Malcolm Roscès

PEST MANAGEMENT IN RICE

Papers presented at the Conference on Pest Management in Rice, held at the Society of Chemical Industry, London, UK, 4–7 June 1990.

PEST MANAGEMENT IN RICE

Edited by

B.T. GRAYSON

Shell Research Ltd, Sittingbourne, Kent, UK

M.B. GREEN

Consultant, Wallington, Surrey, UK

and

L.G. COPPING

DowElanco Ltd, Wantage, Oxon, UK

SCI

FOR THE APPLICATION
OF CHEMISTRY AND
RELATED SCIENCES

Published for the
SOCIETY OF CHEMICAL INDUSTRY
by
ELSEVIER APPLIED SCIENCE
LONDON and NEW YORK

ELSEVIER SCIENCE PUBLISHERS LTD
Crown House, Linton Road, Barking, Essex IG11 8JU, England

Sole Distributor in the USA and Canada
ELSEVIER SCIENCE PUBLISHING CO., INC.
655 Avenue of the Americas, New York, NY 10010, USA

WITH 151 TABLES AND 101 ILLUSTRATIONS

© 1990 SOCIETY OF CHEMICAL INDUSTRY
© 1990 SHELL RESEARCH LTD—pp. 455–464

British Library Cataloguing in Publication Data

Pest management in rice
1. Rice. Pests. Control
I. Grayson, B.T. II. Green, M.B.
III. Copping, Leonard G. IV. Society of Chemical Industry
633.1899

ISBN 1-85166-514-5

Library of Congress CIP data applied for

Printed in Great Britain at the University Press, Cambridge

Preface

The four-day international Conference on Pest Management in Rice, which is the subject of this volume, was the third in an ongoing series of meetings on tropical crops organised by the Pesticides Group of the Society of Chemical Industry, London. The participants came from both the public and private sectors and from many different countries. All the major groups of pests—weeds, microorganisms, arthropods and rodents—were considered, as the organisers believe that it is necessary to address the total pest management problems in each particular growing area, and the variety of the papers indicates the importance of a multi-disciplinary approach to their solution.

Rice is one of the most important world crops and is the major source of food for around 60% of the world's population, with a world production of 500 million tonnes from 150 million hectares of land. Since world stocks amount to only two months supply, many people are at risk from famine. Moreover, it has been esti-mated that the world requirement in 2020 will be about 760 million tonnes, an increase of 50%. This pressure of population on food makes efficient pest management vital and is the reason for bringing together experts from all over the world to this major conference.

Rice is grown under a wide range of farming systems from wet to dry and in a wide variety of environments, from 3000 feet in the Himalayas to sea level in river deltas. Very high levels of management and sophisticated agricultural technology in developed countries such as Japan and the USA contrast with the low manage-ment, low input systems in developing countries in Asia. This is reflected in the use of agrochemicals where the average expenditure per hectare of $680 in Japan contrasts sharply with the average expenditure per hectare of $2 in, for example, Bangladesh. Nevertheless, rice accounts for a total annual global expenditure on agrochemicals of $2400 million, the largest of any crop.

For management of arthropod pests in rice it has become clear that chemical control alone is insufficient and has led to problems of insect resistance and re-surgence due to suppression of predators. An example is the transformation of the brown planthopper in South-east Asia from a non-pest into a major problem as the result of injudicious use of broad-spectrum insecticides. However, the alter-native approach of breeding resistant species has also been proved wanting, and it is now generally agreed that the Integrated Pest Management approach is the right one; this is exemplified in the papers in this volume dealing with insect control. There are many problems to be solved not only in determining the most effective IPM systems for each region but also in deciding how these are to be implemented and whether this is to be by voluntary co-operation between suppliers, farmers and extension services or whether it should be imposed by law. The potential value of computer tools and expert systems to assist implementation of IPM strategies is also discussed.

For management of fungal infections the main technique is the use of chemical control, although some success has also been obtained with biologicals but, so far, to only a limited extent. Natural products also feature in the farmers' armoury, as do compounds which are not directly toxic to the pathogen but which interfere with the infection process. This is exemplified in the papers dealing with fungicides. The importance of learning how to use existing fungicides more effectively by timing applications accurately and by predicting the conditions which favour the more damaging diseases is also addressed.

Weeds can have not only an important direct effect on the reduction of rice yields but also an indirect one by harbouring insects, diseases and other pests. The first line of defence can be the farmer's management practice in the preparation of his land, but almost certainly further control will be required. Different cultivation methods (wet paddy/dry upland, direct seeded/transplanted, multiple cropping) present different weed problems and these problems may change with time. Hand weeding can be and is carried out in some localities, but chemical control is usually more efficient and, with modern selective herbicides, less damaging to the crop. A large number of new herbicides, often for specific use in rice, have been developed recently and some of these are described in the papers dealing with weed control.

Effectiveness and safety of chemical control agents to the operator and to the environment can be increased by improved formulation and by more accurate application. These matters are touched upon in this volume, as is also the severe problem of rat control in rice fields where they cause large losses. The final papers deal with the possibilities for plant breeding and biotechnology in pest management in rice. Cereal crops have proved to be difficult to transform using traditional techniques available to the genetic engineer. Rice, however, is proving to be more amenable than other basic cereal crops. There exist opportunities for introducing new genes to protect the crop from insect and fungal attack. Traits exist in nature which can also be introduced into commercial varieties of rice with a view to overcoming stress-induced yield loss. These and other non-conventional techniques are reviewed.

The introduction of sophisticated pest management systems to rice makes great demands on the time and skill of the growers. The development of expert systems and simulation models may assist growers in developed countries such as the USA and Japan with their decision processes, but it is yet to be seen how these techniques can be applied in developing countries and how IPM strategies can be implemented there. In these countries pesticides have appeared to be an easy option, but in untrained hands they can lead to problems. Attempts to limit their use and to develop and introduce alternative methods of control will make heavy demands upon regulatory authorities, agrochemical manufacturers, research and extension organisations and farmers' skills and, unfortunately, it is in the developing countries that these components of the pest management system are under-resourced or in short supply. It is a challenge to all those involved, especially the wealthier developed countries with their greater financial and technological resources, to try to ensure that the problems of pest management in rice are dealt with and overcome without creating new ones, that production of this vital crop is kept economically viable for the millions who depend upon it for sustenance, and that increases in production needed to feed the increasing populations are achieved. We hope that this conference is a useful contribution to that aim.

B.T. GRAYSON
M.B. GREEN
L.G. COPPING

Acknowledgements

The organisers record their sincere thanks to Shell International Chemical Company for providing the Conference Reception and, at the time of going to press, to the following companies for financial contributions which were used to sponsor the attendance of some speakers and provide bursaries for the attendance of student and senior staff delegates from colleges and universities:

BASF AG
CIBA-Geigy AG
DowElanco Ltd
FMC Corporation
Sandoz Agro Division
ICI Agrochemicals plc
Kureha Chemical Company Ltd
Rhône-Poulenc Agriculture
Schering Agrochemicals Ltd
Takeda Chemical Industries Ltd

B.T.G.
M.B.G.
L.G.C.

Contents

List of Contributors

J. L. Allard, CIBA-Geigy Ltd, Agricultural Division, Research & Development, CH-4002 Basle, Switzerland

J. B. Baker, Louisiana State University Agricultural Center, Baton Rouge, Louisiana 70803, USA

R. P. Bateman, International Pesticide Application Research Centre, Imperial College at Silwood Park, Sunninghill, Ascot, Berkshire SL5 7PY, UK

A. Bhandhufalck, CIBA-Geigy (Thailand) Ltd, PO Box 747, Bangkok 10501, Thailand

J. Cao, Department of Plant Pathology and Crop Physiology, Agricultural Experiment Station, Louisiana State University Agriculture Center, Baton Rouge, Louisiana 70803, USA

D. P. Chakraborty, Hindustan Fertilizer Corporation Ltd, 52A Shakespeare Sarani, Calcutta 700 017, India

J. A. Cheng, Zhejiang Agricultural University, Hangzhou, Zhejiang, China

K. M. Chin, CIBA-Geigy Agricultural Experiment Station, Beg Berkunci, 71309 Rembau, NS, West Malaysia

M. F. Claridge, School of Pure and Applied Biology, University of Wales, Cardiff CF1 3TL, UK

S. H. Crawford, Louisiana State University Agricultural Center, Baton Rouge, Louisiana 70803, USA

R. Fischer, Hoechst AG, R & D Agrochemicals Department, PO Box 800320, D-6230 Frankfurt am Main 80, Federal Republic of Germany

C. D. Forgie, DowElanco Japan Ltd, 1-6-12, Toranomon, Minato-ku, Tokyo, Japan

P. French, ICI Agrochemicals, Jealott's Hill Research Station, Bracknell, Berkshire RG12 6EY, UK

A. Gillespie, Chr. Hansen's Bio Systems A/S, Bøge Alle 10–12, DK-Hørsholm, Denmark

S. S. Gnanamanickam, Reader, Centre for Advanced Studies in Botany, University of Madras, Madras 600 025, India

E. Göbel, Plant Genetic Systems NV, Jozef Plateaustraat 22, B-9000 Gent, Belgium

R. F. S. Gordon, ICI Japan Agricultural Research Station, 780 Kuno-cho, Ushiku, Ibaraki, Japan

G. Görlitz, Hoechst AG, R & D Agrochemicals Department, PO Box 800320, D-6230 Frankfurt am Main 80, Federal Republic of Germany

D. E. Groth, Rice Research Station, PO Box 1429, Crowley, Louisiana 70527, and Department of Plant Pathology and Crop Physiology, Agricultural Experiment Station, Louisiana State University Agriculture Center, Baton Rouge, Louisiana 70803, USA

B. Hanisch, Schering AG, Berlin, Federal Republic of Germany

J. Hayakawa, ICI Agrochemicals, Fernhurst, Haslemere, Surrey GU27 3JE, UK

J. E. Hill, Department of Agronomy and Range Science, University of California, Davis, California, USA

H. Hirata, Shiraoka Research Station of Biological Science, Nissan Chemical Industries Ltd, 1470 Shiraoka, Minamisaitama, Saitama pref., Japan 349-02

J. Holt, Overseas Development Natural Resources Institute (ODNRI), Central Avenue, Chatham Maritime, Chatham, Kent ME4 4TB, UK

J. Jimenez, CAB International Institute of Biological Control, Silwood Park, Ascot, Berkshire, UK

R. G. Jones, Shell International Chemical Company Ltd, Shell Centre, London, UK

G. Kadota, ICI Japan Agricultural Research Station, 780 Kuno-cho, Ushiku, Ibaraki, Japan

J. Kato, DowElanco Japan Ltd, 1-6-12, Toranomon, Minato-ku, Tokyo, Japan

M. Kern, Hoechst AG, R & D Agrochemicals Department, PO Box 800320, D-6230 Frankfurt am Main 80, Federal Republic of Germany

U. Kiessling, BASF Aktiengesellschaft, Agricultural Research Station, D-6703 Limburgerhof, Federal Republic of Germany

W. Knauf, Hoechst AG, R & D Agrochemicals Department, PO Box 800320, D-6230 Frankfurt am Main 80, Federal Republic of Germany

N. Kondo, DowElanco Japan Ltd, 1-6-12, Toranomon, Minato-ku, Tokyo, Japan

T. Konno, Biological Research Center, Nihon Nohyaku Co. Ltd, 4-31 Honda-cho, Kawachi-Nagano, Osaka 586, Japan

K. H. Leist, Hoechst AG, R & D Agrochemicals Department, PO Box 800320, D-6230 Frankfurt am Main 80, Federal Republic of Germany

G. D. Lindberg, Rice Research Station, PO Box 1429, Crowley, Louisiana 70527, and Department of Plant Pathology and Crop Physiology, Agricultural Experiment Station, Louisiana State University Agriculture Center, Baton Rouge, Louisiana 70803, USA

A. Loehken, CIBA-Geigy Ltd, Agricultural Division, Research and Development, Plant Protection, CH-4002 Basle, Switzerland

G. J. Marrs, ICI Agrochemicals, Jealott's Hill Research Station, Bracknell, Berkshire RG12 6EY, UK

N. R. Maslen, Overseas Development Natural Resources Institute (ODNRI), Central Avenue, Chatham Maritime, Chatham, Kent ME4 4TB, UK

T. Matsui, Schering AG, Berlin, Federal Republic of Germany

S. Matsumoto, ICI Japan Agricultural Research Station, 780 Kuno-cho, Ushiku, Ibaraki, Japan

H. Matsuyuki, ICI Japan Agrochemicals Division, PO Box 411, Tokyo 100, Japan

G. A. Matthews, International Pesticide Application Research Centre, Imperial College at Silwood Park, Sunninghill, Ascot, Berkshire SL5 7PY, UK

T. W. Mew, Plant Pathologist, International Rice Research Institute (IRRI), PO Box 933, Manila, Philippines

Y. Miyagi, Biological Research Center, Nihon Nohyaku Co. Ltd, 4-31 Honda-cho, Kawachi-Nagano, Osaka 586, Japan

K. Moody, International Rice Research Institute (IRRI), PO Box 933, Manila, Philippines

S. Nakamura, ICI Japan Agricultural Research Station, 780 Kuno-cho, Ushiku, Ibaraki, Japan

G. A. Norton, Silwood Centre for Pest Management, Department of Biology, Imperial College, Silwood Park, Ascot, Berkshire SL5 7PY, UK

N. M. Pearman, Shell International Chemical Company Ltd, Shell Centre, London, UK

M. Peferoen, Plant Genetic Laboratories NV, Jozef Plateaustraat 22, B-9000 Gent, Belgium

L. G. Peterson, Development Manager/Asian-Pacific Area, DowElanco Pacific Ltd, 40/F Sung Hung Kai Center, 30 Harbour Road, Hong Kong

L. W. Peterson, E. I. Du Pont De Nemours & Company, Stine-Haskell Laboratory, PO Box 30, Newark, Delaware 19711, USA

M. Pfenning, BASF Aktiengesellschaft, Agricultural Research Station, D-6703 Limburgerhof, Federal Republic of Germany

W. T. Reed, E. I. Du Pont De Nemours & Company, Walker's Mill, Barley Mill Plaza, Wilmington, Delaware 19898, USA

R. Rees, Schering AG, Berlin, Federal Republic of Germany

R. Reynaerts, Plant Genetic Laboratories NV, Jozef Plateaustraat 22, B-9000 Gent, Belgium

D. R. Reynolds, Overseas Development Natural Resources Institute (ODNRI), Radar Entomology Unit, RSRE, Leigh Sinton Road, Malvern, Worcs. WR14 1LL, UK

J. R. Riley, Overseas Development Natural Resources Institute (ODNRI), Radar Entomology Unit, RSRE, Leigh Sinton Road, Malvern, Worcs. WR14 1LL, UK

M. C. Rush, Rice Research Station, PO Box 1429, Crowley, Louisiana 70527, and Department of Plant Pathology and Crop Physiology, Agricultural Experiment Station, Louisiana State University Agriculture Center, Baton Rouge, Louisiana 70803, USA

G. Salbeck, Hoechst AG, R & D Agrochemicals Department, PO Box 800320, D-6230 Frankfurt am Main 80, Federal Republic of Germany

D. E. Sanders, Louisiana State University Agricultural Center, Baton Rouge, Louisiana 70803, USA

R. Schaub, Hoechst AG, R & D Agrochemicals Department, PO Box 800320, D-6230 Frankfurt am Main 80, Federal Republic of Germany

R. W. Schneider, Department of Plant Pathology and Crop Physiology, Agricultural Experiment Station, Louisiana State University Agriculture Center, Baton Rouge, Louisiana 70803, USA

U. Schollmeier, Hoeschst AG, R & D Agrochemicals Department, PO Box 800320, D-6230 Frankfurt am Main 80, Federal Republic of Germany

H. H. Schubert, Hoeschst AG, R & D Agrochemicals Department, PO Box 800320, D-6230 Frankfurt am Main 80, Federal Republic of Germany

B. M. Shepard, Resident Director, Clemson University, Coastal Research and Education Center, 2865 Savannah Highway, Charleston, South Carolina 29414, USA

Y. Shirai, Shiraoka Research Station of Biological Science, Nissan Chemical Industries Ltd, 1470 Shiraoka, Minamisaitama, Saitama pref., Japan 349-02

R. J. Smith Jr, US Department of Agriculture, Agricultural Research Service and University of Arkansas Rice Research and Extension Center, Stuttgart, Arkansas, USA

R. R. Stephenson, Environmental & Biochemical Toxicology Division, Shell Research Ltd, Sittingbourne Research Centre, Sittingbourne, Kent ME9 8AG, UK

K. Suzuki, Shiraoka Research Station of Biological Science, Nissan Chemical Industries Ltd, 1470 Shiraoka, Minamisaitama, Saitama pref., Japan 349-02

T. G. Szoke, Rhône-Poulenc Crop Protection Division, Lyon, France

G. H. Toenniessen, Rockefeller Foundation, New York, USA

H. Ueno, ICI Japan Agrochemicals Division, PO Box 411, Tokyo 100, Japan

K. Untung, Department of Entomology, Gadjah Mada University, Yogyakarta 55581B, Indonesia

H. E. van de Baan, Department of Entomology and Pesticide Research Center, Michigan State University, East Lansing, Michigan 48823, USA

W. T. Vorley, Agricultural Division, CIBA-Geigy Ltd, CH-4002 Basle, Switzerland

A. Waltersdorfer, Hoechst AG, R & D Agrochemicals Department, PO Box 800320, D-6230 Frankfurt am Main 80, Federal Republic of Germany

D. R. Wareing, Imperial College, Silwood Park, Ascot, Berkshire SL5 7PY, UK

M. J. Way, Silwood Centre for Pest Management, Department of Biology, Imperial College, Silwood Park, Ascot, Berkshire SL5 7PY, UK

M. O. Way, Texas Agricultural Experiment Station, Route 7, Box 999, Beaumont, Texas 77713, USA

M. E. Whalon, Department of Entomology and Pesticide Research Center, Michigan State University, East Lansing, Michigan 48823, USA

A. T. Woodburn, Director, Allan Woodburn Associates Ltd, 18 Newmills Crescent, Balerno, Edinburgh EH14 5SX, UK

Q. J. Xie, Department of Plant Pathology and Crop Physiology, Agricultural Experiment Station, Louisiana State University Agriculture Center, Baton Rouge, Louisiana 70803, USA

T. Yamaguchi, E. I. Du Pont De Nemours & Company, Walker's Mill, Barley Mill Plaza, Wilmington, Delaware 19898, USA

A. Zoschke, CIBA-Geigy Ltd, Agricultural Division, Research & Development, CH-4002 Basle, Switzerland

RICE PEST MANAGEMENT SYSTEMS - PAST AND FUTURE

G.A.NORTON AND M.J.WAY
Silwood Centre for Pest Management
Depatment of Biology, Imperial College,
Silwood Park, Ascot   SL5 7PY, U.K.

ABSTRACT

In traditional forms of rice cultivation, adaptation to pests was achieved by "naturally selected" cultivation practices and resistant varieties. In recent decades the need for increased rice production throughout Asia has led to the adoption of more intensive production systems. This has produced increases in pest attack and a greater reliance on pesticides as the major form of control. With increasing population, the pressure for more intensive rice production systems will be sustained and risks of serious pest attack will continue. While "breakthroughs" in novel methods of control are possible, what their impact might be in the next decade is uncertain and certainly unproven. At present, there is far more potential to improve pest management by fully utilizing the control methods and practices currently available. What is required is a better understanding of farmers problems that will enable key constraints to be reduced and more appropriate control strategies to be designed.

INTRODUCTION

Rice is one of the oldest cultivated crops: records of its cultivation date back 7,000 years or more in Zhejiang Province, China. It is also the most important crop worldwide. Although world rice production (at 460 million tonnes per year) is less than that of wheat, in Asia, 60% of the world's population cultivates 90% of the total global rice area (Table 1), estimated at 145 million hectares (6 times the total land area of the U.K.).

TABLE 1
Regional distribution of world rice production [1]

Region	% of world rice production
East Asia	45.4
South East Asia	22.2
South Asia	23.5
Latin America	3.9
Africa	2.2
Rest of the world	2.8

In this context, rice pests (by which we mean insect, disease, weed, vertebrate, and other damaging organisms) are important for a number of reasons.

First, unlike other basic food crops, only 2-5% of annual rice production is traded on the world market and world rice stocks are equivalent to just about 17% of annual consumption – enough for about two months. This represents a very risky supply-demand situation which could be threatened by pest outbreaks. For example, ecological stress during the drought of 1987-88 seriously disrupted the security of supply in Asia, threatening the largest proportion of the world's poor and malnourished people.

Second, to meet the food needs of the increasing populations dependant on rice, it is estimated that an increase in production of over 20% is required by 2000, from the 1987 production level of 460 million tonnes; and over 65% by the year 2020 [1]. There are several implications of this for pest management:

*   some of this increased production might be obtained by reducing the losses caused by current levels of pest attack.

*   the changes in production practices, and more intensive rice cultivation that will be necessary to achieve this increase in production, could have an unfavourable effect, both on the status of future pest problems and farmers' ability to deal with them.

It is against this background that the contributions to this Conference should be viewed. In this paper, we set the scene by reviewing the historical development of rice pest problems and their importance today, and express what, in our view, are the prospects for the future.

RICE SYSTEMS, PESTS, AND CROP LOSS

The pest problems of irrigated and rainfed lowland rice, where the crop grows in shallow water, are notably different from the problems of upland and deepwater rice systems, both in terms of pest species, pest management tactics, and the economics of crop protection (Table 2).

TABLE 2
General features of cultural practice, yield, and pest status of upland and irrigated/lowland rices

Features	Upland rice	Lowland rice (irrigated)
Fertilizer/ pesticide use	low	medium-high
Yield potential	low	high
Yield stability	low	high
Role of host plant resistance	low	high
Insect pests		
speciality	mostly polyphagous	mostly monophagous
soil pests	****	0
aerial pests	**	****
Diseases	** to ****	** to ***
Weeds	****	**
Birds	***	*
Rats	*	***(*)

{KEY: 0, *, **, ***, **** = scale of increasing severity}

Since about 90% of the world's rice yield comes from irrigated (72%) and rainfed (19%) lowland rice [1], this paper will concentrate on the pest problems of these two systems.

Some 100 insect species, 74 diseases and 1,800 weed species are recorded as pests in southern and southeast Asia alone.   Of these, 30 insects, 16 diseases and 15 weeds are considered to be economically important (Table 3) [2,3].   Rats and other vertebrates can also be very damaging.

TABLE 3
Economically important pests of tropical rice [3]

Insect pests

Vegetative stage
  Armyworms and cutworms
  Grasshoppers, katydids,
  and field crickets
  Mealybug
  Rice black bugs
  Rice caseworm
  Rice gall midge
  Rice green hairy caterpillar
  Rice green semilooper
  Rice hispa
  Rice leaffolders
  Rice stem borers
      Dark-headed stem borers
      Pink stem borer
      Striped stem borer
      White stem borer
      Yellow stem borer
  Rice thrips
  Rice whorl maggots
  Seedling maggots

Reproductive stage
  Rice brown planthopper
  Rice green leafhoppers
  Rice greenhorned caterpillar
  Rice skippers
  Rice white leafhopper
  Rice whitebacked planthopper
  Rice zigzag leafhopper
  Smaller brown planthopper

Ripening stage
  Rice panicle mite
  Rice seed bugs

Rice Diseases
  Bacterial blight
  Bacterial leaf streak
  Bakanae
  Brown spot
  False smut
  Grassy stunt virus
  Narrow brown leaf spot
  Rice blast
  Rice ragged stunt
  Sheath blight
  Sheath rot
  Stem nematode
  Stem rot
  Tungro virus
  White tip
  Yellow dwarf disease

Weed Pests of Rice
  Commelina   benghalenis
  Cyperus difformis
  Cyperus iria
  Cyperus rotundus
  Dactyloctenium
      aegyptium
  Digitaria
      ciliaris
  Echinochloa
      colona
  Echinochloa
      crus-galli
  Eleusine
      indica
  Fimbristylis
      miliacea
  Monochoria
      vaginilis
  Paspalum distichum
  Portulaca oleracea
  Scirpus martimus
  Sphenoclea zeylanica

Relatively little has been done to assess crop losses, with estimates by Cramer [4] most commonly cited. He gives a loss of some 55%, made up of 35% by insects and 10% each by diseases and weeds. Later evidence does not contradict such estimates. For example, 35-44% general losses from insects [5] and 24% from insects in East and Southeast Asia [6]. Other estimates include those of Litsinger et al, [7] of losses due to non-outbreak pests of 18% and by Moody [8] of from 11-65% from weeds in the Philippines. Locally very severe losses from stemborers (up to 95%) have been recorded in the 1960s and 1970s, as well as leaf and plant hoppers (up to 80%). Recent estimates of losses from particular diseases, notably blast, range from 1-100%, with averages of 5-10% in India and 8-14% in China [9,10].

It seems that pests may approximately halve the potential yield of rice. Thus, alleviation of such losses could make a major contribution to the required target of doubling rice yields by the year 2020.

HISTORICAL DEVELOPMENT OF RICE PEST PROBLEMS

The wild progenitors of rice, Oryza sativa, occurred in a belt in the northern tropics where they grew in shallow flooded valleys in the discrete wet season. Rainfed lowland rice and also irrigated rices are now grown in conditions which are ecologically like the wild habitat. Such conditions possess notable elements of stability in terms of naturally occurring controls. For instance, in irrigated rice in the tropics, there is a remarkable community of natural enemies that develop quickly, thrive in the stable conditions of irrigated rice, and make a notable impact on most insect pests.

For the 9000 years during which rice has been grown as a crop, farmers have selected the best yielding cultivars. Such selection has tended towards strains that show resistance to diseases and insects, and which have the ability to compete with weeds. Indeed, traditional transplanting techniques have evolved largely as a method of combating weeds, fundamentally, the most important pests of rice. Consequently, a combination of cultural practices, "naturally" (unconsciously) selected host plant resistance, and high mortalities from natural enemies has evolved for crops which, although damaged by pests, have produced stable yields on which Asian civilisations have depended.

However, these traditional rice systems were relatively low yielding. With the escalating need for increased food production to meet the needs of an ever increasing population, the situation was transformed by the Green Revolution. A major feature of this revolution was breeding for high yields. However, to achieve the potential of these new varieties, and as a "knock-on" effect of their adoption, several other changes in the rice system occured, with consequent implications for pests and their control.

For instance -

*   the high yielding capability of the early varieties were at the expense of existing natural pest controls.

*   to achieve high yields, nitrogen fertilisers were increasingly used. As with other crops worldwide, this led to a decrease in natural resistance by making rice much more attractive and sensitive to damage by many insect pests and diseases, as well as benefiting weeds.

*   installation and improvement of irrigation schemes constituted another component of this Green revolution which, combined with the use of quick maturing varieties, made possible the growing of three crops a year in some regions. This provided an ideal carry-over system for some pests, for example stemborers and brown planthopper.

These changes continue. For example, in the Muda region of Malaysia, dramatic changes in rice production practices have occurred over the past 15 to 20 years (Fig. 1).

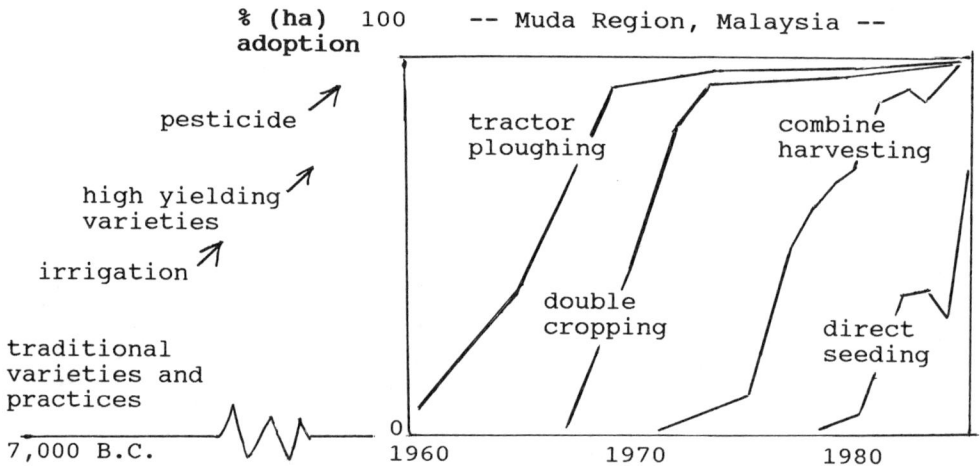

Figure 1. Changes in rice production practices, especially in the Muda region of Malaysia (After [11])

There have been two major implications associated with this development: first, changes in the agro-ecosystem, that can change the status of pest problems. Table 4 provides a summary of expert opinion concerning the impact of new practices on different pest groups.

TABLE 4
The impact of agronomic change on pest impact,
compared with traditional practice (modifed from [11])

| Rice pests | Direct seeding | | Synchronous |
	Broadcast	Drilled .	planting
Leaf folders	+	?	- -
Stemborers	+ +	+	-
GLH/Tungro	+ +	?	- -
Planthoppers	+ + +	+ +	- -
Bugs	+ +	?	- -
Rats	+ +	?	-
Blast	+	?	+ +
Sheath blight	+	?	- -
Weeds	+ +	+	?

Key to Expert opinion:
slight (+), moderate (++), considerable (+++) increase,
slight (-), moderate (--), considerable (---) reduction,
opinion very divided/ "don't know" (?).

The major switch to direct seeding is likely to make
virtually all insect pest problems worse, although increased
weed problems may be of most concern.

Second, while rural labour shortages in the Muda area have
been a major driving force in producing these changes, the
changes themselves have reinforced this labour-saving trend. The
net result is that farmers, in deciding on pest management
strategies, will place considerable emphasis on labour
requirements. Thus, as the intrinsic pest controlling features
of traditional rice systems have been reduced, and pest
problems have become worse, farmers have come to rely more on
pesticides as the major form of control. This, in turn, has
contributed to further development of the rice pest problem - in
particular, in encouraging the three "Rs": Resurgence,
Resistance, and Risks to operators and consumers.

Resurgence of brown planthopper (BPH) provides a typical
example of technology-induced pest problems, in this case making
a species that was a non-pest in tropical S.E.Asia into the most
serious pest of that region. Here the combination of lessened
host plant resistance and sequential crops of well fertilised,
pest-attractive rice provides ideal conditions for rapid build
up of an epidemic pest such BPH in circumstances where other
natural controls fail. Therefore outbreaks of BPH occurred
especially when certain pesticides had a disastrous effect on
the complex of natural enemies that mostly keeps BPH relatively
uncommon, even in Green Revolution conditions. Major outbreaks
which occurred throughout S.E.Asia have now been partly
alleviated by development of resistant strains and, in some
countries such as Indonesia, by banning those chemicals which
induce outbreaks through their unselective action as well as
through apparently subtle effects on the plant that make it more
attractive to the pest [12].

IMPROVING RICE PEST MANAGEMENT

Having looked in some detail at the pest problems associated with rice cropping systems, and how they are changing, we now discuss how to deal with these problems. There are two ways in which pest management can be improved.

(1.) Develop and introduce new technologies and practices, including the breeding of resistant varieties, the development of new insecticides, microbial agents and application equipment, and the development and implementation of new pest monitoring and forecasting systems, and

(2.) Reduce institutional, infrastrucure, information, and knowledge constraints to the use of existing technologies and practices, or modify these technologies and practices to be more appropriate to the target farming system.

New Technologies and Practices

In the past the main emphasis in control of pests of irrigated rice has been the breeding of resistant cultivars, first at the International Rice Research Research Institute (IRRI) and now also by many national organizations. Notable success has been achieved against some pests, particularly diseases and a few insects, though resistance-breaking strains of pests have posed a constant threat. The other main control methods have been the use of pesticides, particularly against insects.

As already mentioned, pests have developed resistance to some of these pesticides, their excessive use has upset biological controls and created new pest problems, they have been harmful to humans and other animals, including important food fish and, in some cases, they have created other environmental problems - notably, inducing development of resistant strains of mosquitoes and other human disease vectors that breed in rice ecosystems.

Thus, increased use of pesticides has led to catastrophe or near catastrophe in some countries. This experience has served to highlight the need for integrated pest management (IPM), defined in strictly practical terms as the farmer's best mix of controls taking into account yields, profits and environmental criteria [13].

In searching for new components of IPM, the only truly novel technology is that of genetic engineering applied to improvement of host plant resistance through intrinsic mechanisms possessed by plants, as well as by insertion of toxins, such as the endotoxin of <u>Bacillus thuringiensis</u>.

Much has been said about the transfer of the BT endotoxin to rice. It is envisaged as protecting rice against important pests such as stem borers and also against other stem and leaf eating Lepidoptera. So far, no strains of BT have been discovered that show adequate toxicity to the target species. However, even if such strains were found and that the endotoxin could be incorporated in the rice plant, two questions still remain - first, whether synthesis of the BT endotoxin will significantly decrease intrinsic yield, and second, how quickly the target pests would develop resistance to BT in these circumstances, where heavy selection pressure is being exerted for resistance. Already strains of several insects have shown high levels of resistance to BT, including genetically engineered BT [14,15].

Another technology, often referred to as new, is the use of pathogens, particularly fungal, as substitutes for synthetic chemicals. The comparative failure of pathogens against insects in general, despite much research over the past 20 years, does not hold out much hope for their use on rice. In particular, such serious disadvantages as lack of fungal survival in store as well as after application, and relatively low toxicity, suggest that they will have comparatively little impact in the future. Genetic engineering is seen as a possible means of introducing competitor non-pathogenic viruses to exclude pathogenic ones. This is an exciting concept, again as yet unproven.

We conclude from this brief summary that the future of pest management in rice, at least for the next 5 to 10 years, will continue to depend overwhelmingly on making the best use of conventional control methods, including conventional pesticides, as part of an integrated pest management approach to the overall pest complex. In view of the unproven value of novel methods, we must accept continued use of pesticides for helping to maintain and increase rice yields.

The range of pesticides is limited, especially as some are being banned because of the harm they can do; for example, certain insecticides, which induce brown planthopper outbreaks in Indonesia. This puts a premium on using available pesticides rationally in an IPM context in order to maintain their efficacy as well as to minimise possible harmful effects.

Reducing Constraints and Modifying Recommendations

A major reason why farmers do not adopt IPM strategies is because the IPM strategy recommended to them is not feasible for a variety of reasons, including such on-farm constraints as an inability to identify pests, lack of time for monitoring, and a lack of management flexibility within the cropping or farming system.

Where this is the case, two means of improvement can be sought.

First, one can attempt to reduce on-farm constraints. A major problem in S.E Asia is a lack of appropriate knowledge at the farm level concerning topics essential for IPM, such as the value of resistant cultivars, pest and symptom identification, and damage thresholds. Recognising this as a major constraint to improved pest management, an FAO Regional programme [16,17] in S.E. Asia aims to -

1) Train farmers in identification of pests and their symptoms, and particularly in distinguishing them from beneficial natural enemy species;

2) Define action thresholds for pesticide usage (Table 5) on which the farmers can develop their own empirical decision criteria.

3) Familiarise farmers in the value of resistant cultivars.

TABLE 5
Thresholds for selected insects in tropical rice (After [3])

Pests	Action threshold
Whorl maggot + green semi-looper & hairy caterpillar	10% damaged leaves
Whorl maggot	0.5 - 1 egg / hill (use of earlier planted fields as indicator)
Stemborers	3 - 8% deadhearts 0.5 - 1 egg mass / m^2
Planthoppers	15 hoppers / hill 0.5 hoppers / tiller (nymphs)

At present, IPM is being practiced on about 4.9 million ha of irrigated rice in Asia, out of a total of 133 million ha. The results are encouraging and the programme is rapidly expanding.

A second approach is to try to reduce the effort involved in IPM. A dilemma that is faced in attempting to implement IPM is to resolve -

* the need for simple decision making rules, that the farmer can readily understand and which are not too demanding of his or her time in data gathering, with

* the need for each IPM strategy to be designed according to the particular problem the farmer faces.

This "farmer specific problem" is defined not only by the biological features of the problem but also the constraints on information gathering, on the options, or the management flexibility that may be peculiar to the particular farming system. Clearly, in attempting to tailor IPM strategies to particular farmers' problems may require considerable decision making and data gathering effort (Fig. 2).

Major rice insect pests (10)

Pest damage symptoms (2)

Plant growth stages (3)
(veg., repro., ripe)

Degree of danger (3)
(high, medium, low)

Yield potential (3)

Value of the crop (3)

Appropriate pesticide (15)

Formulation (2)
(granules, spray)

Dose (3)

Volume (3)

Application frequency (12)

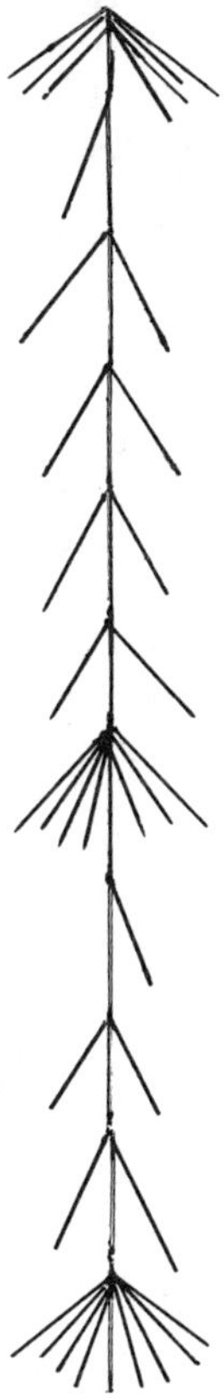

Figure 2. Some of the factors a farmer has to account for in making a decision on the control of one pest in irrigated rice in the Philippines (After [18])

In this context, various decision support systems can be of help. An illustration, concerning the brown planthopper (BPH) is given in a later paper in this volume [19]. A simulation model built for conditions in Zhejiang Province in China has enabled us to conclude that when a pesticide is to be applied against this pest, the most effective time of application is 30 days after transplanting; regardless of transplanting time, the temperature conditions, or whether later sprays are to be applied. The difficult problem is to decide whether a spray needs to be applied at all and here an expert system is proving of value in combining "tailor-made" solutions with a minimum of questions the user has to answer [20].

Although computer-based expert systems are unlikely to be

operational in resource-poor situations for many years, they can be of value in training and in providing a framework for developing operational manuals. At a regional level, data base systems for pest forecasting are more likely to be operational. This type of decision support approach, including multidisciplinary research/extension workshops, that focus on the pest problems of particular rice farmers, is what is required in the future to better match our scientific effort with the real-world problems.

CONCLUSIONS

Rice crop protection systems are the product of many complex factors, operating at different levels (Fig.3).

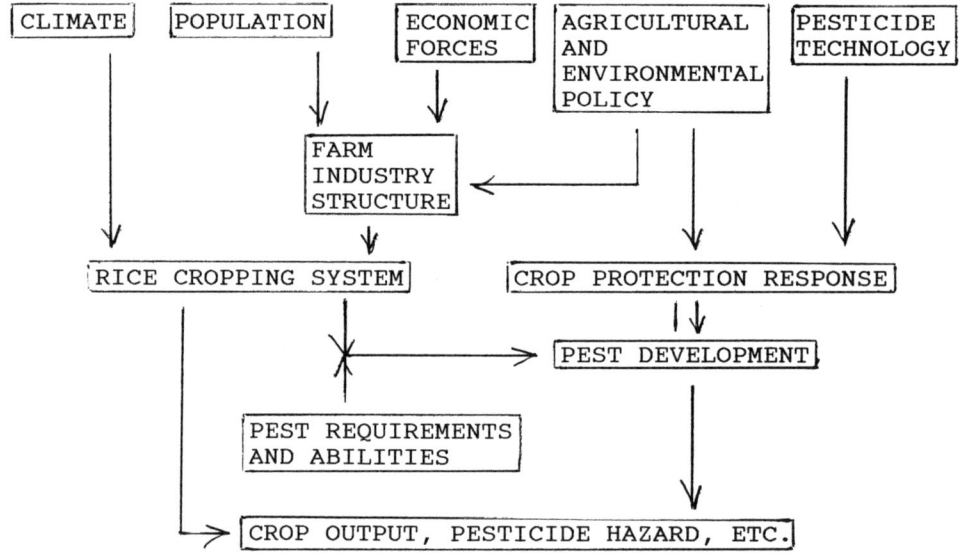

Fig.3 Key factors affecting rice crop protection systems

At a national and regional level, such factors as government policy, input costs and the price of rice, all influence the form of the rice agroecosystem. This, in turn, affects the favourability of the system for different pests and therefore affects the damage caused by these pests. In response to this, farmers attempt to protect their crops, the methods and strategies they use depending on the options available, their performance, and how they fit into the farming or cropping system.

A major objective in the future will be to modify policy, develop technology, provide extension support and train farmers in ways that will encourage and support the development of "sustainable" rice cropping systems - that is, developments that do not lead us into blind alleys, or reduce short term hazards at the expense of much greater, longer term risks. The problem is how to achieve this while, at the same time, meeting the increased production levels that will be required.

In the longer term, technological developments may change the whole situation. Undoubtedly, high investments will continue to be made in the search for such "breakthroughs". However, it would be foolish in the extreme to rely too heavily on future technological change. In the shorter term, we believe that substantial progress can be made by fully utilising existing knowledge and options by first understanding the specific pest problems that particular rice farmers face [18], and then attempting to reduce constraints to improved pest management or to design pest control strategies that are most appropriate to their situation.

REFERENCES

[1]   Anon., _IRRI toward 2000 and beyond_.   International Rice Research Institute, Los Banos, 1989, 66pp.

[2]   Riessig, W.H., Heinrichs, E.A., Litsinger, J.A., Moody, L.F., Mew, T.W. and Barrion, A.T., _Illustrated Guide to Integrated Pest Management in Rice in Tropical Asia_. IRRI, Los Banos, Philippines, 1987.

[3]   Teng, P.S., _Integrated Pest Management in Rice_.   Report to Pest Management Task Force. 1990, 80pp.

[4]   Cramer, H.H., Plant Protection and World Food Production. _Pflanzenschutz-Nachrichten-Bayer_. 1987, 524pp.

[5]   Pathak, P.K. and Dhaliwal, G.S., Trends and strategies for rice pest problems in tropical Asia. _IRRI Research Paper Series_, 1981, No. 64. 1-15.

[6]   Ahrens, C., Cramer, H.H., Mogk, M. and Peschel, H., Economic impact of crop losses. _Proceedings 10th International Congress of Plant Protection_, 1982, 65-73.

[7]   Litsinger, J.A., Canape, B.L., Bandong, J.P., de le Cruz, C.G., Apostel, R.F., Pantua, P.C. and Joshi, R.C., _Rice crop loss from insect pests in wetland and dryland environments of Asia_. International Rice Research Institute, Los Banos, Philippines (Mimeographed report), 1987.

[8]   Moody, L.F., The status of weed control in rice in Asia. _FAO Plant Protection Bulletin_, 1982, _30_, 119-123.

[9]   Padmanabhan, S.Y., Estimating losses from rice blast in India.   In: _The rice blast disease._ Proceedings of a Symposium at the International Rice Research Institute, 1963. John Hopkins Press, Baltimore, 1965, pp. 203-221.

[10] Teng, P.S., Crop loss appraisal in the tropics. _Journal of Plant Protection in the Tropics_, 1986, _3_, 39-50.

[11] Norton, G.A. and Heong, K.L., An approach to improving pest management: rice in Malaysia. _Crop Protection_, 1988, _7_, 84-90.

[12] Kenmore, P.E., Carino, F.O., Perez, C.A., Dyck, V.A. and Gutierrez, A.P., Population regulation of the rice borer planthopper (_Nilaparvata lugens_ Stal) within rice fields in the Philippines. _Journal of Plant Protection in the Tropics_, 1984, _1_, 19-37.

[13] Kenmore, P.E., IPM means the best mix. _Rice IPM Newsletter_, 1987, _1_, 3pp.

[14] McGoughey, W.H., Insect resistance to the biological insecticide, _Bacillus thuringiensis_. _Science_, 1985, _229_, 193-195.

[15] Stone, T.B., Sims, S.R. and Marrone, P.G., Selection of tobacco budworm for resistance to genetically engineered _Pseudomonas florescens_ containing the alpha endotoxin of _Bacillus thuringiensis_ sub-species _Kurstaki_. _J. Invert. Path._, 1989, _53_, 228-234.

[16] Kenmore, P.E., Crop loss assessment in a practical integrated pest control program for tropical Asian rice. In _Crop Loss Assessment and Pest Management_ Ed. P.S.Teng. American Phytopathological Society. St. Paul. 1987. pp. 225-241.

[17] Kenmore, P.E., Litsinger, J.A., Banday, J.P., Santiago, A.C. and Salac, M.M., Philippine rice farmers and insecticides: thirty years of growing dependency and new options for change. In _Management of Pests and Pesticides: Farmers Perceptions and Practices_. Ed. J. Tait & B. Napompeth. Westview Studies in Insect Biology, Bunlder Colorado. 1987. pp. 98-108.

[18] Goodall, G., Challenges to international pest management research in the third world : do we really want IPM to work ? _Bulletin of the Entomological Society of America_, 1984, _30_, 18 - 26.

[19] Wareing, D.R., Holt, J., Cheng, J.A. and Norton, G.A., Use of computer tools for the design of pest management strategies. 1990. (This volume).

[20] Holt, J., Cheng, J.A. and Norton, G.A., A systems analysis approach to brown planthopper control on rice in Zhejiang Province, China. III. An expert system for making recommendations. _Journal of Applied Ecology_, 1990, _27_, 113 -112.

THE CURRENT RICE AGROCHEMICALS MARKET

ALLAN T. WOODBURN
Director
Allan Woodburn Associates Ltd.,
18, Newmills Crescent, Balerno, Edinburgh, EH14 5SX, UK

ABSTRACT

Rice is the single most important crop in terms of the value of consumption of agrochemicals, with a global end-user market value of $2,400 million in 1988. The rice insecticide and rice herbicide sales were both close to $900 million, with rice fungicides at $570 million. The Asian continent, with 90% of the rice producing land, and nearly 92% of the rice production, uses agrochemicals to the value of just over 90% of the global market. However, within this, Japan utilises nearly 60% of the rice agrochemicals, by value, even though it produces only 3% of the world's rice from 1.5% of the land planted to rice. Because of the technical sophistication in Japan, the average expenditure on agrochemicals for rice was $680/hectare in 1988, compared to only around $2/hectare in the underdeveloped Asian countries.

RICE - A STAPLE FOOD CROP

Rice is the most important crop in many areas of the world. The difference between life and death is often dependent on the rice harvests, which in turn can be seriously affected by droughts, floods or excessive pest attacks. In many areas of the underdeveloped world, it is vital to increase rice output, as a means of achieving some degree of self-sufficiency in food supplies and to reduce import bills, or reliance on foreign aid.

As much as 90% of the global rice production is in the hands of small farmers in Asia, who use complex production methods, developed over centuries, to sustain yields in a great diversity of conditions. In all, close on 130 million ha are cultivated in Asia in locations as different as the steep 3000 year old terraces of Fugao in the Philippines, the small intensive fields of Japan and the broad flood plains in the north east regions of the People's Republic of China.

According to the International Rice Research Institute, rice is one of the oldest cultivated crops on earth. A site of sophisticated rice cultivation in southern China is known to be at least 7000 years old.

Rice is probably the world's most versatile crop, growing at more than 3000 metres elevation in the Himalayas and at sea level in the deltas of the great rivers of Asia. Floating varieties grow in water as deep as 4 metres in Thailand, while in Brazil, rice is grown as a dryland crop much like wheat or maize. Transplanted rice requires intensive hand labour in areas like Indonesia, whereas in California, rice is seeded by aeroplane.

Although rice is not an aquatic plant, and in many areas is grown as a rainfed crop, the highest yields are achieved in irrigated paddies. Great ingenuity has been used to develop the system of flooded paddies, the only problem being that it is labour intensive.

In the mid-1970s, the global rice harvests averaged around 360 million tonnes. Ten years later the harvests had risen by 30% so that the average annual production in the 1985 - 1987 period was over 468 million tonnes. There have been three main contributory factors to this dramatic improvement. Firstly, there has been a general increase in the area planted to rice, except in Japan where the government has directed that rice-land be diverted to the growing of other crops. Secondly, the increased use of agrochemicals has led to reduced crop losses through attacks by insects, weeds and diseases. However, over the last ten or fifteen years, perhaps the greatest contribution to increased rice production has been brought about by the introduction of new improved varieties in many irrigated and favourable rainfed areas.

Most of these improvements have been centred on Asia. Production increases have been considerably less dramatic in Latin America, Africa and the Middle East, where rice is also a staple food.

On the basis of IRRI forecasts, global rice consumption will rise from
a level of 420 million tonnes in 1985, through 556 million tonnes at the
end of this century to a staggering 759 million tonnes in the year 2020.
Achievement of these goals will undoubtedly be dependent on advances by the
plant breeders and by the agrochemical scientists, increasingly in co-
operation with each other. Even if these production levels are attained,
because of the anticipated population growth, the average annual rice
consumption per capita will rise by less than 10% between 1985 and 2020.

THE MARKET FOR AGROCHEMICALS IN RICE

Rice is unusual in crop agrochemicals terms since it requires large inputs
of all of the three main agrochemicals categories, insecticides, herbicides
and fungicides. This is shown, on a country-by-country basis, for 1988, in
the accompanying Table 1.

On the basis of these data, rice insecticides accounted for nearly 15%
of the global crop insecticide market value in 1988. The corresponding
proportions for rice herbicides and rice fungicides were just over 9% and
nearly 13.5%, respectively. Overall, with a value of agrochemicals used in
1988 of $2400 million, rice was the single most important crop for
pesticides, shading out both maize and cotton.

The importance of Asia as a rice growing region is well recognised.
Similarly, over 90% of the worldwide rice agrochemicals market value is
located in the Asian region. However, whereas Japan accounts for only
around 3% of the annual global rice production, Japanese rice growers apply
half of the world's rice insecticides and nearly two-thirds of the rice
herbicides and fungicides, by value, annually. The corresponding figures
for the whole of Asia show that the region was responsible for 94.0% of the
insecticides, 81.8% of the herbicides and 97.5% of the fungicides applied
to rice in 1988. The relatively low figure for herbicides reflects the
availability of manual labour throughout the developing nations.

TABLE 1
Rice agrochemicals market values by country - 1988

Country	Herbicides ($m.)	Insecticides ($m.)	Fungicides ($m.)	Others ($m.)	Total ($m.)
Japan	570	455	375	20	1420
S. Korea	48	89	95	3	235
PRC (China)	11	108	35	0	154
Taiwan	26	38	18	5	87
India	18	51	14	2	85
Philippines	17	28	0	3	48
Thailand	17	21	1	0	39
Indonesia	4	24	1	2	31
Bangladesh	3	14	7	0	24
Burma	2	8	4	0	14
Vietnam	2	9	2	0	13
Pakistan	1	3	0	0	4
USA	61	22	4	0	87
Europe	48	24	5	0	77
Brazil	46	1	3	0	50
Rest of World	11	15	6	0	32
Total	885	910	570	35	2400

Sources: AWA Ltd. estimates and Landell Mills Market Research Ltd.

Not surprisingly, there is a very wide difference between the value of agrochemicals applied per hectare of rice in Japan compared to the under-developed areas of Asia. In Japan, where the rice growers tend to have another full-time occupation, the technological inputs in terms of plant varieties, mechanisation and agrochemicals are high. The government guaranteed procurement price for rice is also high. Thus, Japanese rice farmers can afford a total agrochemicals expenditure of over $1400 million on a rice area of just over 2 million hectares. The average expenditure per hectare of rice in 1988 was around $680. In the underdeveloped Asian countries, such as India, Bangladesh, Burma, Vietnam and Pakistan, the average annual expenditure on agrochemicals to protect the rice crop is only between $2 and $3 per hectare.

Elsewhere in Asia, the expenditure on agrochemicals per hectare of rice is also high in South Korea, with a value rapidly approaching $200 per hectare. In the large Chinese rice growing areas, the average annual expenditure on agrochemicals works out at just under $5 per hectare planted, a value slightly above the corresponding figures for both Thailand and Indonesia. The remaining significant Asian rice growing areas, Taiwan and the Philippines, consume proportionately high levels of agrochemicals with over $14 worth of agrochemicals having been applied per hectare of rice in the Philippines in 1988, for example.

The average expenditure on rice agrochemicals in Latin America, Africa and the Middle East is at a similar level to the lowest pertaining in Asia, at around $2.5 per hectare, with the exception of Brazil, where the figure was just over $9 per hectare in 1988, virtually all the requirements being of herbicides. In the relatively small rice growing areas of the United States of America and southern Europe, an average of between $75 and $90 per hectare is the norm for agrochemicals spending.

Rice Insecticide Market

The global rice insecticide market value has been relatively stable in recent years, with any growth in the sector not keeping pace with the overall agrochemicals market value increase. However, the rice insecticide sector is still one of the largest crop insecticide markets with an estimated end-user level value of $910 million in 1988.

Three of the major rice insect pests are the rice stemborer, planthopper and the leafhopper. Experience has shown that the borers are best controlled by organophosphate products, while the carbamate insecticides are more suitable for containing outbreaks of the hoppers. The principal organophosphate insecticides used include:

diazinon, fenitrothion, malathion, phosphamidon, EPN, monocrotophos,

methamidophos, dimethoate, methyl parathion, azinphos and phenthoate,

while the leading carbamate insecticides for use on rice include:

cartap, MTMC, MPMC, isoprocarb (MIPC), BPMC, carbofuran, carbaryl and

benfuracarb.

The range of insecticidal products which are used on rice is fairly large, and, as a result, no single agrochemicals producer holds a dominant share of the market. The Japanese companies, such as Sumitomo, Kumiai, Hokko, Mitsubishi and Takeda are all well represented in their local market as well as in the southeast Asian markets generally. Other companies which have significant presences in the rice insecticide sector include Ciba-Geigy, Bayer, Shell and FMC.

The application of insecticides over the years has been proven to provide rice growers with considerable benefits. For example, it has been estimated in Japan that the annual rice yields would be reduced by at least 25% if no insecticides were applied. Additionally, the value of rice protected has been shown to be seven times the expenditure involved in purchasing and applying insecticides to rice.

As previously indicated, it is estimated that 94% of the total market for rice insecticides in 1988 was in Asia, with the remainder having been applied in areas where high quality rice is produced, such as in the United States of America, Spain, Italy and to a lesser extent in the Soviet Union.

Japan, with rice insecticide sales totalling $455 million in 1988, accounted for half of the worldwide sector value. Over the past forty years, Japanese growers have made great technical advances in the planting and growing of rice such that there is currently an over-production situation. As the improvements in the rice production methods were introduced, with the associated changes in the environment around the paddy fields, came a greater occurrence of insect pests. Early control was achieved through the use of DDT, BHC, parathion and EPN. These products were virtually banned in Japan in the late 1960s and replaced by less toxic products such as malathion, phenthoate, trichlorfon and diazinon. At about the same time carbamate insecticides were introduced to control hopper infestations, but the onset of resistance to these products led to a resurgence in the use of organophosphates, both alone and in mixtures with carbamates.

In Japan, several species of leafhoppers and planthoppers are generally acknowledged as serious rice pests, because they significantly damage rice production, either by transmitting virus pathogens or by feeding on the plants. Further, the development of insecticide resistance among these pests has become a major problem.

An unusual feature of the rice agrochemicals market, and one where the pioneering work was done in Japan concerns the use of insecticide / fungicide mixtures. Nearly one-quarter of the rice insecticide applications in Japan are in the form of these mixtures with fungicides. They are mainly used for simultaneous control of insect pests and sheath blight, or blast, greatly contributing to labour savings in the middle and late stages of paddy rice cultivation.

A recent significant problem has arisen in Japan through the accidental introduction, believed to have been in imported hay from California, of the rice water weevil, in 1976. Within ten years, the pest had spread throughout all of the prefectures, and infested an area equivalent to two-thirds of the total planted area in 1988. Over the past few years, in excess of Yen20,000 million has been spent annually on the control of the rice water weevil, i.e. over $160 million each year. To put this expenditure into perspective, the control of this one insect pest in Japan costs 50% more than any other country spends on its total insect control programmes in rice currently.

Incidence of the rice water weevil was around 20% higher in Japan in 1988 resulting in carbosulfan becoming one of the country's largest selling rice insecticides. Other recent introductions which have also performed well include benfuracarb and buprofezin. However, earlier introductions such as diazinon, disulfoton, BPMC and cartap still remain in significant demand.

In the second largest rice insecticide market, the People's Republic of China, the most important rice insecticides in 1988 were the organophosphates, methamidophos, with over 40% of the sector's share, and dimethoate, with a further 20%.

The wide diversity of insecticides which are used to control insect pests in rice is amply illustrated by the number of different products which command the largest market shares in the other major country markets. For example, in South Korea, the largest selling rice insecticides are cartap, fenobucarb and carbofuran, in India, endosulphan, quinalphos and phosphamidon, in Taiwan, carbofuran and buprofezin while in the Philippines the organophosphates, monocrotophos, methyl parathion and azinphos-ethyl are the most important rice insecticides. However, in Indonesia, the organophosphorus insecticides were banned in November, 1986, since widespread resistance to these products was evident in the brown planthopper. As a result, the carbamate insecticides, carbofuran and fenobucarb, between-them, accounted for 75% of the country's rice insecticide sales in 1988.

Research and development targets within the rice insecticides sector still remain the planthoppers and leafhoppers, with the rice leafroller and, more recently, the rice water weevil being other important insect pests requiring control measures.

Rice Herbicide Market

The rice herbicide market, like the rice insecticide sector, has not kept pace with the overall agrochemical market value increase in recent years. There are two principal reasons for this, namely that the developing nations continue to utilise cultural methods to control weed populations, augmented by hand weeding, while the recent Japanese acreage reductions directly impact herbicide demand levels.

Nonetheless, the sector, with an end-user market value of $885 million in 1988 is the fifth largest crop herbicide market. As the data in Table 1 indicate, Japan accounted for 64.4% of the global rice herbicide market in 1988, the rest of Asia only 17.4% and the remainder of the world, principally the United States of America, Europe and Brazil, 18.2%. The Japanese rice grower demands the highest technical profile and the maximum level of control from his herbicides, no doubt influenced by the guaranteed producer price, which is around six times that paid to Californian growers.

A great number of cultural methods have evolved to suppress the growth of weeds which compete with rice. These include burying the weeds as part of the preparation of the paddy, control of the water depth to drown the weeds, hand weeding and transplanting rice seedlings to give the crop a start over the weeds. However, all of these options are fairly expensive, both in terms of time and money, especially where the availability of labour is limited. The task is further complicated by the fact that some weeds, such as the various species of barnyard grass, are difficult to distinguish from rice in their early stages.

Rice herbicides, initially 2,4-D, were introduced into Japan over forty years ago, and proved their worth by significantly reducing the labour requirements in paddy rice cropping. However, a significant increase in perennial weeds infestations resulted, a problem which had not been severe in the era of manual weeding. This aspect, coupled with the ever present threat of pollution from the drainage systems, necessitated the development of extremely safe herbicides, and a desire to reduce the quantities applied to as low a level as possible. The agrochemicals industry has generally succeeded in meeting these grower and environmental needs, since, with only a few exceptions, all of the rice herbicides currently in use in Japanese paddy fields have been introduced in the last twenty years.

Until the introduction of the one-shot herbicide, a fairly sophisticated weeding procedure for Japanese wetland rice had evolved, as follows:

a) first weeding, which takes place between the initial ground preparation for planting and 4-5 days after planting. Products used in this first weeding include oxadiazon, butachlor, CNP, thiobencarb and chlormethoxynil.

b) second weeding, between 10 and 25 days after planting. Herbicidal mixtures are extremely important at this stage, specifically those containing MCPB and simetryne with either molinate or thiobencarb.

c) third and final weeding, at 30 to 35 days after planting. Primarily to control perennial weeds, and bentazone is often the product of choice.

Under this scheme, the average number of herbicide treatments to rice in Japan rose to nearly 2.5 annually, although differences in growing conditions meant that the actual number of applications varied considerably from region to region. Pressure was then exerted to develop new herbicides, effective against both annual and perennial weeds, which could reduce the spray frequency from two to three times to only once per season. Such herbicides would also further reduce labour costs and the quantities of agrochemicals applied.

The initial one-shot herbicides were introduced in 1982, and are used as early-stage herbicides, applied shortly after the rice plants have been transplanted. The growers can now select the herbicide most suitable for their particular planting conditions and the species of weeds present. Follow-up treatments may still be required in cases where perennial weed infestations are either severe or seasonally variable. One of the effects of the increasing use of these one-shot herbicides has been the reduction in the average number of rice herbicide treatments in Japan to 2.2 in 1988.

One-shot products were applied to 58% of the Japanese paddy rice acreage in 1988, and although one of the original introductions, Kusakarin, a butachlor pyrazolate mixture, was still the largest selling rice herbicide in Japan, its lead position was being seriously threatened by the more recently launched one-shot products based on bensulfuron-methyl, specifically in mixture with either thiobencarb or dimepiperate.

Because of the reduced technical sophistication of the grower in the rest of Asia, allied to the high price of these one-shot products, the rice herbicide markets are totally different in these countries. For example, in South Korea, butachlor, either alone, or in mixtures with naproanilide usually, accounted for over 70% of the market in 1988. In Taiwan, the major rice herbicides are glyphosate, paraquat and butachlor mixtures, while in India, butachlor accounts for over 90% of the rice herbicide market. 2,4-D and pretilachlor are important in both the Philippines and Thailand, as is fenoxaprop-ethyl in the latter country. Rice herbicide usage in China is low and is satisfied from local production of 2,4-D, trifluralin and MCPA.

Outwith Asia, the major markets are in the USA, Brazil and Europe. In the United States, propanil is the leading rice herbicide followed by bentazone and thiobencarb. Molinate is the market leader in the west European rice growing regions of Italy and Spain, while propanil and molinate, alone or in combination with each other, commanded nearly 70% of the Brazilian market for rice herbicides in 1988.

Not surprisingly, most of the research and development effort in the rice herbicide sector is directed at improving the Japanese one-shot product range, through broadening the weed spectrum and lengthening the application periods. This will remain the prime focus, until the other Asian nations can afford the social consequences of increased chemical control of weeds in rice.

Rice Fungicide Market

Rice is attacked by a number of fungal diseases, the most important of which is blast. Although rice blast is a serious problem in most rice growing countries, it is largely in Japan that progress on fungicidal control has been achieved. The use of fungicides is only one factor in controlling the incidence of blast disease, and their timely applications must be employed in conjunction with the use of healthy seeds, early removal of seedlings for complementary planting and application of the proper amount of fertilisers.

In Japan especially, the changes in rice cultivation, such as high density planting and application of large amounts of fertilisers to increase the yield, resulted in the onset of other disease problems as well as blast. For example, the increase in early season rice cultivation through the use of vinyl film in nurseries in the late 1950s caused the incidence of sheath blight to reach serious levels. Virus diseases such as stripe, dwarf, black-streaked dwarf and yellow dwarf, all of which are transmitted by leafhoppers and planthoppers, also increased rapidly with the propagation of early season rice cultivation in Japan. Thus, the type of problems which have beset west European cereals farmers for the past ten or fifteen years have been common to Japanese rice growers for much longer.

The increased planting of IRRI varieties in recent years in southeast Asia has led to much greater disease incidence in these areas, but, since the varieties used in Japan are japonica, these extra problems have not invaded Japan.

In view of the importance of disease attacks on the rice yields which can be achieved, it is somewhat surprising that only five country markets accounted for over 94% of the global rice fungicide market in 1988. The Japanese market alone was responsible for close on two-thirds of the total world value of $570 million.

On average, in Japan, the number of applications to try to control rice blast is 2.5 annually. Early control of this disease was effected by the use of organomercury compounds, although, since the late 1960s these have been banned in Japan. A number of highly effective antibiotics, organophosphates and organochlorine products were subsequently introduced, but resistance problems have been increasingly encountered.

Several application methods have been established, such as granule application of systemic fungicides to nursery boxes immediately before transplanting, submerged application of systemic products in the field to prevent leaf or panicle blast, and foliar spraying and dusting by ground and aerial applications. Methods and timing of applications vary in the regions as the climatic and cultural practices are different.

Throughout the 1980s, the incidence of rice blast has been falling in Japan, but whether this has been due to the various methods which have been increasingly employed, or whether to advantageous weather conditions, or a combination of both factors, has yet to be established.

Sheath blight has been treated for many years with organoarsenic compounds with assistance more recently from polyoxin and validamycin antibiotics. In Japan, the area affected by sheath blight is on a long-term increasing trend, and the only satisfactory control method is by fungicidal applications.

It has been estimated that annual crop losses in Japan attributable to sheath blight have been in the region of 500,000 tonnes annually over recent years, a more than 50% decrease on the losses which were the norm throughout the 1960s. The damage prevention effect has been as a result of the chemical control methods practised. In this respect, growers have gained access to three new sheath blight control fungicides, pencycuron, flutolanil and diclomezine in the past four seasons.

A number of economically important diseases do not have satisfactory control measures, including bacterial leaf blight and virus diseases, while seedling blight remains a problem. Additionally, the "Bakanae" disease, a typical seed-borne disease that occurs from nursery to paddy, has re-emerged as a major problem in the past eight to ten years as a result of the onset of resistance to benomyl preparations.

Because of the intensive nature of rice cultivation in Japan, a number of fungicides are applied annually. It is estimated that the principal products which were used in 1988 were probenazole and isoprothiolane. The situations in the second and third largest rice fungicides markets, South Korea and the People's Republic of China, closely resemble Japan, although the changes which have taken place in Japan have been adopted at a much slower rate. In South Korea, the major products used are neo-asozin, edifenphos and tricyclazole, while around 80% of the Chinese market is satisfied by three fungicides, iprobenfos, isoprothiolane and edifenphos.

In Taiwan, rice blast is preferentially controlled with fthalide and tricyclazole, with pencycuron being widely used to effect sheath blight control. Between them, these three products accounted for nearly 60% of the rice fungicide market in Taiwan in 1988. The other major market, India, has edifenphos, carbendazim and iprobenfos as the principal rice fungicides, with over 75% of the market between them in 1988.

The major target diseases which affect rice, and yields thereof, which remain the principal foci for research and development are rice blast, sheath blight, bacterial leaf blight and "Bakanae" disease, with the Japanese market being the main testing ground.

Other Agrochemicals Used on Rice

It has been estimated in Table 1, that over 98.5% of the agrochemicals applied to rice were either insecticides, herbicides or fungicides in 1988. The remainder of the market is in plant growth regulators and molluscicides, although both sectors are currently small in value terms.

Especially in Japan, the development of chemical products that could prevent rice products from lodging has been eagerly awaited. The first specific anti-lodging agent for paddy rice, inabenfide, was registered in Japan in October, 1986. The initial reactions from the growers have been favourable, although the product's application rate is high and the application must be made 40 to 60 days before heading. More recently, anti-lodging products which are active at significantly lower dosage rates have gained approval, and this sector of the rice agrochemicals market is expected to witness explosive growth over the next few years, primarily in Japan, but to a lesser extent in South Korea and Taiwan also. This will be especially so if consistent yield enhancement is achieved.

The application of molluscicides to paddy rice is largely confined to Taiwan and the Philippines, and has arisen because attempts to introduce snails into the local diets were not welcomed by consumers. The unwanted snails were then dumped in irrigation ditches and public waterways with disastrous consequences for farmers.

AGROCHEMICALS USAGE IN RICE - THE FUTURE

It has been estimated that the world's annual rough rice production must increase from the recent levels of around 470 million tonnes to 560 million tonnes by the turn of the century, and to 760 million tonnes by the year 2020, just to keep pace with population growth. In effect, the increases in rice production which have been achieved in Asia in the past twentyfive years must be repeated over the next three decades, i.e. a further doubling of the Asian rice harvests.

In view of the constraints on land resources, a significant portion of the increased production will have to come from yield increases, an area where agrochemicals can make a very significant contribution. If the Asian rice producing area increases by 10% over the coming thirty years, to over 142 million hectares, and production rises by over 60% to 690 million tonnes to keep pace with population growth and allow exports to be made to accrue some foreign currency, then yields must rise to nearly 4.85 tonnes per hectare, up a total of 47%, or by an average minimum of 1.2% per annum.

The great production gains of the 1960s and 1970s were largely achieved through the introductions of improved varieties on the irrigated and favourable rainfed areas. Plant breeders have now turned their attentions to a new range of improved rices for environments which are plagued by drought, flood, low temperature and saline soils. Such successes will result in rice production increases, but judicious use of agrochemicals in the under-developed countries is now essential to safeguard the productivity gains which have already been engineered by modern agriculture.

In all probability, there will be at least a further 20-25% reduction in the area grown to rice in Japan by the year 2020. This will lead to a corresponding decrease in the value of the market for rice agrochemicals there. Indeed, if the authorities decide to use the producer procurement price as an anti-production weapon, the reduction in rice agrochemicals usage will be even greater. It is possible that the introduction of good tasting cultivars of rice plants, which will be more susceptible to disease attacks and which tend to be tall growing, will increase the necessity for fungicide and growth regulator sprays.

However, even if antiviral agents and/or microbial products are developed to control virus diseases in rice, unless they are sufficiently cheap to allow them to be applied in the under-developed nations, then the long-term outlook for the rice agrochemicals market is pessimistic, although there may well be some short-term demand increases. Unfortunately, because of limitations on the supply of money to purchase yield-increasing agrochemicals, the long-term rice production goals may not be achievable.

Finally, Integrated Pest Management (IPM) techniques have been proposed to overcome some of the problems associated with widespread agrochemicals usage. However, experience in other crops has shown that these concepts, while often applicable on a limited geographic basis, are generally not transferrable to other regions. Of more significance in the future will be the quick and widespread adoption of Resistance Management Programmes. When combined with the introduction of new, environmentally more acceptable agrochemicals, these programmes will provide the greatest prospects of achieving the required rice yield targets, while limiting to a minimum the adverse effects of agrochemicals usage.

FOLIAR FUNGICIDES FOR CONTROL OF RICE DISEASES IN THE UNITED STATES

D. E. Groth, M. C. Rush and G. D. Lindberg
Rice Research Station, P. O. Box 1429, Crowley, LA 70527
and Dept. of Plant Pathology and Crop Physiology,
Agricultural Experiment Station,
Louisiana State University Agriculture Center
Baton Rouge, LA 70803, USA

ABSTRACT

The most common and important foliar diseases of rice in the southern United States are sheath blight, blast, stem rot, brown leaf spot, narrow brown leaf spot and leaf smut. Fungicides have been evaluated for more than 20 years for efficacy against these diseases at the Rice Research Station in Crowley, Louisiana, USA in small plot experiments. Five fungicides are presently registered for commercial use. The commercial fungicides benomyl, propiconazole, iprodione, copper plus sulfur, and thiabendazole have received registrations since 1976. Benomyl, propiconazole, and Iprodione have similar activity against sheath blight but thiabendazole and copper plus sulfur are not recommended because of poor performance. Benomyl also has activity against blast and narrow brown leaf spot. Propiconazole gives excellent control of narrow brown leaf spot and leaf smut. Iprodione suppresses the brown leaf spot disease. Several new fungicides which are narrow spectrum and specific for sheath blight control will soon be registered.

INTRODUCTION

Damage caused by diseases on rice in the United States can greatly impair productivity and in some cases completely destroy a crop. Some of the most important and common foliar diseases in Louisiana include sheath blight (_Rhizoctonia solani_ Kuhn), blast (_Pyricularia oryzae_ Cav.), stem rot (_Magnaporthe salvinii_ (Catt.) Krause & Webster), brown leaf spot (_Cochiobolus miyabeanus_ (Ito & Kur. Drech.), narrow brown leaf spot (_Cercospora oryzae_ Miyake), and leaf smut (_Entyloma oryzae_ H. & D. Sydow) (1).

Sheath blight is the most important disease causing significant losses each year (2,3). This disease is favored by the high temperatures, humid conditions, high fertility levels, thick stands and dense crop canopies which typically exist in the rice areas of the southern United States. Small dark-brown sclerotia survive between crops on straw and in the soil. These sclerotia float in the flood water, come in contact with rice tillers, and infect plants at or near the water line. The disease progresses up the plants, forming a "snakeskin" pattern of banding on the stem and leaf surfaces. The disease spreads from plant to plant outward through the field, causing circular areas of dead and collapsing rice plants. The disease develops rapidly at the boot and heading stages. Medium-grain varieties are more resistant than long-grain varieties.

Blast is the second most important disease of rice in the United States. This disease is more sporadic in occurrence than sheath blight. This is probably due to higher levels of resistance and generally unfavorable weather conditions. The blast fungus overwinters in rice straw, stubble, and on seeds. The disease spreads rapidly in the field by means of airborne spores. In the vegetative stage of rice, elongated, spindle-shaped lesions with brown borders appear on the leaves. Severe infestations can lead to large areas of dead plants. Leaf blast development is usually associated with the loss of flood or prolonged delay of flooding (4). Excessive N levels may also increase disease severity. Correct water management and application of a foliar fungicide are the most important control measures at this stage. After heading, brownish lesions can develop on the node at the base of the head, causing sterile florets or "blasting" followed by breaking over of the head to produce the "rotten-neck" symptom. Symptoms also occur on the nodes of the stem and at the base of the flag leaf blade. Preventative fungicide sprays at boot and heading can suppress rotten-neck symptoms, although the primary control method used is disease resistance. Cultivars differ greatly in their level of resistance and selection of a resistant cultivar is the most important decision a farmer makes.

Stem rot has been a major disease in the United States but has become a minor problem in the last few years. The fungus survives as pin-head size fungal sclerotia in rice straw and in the soil. Sclerotia float in the flood water, attach to the plant, germinate and cause infection. Black angular lesions with a yellowish border develop on the leaf sheath near the water surface. The infection progresses into the culm and may lead to breaking or collapse of the culms, causing lodging. Sclerotia develop inside the leaf sheaths and stems before and after maturity. Stem rot is most severe when rice is grown under low potassium levels but this seldom occurs because of extensive fertilization programs.

Brown spot can be an important disease which is associated with poor growing conditions, including low fertility and other stressful conditions. The pathogen can be seedborne but can also survive in rice straw or stubble. It is spread by wind blown spores. The disease is characterized by circular to slightly elongate reddish brown lesions on the leaf, often surrounded by a yellow or gold halo. Under severe conditions, spots can have grey necrotic centers. Most varieties have a good level of resistance, but the major control practice is to avoid stressing the plant.

The narrow brown leaf spot fungus overwinters in rice straw and stubble. The fungus spreads by wind-blown spores. Symptoms appear late in the season and appear as either linear brown lesions on the leaves, leaf sheaths and floral parts or on the panicle causing damage similar to rotten-neck blast except that the internode instead of the node is affected. Net blotch-like symptoms appear on lower leaf sheaths at or below the collar causing death of the leaf blade. Premature ripening, lodging and yield loss may occur. New cultivars are usually released as resistant, but after 3 to 4 years they become susceptible.

Leaf smut is extremely common in the United States but has not been shown to cause significant damage. The main problem is cosmetic and farmers tend to be overly concerned by this disease. The fungus overwinters in soil and dead leaves. The fungus is spread by airborne spores. Typical small, slightly raised black spots appear on both sides of the leaves and sometimes on the sheath. Leaf smut occurs late in the season and no control measures are currently recommended.

Several fungicides have been tested in previous years for blast control. Most of the effective agents were heavy metals including phenyl mercuric acetate, triphenyltin hydroxide (Duter) and the copper containing compounds including Bordeaux mixture (5, 6, 7). Several antibiotics have also been tested. Tin compounds were only granted 2 years of emergency registration. The copper containing compounds tend to be phytotoxic on rice and if disease is not serious, cause more damage than good. Benomyl was the first modern fungicide registered on rice in the United States (8).

Ultimately, control of rice diseases will involve an integrated disease management system using a blend of pesticides, disease resistance, cultural practices, biological control and regulatory procedures. To control these rice diseases the farmer first must know how to identify the major diseases and scout fields regularly for their occurrence. Control measures presently used include the use of resistant cultivars, moderately effective fungicides and cultural practices that reduce disease and lower inoculum survival rate. Resistant, commercially acceptable cultivars are not always available for all of the diseases and often cultural management is ineffective and impractical, so the rice producer must rely on the use of fungicides as his main line of defence. Therefore, it is even more important that safe effective fungicides are found, developed, and made available to Louisiana rice growers for use in disease management programs. The objective of the rice pathology program at the LSU Agricultural Center has been to evaluate new and experimental fungicides for effectiveness in controlling various rice diseases and improving yields and to determine the most effective timing of application of these fungicides.

Descriptions of the major diseases of rice and general control information has been published and recently updated by Oh (9).

MATERIALS AND METHODS

Fungicide evaluation trials have been conducted at the Rice Research Station in Crowley, LA for the last 18 years and on the Errol Lounsberry Farm in Lake Arthur, LA for the last 3 years (10 through 41). Various cultivars have been used including Lemont, Labelle, Saturn, M201, and Early Calusa #2. Plots consisted of seven rows 4.9 to 7.6 m long with 18 cm spacing. Experiments were arranged in randomized complete block designs with four to six replications. Standard agronomic practices were used to manage the tests (40). Sheath blight inoculated plots had a moist, autoclaved rice grain:rice hull (1:2) medium culture of R. solani applied at 5-7 weeks after emergence (shortly before the first internode elongation stage of growth) at the rate of 80-100 ml of inoculum per plot to produce uniform disease development. Fungicides were applied in water at 93 to 280 l X ha^{-1} with a CO_2 pressurized backpack type sprayer equipped with 2-4 nozzle boom with flat fans or cone tips at various growth stages of the rice plant. These growth stages included green ring (GR)(first internode starting to elongate), panicle differentiation (PD)(panicle 2 mm), booting (B)(panicle 1-5 cm) and heading (H)(70-80 percent of panicles exerting). Fungicides were applied singly, in combinations, or sequentially to plots. In general these treatments were applied between June 15 and August 15. Disease ratings were taken at 2 to 14 days before harvest using 0 to 9 severity ratings scales for sheath blight (Table 1), brown leaf spot (Table 2) narrow brown leaf spot (Table 3) and leaf smut (Table 4). In addition, infestation levels of sheath blight (percent tillers infected and percent tillers dead at maturity) and blast (percent rotten-neck, percent panicles infected) were recorded. Sheath blight and blast control were evaluated in separate trials on different cultivars at the same location. The center four rows were harvested with a small plot combine. Sample weight and moisture were determined, and rough rice grain yields were calculated as kg X ha^{-1} at 12% moisture for each treatment.

RESULTS

At present there are five fungicides registered for rice in the United States. These are benomyl (Benlate 50WP & 50DF), iprodione (Rovral 4F & 50WP), propiconazole (Tilt 3.6EC), thiabendazole (Folatec 3.8F), and copper plus sulfur (Top-Cop).

Table 1
Rating system for determining sheath blight severity on rice

Disease rating	Sheath blight development
0	Plants healthy, no symptoms.
1	Restricted oval lesions at water-line or inoculation points, lesion centers grey-green to nearly white, margin of lesion a broad red-brown or purple-brown border usually broader than necrotic center, less than 2.5% of tissues affected.
2	Few oval or coalesced lesions on lower sheaths or at infection points, lesions with broad red-brown border, 5% or less of tissues affected.
3	Lesions on lower leaf sheaths or at inoculation points, lesions with narrow red-brown border, coalescing, less than 10% of tissues affected.
4	Lesions mainly restricted to sheaths on lower third of plant, lowest leaves, or inoculation points, lesions discrete or coalescing with narrow red-brown border, 10 to 15% of leaf and sheath tissues affected.
5	Lesions mainly restricted to sheaths and leaves of lower half of plants, lesions usually coalescing with large necrotic centers and narrow red-brown borders, 15 to 25% of tissues affected, culm not injured.
6	Lesions usually coalescing and affecting lower 2/3 of sheath area of plant, lesions extending to blades of lower leaves or lower leaves killed by injury to sheath, 25 to 40% of tissues affected, culm of infected tillers usually not affected.
7	Lesions usually coalescing and affecting lower 3/4 of sheath area of plant, lesions extending to leaf blades of lower 2/3 of plant, 40 to 60% of tissues affected, outer portion of culm may be brown or have brown streaks near water-line.
8	Lesions reaching to flag leaf, lower sheaths with coalesced lesions covering most of tissue, lower and middle leaves dead or dying, 60 to 80% of tissues affected, culms with brown streaks or turning light brown to center and water-soaked, severely affected tillers lodging, florets in lower 1/3 of panicle often not filling.
9	Lesions reaching to flag leaf, lower leaves mostly dead, sheaths dried, culms brown, water-soaked or collapsing, most of tillers lodged, florets in lower 1/3 to 1/2 of panicle not filling.

Table 2

Rating system for determining brown spot (Cochiobolus miyabeanus) severity on rice

Disease Rating	Brown spot development
0	Plants healthy, no symptoms
1	Few to many dark specks on pin-head size, no necrosis (collapsed cells).
2	Dark brown specks, 0.5 to 1 mm in diameter, no necrosis.
3	Small round or oval brown spots, 1 to 2 mm in diameter, no necrosis or grey in centers.
4	A few (10/leaf or less) dark brown spots, 2 to 3 mm in diameter, grey necrotic area in center.
5	Less than 15 lesions/leaf of typical circular or oval spots, spots 2 to 4 mm with grey necrotic centers and brown margin, many have a chlorotic halo around spot.
6	Fifteen to 25 typical brown spot lesions/leaf; circular, oval,or sometimes linear, 3 to 5 mm in diameter or length, with large necrotic center and brown margin, often with yellow or gold halo around lesion.
7	Less than 50 lesions/leaf, lesions oval to elongated, 4 to 6 mm in diameter or length, Lesions mainly with grey necrotic centers, narrow brown margin.
8	Many (50 to 75) lesions/leaf of group 7 size or larger, less than 25% of leaf area killed by coalescence of lesions.
9	More than 75 lesions/leaf of group 7 size or larger (usually 6 to 10 mm) more than 25% of leaf area killed by coalescence of lesions.

Table 3
Rating system for determining narrow brown leaf spot
(Cercospora oryzae) severity on rice

Disease rating	Narrow brown leaf spot development
0	Plants healthy, no symptoms.
1	Few to many dark specks of pin-head size, no necrosis (collapsed cells).
2	Dark brown specks, 0.5 to 1 mm in diameter, no necrosis.
3	Small linear reddish-brown spots, 1 to 2 mm in length, about 0.5 mm wide.
4	A few (15/leaf or less) reddish-brown spots, 3 to 4 mm in length, less than 1 mm wide.
5	Less than 25 lesions/leaf of typical linear spots, spots 5 to 6 mm long and less than 1 mm wide.
6	Twenty-five to 50 typical narrow brown spot lesions/leaf; linear, 7 to 10 mm in length, 1 mm wide.
7	Less than 75 lesions/leaf, lesions 4 to 6 mm in length, some lesions may have grey necrotic centers with narrow brown margin, 1 to 1.5 mm wide.
8	Many (75 to 100) lesions/leaf of group 7 size or larger, less than 25% of leaf area killed by coalescence of lesions.
9	More than 100 lesions/leaf of group 7 size or larger (usually 10 to 20 mm) more than 25% of leaf area killed by coalescence of lesions.

Two fungicides, flutolanil (Moncut 50WP) and pencycuron (Monceren 75WP), are close to being registered. Most of the results and discussion will be based on these fungicides but numerous other compounds have been extensively tested. A list of these fungicides and their performance are presented in Table 5.

Table 4
Rating system for determining leaf smut (<u>Entyloma</u> <u>oryzae</u>) on rice

Rating	Leaf smut development
0	No evidence of infection.
1	Only one to three lesions of minute size, less than 0.5 mm in length and 0.1 mm in width, produced on the flag leaf.
2	About 4 to 10 lesions of minute size, less than 0.5 mm in length and 0.1 mm in width, produced on the flag leaf and distributed over 1.0 to 2.0% of the total flag leaf area.
3	Not more than 100 sori of about the same size as class 2, and distributed over to 10% of the total flag leaf area.
4	More than 100 sori of the size 1.0 mm in length and 0.5 mm in width distributed over about 10 to 30% of the total flag leaf area.
5	Medium size sori, about 1.5 mm in length and 1.0 mm in width, distributed over about 30 to 50% of the total flag leaf area.
6	Larger sori, about 2.0 mm in length and 1.0 mm in width, distributed over about 50 to 60% of the total flag leaf area.
7	About the same size of sori as class 6 but distributed over 60 to 70% of the total flag leaf area. Ten to 25 percent of the flag leaf turns necrotic from the blade tips.
8	Large sori, 2.5 to 3.5 mm in length and 2.0 mm in width distributed over 70 to 80% of the total flag leaf area. About 25-30 percent of the flag leaf turns necrotic from the tips.
9	Almost all sori coalesce and the size of the lesion is not discernible. The flag leaf area is covered by sori and 30 to 60 percent of the leaf turns necrotic from the tip.

Table 5
Fungicides that gave significant yield increases when tested at the LSU Rice Research Station, Crowley, Louisiana, USA for their effectiveness against rice diseases

CHEMICAL	COMPANY	LFST	BRSP	NRBR	SHBT	BLST
OAC-2481	Olin	NS	NS	NS		+
DU-TER	Griffin	+	+	+	+	+
MANZATE 200	Dupont		NS	NS	+	
BRAVO 6F	Diamond Sham.	+	-	+		
RH-3928	Rohm & Haas		-	+		+
HINOSAN	Mobay	+				+
DIFOLATAN	Valent	+	NS	+		
EL-291	Elanco	+	NS	+		+
RH-2161	Rohm & Haas	+	+	+		+
DITHANE M45 F	"	+	+	+		+
BEAM	Elanco	+	+	+		+
VITAVAX	Uniroyal				+	
SUPERTIN 4L	Griffin	+	+	+	+	+
TOPSIN-M	Penwalt	+	+	+	+	+
BAYCOR	Bayer	+	+	+	+	
EL 236	Elanco	+	+	+	+	
OAC-3910	Olin				+	
FUJI-ONE	-				NS	+
FOLICUR	Mobay	+	+	+	+	NS
PP523	PPG	+	+	+	NS	
EL-228	Elanco		+	+		-
ORTHOCIDE	Chevron	+	NS	NS		
BAYLETON	Mobay	+	NS	+	+	
GLYOCEX	-	+	NS	+		
DPX 14	Dupont	+	NS	+		
DPX770	"				+	
DPX H6573	"	+	+	+	+	
DPX 965	"				+	+
LY156236	-	NS		NS	+	
ME 227	-		NS	+	+	
PROCHLORAZ	Nor Am	+	NS	+	+	+
Mon-24077	Monsanto	+			+	
BAS-480 00F	BASF	+			+	
SAN 619F	Sandoz	+			+	
CGA-455	Ciba-Geigy	+			+	
A-1055		+	+	+	+	+
SDS-45037		NS	NS	+	NS	
EF-705		NS	+	NS	+	

+ = Significant (P=05) decrease in disease or increase in yield, - = significant increase in disease, ns = no significant effect. LFST = leaf smut, BRSP = brown spot disease, SHBT = sheath blight, and BLST = blast disease

YIELD

Benomyl has been registered for rice in the United States since 1976. In 1987 the 50DF (dry flowable) formulation was made available. The use of this product has resulted in a 550 to 700 kg X ha^{-1} average yield response (Fig 1 & 2). Benomyl is broad-spectrum in its activity, but has the drawback of only suppressing sheath blight (Fig 1). Benomyl is applied as two 1.1 kg X ha^{-1} applications, one at the early boot (1-5 cm panicle in the boot) and the second at 70-80% of the panicles emerging.

Another registered fungicide is thiabendazole. This fungicide has consistently been less effective than benomyl in disease control and yield response (Fig. 1). Thiabendazole is not recommended because of its poor and erratic performance (Fig. 2). One advantage of this fungicide is that it is relatively inexpensive and has some activity against both blast and sheath blight. High rates of this fungicide must be used and even then disease control may not be satisfactory.

A recently registered fungicide is propiconazole which was registered in 1987. Propiconazole is applied either as two 0.45 l X ha^{-1} applications at green ring (first joint elongating) and early boot or as a single 0.741 l X ha^{-1} application at early boot (panicle 0.5 to 5 cm long). If disease starts to develop again after these treatments, another registered fungicide will have to be used at heading since propiconazole can not be applied to an exposed panicle because of registration restrictions. The major limitation of propiconazole is that the low rates allowed by the registration are the minimum effective rates for disease control. Fifty to 100 percent increases in use rates would be more effective. If the timing is right, that is when the disease just begins to develop upward during the late jointing stages of growth, and environmental conditions are not too conducive for disease development, high yield increases are possible for these use rates (Fig. 2). Significant yield increases are not achieved when disease pressure is high (Fig. 2, 1988).

The most recently registered fungicide (1989) is iprodione. Iprodione is similar to benomyl in yield response (Fig 1). Iprodione is applied as two 0.55 ai kg X ha^{-1} applications at early boot and heading. Yield increases are not always as high with this compound as would be expected from the amount of disease control given (Fig. 1). However, iprodione is a very consistent performer and the yield increases over the unsprayed check have been the most stable of any of the fungicides registered (Fig 2).

Pencycuron is an experimental fungicide that gives excellent yield responses when sheath blight is the primary disease (Fig. 1). Pencycuron will probably be applied as two 0.28 kg X ha^{-1} applications at panicle differentiation (PD) and boot or as a boot application followed by a broad spectrum fungicide such as benomyl (Table 6). The advantage of this sequential spray is it combines a broad spectrum fungicide with a specific fungicide to control several diseases that may limit the yield response due to sheath blight control, especially blast. A sequential spray also reduces the possibility of resistance to one fungicide. A registration is expected in the next few years.

Table 6.
Effect of single and sequential applications of fungicides on
sheath blight development and yield performance over the
unsprayed inoculated check. Data are the average of 4
years/eight tests conducted at the Rice Research Station,
Crowley, Louisiana, USA, 1986-1989

Fungicide(s) [a] used	Rate Kg ai X ha^{-1}	Timing[b] of Application	Sheath Blight Rating (0-9)	Yield Increase (Kg X ha^{-1})
benomyl	0.55	B+H	5.4	732
propiconazole	0.4-0.5	B	5.6	803
propiconazole/ benomyl	0.4-0.5 0.55	B H	4.7	732
pencycuron	0.28	B+H	3.1	1039
pencycuron/ benomyl	0.28 0.55	B H	3.4	1032

[a] / = a sequential spray of two different fungicides
[b] B = booting growth stage and H = heading growth stage

Flutolanil (Moncut) is very similar to pencycuron in
specific activity to sheath blight (Fig. 1). Yield increases
due to sheath blight control have been lower than pencycuron
and more erratic (Fig 2). Again sequential sprays with a
broad spectrum fungicide have been very effective. A full
Federal registration is expected in the next few years.

SHEATH BLIGHT

Benomyl has fair activity against sheath blight in reducing
the disease ratings, percent tillers infected, and maintaining
more live tillers in the ratoon crop after harvest (Fig. 1).
This has a major effect on the yield performance of the second
or ratoon crop. Thiabendazole is weak on all aspects of
sheath blight control. Propiconazole, iprodione and
flutolanil all have activity against sheath blight similar to
benomyl. Propiconazole is applied early in the season and its
residual activity is not long enough. Pencycuron has
excellent activity against sheath blight reducing both disease
ratings and infestation levels.

There is consistently a clear positive trend between
evaluations of sheath blight as expressed as infestation
levels, disease ratings, and dead tillers (Table 7). There
is often a strong correlation between the three disease rating
systems and grain yields (Table 7). Some years though
correlations between disease ratings and yield are not
significant because there are experimental fungicides in the

SB INFESTATIONS

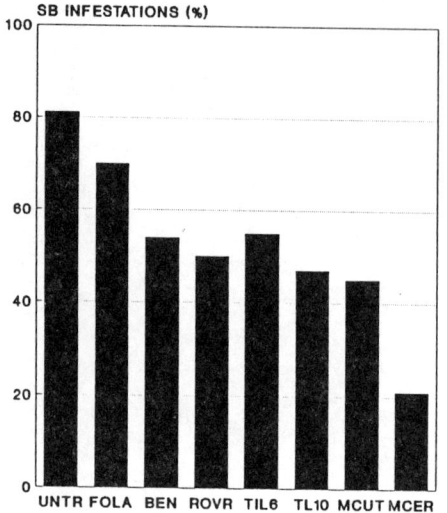

SB RATINGS

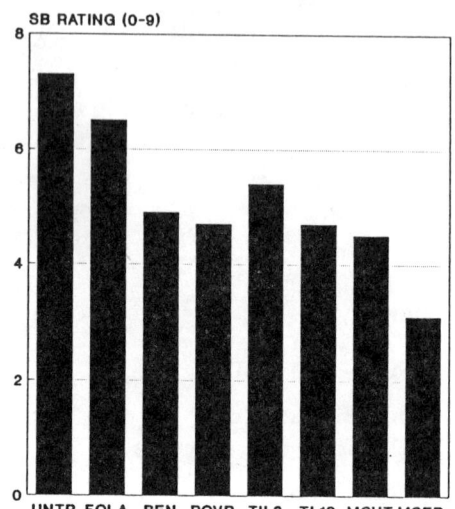

YIELD RESPONSE

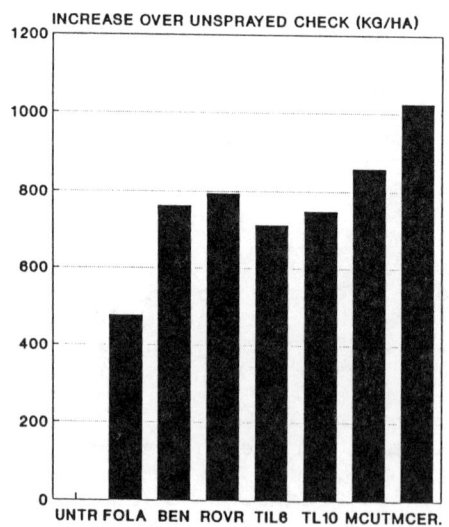

DEAD TILLERS

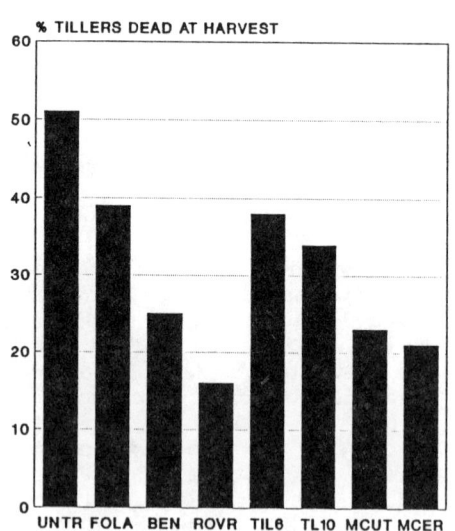

Figure 1. Performance of Benlate (BEN), Folatec (FOL), Rovral (ROV), Tilt as either two 6 oz applications (TIL6) or one 10 oz application (TL10), Moncut (MCUT) and Moncerene (MCER) on sheath blight control and yield performance as compared to the untreated check (UNTR). Data is the average of nine years/15 tests conducted in Louisiana, USA, 1981-1989.

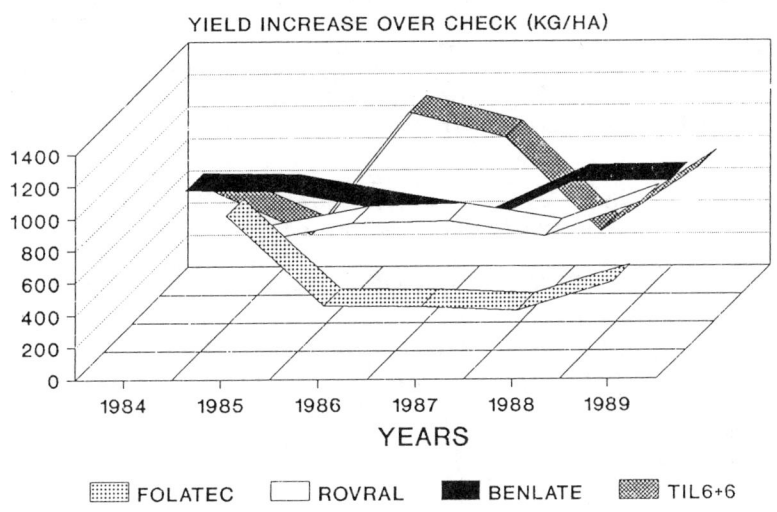

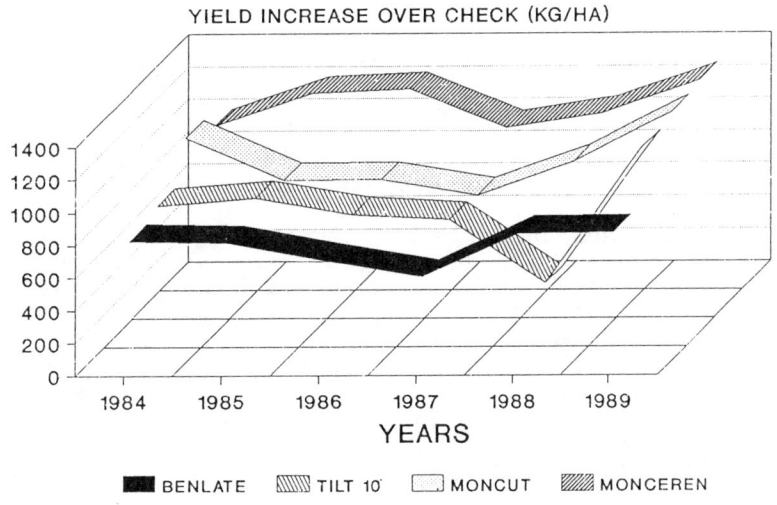

Figure 2. Yield increase due to sheath blight control by Benlate (BEN), Folatec (FOL), Rovral (ROV), Tilt as either two 6 oz applications (TIL6) or one 10 oz applications (TL10), Moncut (MCUT) and Moncerene (MCER) on yield performance over the untreated check (UNTR). Data for each year is the average of all tests conducted in Louisiana, USA.

test which do not control sheath blight but do give a yield response, some that control sheath blight but do not increase yields due to phytotoxic effects, and some fungicides actually increase sheath blight probably due to eliminating antagonistic microorganisms (Table 5). Since disease ratings are well correlated, any one of the rating scales could be used if time is limited, but with a lesser degree of accuracy.

Table 7
Correlation of yield with sheath blight ratings and correlation among rating types

	Yield X Rating Methods[a/]			Rating method X Rating method		
Year	Yield X Infect	Yield X Killed	Yield X 0-9	Infect X Killed	Infect X 0-9	KilledX 0-9
1977	-0.71**	-0.79**	-0.77**	0.92**	0.97**	0.97**
1978	-0.39ns	-0.05*	-	0.87**	-	-
1979	0.27ns	-	-	-	-	-
1980	-0.11ns	-0.13ns	-	0.97**	-	-
1981	-0.71**	-0.80**	-0.65**	0.86**	0.96**	0.87**
1982	-0.56**	-0.80**	-0.65**	0.78**	0.96**	0.88**
1983	-0.47**	-0.49**	-	0.98**	-	-
1984	-0.89**	-0.85**	-0.84**	0.91**	0.95**	0.95**
1985	-0.80**	-0.82**	-0.79**	0.93**	0.95**	0.95**
1986	-0.18ns	-0.23**	-	0.82**	-	-
1987	0.13ns	0.07ns	0.34	0.79**	0.92**	0.92**

[a/] Rating methods were Infected = percent of tillers infected at maturity, Killed = percent tillers dead at maturity, and 0-9 = the 0 to 9 rating scale listed in Table 1. Mean yields for each treatment in a test correlated with each method of measuring disease on each treatment.

In years when there is little natural sheath blight there has been a 1434 kg X ha^{-1} yield increase in the unsprayed uninoculated check over the unsprayed inoculated check (Table 8). Benomyl increased yields an average 805 kg X ha^{-1} and pencycuron increased yields 1032 kg X ha^{-1} which represents 59 and 75 percent returns of the yield lost. The yield increase from using benomyl is probably due to control of other diseases, as well as sheath blight control. The pencycuron treatments lose some yield potential to these other diseases. Using an estimated value of $0.09 dollars per kg of rough rice in the United States, the benomyl sprays returned an average $72.45 per ha and pencycuron returned an average $92.88 per ha. The cost of applying benomyl is about $48.00 per hectare.

Table 8
Comparison of uninoculated, unsprayed; benomyl, and pencycuron
treatments for control of sheath blight on rice and for
increased yield over inoculated, unsprayed plots

Years	Inoculated Rating (0-9)	Yield (Kg X ha^{-1})	Uninoculated Rating (0-9)	Yield (Kg X ha^{-1})	benomyl Rating (0-9)	Yield (Kg X ha^{-1})	pencycuron Rating (0-9)	Yield (Kg X ha^{-1})
1984	8.3	5326	1.9	1103	5.0	847	2.7	787
1985	6.6	6282	2.4	1135	4.6	768	2.6	1154
1986	7.0	7003	1.0	1627	5.0	683	3.5	1610
1988	6.8	5373	3.0	1515	5.0	861	3.5	1162
1989	6.2	5399	0.8	1788	5.0	1044	2.8	447
Mean	7.0	5985	1.8	1434	4.9	805	3.0	1032
Mean % return of uninoculated plot yield						(59%)		(75%)

BLAST

Benomyl has the best activity against blast of the registered
fungicides (Fig. 3). It is able to suppress disease levels
but not completely control rotten-neck blast. Benomyl is also
effective for leaf blast control, but reestablishing the flood
on a drained field often has as much if not more effect on
reducing disease (4). Sprays for rotten-neck blast have to
be preventative since there are no good prediction systems
available at this time. The other registered fungicide with
activity against blast is thiabendazole. It has been
consistently less effective than benomyl and is not
recommended in Louisiana. Propiconazole, iprodione,
flutolanil and pencycuron have little or no effect on blast.
The fungicide tricyclazole (Beam) has performed very well
against blast in our tests, but has not been registered in the
United States. Numerous other fungicides have been tested for
blast control with varying levels of activity (Table 5), but
none are approaching registration at this time.

BROWN SPOT

Of the presently labeled fungicides, only iprodione has
significant activity against the brown spot disease (Fig. 3).
The other fungicides suppress brown spot slightly but not
enough to be significant. Under normal (unstressed rice)
conditions in the field, brown spot is not a problem and the
poor activity of these fungicides should not be a deterrent
to their use.

NARROW BROWN LEAF SPOT

Benomyl and propiconazole treatments have excellent activity
against narrow brown leaf spot (Fig. 3). This disease can be
very severe and a portion of the yield increases these

NECK BLAST

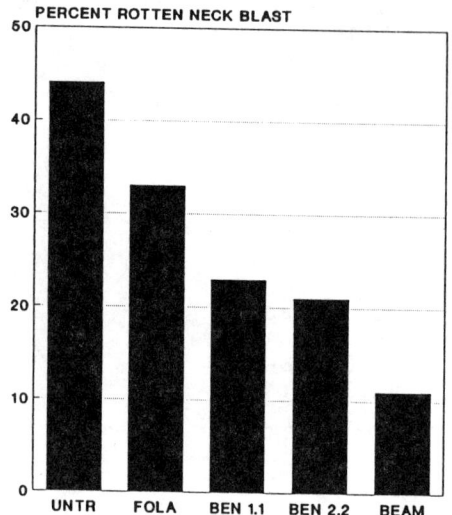

BROWN SPOT

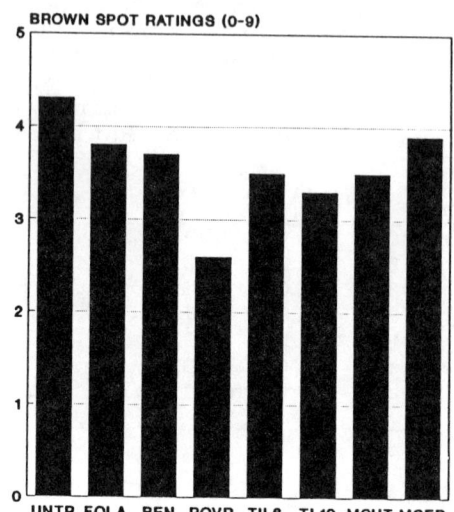

NARROW BROWN SPOT

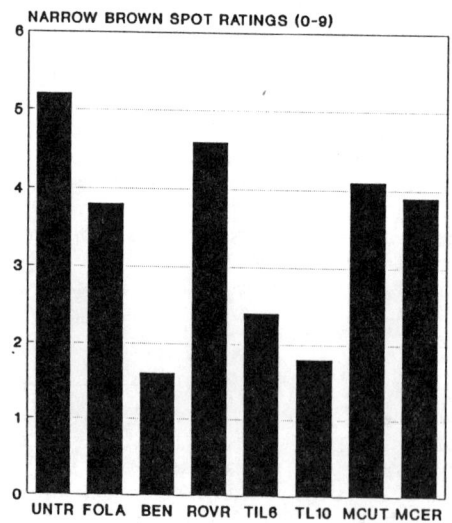

LEAF SMUT

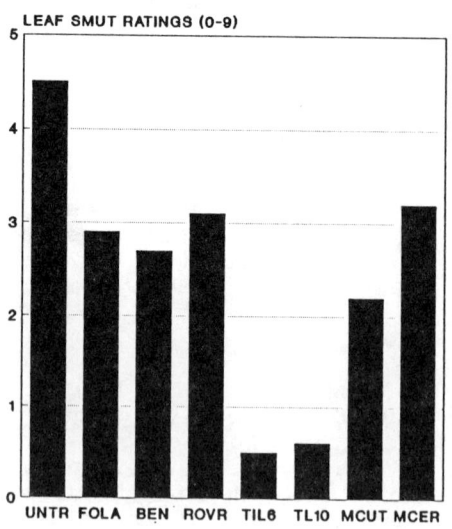

Figure 3. Performance of Benlate (BEN), Folatec (FOL), Rovral (ROV), Tilt as either two 6 oz applications (TIL6) or one 10 oz application (TL10), Moncut (MCUT) and Moncerene (MCER) on rotten-neck blast, brown spot, narrow brown leaf spot and leaf smut control compared to the untreated check (UNTR).

fungicides produce is probably due to control of this disease. Lack of activity against narrow brown leaf spot can also explain poor yield response in some cases of fungicides that control sheath blight, but do not control this disease.

LEAF SMUT

All of the fungicides registered have some activity against leaf smut (Fig. 3). Iprodione and pencycuron have the weakest activity, whereas propiconazole almost eliminates leaf smut. Leaf smut is not considered an important disease in the United States and lack of control may not be important in recommending a fungicide.

DISCUSSION

Fungicides must be evaluated using several years of data because of environmental variation from plot to plot and between years. Fungicides are also seldom 100 percent effective in controlling diseases. Many factors affect efficiency including timing, cultural practices, inoculum levels, weather patterns, varietal response, amount of spray volume, and the adjuvants that were used. One of the most important factors is not to allow the disease to get out of control before the fungicide is applied. Salvage spray applications in general do not give the yield response needed because the damage has already taken place. If a salvage spray is required, higher than recommended rates may be necessary to get a good yield response. The best way to avoid a salvage spray is to scout rice fields on a seven to 10 day schedule starting at the green ring stage and continuing through the panicle emergence stage. Recommendations for sheath blight in the United States are to apply a fungicide when 5 to 10 percent of the rice tillers of a susceptible cultivar are infected at the early jointing stages of growth. With a less susceptible variety the treatment threshold is 10 to 15 percent of tillers infected at the second or third joint stages of growth.

The economics of using fungicides in Louisiana is very important. In some cases a fungicide application is not justified because the farmer will not receive an adequate return on his investment. This usually occurs when the field is under a share cropping situation where the land owner receives 25 to 30 percent of the yield as payment for the land and water. As the land owner usually does not provide funding for the cost of chemicals the grower is reluctant to use any pest management chemicals including fungicides as 25 to 30 percent of the yield increase will go to the landlord. Often rice diseases and other pests will only reduce yield and grain quality and not completely destroy the crop. Share cropping farmers sometimes prefer to have reduced inputs and reduced yields rather than shoulder the entire cost of pesticides. There are exceptions to this, especially with blast, where near complete crop failures occasionally occur in isolated fields. In general, the farmer will receive a high enough yield increase and improvement in harvestability and grain quality to justify applying a fungicide when disease conditions exist.

Often farmers are not satisfied with the performance of their fungicide for sheath blight control because symptoms of the disease show up late in the season. In small plot work, even though the disease starts to develop late in the season after the fungicide has weathered or broken down, consistently high yield increases have been obtained. Apparently this late disease development occurs after grain filling is essentially complete, and the disease is suppressed long enough to increase yields over the unsprayed control. The other problem is that the fungicides registered are only moderately effective against sheath blight at the registered rates and more effective fungicides are needed if sheath blight is to be completely controlled. Higher rates of registered fungicides are more effective but cost and registration restrictions prevent their use.

A major problem with fungicides is that aerial applications do not always place the fungicides on target or environmental conditions limit fungicide effectiveness. When fungicides are applied by air, most of the material is deposited on the upper third of the canopy. It then requires weathering and redistribution to move the fungicide into the lower canopy. At the time when a fungicide should be applied, most of the disease pressure is on the lower third of the plant. This situation makes redistribution into the lower canopy by weathering necessary as the labeled fungicides are not systemic or the systemic fungicides are only locally systemic and move toward the tip of leaf blades. Weathering at the same time is detrimental since fungicidal activity is lost. Other conditions that limit deposition at application include drift, volatility, and calibration errors with spray equipment.

In the near future several fungicides will probably be registered on rice in the United States. Development of improved disease prediction or scouting techniques along with improvement of application technologies for existing compounds will complement chemical disease control methods. Major emphasis will be devoted to these topics in the United States.

LITERATURE CITED

1.   Groth, D. E. and Hollier, C. A., A survey of rice diseases in Louisiana. Louisiana Agriculture, 1986, 29, 10-12.

2.   Marchetti, M. A., Potential impact of sheath blight on rice yield and milling quality of short statured rice lines in the southern United States. Plant Disease, 1983, 67, 162-165.

3.   Lee, F. N. and Rush, M. C., Rice sheath blight: A
        major rice disease. Plant Disease, 1983, 67, 829-
        832.

4.   Kim, C. Effect of water-management on the etiology
        and epidemiology of rice blast caused by Pyricularia
        oryzae Cav.,   Dissertation,   Louisiana   State
        University, 1986, 170pp.

5.   Ashrafuzzaman, M. H. and Frederiksen, R. A. Chemical
        control of rice blast caused by Pyricularia oryzae.
        Phytopathology, 1967, 57, 457.

6.   Lindberg, G. D.   Chemical control of blast in
        Louisiana. Rice J., 1967, 70(7), 62.

7.   Lindberg, G. D. and Atkins, J. G., A progress report
        on blast in Louisiana and Texas. Proc. Rice Tech.
        Working Grp., 1961, 1, 22.

8.   Rush, M. C., Lindberg, G. D., and Whitem, H. K.,
        Benomyl: new fungicide for foliar diseases in rice.
        Louisiana Agriculture, 1977, 20, 10-11.

9.   Oh, S. H.,   Rice diseases 2nd Ed.   Common.
        Mycological Inst., Kew, Surrey, England, 198 , pp.

10.   Rush, M. C. and Lindberg, G. D., Fungicidal control
        of rice foliar diseases. Ann. Prog. Rept., Rice Res.
        St.,   La. Agri. Exp. Stn., L. S. U. Agricultural
        Center, 1974, 66, 181-183.

11.   Lindberg, G. D.,   Chemical control of blast in
        Louisiana. Ann. Prog. Rept., Rice Res. St.,   La.
        Agri. Exp. Stn., L. S. U. Agricultural Center, 1974,
        66, 184-186.

12.   Rush, M. C., Lindberg, G. D. and Shahjahan, A. K.
        M., Fungicidal control of rice foliar and stem
        diseases. Ann. Prog. Rept., Rice Res. St.,   La.
        Agri. Exp. Stn., L. S. U.
        Agricultural Center, 1975, 67, 197-199.

13.   Lindberg, G. D., Chemical control of blast in
        Louisiana. Ann. Prog. Rept., Rice Res. St.,   La.
        Agri. Exp. Stn., L. S. U. Agricultural Center, 1975,
        67, 200-202.

14.   Rush, M. C., Lindberg, G. D., and Shahjahan, A. K.
        M., Fungicidal control of rice foliar and stem
        diseases. Ann. Prog. Rept., Rice Res. St., , La.
        Agri. Exp. Stn., L. S. U. Agricultural Center, 1976,
        68, 124-221.

15.   Lindberg, G. D., Chemical control of blast in Louisiana. Ann. Prog. Rept., Rice Res. St., La. Agri. Exp. Stn., L. S. U. Agricultural Center, 1976, 68, 222-224.

16.   Rush, M. C., Lindberg, G. D., and Morgan, A. L., Fungicidal control of rice foliar and stem diseases. Ann. Prog. Rept., Rice Res. St., La. Agri. Exp. Stn., L. S. U. Agricultural Center, 1977, 69, 233-243.

17.   Lindberg, G. D., Chemical control of blast in Louisiana. Ann. Prog. Rept., Rice Res. St., La. Agri. Exp. Stn., L. S. U. Agricultural Center, 1977, 69, 244-245.

18.   Rush, M. C., Lindberg, G. D., and Morgan, A. L., Fungicidal control of rice foliar and stem diseases. Ann. Prog. Rept., Rice Res. St., La. Agri. Exp. Stn., L. S. U. Agricultural Center, 1978, 70, 279-284.

19.   Lindberg, G. D., Chemical control of blast in Louisiana. Ann. Prog. Rept., Rice Res. St., La. Agri. Exp. Stn., L. S. U. Agricultural Center, 1978, 70, 285-287.

20.   Rush, M. C., Lindberg, G. D., Morgan, A. L., and Ravenhorst, D., Fungicidal control of rice foliar and stem diseases. Ann. Prog. Rept., Rice Res. St., La. Agri. Exp. Stn., L. S. U. Agricultural Center, 1979, 71, 215-219.

21.   Lindberg, G. D., Chemical control of blast in Louisiana. Ann. Prog. Rept., Rice Res. St., La. Agri. Exp. Stn., L. S. U. Agricultural Center, 1979, 71, 224-277.

22.   Rush, M. C., Lindberg, G. D., and Ravenhorst, D., Fungicidal control of rice foliar and stem diseases. Ann. Prog. Rept., Rice Res. St., La. Agri. Exp. Stn., L. S. U. Agricultural Center, 1980, 72, 273-277.

23.   Lindberg, G. D., Chemical control of blast in Louisiana. Ann. Prog. Rept., Rice Res. St., La. Agri. Exp. Stn., L. S. U. Agricultural Center, 1980, 72, 278-279.

24.   Lindberg, G. D., Chemical control of blast in Louisiana. Ann. Prog. Rept., Rice Res. St., La. Agri. Exp. Stn., L. S. U. Agricultural Center, 1981, 73, 250-253.

25. Rush, M. C., Zhuo-Tong, Y., Lindberg, G. D., McLeod, J. M., and Castro, A., Chemical control of rice foliar and stem diseases. *Ann. Prog. Rept.*, Rice Res. St., La. Agri. Exp. Stn., L. S. U. Agricultural Center, 1982,74, 339-338.

26. Lindberg, G. D., Chemical control of blast in Louisiana. *Ann. Prog. Rept.*, Rice Res. St., La. Agri. Exp. Stn., L. S. U. Agricultural Center, 1982, 74, 339-341.

27. Rush, M. C., Lindberg, G. D., and Scott, J. M., Fungicidal control of rice foliar and stem diseases. *Ann. Prog. Rept.*, Rice Res. St., La. Agri. Exp. Stn., L. S. U. Agricultural Center, 1983, 75, 271-279.

28. Lindberg, G. D., Chemical control of blast in Louisiana. *Ann. Prog. Rept.*, Rice Res. St., La. Agri. Exp. Stn., L. S. U. Agricultural Center, 1983, 75, 280-281.

29. Rush, M. C., Acedo, J. R., and Groth, D. E., Fungicidal control of rice foliar and stem diseases. *Ann. Prog. Rept.*, Rice Res. St., La. Agri. Exp. Stn., L. S. U. Agricultural Center, 1984, 76, 268-273.

30. Lindberg, G. D., and Rush, M. C., Chemical control of blast in Louisiana. *Ann. Prog. Rept.*, Rice Res. St., La. Agri. Exp. Stn., L. S. U. Agricultural Center, 1984, 76, 291-292.

31. Rush, M. C., Lindberg, G. D., and Groth, D. E., Fungicidal control of rice foliar and stem diseases. *Ann. Prog. Rept.*, Rice Res. St., La. Agri. Exp. Stn., L. S. U. Agricultural Center, 1985, 77, 278-283.

32. Lindberg, G. D. and Rush, M. C., Chemical control of blast in Louisiana. *Ann. Prog. Rept.*, Rice Res. St., La. Agri. Exp. Stn., L. S. U. Agricultural Center, 1985, 77, 298-299.

33. Groth, D. E., Commercial rice fungicide trial. *Ann. Prog. Rept.*, Rice Res. St., La. Agri. Exp. Stn., L. S. U. Agricultural Center, 1986, 78, 174-176.

34. Rush, M. C., Lindberg, G. D., and Groth, D. E., Fungicidal control of rice foliar and stem diseases. *Ann. Prog. Rept.*, Rice Res. St., La. Agri. Exp. Stn., L. S. U. Agricultural Center, 1986, 78, 179-184.

35.   Lindberg, G. D., and Rush, M. C.,   Chemical control
        of blast in Louisiana. Ann. Prog. Rept., Rice Res.
        St.,   La. Agri. Exp. Stn., L. S. U. Agricultural
        Center, 1986, 78, 206-207.

36.   Groth, D. E., Off-station commercial rice fungicide
        trial. Ann. Prog. Rept., Rice Res. St.,   La. Agri.
        Exp. Stn., L. S. U. Agricultural Center, 1987, 79,
        208-209.

37.   Rush, M. C., Lindberg, G. D., and Groth, D. E.,
        Fungicidal control of rice foliar and stem diseases.
        Ann. Prog. Rept., Rice Res. St.,   La. Agri. Exp.
        Stn., L. S. U. Agricultural Center, 1987, 79, 214-
        224.

38.   Lindberg, G. D., and Rush, M. C.,   Chemical control
        of blast in Louisiana. Ann. Prog. Rept., Rice Res.
        St.,   La. Agri. Exp. Stn., L. S. U. Agricultural
        Center, 1987, 79, 235-236.

39.   Groth, D. E., Frey, M., Nugent, J., and Rush, M. C.,
        Commercial rice fungicide trial. Ann. Prog. Rept.,
        Rice Res. St.,   La. Agri. Exp. Stn., L. S. U.
        Agricultural Center, 1988, 80, 262-265.

40.   Rush, M. C., Lindberg, G. D., and Groth, D. E.,
        Fungicidal control of rice foliar and stem diseases.
        Ann. Prog. Rept., Rice Res. St.,   La. Agri. Exp.
        Stn., L. S. U. Agricultural Center, 1988, 80, 288-
        298.

41.   Lindberg, G. D., and Rush, M. C.,   Chemical control
        of blast in Louisiana. Ann. Prog. Rept., Rice Res.
        St.,   La. Agri. Exp. Stn., L. S. U. Agricultural
        Center, 1988, 80, 282-283.

CHEMICAL CONTROL OF SEEDLING DISEASES OF RICE IN LOUISIANA

M. C. RUSH AND R. W. SCHNEIDER
Department of Plant Pathology and Crop Physiology
Agricultural Experiment Station
Louisiana State University Agricultural Center
Baton Rouge, Louisiana   70803, U.S.A.

ABSTRACT

Rice is a major crop in the state of Louisiana, USA, with 210,000 hectares produced in 1988 having a farm value of 192 million dollars. The water-mold disease in water-planted rice is a major problem leading to replanting or losses due to nonuniform stands. The disease is primarily caused by species of Achlya and Pythium. Seed-protectant fungicides have been field-tested since 1972. These tests are summarized and the labeled seed-protectant fungicides available in 1990 are listed. Seed-protectants gave about a 30 percent increase in stand over that produced by untreated seeds. Treatment of flood water or soil with metalaxyl significantly increased stands by controlling Pythium spp. Seedling blight caused by seedborne and soilborne fungi can also be controlled by fungicides.

INTRODUCTION

In the state of Louisiana in the United States, rice is the fourth most important field crop with an estimated farm income of 193 million dollars from 520,000 acres of rice in 1988. The seed-rot, water-mold complex and seedling blight are common diseases affecting rice in Louisiana (1). These diseases cause stand reduction and stand irregularities in water-planted and, to a lesser extent, drill-planted rice. About 70 percent of the rice in Louisiana is normally planted by airplane into a 10 to 15 cm flood in the field using presprouted or dry seeds. About 30 percent of the rice acreage is planted with grain-drills into a dry seedbed. Losses to seedling diseases are incurred directly through the costs of replanting or through loss of yield from thin stands, and indirectly through the inefficient use of space and light in irregular stands and by increased competition with weeds or increased weed control costs. Also, if it is necessary to replant or overseed, the rice crop may mature late in the season when conditions are less favorable for high yields because of tropical storms and high disease pressure from foliage diseases.

Rice seed germination and growth of seedlings is often poor under the weather conditions normally encountered in March and early April in the southern rice area of the United States. This area includes Arkansas, Louisiana, Mississippi, and Texas. Considerable damage to water-seeded rice is encountered under these cool weather conditions regardless of the seed-protectant fungicides used. The fungi causing seed-rot and water-mold disease in water-seeded rice are primarily species of Pythium, Achlya, and Fusarium. Drill-seeded rice can be injured by several seedborne fungi including Helminthosporium Bipolaris oryzae, Curvularia lunata, Alternaria padwickii, Fusarium roseum, and Fusarium moniliforme. The soilborne fungi Rhizoctonia solani, Sclerotium rolfsii, and Pythium spp. can also damage drill-seeded rice (1,2).

Theoretically, seed-rot and seedling diseases of rice should be economically controlled with seed-protectant fungicides. For this reason a program to develop more effective fungicidal seed treatments has been conducted at the Louisiana State University Rice Research Station since the early 1970's. Fungicides or formulations of fungicides are field screened in experimental tests each year. Those fungicides performing well in the experimental tests are advanced to further field testing in subsequent years.

SEEDLING DISEASES IN DIRECT SEEDED RICE

With the increase in the use of water-seeding as a method for planting rice, it has become more difficult to obtain uniform stands of sufficient density to obtain maximum yields. The most important biological factor contributing to this situation is the water-mold or seed-rot disease caused primarily by water-mold fungi in the genera Achlya and Pythium. Recently, certain Fusarium spp. also were found associated with molded seeds and the disease is caused by a complex of these fungi (3,4,5,6,7,8,9,10). The severity of this disease is more pronounced when water temperatures are low or unusually high. Low or high water temperatures slow the germination and growth of rice seedlings, but do not affect growth of the pathogens. In surveys conducted in Louisiana during the early 1970's, an average of 45% of the water-planted seeds were lost to water-mold. This means that about 29 kg of seed out of 66 kg planted were lost to the disease.

Water-mold may be observed through clear water as a ball of fungal strands surrounding seeds on the soil surface. After the seeding flood is drained, seeds on the soil surface are typically surrounded by a mass of fungal strands radiating out over the soil surface from the affected seeds (Fig. 1). The result is a circular copper-brown or dark green spot about 2 cm in diameter with a rotted seed in the center. The color results from the growth of bacteria and green algae which are mixed with the growing water-mold hyphae. Achlya spp. typically have coarse hypha visible to the eye without magnification (Fig. 2). Pythium spp. have a fine mycelium visible in mass as a gelatinous matrix around seeds.

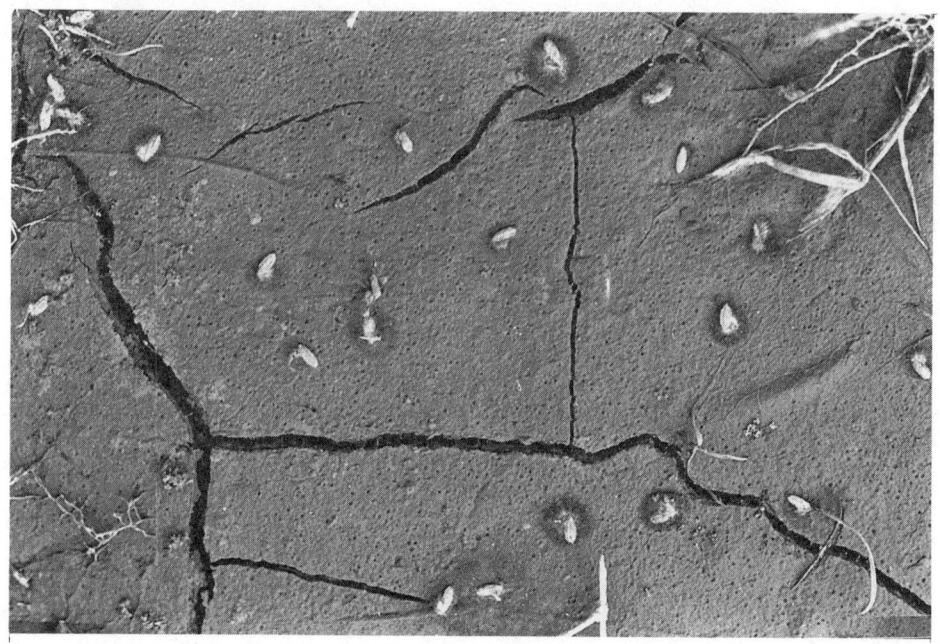

Figure 1. Water-mold symptoms in the field after seeding flood is removed.

Achlya spp. normally attack the endosperm of germinating seeds. They destroy the food source for the growing embryo, and then the fungi may attack the affected embryo or small seedlings. Pythium spp. usually attack the developing embryo directly. When the seed is affected by the disease, the endosperm becomes liquified and oozes out as a white, thick liquid when the seed is pressed between your fingers. The embryo turns yellow-brown and finally dark brown. If the affected seed germinates, the small shoot and root are attacked and the seedling is stunted. When infection takes place after the seedling is established, the seedling is stunted, turns yellow and grows poorly. If the weather is favorable for plant growth, seedlings often recover and are not severely damaged.

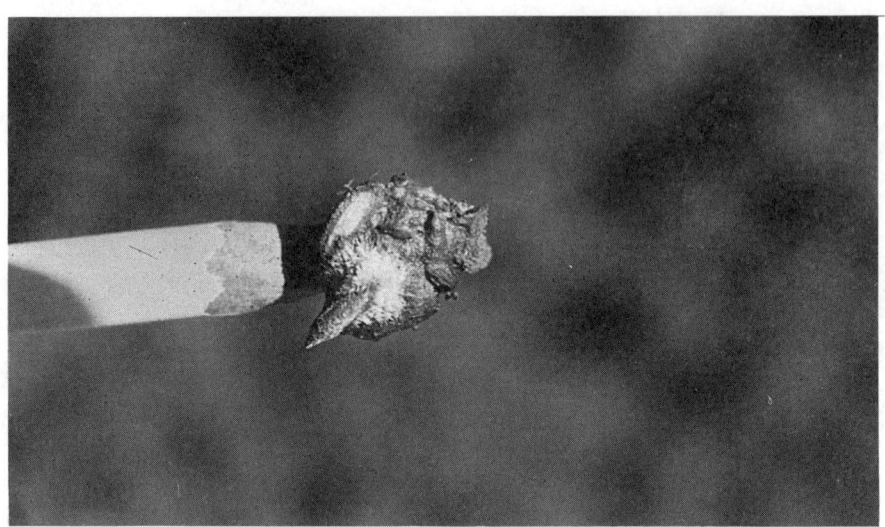

Figure 2. Rotted rice seed with coarse mycelium of <u>Achlya</u> sp. radiating from seed.

The disease is controlled or is less severe when rice is water-seeded in April or early May when weather conditions favor seedling growth. When the mean of the high and low temperatures is 18°C, conditions begin to favor seedling growth and water-mold is less severe. Night temperatures should be above 10-13°C. Seeds should be vigorous and should have a high germination percentage.

The fungal species normally isolated from rotted seeds in water-planted rice in Louisiana include <u>Achlya</u> <u>klebsiana</u> Pieters, <u>Achlya</u> <u>conspicua</u> Coker, <u>Pythium</u> <u>dissotocum</u> Drechsler, and <u>Pythium</u> <u>spinosum</u> Sawada. In pathogenicity tests, <u>A.</u> <u>klebsiana</u> and <u>P. dissotocum</u> were highly virulent (4). Disease damage was most severe when germinating seeds were simultaneously inoculated with zoospore suspensions of <u>A.</u> <u>klebsiana</u>, <u>A.</u> <u>conspicua</u>, <u>P.</u> <u>dissotocum</u>, and <u>P.</u> <u>spinosum</u> confirming that these fungi act in concert in the field to cause the water-mold disease complex in Louisiana. Light and electron microscopic observations of infected rice seed confirmed that <u>Achlya</u> species attack and destroy the seed endosperm while <u>Pythium</u> species infect the embryos and radicles of germinating seeds (4).

Seedling blight is a disease complex caused by several different seedborne and soilborne fungi including species of <u>Bipolaris</u>, <u>Curvularia</u>, <u>Sarocladium</u>, <u>Fusarium</u>, <u>Rhizoctonia</u>, <u>Pythium</u>, and <u>Sclerotium</u> (2). Typically, rice seedlings are weakened, stunted, or killed by the fungi causing stands of rice to be irregular and thin from the time they are established.

The fungi involved directly penetrate the germinating rice seed or young seedling and either kill or injure it. If blighted seedlings emerge from the soil, they are usually weak and chlorotic.

The severity of seedling blight depends mainly upon three factors: the percentage of seed that is infected by blight fungi, soil temperature, and soil moisture content. Seedling blight is more severe on rice that has been seeded early when the soil was cold and damp. This disadvantage of early seeding can be partially overcome by seeding at a shallow depth. Conditions that tend to delay seedling emergence from the soil often favor seedling blight.

Seeds that carry blight fungi frequently have spots or discolorations on the hulls. However, seed can be infested and still appear to be clean. Helminthosporium (Bipolaris) oryzae is one of the chief causes of seedling blight and is seedborne. A seedling attacked by this fungus has dark areas on the basal parts of the first leaf, the crown area, and roots. The soilborne blight fungus, Sclerotium rolfsii, sometimes kills or severely injures large numbers of rice seedlings after they emerge if the weather at emergence time is moist and warm. A cottony white mold develops on the lower parts of affected plants. This type of blight can be checked by flooding the land immediately (1).

Water- and soil-borne fungi in the genus Pythium may attack and kill seedlings from germination to about the three-leaf stage of growth. Roots may be discolored brown or black with the shoot suddenly drying and turning straw-colored. This disease is most common in water-seeded rice after the field is drained. It may also occur in drill-seeded rice after the field is flushed or in prolonged wet, rainy periods.

Treatment of seed with a fungicide is recommended to improve or ensure stands (10,11,12,13). Proper cultural methods for rice production, such as proper planting date or shallow seeding of early planted rice, will help control the seedling blight fungi.

A possible alternative to seed-protectants to control water-mold in water-seeded rice is the use of systemic biocidal fungicides in treating soil before flooding or treating the flood water in a field prior to seeding. Recent research to demonstrate the presence of feeder root necrosis in rice with treatment of the soil or water with metalaxyl (Ridomil 2E) demonstrated that Pythium species causing water-mold could be controlled in the seeding flood resulting in significant increases in stand (14). This research is continuing with a reorientation toward controlling seedling diseases.

PROCEDURES USED TO EVALUATE FUNGICIDES

Seed-Protectants

Seeds from a single lot of a current commercial cultivar were treated in February or the first 2 weeks of March each year for planting in April and May (15,16). The formulated fungicides were placed into water (2 ml per 0.114 kg seed) in 1 liter glass jars, stirred with a magnetic stirring bar until suspended, and the suspended fungicide was used to coat the inside of the jar. Preweighed seeds were poured into the coated jar immediately, the jar was shaken by hand for 60 seconds, and the jar was placed onto a jar mill to revolve for 10 minutes or until seeds were uniformly coated with the fungicide. Treated seeds were stored at room temperature (22°C) in heavy manila seed envelopes until counting for planting. Seeds were counted with an electronic counter.

Simulated water- and drill-seeding methods were used to plant treated seed at the Rice Research Station at Crowley, Louisiana. In the water-seeded tests, five or six replications of 200 presprouted seeds per plot were planted. Seeds were soaked in five volumes of tap water in 15 mm X 150 mm culture tubes for 24 to 30 hours before planting. Plots consisted of cylinders 30-cm high and 60-cm in diameter made of 0.3 cm mesh galvanized hardware cloth placed into a leveed plot area in a randomized complete block design (Figure 3). The cylinders were placed 60 cm apart within blocks, with a 2 meter alley between blocks. The plot area was flooded 10 to 15 cm deep and the treated seeds from each culture tube were scattered into the water in each cylinder. After planting, the cylinders were covered with 0.6 cm mesh galvanized hardware cloth to keep out rodents and birds. Two individually leveed plot areas were planted at each planting date using identical procedures. In one plot area the seeding flood was removed after 3 days and the plots were flushed as needed to maintain growth. In the other plot area the seeding flood was maintained at 5 to 10 cm depth until 3 days before stands were counted.

Drill-seeded plots consisted of the same treatments, planted at the same time, in 3-meter rows with 200 seeds per row. Plots were replicated five or six times using the same randomized complete block design as in the water-seeded tests. Rows were spaced 25 cm apart within blocks with 2 meter alleys between blocks. Seeds were planted with a hand-drop seeder. The drill-seeded plot area was flushed; i.e. flooded and drained, to start germination at the same time the water-seeded tests were planted.

Stand (number of plants per plot) was counted at the 3 to 4 leaf stage or 3 to 4 weeks after seeding. This is the stage when the permanent flood would normally be placed onto the field.

Figure 3. Planting water-seeded seed-protectant fungicide test.
Flushed drill-seeded test in background.

Soil or Water Treatments

Field experiments were conducted in 1987 with metalaxyl (Ridomil
2E - Ciba-Geigy Corporation) and metam sodium (Busan 1020 -
Buckman Laboratories, Inc.) to determine the effects of a
Pythium specific fungicide and a biocide on stand establishment
under water-seeded conditions.

Plots were individually leveed so that chemical treatments
and irrigation water could be controlled (Figure 4). The
cultivar Lemont was used for these tests. All plots were water-
seeded at 57 kg seeds per hectare on May 13. Fertilizer was
applied by broadcasting into the permanent flood at 15 days
after planting as 668 kg of 20-10-10 (N, P, K) per hectare. An
application of 89 kg per hectare of ammonium sulfate (21% N)
was broadcast over the plots 35 days after planting. Weeds were
controlled by draining the plots at 12 days after planting,
applying Propanil (3.56 lb X ha^{-1} ai) at 13 days after planting,
and reflooding at 15 days after planting. Furadan was applied
at 0.45 kg X ha^{-1} at 35 days after planting by broadcasting 3%
granules over each plot.

The objectives of the experiment were to compare liquid
metalaxyl (Ridomil 2E), granular metalaxyl (Ridomil 5G), and
metam sodium for their effects on stand establishment in water-
seeded rice. The treatments used in this test are listed in
Table 1.

Figure 4. Overview of field tests with metalaxyl and metam sodium for seedling disease control showing individually leveed plots.

Table 1
Treatments used to determine the effects of metalaxyl and metam sodium on stand establishment in water-seeded rice.

Treatment	Rate (Units/ha)	Method of Application	Water Management[a]
1 Control	--	flooded 9 days before planting	1
2 Metalaxyl (2E)	1.04 liters	9 days pre-plant in flood water	1
3 Metalaxyl (2E)	2.08 liters	"	1
4 Metalaxyl (2E)	4.16 liters	"	1
5 Metalaxyl (2E)	8.32 liters	"	1
6 Metalaxyl (2E)	1.04 liters	sprayed on soil before flooding	2
7 Metalaxyl (2E)	2.08 liters	"	2
8 Metalaxyl (5G)	0.25 kg	granules broadcast on soil before flooding	2
9 Metalaxyl (5G)	0.5 kg	"	2
10 Metam sodium	83.4 liters	9 days preplant in flood water	3

Table 1 (cont'd)

Treatment	Rate (Units/ha)	Method of Application	Water Management[a]
11 Metam sodium	166.8 liters	9 days preplant in flood water	3
12 Metam sodium	333.5 liters	"	3
13 Metam sodium	667.0 liters	"	3

[a] Water management systems were 1 = maintained continuous flood (10-15 cm) for 9 days before planting, drained 1 week after planting, flushed at 9 days after planting, reflooded 14 days after planting; 2 = Flooded at planting, drained after 8 days, reflooded 14 days after planting; 3 =maintained continuous flood for 9 days before planting then planted and drained, flushed 6 days after planting, reflooded 14 days after planting.

RESULTS AND DISCUSSION

Seed-Protectant Fungicides

In 19 years of field testing seed-protectant fungicides, effects on stand ranged from significantly less than untreated seeds (phytotoxic) to as much as a 300 percent increase. Water-mold disease was present in our field tests (Fig. 5) in amounts comparable to commercial fields. More than 750 fungicides, fungicide combinations, and treatment rates were tested. During this period 23 seed-protectant fungicides were registered for use on rice seeds, many of them based on the results of our field trials (Table 2). Through 1989, more than 90 percent of the rice seeds planted in Louisiana were treated with fungicides. For the 1990 season, only 10 materials remain as many companies refused to reregister their products for rice seed-treatment use (Table 3). Captan and captafol products were not reregistered for rice in 1990. These products, along with etridiazole, which is no longer being sold for rice, have been lost as seed-protectants for rice.

Table 2
Seed-protectant fungicides registered for rice
in the period 1970-1990.

Common Name	Trade Name	Rate per 100 lbs seed	Company
captan	Captan 80 WP	2.25 oz.	Stauffer
captan	Gustafson Captan 30-DD	2.7 fl. oz.	Gustafson
captan	Gustafson Captan 400-DD	3-6 fl. oz.	Gustafson
captan	Orthocide 4 Fl	3.4 fl. oz.	Chevron

Table 2 (cont'd)

Common Name	Trade Name	Rate per 100 lbs seed	Company
captan + methoxychlor	Orthocide + Methoxychlor 75-5 WP	2.25 oz.	Chevron
captafol	Difolatan 4 Fl	4-6 fl. oz.	Chevron
mancozeb	Dithane M-45	4-6 oz	Rohm & Haas
mancozeb	Dithane F-45	2.2-6.4 fl oz.	Rohm & Haas
mancozeb	Dithane DF	3.1-4.3 fl oz.	Rohm & Haas
thiram	Arasan 70S-Red	2.35 oz.	Dupont
thiram	Arasan 42S	3.3 fl. oz.	Dupont
thiram	Gustafson 42S	3.3 fl. oz.	Gustafson
thiram	Gustafson Thiram 30 Fl	3.38-8.78 fl oz	Gustafson
copper hydroxide	Kocide SD	4-8 fl. oz.	Kocide
copper hydroxide	Champion SD	4-8 fl. oz.	Agtrol
terrachlor + terrazole + zinc omadine	TerraCoat ZN 2055 WP	5.0 oz.	Olin
mancozeb	Manex II	3.4-6.7 fl oz.	Griffin
mancozeb	Manzate 200DF	2-4 oz.	Dupont
metalaxyl	Apron Fl	0.75-1.5 fl oz.	Gustafson
metalaxyl	Apron 25 WP	1-2 oz.	Ciba-Geigy
carboxin + thiram	Vitavax 200 FF	4 fl. oz.	Gustafson
carboxin + thiram	Vitavax R	4 fl. oz.	Uniroyal

Table 3
Seed-protectant fungicides available for rice in 1990.

Chemical	Trade Name(s)	Rate per hundred lbs seed	Company
metalaxyl	Apron Fl	0.75-1.5 fl oz	Gustafson
	Apron 25 WP	1-2 oz.	Ciba-Geigy
copper hydroxide	Champion SD	4-8 fl. oz.	Agtrol
	Kocide SD	4-8 fl. oz.	Griffin

Table 3 (cont'd)

Chemical	Trade Name(s)	Rate per hundred lbs seed	Company
thiram	Gustafson 42-S	3.3 fl. oz.	Gustafson
mancozeb	Dithane F-45	2.2-6.4 fl. oz.	Rohm & Haas
	Dithane DF	2.1-4.3 oz.	Rohm & Haas
	Manex II	3.4-6.7 fl. oz.	Griffin
	Manzate 200 DF	2-4 oz.	Dupont
carboxin + thiram	Vitavax 200 FF	4 fl. oz.	Gustafson

Table 4 gives the mean percent increase in stand over that from untreated seeds for each of the fungicides available in 1990 in tests conducted over the years 1976 to 1989.

Figure 5.   Plot in water-seeded, seed-protectant fungicide test showing typical water-mold symptoms.

Table 4
Seedling disease control given by seed-protectant fungicides
available in 1990 in tests conducted over the period 1976-1989.

Fungicide[a]	Mean Percent increase in stand over untreated seeds			Number of tests
	DS	WS-D	WS-C[b]	
metalaxyl (Apron 25 WP, Apron FL and CGA 48988)	12.2	20.1	41.7	45
copper hydroxide (Kocide SD, Champion SD)	16.9	32.5	47.3	71
thiram (Arasan 70S Red, Arasan 42S and Gustafson 42S)	28.1	28.8	40.5	42
mancozeb (Dithane M-45, Dithane F-45, Dithane DF Manex II, and Manzate 200)	18.4	28.6	37.2	99
carboxin + thiram (Vitavax 200 FF, Vitavax R)	14.4	18.9	18.8	105
Overall Means	18.0	25.8	37.1	

[a] All formulations and rates tested of the same chemical were combined.

[b] DS = drill-seeded, WS-D = water seeded, drained in 3 days and flushed as needed, WS-C = water-seeded, maintained as a continuous flood.

The continued use of presprouting fungicide treated seeds before water-seeding has recently been questioned by the Louisiana Department of Agriculture and by the Louisiana Department of Environmental Quality because the remaining fungicide-contaminated soak water is now considered hazardous waste. One seed company that presprouts seeds for its customers has estimated that it must dispose of up to 190,000 liters of this fungicide-contaminated water. In the past, this material has been released into streams and bayous. This practice will not be allowed from the 1990 season onward. It appears that the problem will be addressed during the 1990 season by dumping the waste water into rice fields planted with treated seeds, by water-seeding with dry fungicide treated seeds, and by increasing the amount of drill-seeding. Experiments are in progress to develop a system to presprout seeds without having excess, contaminated water. If metalaxyl can be registered and used to treat soil or the seeding flood before water-seeding rice, this will greatly reduce the need for treating seeds with fungicides.

Soil or Water Treatment

Metalaxyl, applied in the flood at 9 days before planting at approximately 1, 2, 4, and 8 liters per hectare (Table 1) significantly increased stand, shoot weight, and yield over the untreated plots (Table 3).

Table 5
Effects of metalaxyl and metam sodium on
stand, shoot weight, and yield of rice.

Treatment No. [a]	Stand (plants/sq. ft.)	Shoot wt. (g/5 plants)	Yield (kg X ha^{-1})
1	14.5	8.5	3953
2	22.3	8.8	4413
3	26.3	8.9	4706
4	25.4	13.4	4381
5	27.1	9.0	4309
6	26.5	12.8	4977
7	20.1	12.1	4576
8	26.6	11.3	4640
9	20.8	14.9	4766
10	19.2	17.5	4552
11	29.0	13.2	4375
LSD (P = 0.05)	4.2	2.8	408

[a] Treatments shown in Table 1.

Stand increases ranged from none to 100% over the stand in untreated plots. Yields were increased by up to 28%. In the metalaxyl treatments, the 0.23 kg X ha^{-1} ai granular treatment was about equal to the 0.23 kg X ha^{-1} ai liquid treatment (Treatment 3) for stand, shoot weight, and yield (Table 5). Metalaxyl appeared to work about equally well whether it was added to the flood water 9 days preplant, sprayed onto the soil just prior to flooding the plots, or applied to the soil as a granule at preflood (Figures 6 & 7). As metalaxyl is specific in its activity against _Pythium_ spp. and with limited activity against _Achlya_ sp. (4), it suggests that _Pythium_ spp. have a major role in the water-mold disease.

Figure 6. Field plot treated with metalaxyl (Ridomil 5G) and
untreated control plot 15 days after seeding. A) Control plot
without metalaxyl 15 days after seeding. B) Metalaxyl treated
plot after 15 days

Figure 7. Overview of field tests with metalaxyl and metam sodium 42 days after seeding. A) Control plot after 42 days. B) Metalaxyl plot after 42 days.

The metam sodium treatments also significantly increased stand, shoot weight, and yield over that in the control plots (Table 5). This chemical is a water-soluble fumigant with biocidal properties. It appeared to control the water-mold fungi and nitrogen appeared to be conserved in these plots.

CONCLUSIONS

Seed-protectant fungicides that are effective can increase stands up to 300 percent with the normal increase being about 30 percent. This would mean that a potential stand of 108 plants per square meter is increased to 140 plants per square meter. The first stand would be unacceptable to most growers and the field would have to be replanted. The second stand is rather low but a crop can be made from that stand. In water-seeded rice in Louisiana a stand averaging 160-215 plants per square meter would be considered optimal.

A serious problem is developing with disposal of the waste water from presprouting fungicide treated seeds. Most of the presently used fungicides do not have a label that would allow this water to be released into planted fields. Dumping the water into a pond or similar disposal site would create a hazardous waste dump. It can cost millions of dollars per hectare to decontaminate these areas. Water-seeding dry seeds creates a condition where seeds are easily moved on the soil surface in response to wave action created by winds. The seeds gather into low areas or are washed onto the edges of levees. This creates nonuniform stands. In addition, the added time before draining the flood water, so that the seeds can sprout, means more time is available for the fungicide to wash off the seeds and for the seeds to be exposed to the water-mold fungi.

It will be necessary to develop a method to presprout seeds that does not create fungicide-contaminated waste water or to develop a procedure to groove the field so that water-seeded dry seeds are caught in rows in the field. A grooving machine has been developed and tested at the Rice Research Station at Crowley.

A final solution to these problems would be the registering of a fungicide to treat the soil or flood water to destroy the water-mold fungi. Steps are being taken to develop metalaxyl (Ridomil 2E) for this purpose. The present limitations are the high costs of registering a pesticide for use in water and the cost of Ridomil 2E which would lead to a cost of about $35 per hectare to prevent seedling disease and seed-rot in water-seeded rice. This cost would be offset by the reduction in seeding rate, thereby reducing the cost of seed. It appears that metam-sodium will not be a solution to the problem as the high application rates required (about 250 liters per ha) are not cost effective.

Resistance to metalaxyl has developed in several pathogens including <u>Phytophthora</u> <u>infestans</u> in potato (17). This suggests that widespread use of metalaxyl could lead to resistance in the water-mold <u>Pythiums</u>.

Control of the water-mold complex in water-seeded rice will be accomplished by a disease management system that includes planting when the water and soil temperature favors rapid germination of seeds and seedling growth, proper water management, and the use of fungicides. As the disease is the result of a complex of several fungi acting in concert, disease resistance will be difficult to develop. However, seedling vigor is a heritable characteristic and breeding lines can be selected for fast growing, vigorous seedlings. This characteristic would help reduce water-mold. Seeding blight and preemergence rotting of rice seeds are not a serious problem under normal planting conditions, and seed-protectant fungicides can be readily used to control fungi causing this problem without the concern of developing a hazardous waste.

References

1. Anonymous., _Rice Production Handbook_, Bulletin 2321, Louisiana State University Agricultural Center, Baton Rouge, 1987, pp. 36-46.

2. Oh, S.H., _Rice Diseases_, Commonwealth Mycological Institute, Kew, Survey, 1985, pp. 301-305.

3. Cools, W.G., _Achlya and Pythium Species Associated with the Water-Mold Disease of Rice (Oryza sativa L.) in Southwestern Louisiana_. M.S. Thesis, Louisiana State University, Baton Rouge, 1972.

4. Krishna, P.G., _Etiology and Control of the Rice Water-Mold Disease Complex in Louisiana_, M.S. Thesis, Louisiana State University, Baton Rouge, 1983.

5. Krishna, P.G., and Rush, M.C., The role of _Pythium spinosum_ in the complex causing water-mold disease of water-seeded rice. _Phytopathology_, 1983, 73, 502-503.

6. Webster, R.K., Hall, D.H., Heeres, J., Wick, C.M., and Brandon, D.M. _Achlya klebsiana_ and _Pythium_ species as primary causes of seed-rot and seedling diseases of rice in California. _Phytopathology_, 1970, 60, 964-968.

7. Rush, M.C., Marchetti, M.A., and Adair, C.R., Stand establishment in early seeded rice in the South. _Rice Journal_, 1972, 175, 32-34.

8. Rush, M.C. and Gifford, J.R., Enhancement of seed-rot and seedling diseases of rice by an insecticidal seed treatment. _Plant Dis. Reptr._, 1972, 156, 154-157.

9. Ito, T.S. and Nagai, M., On the rot disease of the seeds and seedlings of rice plants caused by some aquatic fungi. _Jour. Fac. Agr. Hokfacido Imp. Univ._, 1931, 32, 45-69.

10. Rush, M.C., Improving stands of early-seeded rice. _Louisiana Agriculture_, 1973, 116, 10-11.

11. Rush, M.C., Controlling seed-rot and seedling diseases in rice. Phytopathology, 1985, 175, 1329.

12. Webster, R.K., Hall, D.H., Wick, C.M. and Brandon, D.M., Seedling disease and its control in California rice fields. Rice Journal, 1970, 73, 14-17.

13. Webster, R.K., Hall, D.H., Wick, C.M., and Brandon, D.M., Chemical seed treatment for the control of seedling disease of water-sown rice. Hilgardia, 1973, 141, 689-698.

14. Schneider, R.W., Rush, M.C., and Pillay, M., Feeder-rot necrosis recognized as a serious disease in rice. Louisiana Agriculture, 1988, 131, 5-7.

15. Rush, M.C., Chemical control of rice seedling diseases. Ann. Prog. Rpt., Rice Res. Sta., LA Agr. Expt. Sta., LSU Agr. Center, 1988, 80, 299-320.

16. Rush, M.C., Chemical control or rice seedling diseases. Ann. Prog. Rpt., Rice Res. Sta., LA Agr. Expt. Sta., LSU Agr. Center, 1987, 79, 225-234.

17. Kadish, D. and Cohen, Y., Estimation of metalaxyl resistance in Phytophthora infestans. Phytopathology, 1988, 78, 915-919.

THE BENEFITS OF SEED/EARLY SEASON FUNGICIDE APPLICATION FOR THE MANAGEMENT OF RICE BLAST (*PYRICULARIA ORYZAE*)

AXEL LOEHKEN
CIBA-GEIGY LIMITED
Agricultural Division
Research and Development
Plant Protection
4002 Basle, Switzerland

ABSTRACT

The benefits of seed/early season fungicide application for the management of rice blast (*Pyricularia oryzae*) are demonstrated based on results from trials for the development of the systemic blasticide pyroquilon. In crops with an effective control of leaf blast up to the end of tillering phase, during which rice is very susceptible to this disease and blast attacks can cause severe damage to plant development, panicle emission and ripening is more uniform and more and heavier panicles/hill are produced. The rice crop protected from leaf blast is more vigorous throughout the whole growing period resulting in more productive tillers and in significantly higher yield.

INTRODUCTION

The importance of blast and of its control to the rice culture is emphasized by so many authors that there is no need to repeat it. Especially important for this paper is the direct effect of leaf blast infection in reducing rice crop yields. Early leaf blast can cause stunting of the rice plants, reduced tillering, a prolonged ripening phase, less panicles per hill and reduced 1000 grain weight [1,2,3]. These factors cause yield reductions, even if no panicle infections occur.

Also relevant is the generally accepted fact that panicle infections are often preceded by leaf blast attacks [1] and that the lesions on the upper five leaves could produce spores for infection of panicles at the initial heading stage [4]. Effective control of leaf blast therefore reduces the quantity of inoculum available for the infection of panicles.

Effective control of early leaf blast is best achieved by applying systemic blasticides in a way they can be taken up by the oot system. Foliar sprays at this early stage of crop development are less effective because leaves at this stage are too small to take up enough of the chemical and new leaves are rapidly unfolding.

Pyroquilon is such a systemic blasticide and the results gathered during the development of this compound serve as the experimental base for this paper.

MATERIALS AND METHODS

Pyroquilon

Pyroquilon is a purely preventive acting compound. Pyroquilon does not inhibit growth of this fungus in culture or conidial germination and appressorial formation on the plant but blocks melanin biosynthesis in the appressoria and thus prevents penetration of the epidermal walls. Successful penetration is observed only when appressoria are melanized [5].

Pyroquilon is sold under the trade names "FONGORENE®" for the use as seed treatment and foliar spray and "CORATOP®" for the use as an into-water applied granule [6].

This compound has to be used in a protective schedule, before blast occurs in the field. The easiest ways to do this are a seed treatment for dry seeded rice and early into water granule applications into the seedling box, seedling bed or rice field for direct wet sown or transplanted rice to control early leaf blast.

Seed treatment of wet sown rice is not as effective as the treatment of dry seed because the compound is washed-off and diluted.

Evaluations

Leaf blast: The percentage of green leaf area reduction by blast was estimated visually and called " % damage by leaf blast". Factors taken into account were diseased leaf area, dead leaves, stunting of plants and number of tillers and leaves. Values given represent the average damage on the whole plot.

Panicle/neck blast: In Brazil an overall estimation of blast diseased panicles and necks in treated and untreated was done.

Yield factors: numbers of tillers/plot, numbers of panicles/plot, grains/panicle were counted in representative samples.

Yield measurements: Whole plots were harvested. In Thailand hills were cut by hand and threshed in special machines for experimental trials; on-farm trials in Brazil and Colombia were yielded with combine harvesters. Grain weights were all converted to 14 % moisture content.

Trial Details

Seed treatment trials RI 1 and RI 2 (Figures 1 and 2): These trials were conducted in upland rice nurseries on the CIBA-GEIGY research station in Cikampek, West Java in 1984 and 1985 respectively. Plot size was 2x2m with 4 replicates (RCBD) and inoculum provided by infector rows along both sides of each plot. Disease was promoted by frequent sprinkler irrigations. The variety used was Cimandiri.

In the trial RI 1 (Figure 1) the disease started to develop more slowly and foliar applications were carried out 14 days after seeding (DAS), followed by sprays every week.

In trial RI 2 (Figure 2) foliar applications were started 7 DAS because blast came in very aggressively soon after seeding.

For the seed treatment in both trials pyroquilon at 4 g a.i. was used as a slurry with 10 ml of water kg^{-1} of seed.

Foliar treatments were 500 g a.i. ha^{-1} of fthalide at each weekly application.

Seed treatment trial CO 1 (Figure 3): This trial was carried out in the Llanos Orientales, Colombia in 1983 to test the efficacy of the seed treatment in a practical field trial with natural infection. The variety CICA 4 was treated with 4 g a.i. kg^{-1} seed. A seed rate of 188 kg ha^{-1} was used in one hectare plots without replication. Treated and untreated plots were adjacent. Farmers' practice for irrigation (frequent floodings of the field) was followed.

Seed treatment demonstration trials in Brazil (Tables 1 and 2): From 1986 to 1989 a series of demonstration trials was conducted on upland rice farms in Central Brazil with natural infection. Trials were seeded and managed following local farmers' practice using a seed rate of 40-50 kg ha^{-1}. Seed was pre-treated with carbofuran at 3.5 g a.i. kg^{-1} seed followed by 4 g a.i. kg^{-1} pyroquilon in rotating drums. In 1988 an Amazone Trans-Mix machine was used for treatment. Plot size was of 3 hectares without replicates, treated and untreated plots were adjacent. Leaf blast attack was usually rated twice and in the trials from 1988/89 (Table 2) three times.

Additional observations such as number of panicles per plot, number of tillers and grains per panicle were counted in samples taken in some of these trials.

Into water granule trials in Thailand (Figures 4 and 5, Table 3): In the Chachoengsao area near Bangkok, Thailand, rice is directly seeded into flooded fields after pre-germinating the seeds. Leaf blast usually starts to appear around 4 weeks from seeding in the dry season and is a major constraint to the growing of high yielding varieties in this region. In 1986 trials on 5x5 m plots in 4 completely randomized replicates with early applied pyroquilon granules at 1.5 kg a.i. ha⁻¹ were carried out. 20-25 days after seeding is the most appropriate timing for such an application. Plots were separated by plastic dikes and disease promoted by excessive nitrogen fertilization.

In trials 01-04 (Table 3) the pyroquilon protective treatment was reinforced by a follow-up granule treatment at 55 days after seeding. In trials 05-08 this additional treatment was omitted. The farmers' practice is to apply a foliar spray of edifenphos at 500 g a.i. ha⁻¹ curatively (some days after first lesions are seen followed by another spray a week later). This practice was followed and for comparison reasons one into water application of pyroquilon at the same timing of the first standard foliar application was included.

In 1987 the same trial protocol as in 1986 was conducted but a granule treatment of pyroquilon at 500 g a.i. ha⁻¹ at 25 days after seeding was added. One typical trial of this project is presented.

RESULTS

Comparison of Seed Treatment to Foliar Sprays for the Control of Early Leaf Blast in Upland Rice

The disease development and control given by treatments in trials RI 1 and RI 2 are presented in the following two figures. Disease ratings in percent damage caused by leaf blast are given on the vertical axes and evaluation dates in days after seeding on the horizontal axes. Weekly foliar sprays with the standard fthalide were in one trial equal to one seed treatment application with pyroquilon under heavy disease pressure in upland rice nursery, but in the second, when the epidemic was earlier and more rapid, the seed treatment was significantly superior to the foliar spray.

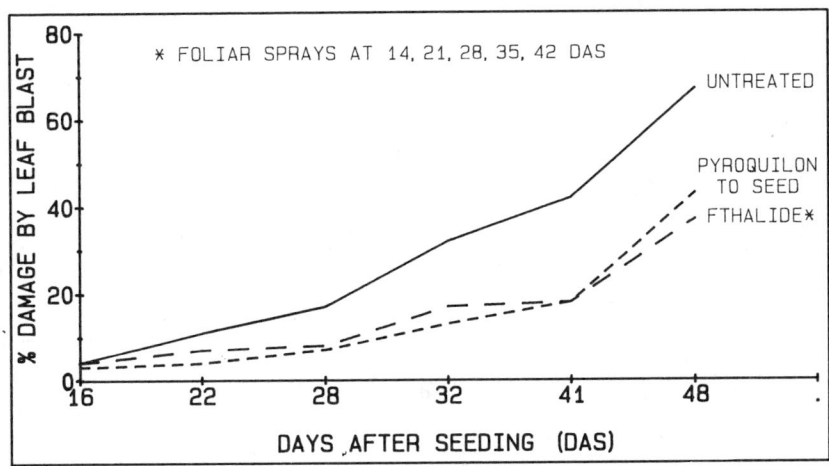

Figure 1. Trial RI 1 (Indonesia)

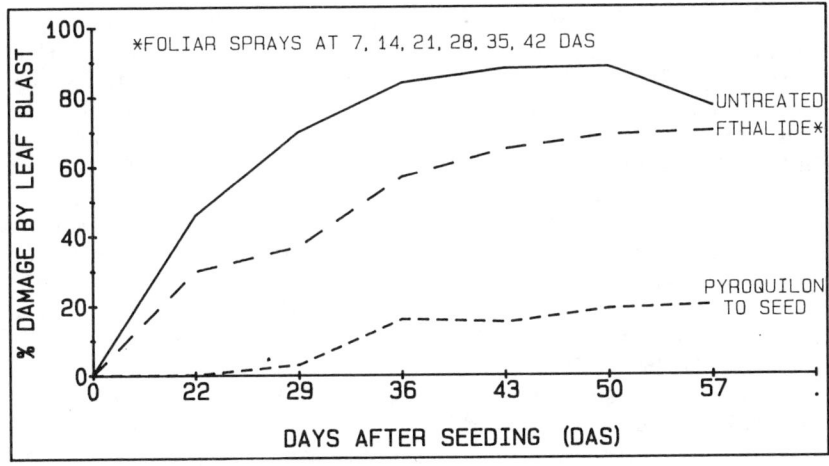

Figure 2. Trial RI 2 (Indonesia)

<u>On</u> <u>Farm</u> <u>Large</u> <u>Plot</u> <u>Seed</u> <u>Treatment</u> <u>Demonstration</u> <u>Trials</u> <u>in</u> <u>Upland</u>
<u>Rice</u>
In.trial CO 1 damage by leaf blast reached 25 % in the untreated
field and yield was 3889 kg ha^{-1}. In the field sown with
pyroquilon treated seed, disease damage was kept below 5 % during
the whole growing season and yield was 4494 kg ha^{-1}, 605 kg ha^{-1}
more than in untreated.

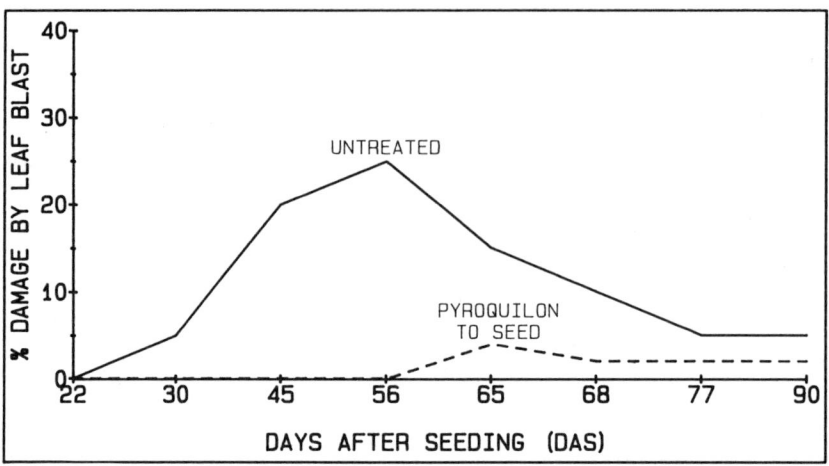

Figure 3. Trial CO 1 (Colombia)

Numerous demonstration trials were carried out in Central Brazil from 1986-1989 on upland rice to compare seed treated plots to untreated (Tables 1 and 2). The general observation throughout the trials was a more uniform crop development (tillering, panicle emission and ripening) in seed treated plots. The untreated plots presented an average of only 70 % of the productive tillers of treated plots and 20 % reduction in plant height. And 5-10 % of killed plants against none in the treated fields. Other differences noticed were 20-30 % longer flag leaves and 20 % more seeds/panicle in treated plots. The 1000 grain weight did not differ significantly, probably because plots were yielded with combine harvesters and lighter grains lost during the cleaning process. All these factors resulted in an average of 30-40 % yield increase in seed treated fields (Tables 1 and 2). Farmers noticed the differences resulting from seed treatment at the vegetative phase by just comparing the crop vigour or pulling out plants in treated and untreated fields, where the former presented a visibly longer and better developed root system.

TABLE 1
Seed treatment demonstration trials in Central Brazil
(1986 ~ 1988)

location	season	variety	T**	% blast on leaves 45-50 DAS	60-70 DAS*	yield kg ha^{-1}	% yield gain in treated
Goianésia	86/87	IAC25	T	5	40	1200	33
			C	40	50	900	
Faz.Ipê	86/87	IAC47	T	3	20	3804	93
			C	20	40	1965	
Faz.Ipê	87/88	CNA4120	T	2		1782	31
			C	15		1362	
Faz. Lagares	87/88	IAC47	T		10	2387	29
			C		35	1847	
Faz.J.Riva	87/88	CNA4120	T		2	2998	23
			C		20	2438	
Faz. Florismar	87/88	IAC47	T		40	2663	216
			C		70	843	
Faz. Planagri	87/88	CNA4120	T	15		2370	68
			C	35		1410	
Faz. Transal	87/88	IAC165	T	10		1568	41
			C	35		1108	
Faz. Marcelino	87/88	IAC47	T		20	3202	36
			C		60	2352	

* DAS (days after seeding)
** T=treated; C=untretaed check

TABLE 2
Seed treatment demonstration trials in Central Brazil
(1988/1989)

location	variety	T**	% blast on leaves 30 DAS	45 DAS	60 DAS*	yield kg ha^{-1}	% yield gain in treated
Faz. Ouro Quente	IAC25	T	0	5	30	2088	30
		C	20	45	60	1610	
Faz. Calcareo	Guarani	T	0	0	10	2300	92
		C	0	20	40	1200	

TABLE 2
(continuation)

| location | variety | T** | % blast on leaves | | | yield | % yield gain |
			30 DAS	45 DAS	60 DAS*	kg ha⁻¹	in treated
Faz.Ipê	IAC25	T	0	5	20	1431	16
		C	20	30	40	1230	
Faz. Santa Rita	IAC25	T	0	20	50	1924	30
		C	20	50	70	1480	
Faz. Felicidade	IAC165	T	0	0	10	2005	32
		C	10	30	40	1515	
Faz. Perpetuo Socorro	IAC165	T	0	5	20	1675	29
		C	20	50	60	1300	
Faz. Ferradura	IAC47	T	0	5	20	2430	35
		C	20	40	60	1800	

* DAS (days after seeding)
** T=treated; C=untreated check

Into Water Granule Application to Control Early Leaf Blast in Direct Seeded Flooded Rice

In the following two figures, results from trial TF-86-07 of a project with 8 trials carried out in Thailand in 1986 are presented. Pyroquilon granules applied at 25 days after seeding (before any blast occurred) were compared to an application at 34 days after seeding (soon after first blast lesions were visible) and to foliar sprays with 500 g a.i. ha⁻¹ of edifenphos at 34 and 41 days after seeding. The disease level reached 70 % in the untreated check 55 days after seeding. The preventively applied pyroquilon granule gave excellent disease control whereas the curative application of pyroquilon granules (at 34 DAS) was less effective. The standard foliar spray with edifenphos only gave a disease control of ca. 50 % (Figure 4). The corresponding yield results are presented in Figure 5. The better the disease control the higher were the yield responses.

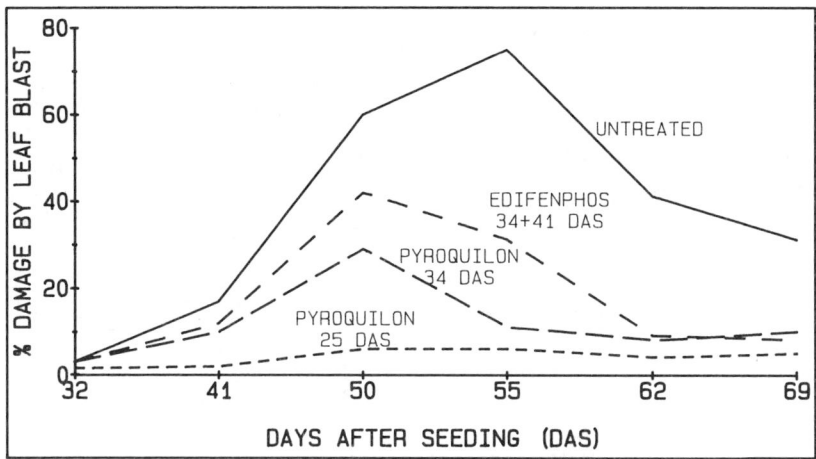

Figure 4. Trial TF-86-07: Leaf blast ratings.

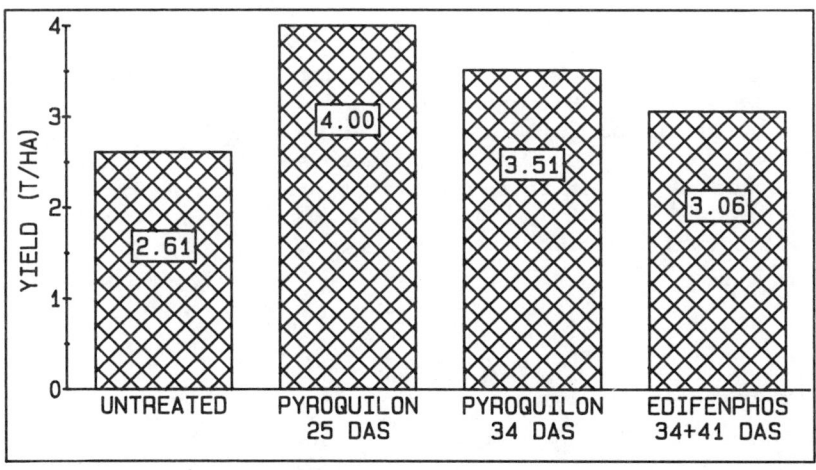

Figure 5. Yield results from trial TF-86-07.

Yield results from all the 8 trials from the Thailand project 1986 are presented in Table 3. Trials 01-04 received 2 applications at 25 and 55 days after seeding, and trials 05-08 received only one application at 25 days after seeding. The additional treatment at 55 DAS did not improve the performance of the protective pyroquilon treatment. The protective treatment resulted in higher yield response than the curative treatments in all trials.

TABLE 3
(yield in t ha^{-1})

treatment/trial	01	02	03	04	mean	05	06	07	08	mean
untreated	0.4	1.2	2.9	2.6	1.8	2.0	1.5	2.6	2.4	2.1
pyroquilon protective	3.6	3.5	4.0	3.9	3.8	3.9	3.5	4.0	3.9	3.8
pyroquilon curative *	0.8	2.7	3.5	3.0	2.5	3.4	3.1	3.5	3.1	3.3
edifenphos curative *	0.8	2.5	3.0	3.0	2.3	3.2	2.8	3.1	2.8	3.0

* Pyroquilon curative was applied soon after the first lesions were noticed, only once and edifenphos was applied at this same timing for the first application followed by a second, one week later.

Follow-up trials in the same region to compare a lower rate (500 g a.i. ha^{-1}) for the granule application at 25 days after seeding (DAS) to the previously tested high rate of 1500 g a.i. ha^{-1} and to the local foliar standard edifenphos were carried out in 1987. Disease ratings (in % damage caused by leaf blast) and yield figures of one of these trials is presented below (trial TF-87-02 in Bam Po). The lower rate was sufficiently effective for disease control and yield responses.

TABLE 4
(% damage caused by leaf blast)

treatment/DAS	35	40	45	50	55	60	64	yield (t ha^{-1})
untreated	0.2	5	18	74	80	69	68	2.4
pyroquilon 500 g a.i. ha^{-1}	0.2	1	2	7	7	5	3	5.3
pyroquilon 1500 g a.i. ha^{-1}	0.2	1	1	3	3	2	1	5.2
edifenphos 500 g a.i. ha^{-1} at 35 + 42 DAS	0.2	3	10	43	41	33	29	3.8

Emission of New Leaves in Upland Rice

In detailed studies with Indonesian rice variety Cimandiri, used
in the trials in Indonesia, leaves/hill (healthy hills) were
counted up to 42 days after seeding. The emission of new leaves
is very fast during the early development phase of rice (Table 5).

TABLE 5
(leaf emission in upland rice)

days after seeding	leaves/hill
10	12
14	17
21	31
28	56
35	127
42	174

DISCUSSION

Importance of Leaf Blast to the Rice Crop

Early crop establishment and plant development during the tillering phase of rice can be severely affected by blast. It does not
need to be a very strong attack. If it occurs early in the
development of the rice plant, before or at initial tillering,
disease levels which would just by their quantity of symptoms not
alert a farmer (with the exception of the very careful farmers in
the Far East), do already cause significant damage to the rice
crop. Such damage will later reflect in yield losses (Figure 3).
Early leaf blast can also be of such a severity that entire tillers
or even hills are destroyed [1]. In extreme cases farmers would
have to reseed.

Rice crops can also be infected after the tillering phase but
such late leaf infections are much less aggressive and cause less
crop damage. Reports of severe late leaf infections are seldom
found. This fact can be explained by the increased resistance of
older rice leaves towards *P. oryzae* (age resistance) [7].

The importance of early blast infections to the establishment, development and finally productivity of rice crops shows
that only early applications with blasticides can provide an
efficient management of blast in such situations (Figures 4 and
5).

Control of Leaf Blast in Dry Seeded Rice

Dry seeded rice is the most vulnerable to economical damage to the
crop by leaf blast. In most situations dry seeded rice corresponds to upland cropping. The most obvious way to control leaf

blast in such a growing system is a seed treatment or a foliar spray, and the results presented in Figures 1 and 2 indicate that. Seed treatment is much more efficient in controlling early leaf blast (Figures 1 and 2) as it offers a way to apply the optimum amount of compound without losses of chemical. It is quickly taken up by the seed and roots and translocated to the new growth with efficacy lasting over the first 40-50 days after seeding [8,9]. This is exactly the time during which plants are most susceptible to leaf blast and during which this disease causes severest damage to the young crop. A foliar spray cannot be as effective because the leaf area is still very small when first infections occur and, therefore, only limited amounts of fungicide find their target. New growth at this early development stage is very fast (Table 5) and because of this foliar sprays have to be repeatedly in order to achieve effective disease control.

Compounds to be used for seed treatment of rice must be ystemic with a long-lasting activity and be safe to the young rice plants. It is extremely important, in the case of blast, that the compound does not only disinfect the seed by inactivating conidia or mycelium existing on the rice grains used for seed, but that such a compound protects the rice plant from airborne inoculum through the tillering phase. Airborne inoculum is the most important source of disease for spreading blast [1,3]. Tillering is a phase during which the rice plant is highly susceptible to *Pyricularia oryzae* [10,11].

Control of Leaf Blast in Direct Wet Sown Rice
The treatment of wet sown rice seed is less reliable because the compound is washed-off and diluted. But also in wet sown rice it is very difficult to time a foliar spray because it should coincide with the first appearance of the disease. Only strongly curative fungicides would allow the delay of an application to a time when disease is already visible in the field. When some lesions are found, a high infection with latent lesions may be present in the crop thus compromising the success of an application with a protective compound. If a foliar spray is applied too early it will not protect the new growth, as significant amounts of a chemical taken up by a leaf are only redistributed within this same leaf and new leaves are formed as quickly as in upland rice (Table 5).

An into-water application with systemic blasticides taken up by roots offers a much better distribution into new leaves for a similar period as a seed treatment in upland rice. Three weeks after seeding or transplanting is an optimum timing for such applications and season-long protection for leaf blast can be expected from one single granule application (Figures 3 and 4, Tables 3 and 4).

To achieve the same level of control, depending on the conditions and availability of inoculum, 2-4 foliar applications may be needed.

Control of Leaf Blast in Transplanted Rice

In transplanted rice leaf blast is generally less severe than in direct seeded rice. At transplanting the nutrient level in the seedling changes and the old (sometimes infected) leaves die and drop. The epidemic is delayed and is seldom severe [1].

In transplanted rice blast may occur in the seedling boxes/beds. It can be very damaging and be controlled in the same way as direct seeded rice. To reduce the inoculum source coming from seedling boxes/beds one pre-plant incorporation or early granule application (before first infections take place) are the best recommendation.

The advantages of early fungicide applications for the control of blast in transplanted rice are less pronounced than with direct seeded rice, but in such a system it is also possible to apply the blasticide into the seedling box/bed any time before transplanting, depending on the characteristics (phytotoxicity, lasting of activity, systemicity) of the compound and the timing of blast appearance.

Advantages of Early Treatments for the Management of Rice Blast

The most recognized and feared phase of blast is panicle- and specially neck blast. Therefore, the efficacy of early season treatments against this disease are often questioned. But, before blast causes direct damage to the panicles, this disease can reduce the yield potential of a rice crop tremendously by interfering with crop establishment and the formation of productive tillers when attacking before or during the tillering stage. In many areas early leaf blast is more important than panicle blast or a predisposing factor for the quantity of panicle blast infections by producing the inoculum. It is not common for panicles to be attacked without the previous occurrence of leaf blast [1]. If the crop does not reach the reproductive phase in a good condition the importance of neck blast is only relatively minor. Therefore, the first thing needed to be controlled in endemic blast areas is early leaf blast.

Beside the very obvious differences (like panicles per plot, size of panicles, size of plants) the more uniform panicle emission and ripening are of great importance for improving yield in treated plots as farmers can time the harvesting more precisely, because more panicles are at the optimum ripening stage than in the untreated. Also the time of exposure to neck blast in treated plots is shorter when panicle development is more uniform. Rice fields affected by early leaf blast start the heading process earlier for the fewer main tillers but have many more latecomers, prolonging the risk for panicles available to be infected by blast. In areas where panicle/neck blast usually occur, the timing for an application with chemicals at the heading stage to control this phase of the disease is much easier to be defined in uniformly heading fields. If large enough areas are treated, the quantity of inoculum available for neck blast infections is reduced to such an extent that severe damage by panicle/neck blast may not occur.

Another important use of the early control of blast could be the treatment of newly released resistant or partially blast resistant varieties in endemic areas [12,13]. In blast endemic areas resistance breaks down after 2-3 [14] years and it is more and more difficult to produce new resistant varieties. The life-time of these varieties could be prolonged by controlling early eaf blast as less blast conidia are exposed to the host plant resistance selection pressure (14).

Seed treatments in general offer advantages of being relatively easy to apply and of lower risk to the environment. The same can be said for seedling box applications.

CONCLUSIONS

Early infections of rice crops with leaf blast cause damage to the optimum development of the plants and as a consequence cause significant yield reductions. Conidia deriving from early attacks are an important source of inoculum for surrounding fields. If the blast epidemic is not stopped, conidia from the upper leaves will infect panicles.

Effective control of early leaf blast resulted in significant yield improvements in upland rice in field trials in Central Brazil, where a simple seed treatment with a systemic blasticide reduced yield losses by 30 to 40 % through keeping the leaf blast infection low up to 40-60 days after seeding. Probably the benefits would be even greater if larger areas were treated for several years to reduce drastically inoculum in a region.

The most important advantages of controlling early leaf blast are the good response of the crop in reaching the heading stage with an optimum of vigour and uniformity thus laying the basis for maximum productivity.

The benefits of early leaf blast control can be shown in any leaf blast endemic area, disregarding the cropping system, if the right means for control are available and if the epidemiology, especially for the timing of first infections, is considered correctly.

If the disease occurs early, a preventive application is more effective than a curative one with today's available compounds. As the leaf area is very small during early growth stages and only little of the sprayed compound can find the target, a treatment taken up by the root system is more effective at this stage. Early treatments, especially a seed treatment, are very localized applications with a minimum of risk to the environment.

The control of leaf blast by seed/early season fungicide applications could be used as an important component of blast management, that prevents the vegetative phase damage and helps to keep inoculum pressure down for the subsequent danger of panicle blast infections and for the spread of blast to the surrounding rice fields. It should be used in combination with other blast management means, such as the use of resistant varieties, in order to make them more effective and durable.

REFERENCES

1. Ou, S.H., 1972: Rice diseases. Commonwealth Mycological Institute, Kew. Surrey, England, pp. 97-184.

2. Goto, K., 1965: Estimating losses from rice blast in Japan. The rice blast disease. Proceedings of a symposium at IRRI, July, 1963, pp. 195-202. Baltimore, Maryland, John Hopkins Press.

3. Suzuki, H., 1975: Meteorological factors in the epidemiology of rice blast, Ann. Rev. Phytopathology 13, pp. 239-256.

4. Kato, H., Sasaki, T. and Koshimizu, Y., 1970: Potential for conidium formation of *Pyricularia oryzae* in lesions on leaves and panicles of rice. Phytopathology 60, pp. 608-612.

5. Woloshuk, C.P. and Sisler, H.D., 1981: Tricyclazole, pyroquilon, tetrachlorphthalide, PCBA, coumarin and related compounds inhibit melanization and epidermal penetration by *Pyricularia oryzae*. J. Pesticide Sci. 7, 1982, pp. 161-166.

6. Nakamura, M.: CORATOP®/FONGORENE® (Common Name: Pyroquilon), A new systemic blast fungicide in rice, Japan Pesticide Information No. 48, 1986, pp. 27-30.

7. Bernaux, P., 1981: Evolution de la sensibilité des glumelles due riz à *Pyricularia oryzae* Cav. et à *Drechslera oryzae* (Br. de Haan), Sub et Jain: consequences pour la transmission des maladies, Agronomie 1 (4), pp. 261-264.

8. Prabhu, A.S. 1985: Evaluation of pyroquilon seed treatment for blast (Bl) control in upland rice, IRRN 10: 1, 1985, p. 13.

9. Guyer, R. and Marjudin, K., 1986: Effective control of rice blast in the tropics with FONGORENE® . Proceedings of the Second International Conference on Plant Protection in the Tropics, p. 371 (abstract).

10. Hashioka, Y., 1950: Studies on the mechanism of prevalence of the rice blast disease in the tropics, <u>Tech. Bull. Taiwan Agric. Res. Inst.</u> <u>8</u>, p. 237.

11. Andersen, A.L., Henry, B.W. and Tullis, E.C., 1943: Factors affecting infectivity, spread and persistance of *Pyricularia oryzae* Cav. <u>Phytopathology</u> <u>57</u>, pp. 237-241.

12. Filippi, M.C. and Prabhu, A.S., 1989: Controle da Brusone com tratamento de sementes e seu efeito sobre Brusone nas paniculas em arroz de sequeiro, <u>Fitopatologia</u> <u>Brasileira</u> <u>14</u>, p. 149 (abstract).

13. Williams, R. J., Marjudin, K., and Guyer, R., 1985: Rice blast control by seed treatment with pyroquilon, <u>Phytopathology</u> <u>75</u>, p. 1383.

14. Ahn, S.W.,1981: Slow Blasting Resistance in <u>Report</u> <u>and</u> <u>Recommendations</u> <u>from</u> <u>the</u> <u>Meeting</u> <u>for</u> <u>International</u> <u>Collaboration</u> <u>in</u> <u>Upland</u> <u>Rice</u> <u>Improvement</u>, <u>Brazil</u> <u>1981</u> <u>(IRRI)</u>, pp. 23-25.

BIOLOGICAL CONTROL OF RICE DISEASES (BLAST AND SHEATH BLIGHT) WITH BACTERIAL ANTAGONISTS : AN ALTERNATE STRATEGY FOR DISEASE MANAGEMENT

S.S. GNANAMANICKAM[1] and T.W MEW[2]

[1]Reader, Centre for Advanced Studies in Botany
University of Madras, Madras 600 025, India and
[2]Plant Pathologist, International Rice Research Institute
(IRRI), P.O. Box 933, Manila, Philippines

ABSTRACT

As part of efforts to develop alternate strategies for rice disease management, more than 400 strains of bacteria isolated from IRRI rice fields were screened in the laboratory for antagonism towards Pyricularia oryzae and Rhizoctonia solani, the fungal pathogens of blast and sheath blight respectively. On the basis of growth-inhibition in plate assays, 9 bacterial strains which included 3 strains of Pseudomonas fluorescens, 5 strains of Bacillus spp and a strain of Enterobacter were identified. Diameter of inhibition zones ranged from 20-40 mm against P. oryzae and from 20-31 mm against R. solani. These bacterial strains also suppressed sheath blight (Sh-B) in detached rice leaves. They were further evaluated in field tests for suppression of blast and Sh-B when applied as seed treatments or as seed treatments plus sprays. In three field experiments conducted in the IRRI farm, bacterial treatments afforded significant Sh-B control and performed better than validamycin, the fungicide routinely used for Sh-B control. Suppression of Sh-B was more pronounced in direct-seeded rice than in tranplanted rice. Bacterial treatments reduced leaf blast severities by 50-73% in var-IR50 and 34-80% in var C-22 in lowland experiment and 47-57% reduction in var. UPLRi-5 in an upland experiment. Further investigations with a Pseudomonas strain, 7-14 have shown that it produces 2 fluorescent antiblast antibiotics (AB_1 and AB_2) in culture which at 1.0 ppm inhibit conidial germination of P. oryzae and afford 90-92% reduction in leaf blast development in treated rice (var. IR 50) seedlings. Therefore, biological control of blast by Pseudomonas is antibiotic-mediated.

INTRODUCTION

Rice is the staple food for 2.7 billion people in Asia where 90% of the world's rice is grown and eaten (1). Blast disease caused by <u>Pyricularia oryzae</u> cav. is a severe production contraint in dryland and unbunded upland rice cultures where rice farmers have no access to nor can they afford fungicides. Because of the instability of the rice blast fungus and the marked variability in pathogenicity found in different strains of this organism, control through breeding for resistance has met with only partial success. Likewise, sheath blight (Sh-B) of rice caused by an aerial form of the fungus, <u>Rhizoctonia solani</u> (Kuhn) (<u>Thanetophorus cucumeris</u> (Frank) Donk) has emerged as one of the most important diseases in modern rice production in both the temperate and tropical rice growing countries causing serious crop losses (2). In recent years near total crop failures have occurred in Vietnam where farmers use very high seeding rates (300-400 kg ha^{-1}) to raise direct seeded rice crops. Sh-B control by the use of resistance cultivars has not been very successful because adequate level of host resistance has not been found. Fungicide application is effective but is limited to the more affluent countries (3). Therefore, alternate disease management strategies are sought that are ecology sound and cost effective.

Recent studies conducted in Australia, Canada, China, India, the Netherlands, Philippines and the United States reveal the potential of root colonizing microorganisms to inhibit or displace soilborne pathogens at the root soil interface and thereby protect the root health of perennial and annual plants (4) such as cotton, potato, tobacco, flax raddish, cucumber, wheat and rice (5,6,7,8,9,10,11,12,13). These microorganisms inhibit pathogens by producing antibiotics, siderophores (compounds that chelate biologically available iron), and possibly substances that stimulate plant growth (14, 15, 16, 17). This paper describes new evidence on the use of antagonistic bacteria for the control of rice blast and Sh-B and presents evidence for the basis for such antagonism in a fluorescent <u>Pseudomonas</u> strain.

disease caused by <u>Pyricularia</u> <u>oryzae</u> cav. is a severe production contraint in dryland and unbunded upland rice cultures where rice farmers have no access to nor can they afford fungicides. Because of the instability of the rice blast fungus and the marked variability in pathogenicity found in different strains of this organism, control through breeding for resistance has met with only partial success. Likewise, sheath blight (Sh-B) of rice caused by an aerial form of the fungus, <u>Rhizoctonia</u> <u>solani</u> (Kuhn) (<u>Thanetophorus</u> <u>cucumeris</u> (Frank) Donk) has emerged as one of the most important diseases in modern rice production in both the temperate and tropical rice growing countries causing serious crop losses (2). In recent years near total crop failures have occurred in Vietnam where farmers use very high seeding rates (300-400 kg ha^{-1}) to raise direct seeded rice crops. Sh-B control by the use of resistance cultivars has not been very successful because adequate level of host resistance has not been found. Fungicide application is effective but is limited to the more affluent countries (3). Therefore, alternate disease management strategies are sought that are ecology sound and cost effective.

Recent studies conducted in Australia, Canada, China, India, the Netherlands, Philippines and the United States reveal the potential of root colonizing microorganisms to inhibit or displace soilborne pathogens at the root soil interface and thereby protect the root health of perennial and annual plants (4) such as cotton, potato, tobacco, flax raddish, cucumber, wheat and rice (5,6,7,8,9,10,11,12,13). These microorganisms inhibit pathogens by producing antibiotics, siderophores (compounds that chelate biologically available iron), and possibly substances that stimulate plant growth (14, 15, 16, 17). This paper describes new evidence on the use of antagonistic bacteria for the control of rice blast and Sh-B and presents evidence for the basis for such antagonism in a fluorescent <u>Pseudomonas</u> strain.

MATERIALS AND METHODS

Bacterial strains. Isolation of fluorescent and nonfluorescent strains of bacteria from rice field soil, water, rice tissues and other sources have been described previously (10). These strains were available in IRRI's culture collection.

Additional isolations were made for root-colonizing bacteria from the rice rhizosphere. A total of 267 samples of rice roots (ca. 10 g each) were collected with adhering soil from different blocks of the IRRI farm for lowland rice during July 1987 to June 1988. Samples were washed in running tap water for 10 min. and blotted dry. Twenty five root tips each 20-25 mm long were removed at random and were imprinted by gentle pressing for 30 sec on dry surface of either King's medium B (KB) or yeast extract peptone dextrose agar (YPDA) for the detection of fluorescent and nonfluorescent strains respectively.

The remainder of each sample was surface sterilized by washing serially in 10% sodium hypochlorite (5 min), 70% ethyl alcohol (20 sec) and three changes of sterile distilled water. After drying, five smaller samples of 1.0 g each were ground asceptically in a pestle and mortar with 9.0 ml sterile water. From a serial dilution prepared, 0.1 ml of each dilution was plated onto six replicate plates of KB and YPDA by following the standard dilution plating techniques (10,18). Fluorescent colonies were detected by viewing KB plates under UV light after 48 h incubation at 28°C. Non-fluorescent colonies that appeared on KB and YPDA plates were restreaked on KB and nutrient broth yeast extract (NBY) agar plates and marked as nonfluorescent if there was no pigment production on both media. Single colony isolates were stored in NBY and YPDA slants.

Test for in vitro antibiosis. Four hundred strains were screened in the laboratory for antagonism towards the blast fungus, P.oryzae. Of these, 200 strains were from the existing culture collection assembled during earlier studies (10) and another 200 strains were from isolations made during the present study. Strains were routinely screened for mycelial growth inhibition in P.oryzae and R.solani by dual plate tests (10,11). Diameter of inhibtion zones were measured after 3 and 5d incubation at 28°C. Only those strains which induced inhibition zones of more than 20 mm dia were retained for further evaluation.

Production of fluorescent pigment (siderophore) in plain (KB) and in FeCl$_3$ (1.0 to 10 nM)-amended KB was also tested for two of the fluorescent strains (4-15 and 7-14) which were strongly antagonistic to P. oryzae.

Strains which were strongly positive for in vitro antibiosis towards R. solani were screened for Sh-B reduction on detached rice leaves (var. IR58, 35 d old). Leaves at the same growth stage were removed from screenhouse grown plants, washed in running tap water and cut into 5-7 cm long pieces. Five such pieces were arranged in each water agar plate (9 cm dia). Five laboratory produced sclerotia soaked or pretreated either in bacteria or water were placed on the centre of the leaf (one sclerotia per leaf) surface in 6 replicate plates for each strain evaluated. Plates were observed every 24 h for mycelial growth in water agar. Sh-B development was recorded in checks and bacteria treatments after 7 d incubation at 28°C by measuring the average length of Sh-B lesions.

Identification of bacterial strains

Two fluorescent (4-15 and 7-14) and two nonfluorescent (33 and 4.03) strains which showed strong antibiosis, towards P. oryzae were characterized on the basis of results of conventional bacteriological tests described under the determinative scheme of Palleroni (19) for fluorescent strains and the API system (20) for nonfluorescent strains.

Using standard procedures, mutant strains for these four bacteria were generated by incorporating resistance to rifampicin (R) or rifampicin and nalidixic acid (RN) (100 ppm). The mutants were checked for retention of antibiosis towards P. oryzae.

Likewise, three fluorescent (4-15, 7-14 and 784) and six nonfluorescent strains (1-14-1, 17, 33, 4-03, 591 and 655) that showed strong in vitro antibiosis towards R. solani and reduced Sh-B in detached rice leaves were characterized.

Seed inoculation and bacteria treatments for field experiments for blast and Sh-B control

1.   **Blast** : Seeds of rice varieties were coated with bacteria by following a simple scheme : Bacterial cells of test strains from 24-h-old cultures were scraped into a 1% carboxymethylcellulose buffer. Seeds (1.3 kg/ treatment) were treated with the bacterial suspension and incubated overnight in polythene bags at 25°C. Excess buffer-bacteria mixture was drained and seeds were dried in sterile air for 12 h before sowing in 4x5 m plots. At sowing, seeds had 10^9 colony-forming-units (cfu) g^{-1}. Seeds coated with a fungicide, pyroquilon (CGA 49104) (8.0 g kg^{-1} seed) and untreated seeds were sown as checks.

Two field experiments were conducted. One was at the Blast Nursery of the IRRI farm during dry season (DS) 1989. Seed inoculation of rice (var IR50 and C22) with strains, 4-03R, 33R, 4-15 R and 7-14 RN, pyroquilon and a nontreated check formed the six treatments for this randomized complete block (RCB design) experiment which had 4 replications for each treatment. Besides seed treatment, the IR50 and C22 rice crops received four bacterial (10^8 cfu ml^{-1}) sprays at 10, 25, 40 and 55 days after planting.

Another field experiment was conducted at the IRRI site for Upland Rice Research for Acid soils at Cavinti in the Philippines where there is usually heavy incidence of blast. The rice var. UPLRi-5 was seed inoculated and sprayed (exactly as described for the other experiment) with the same four bacterial strains.

In both experiments, bacterial multiplication was monitored from root and shoot samples removed at 10 d intervals on NBY amended with R or RN. Leaf blast and neck blast were assessed.

2. **Sh-B** : Three field experiments were conducted in the IRRI farm during the dry season (DS) (Jan-May, 1988), wet season (WS) (July-December), 1988 and DS 1989. The crop was raised from inoculated seed or from nontreated (check) and fungicide-treated seeds. Fungicides and checks were included to allow for comparisons on the efficiency of bacterial treatments to be made. Application of fertilizers at standard rates and other agronomic practices were followed. For raising a transplanted crop seeds were first sown in a raised wet seedbed and seedlings of 21 d were transplanted in field plots of 5.4 x 3.0 m. For each treatment, there were 4 replications arranged in a split-plot design. Seeds were sown directly in field plots in the case of a direct-seeded crop.

In the first experiment (DS 1988) bacterial treatments were limited to seed inoculation while in the next two experiments, additional sprays (2 during WS 1988 at 25 and 40 days after planting and 3 during DS 1989 at 25, 40 and 55 DAP) were also made with bacterial cell suspensions that contained 10^8 cfu ml^{-1} by using knapsack sprayers. Fungicides were sprayed at recommended concentrations.

Treatment for DS 1988 experiment included 5 bacteria of which 3 (In-b-33, In-b-3, and In-b-655) were nonfluorescent strains, one (In-b-784) fluorescent strain and a mixture of In-b-33 + In-b-784. Two fungicides, quintozene (2 g kg^{-1} seed) and validamycin (3% a.i) were used together with bacterial treatments on both direct-seeded and transplanted crops of varieties, IR58, and IR62.

In experiment 2 (WS 1988), five bacterial treatments (2 fluorescent strains, In-b-784 and In-b-7-14 and 2 nonfluorescent strains, In-b-33 and In-b-1-14-1 and a mixture of In-b-33 + In-b-1-14-1) and validamycin were tested on direct-seeded and transplanted crop var. IR58. In the DS 1989 experiment, the same bacterial treatments with an additional treatment made up of a mixture of In-b-7-14 + In-b-784 (both fluorescent) strains were tested on a transplanted crop of IR 58.

Basis for biological antagonism in a Pseudomonas strain antibiotic production

Strain 7-14 was grown in 200 ml of potato glucose broth in 500 ml Erlenmeyer flasks as shake culture for 5 d. Culture conditions for production of antibiotic and its isolation were those used by Gurusiddaiah et al (21) From a batch of 10 l broth an oily residue of 100-350 mg l^{-1} was obtained after solvent extraction with methylene chloride and concentration. The residue was further purified through sephadex LH20 column chromatography twice. A partially purified substance obtained by selective elution of the column with chloroform : acetone was chromatographed on tlc (Merck silica gel, F254S 0.2 mm) plates and developed in chloroform : methylene chloride (9:1). The fluorescent antibiotics were further characterised by their R_f values, uv absorption spectra and biological activity.

Tests for biological activity of the antibiotic bands

The two substances (AB$_1$, AB$_2$) were separately tested for inhibition of conidial germination in P. oryzae. Conidia of

P. oryzae were made up in sterile water @ 25,000 conidia ml^{-1}
by following procedures described by Mackill and Bonman (22).
Their germination was assessed in slide germination tests in
the presence of 0(check) to 1000 ppm concentration of the
antibiotic after 1 to 6 h of incubation at 28°C. There were 6
replications for each concentration tested. Biological
activity of the antibiotics was also assessed in three
separate glasshouse tests. Seedlings of IR50 rice (3-4 leaf
stage) were spray inoculated with P. oryzae conidia (50,000
ml^{-1}) suspended either in water or aqueous solution (1.0 ppm)
of the two antibiotic substances together. Leaf blast lesions
were counted 5 d after inoculation of 400 leaves.

RESULTS

Identification of bacterial strains

Amongst the four strains which were considered most
efficient antagonists for P. oryzae the two fluorescent
strains, 4-15 and 7-14 were identified as strains of
Pseudomonas fluorescens and the nonfluorescent strains, 33 and
4-03 as Bacillus spp. Species level identification of the
Bacillus strains is incomplete.

Among the efficient antagonists for R. solani, the 3
fluorescent strains (In-b-784, In-b-7-14 and In-b-4-15) were
identified as Pseudomonas fluorescens. The nonfluorescent
strains In-b-1-14-1, In-b-4-03, In-b-17, In-b-33, and In-b-655
were species of Bacillus while In-b-591 was an Enterobacter
sp. More data are being gathered to complete the species
level identification.

In vitro **antibiosis.** In plate tests, Pseudomonas
fluorescens strains 4-15 and 7-14 and Bacillus strains, 4-03
and 33 caused maximum inhibition of the blast fungus, P.
oryzae. The average diameter of the inhibition zones was 38.5
mm for strain 7-14 (Fig 1), 30.4 mm for strain 4-15, 26.3 mm
for strain 33 and 21.1 mm for strain 4-03. The rifampicin (R)

or rifampicin and nalidixic acid (RN) resistant mutants of these strains retained their ability to inhibit P. oryzae.

Further, in plate tests, inhibition of P. oryzae by strains 7-14 and 4-15 was not reversed by amendments with FeCl$_3$ (upto 10 mM) in King's medium or potato dextrose agar.

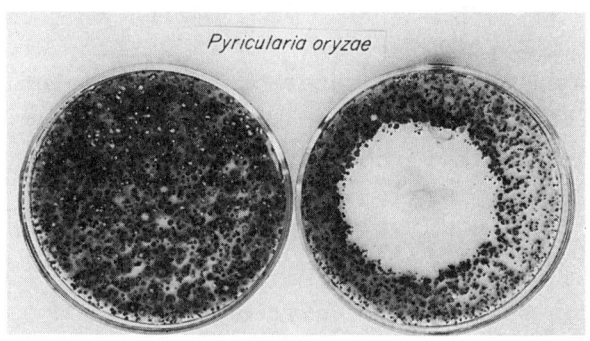

Figure 1. Inhibition of <u>Pyricularia</u> <u>oryzae</u> in potato dextrose agar plate by fluorescent pseudomonad strain, 7-14, The plate on the right was inoculated in the centre with loopful of cells of 7-14 after the agar surfaces of both plates were spread with conidial suspension of P. oryzae. The plate on the left (check) did not receive bacterial inoculation. Fungal inhibition was observed after 4 days of incubation.

One hundred and sixty-eight of the total strains (42%) screened <u>in</u> <u>vitro</u> showed antagonism towards <u>R.</u> <u>solani</u> which was indicated by the appearance of mycelial growth-inhibition zones of 2.0 to 31.0 mm dia. Only 37 strains consistently caused inhibition zones of 20-31 mm and the nonfluorescent strains (26 strains) were predominant amongst them. When these 37 strains and 15 other strains which induced smaller zones were evaluated in the detached leaf tests only 9 strains caused Sh-B suppression ; all were from among the 37 strains. Sh-B suppression caused by these strains (3 fluorescent and 6 nonfluorescent) were 91% by strain 1-14-1 (NF-nonfluorescent), 88% by strain 591 (NF), 87% by strain 7-14 (F = fluorescent),

72% by strain 33 (NF), 66% by strain 4-03, 59% by strain 17 (NF), 56% by strain 784 (F) and 39% by strains 4-15 (F) and 655 (NF).

Biological control of leaf and neck blast severities

Seed treatment and spray application with bacteria led to reductions in leaf blast severity in both IR50 and C22 rice crops (Table 1) in the experiment conducted at the blast nursery. The fungicide, pyroquilon was the most effective of all the treatments. Amongst the bacterial treatments, strains, 4-03 and 7-14 afforded 73 and 62% reductions in leaf blast in var IR50 and strains, 33 and 34-03 afforded 80 and 66% reductions in leaf blast in var C22. Reductions in neck blast severity afforded by bacteria were not significant (Table 1).

TABLE 1
Effect of seed treatment and spray application with antagonistic bacteria on leaf blast severity, Blast Nursery, IRRI, DS 1989

| Treatment | Leaf Blast Severity | | | | Neck Blast Incidence | | | |
	IR50	Disease Control (%)	C22	Disease Control (%)	IR50	Disease Control (%)	C22	Disease Control (%)
Bacterial strain								
4-03 R	0.92	73.4	0.19	66.1	67.42	20.6	21.32	22.2
33 R	1.63	52.9	0.11	80.4	47.63	35.7	15.20	44.5
4-15 R	1.74	49.7	0.37	33.9	79.13	-2.3	27.03	1.4
7-14 RN	1.31	62.1	0.31	44.9	74.40	3.8	22.03	19.6
Pyroquilon	0.58	83.2	0.00	100.0	28.20	63.6	9.53	65.2
Check	3.46	-	0.56	-	77.37	-	27.4	-

$$^a \text{ Leaf Blast Severity Index} = \frac{n(1) + n(2) \ldots n(9)}{\text{Total } n} \times 100$$

where $n(1)$, $n(2)$,...$n(9)$ are number of tillers with disease score 1,2,..,9 in a 1-9 scale of IRRI's standard Evaluation System (SES). Scores 1-3 indicate a resistant (R) reaction, 4-5 a moderately resistant (MR) and 6-9 a susceptible (S) reaction.

$$^b \text{ Percent Disease Control} = \frac{\text{Control} - \text{Treatment}}{\text{Control}} \times 100$$

Bacterial populations monitored during this experiment showed fluctuations. In IR50, strains 33 and 4-03 had higher populations of 10^5, 10^7, 10^7 (cfu g^{-1} tissue at 20,35, 50 and 80 d after seeding than strains 7-14 and 4-15 which had 10^4, 10^5, 10^5 and 10^3 cfu g^{-1} tissue for the corresponding periods. Nontreated checks had other bacteria whose populations never exceeded 10^2 cfu g^{-1} tissue during the cropping period. In var-C22, strain 33 had the highest population of 10^3, 10^8, 10^6 and 10^5 cfu g^{-1} tissue at 25, 35, 50, 65 and 80 d after seeding. This was followed by strains 4-03, 7-14 and 4-15 in descending order.

In the second field experiment conducted at Cavinti, Philippines, strains 4-15 R and 7-14 RN of Pseudomonas fluorescens afforded 59 and 47% leaf blast reduction in rice var. UPLRi-5. The Bacillus strains, 4-03 R and 33 R gave 46 and 44% protection (Table 2). All these were significant reductions. Of all the treatments, pyroquilon was most effective. Strain 7-14 was the most effective of all treatments (including pyroquilon) for neck blast reduction (Table 2). There were also small increases in grain yield due to bacteria and pyroquilon treatments.

P. fluorescens strains had populations of 0.5×10^5 cfu g^{-1} tissue upto 40 d after seeding and were not detected in subsequent samples. The Bacillus strains, however, had higher populations of 0.9×10^6 cfu g^{-1} tissue at 30 d and 1.0×10^5 cfu g^{-1} tissue at 60 and 110 d after seeding.

TABLE 2

Effect of seed inoculation and sprays with antagonistic bacteria on rice blast in cv. UPLRi-5 Cavinti, 1988 WS.

	Blast severity		Grain
Treatment	Leaf blast	Neck blast	yield (g)
Bacterial strain			
4-03 R	3.32	2.96	100.45
33 R	3.49	3.70	95.53
4-15 R	2.57	2.95	92.53
7-14 RN	3.29	2.75	102.65
Pyroquilon	1.95	3.68	96.33
Check	6.27	3.77	96.33
LSD (0.05)	2.22	1.79	43.55

[a] Number of blast lesions cm^{-2} leaf area.

[b] Neck blast severity index = $n(1)+n(2)+n(3)...n(9) \times 100$ $n(1)$, $n(2)$, etc are number of tillers with disease score 1,2, or 9 in a 1-9 scale of IRRI's Standard Evaluation System (SES). A score of 1 indicates no neck blast and score of 9 indicates maximum neck blast severity.

[c] From 100 panicles per plot.

Suppression of Sh-B in field plots by bacterial treatments Experiment I (DS 1988). When 5 bacterial treatments were compared with nontreatment (check) and fungicide treatments on direct-seeded and transplanted IR58 and IR62 rice crops, significant Sh-B reductions were seen in all treatments in direct-seeded IR58 and transplanted IR62 crops (Table 3).

Fluorescent strain 784 and a mixture of 784 + nonfluorescent strain 33 afforded disease reductions of 66 and

62% while quintozene and validamycin treatments caused 68 and 42% respective reductions in direct-seeded IR58. In transplanted IR58, both the fungicides and strain 4-03 treatments alone caused significant Sh-B reductions.

In direct-seeded IR62 rice, all bacterial treatments caused significant reductions of 71-81% while quintozene and validamycin afforded 68 and 71% reductions respectively. In transplanted IR62, neither the bacteria nor the fungicides caused significant disease suppression. Grain yields were not affected by bacteria or fungicide treatments.

TABLE 3

Effect of seed inoculation in direct-seeded and transplanted rice (cv. IR58 and IR62) on the incidence of sheath blight. Dry season 1988. IRRI, 1988

| Seed Treatment | Sh-B severity[a] | | | |
| | Direct-seeded | | Transplanted | |
	IR58	IR62	IR58	IR62
1. Quintozene	1.62	0.98	1.93	1.03
2. Validamycin	2.94	1.29	1.93	0.79
3. In-b-33	2.95	1.27	2.62	0.77
4. In-b-4-03	2.27	1.90	1.13	0.78
5. In-b-784	1.72	1.59	2.48	0.86
6. In-b-655	2.06	1.58	3.08	0.67
7. In-b-33 + 784	1.88	2.08	2.70	0.63
Check (non-inoculated, pathogen inoculated)	5.00	1.42	2.64	3.11
LSD (0.05)	0.83		0.64	

[a] Average of 4 replicates.

Experiment II (WS 1988) This experiment is a comparison between direct-seeded and transplanted IR58 crops (Table 4). Sh-B development was 13.4% in checks of the former and 12.8% of the latter. Validamycin and all 5 bacterial treatments caused significant Sh-B reduction in all the direct-seeded crop while validamycin and 4 of the 5 bacterial treatments caused Sh-B reductions in the transplanted crop (Table 4).

In direct-seeded IR58 Sh-B reductions afforded by bacterial treatments ranged from 43 to 98% while that of validamycin was 63%. Strains 1-14-1 (nonfluorescent) and 7-14 (fluorescent) caused Sh-B reductions better than validamycin. In the transplanted crop, however, validamycin was most effective and caused a 78% Sh-B reduction. Bacterial treatments afforded 21 to 66% reductions (Table 4). Yield increases were not significant.

TABLE 4

Effect of seed inoculation and two sprays with bacteria on sheath blight severity in direct-seeded and transplanted IR58 (WS 1988). IRRI, 1988.

| | Sh-B severity[a] | |
Treatment	Direct-seeded	Transplanted
Validamycin	4.99 bc	2.76 c
In-b-33	6.36 bc	9.27 ab
In-b-1-14-1	0.19 c	5.99 bc
In-b-7-14	2.11 c	5.02 bc
In-784	5.09 bc	4.43 bc
In-b-33 + 1-14-1	7.74 b	4.98 bc
Check	13.39 a	12.83 a

[a] Averages of 4 replicates. Means followed by the a common letter are not significantly different from each other at the 5% level by DMRT.

Experiment III (DS 1989) This experiment was designed to examine the effect of bacteria applied as seed treatments and three additional sprays for Sh-B reductions in transplanted IR58 rice crop (Table 5). The nontreated checks had 59.8% Sh-B incidence while validamycin and all 6 bacteria treatments caused significant suppression of Sh-B. Disease reductions caused by validamycin was 93.8% along with significant increase in grain yield. Bacterial treatments caused 61-74% Sh-B reductions but no significant yield increase.

TABLE 5

Effect of seed inoculation and three sprays with bacteria on sheath blight severity in transplanted IR58 (DS 1989) IRRI, 1989

Treatment	Sheath blight severity		Yield	
Validamycin	3.68	c	174.60	b
In-b-33	15.58	bc	159.00	ab
In-b-1-14-1	19.32	b	150.30	a
In-b-33 + 1-14-1	20.76	b	149.80	a
In-b-7-14	19.49	b	151.00	a
In-b-784	21.94	b	144.60	a
In-b7-14 + 784	23.19	b	139.10	a
Check	59.84	a	130.20	a

[a] Averages of 4 replicates. Means followed by a common letter are not significantly different at the 5% level by the DMRT.

Characteristics of antibiotics produced by Pseudomonas fluorescens strain, 7-14.

A sample of 6.0 g of crude extract made from 10 1 of culture fluids of strain 7-14 yielded 2.8 mg of partially purified antibiotic at the end of two rounds of Sephadex LH 20 column chromatography. On tlc this substance yielded a major UV fluorescent (266 nm) blue band designated as AB_1 and also a minor blue fluorescent band (AB_2). The major principles (AB_1) moved to Rf values of 0.345 in $CHCl_3$- acetone (4:1), 0.240 in $CHCl_3$-acetone (9:1), and 0.048 in CH_2Cl_2 (100%). Both AB_1 and AB_2 had similar UV-spectra and had a major absorption peak at 220 nm and minor absorption peak at 281 nm.

At 1.0 ppm concentration these antiblast antibiotics afforded 70-100% inhibition of conidial germination in P. oryzae (Fig. 2a and 2b). They also protected IR50 rice seedlings from infection by P. oryzae. In the three glasshouse tests made, there was an average of 23 blast lesions per leaf in the plants sprayed with conidia in water (check) while those seedlings sprayed with antibiotic treated conidia had an average of 1.9 lesions per leaf (Fig. 3).

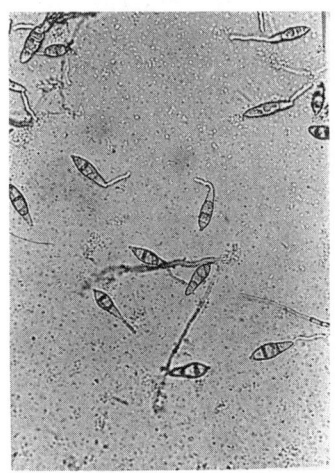 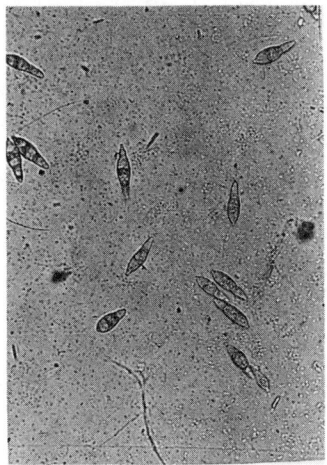

Figure 2a,b. Inhibition of conidial germination in Pyricularia oryzae by antiblast substances produced by Pseudomonas strain, 7-14. Conidia were suspended either in water (2a) or in an aqueous solution of AB1 and AB2 at 1.0 ppm concentration (2b). Germination counts were made after 6 h. Antiblast substances inhibited conidial germination that ranged from 70 to 100%.

Fig. 3. Protection of IR50 rice seedlings from leaf blast development by treatment with <u>Pseudomonas</u> antibiotic. Leaves on the right are from seedlings that were spray inoculated with conidia of <u>Pyricularia</u> <u>oryzae</u> suspended in water (50,000 ml^{-1}). Leaves on the left showing a fewer number of blast lesions are from seedlings that were spray-inoculated with the same level of conidia suspended in a 1.0 ppm aqueous solution of antiblast substances (AB_1 & AB_2). Lesions were counted from 400 leaves for each treatment on the 5th day after inoculation. In the three glasshouse tests, there were 93, 91 and 90% protection due to antibiotic treatment.

DISCUSSION

In this study it has been demonstrated that bacterial strains isolated from rice fields and selected on the basis of <u>in</u> <u>vitro</u> tests for antibiosis can be useful for reduction of leaf and neck blast severities when used as seed treatments and sprays. Although the reductions were not as good as those afforded by pyroquilon, reductions that ranged from 50 to 73% and 34 to 80% were obtained through bacterial treatments (Table 1) in varieties IR50 and C22 respectively.

At least in the first field experiment conducted at the Blast Nursery, reductions in leaf blast severities caused by bacterial treatments correlated with the strain's ability to colonize and survive in the rice plant. This is evident from the observation that Bacillus strains 33 and 4-03, and in particular strain 33, caused maximum leaf blast reductions (80% reduction caused by strain 33) in var C22 in IR50 and C22 rice crops (Table 1). Whereas the data obtained on leaf blast reductions by the same set of bacteria treatments in experiment 2 conduct at the Upland Acid Soil site at Cavinti did not show a correlation between percent leaf blast reduction and bacterial populations (Table 2). It is noteworthy that the P. fluorescens strains (7-14 and 4-15) in spite of their lower population were more effective in reducing leaf blast and neck blast than the Bacillus strains. Therefore, on the basis of performance in the Upland Site where blast is usually a severe production constraint, strain 7-14 was investigated further to understand the mechanism(s) involved in biological suppression of blast.

Circumstantial evidence suggested that siderophore production, one of the important mechanisms known to mediate bacterial antagonism in fungal inhibition is unlikely to be involved in highly acidic soils (23) such as Cavinti soil pH 4.0 and with an Fe-content that ranges from 3.7 to 4.6% (24). Further, in plate tests, inhibition of P. oryzae by 7-14 was not reversed by amendments with $FeCl_3$ (upto 10 mM) in King's medium B or potato dextrose agar.

Similar evidence is also available for Sh-B suppression. Results obtained in glasshouse experiments suggest that soil pH that is optimal for lowland rice production (pH 5.5 to 6.5) and acid pH (5.0) are suitable for obtaining significant results on Sh-B reduction by bacterial treatments (25). The soils used in these experiments had active Fe-contents of 1.3 to 2.8%. It is known that fluorescent siderophores (FS) are not produced in the presence of moderate amounts of Fe (9,23).

Therefore, searching for antibiotics to explain the antiblast activity in strain 7-14 appeared as a reasonable option.

In the field, Sh-B reduction due to bacterial treatments was more pronounced in the direct-seeded IR58 crops in the DS and WS 1988 experiments (Table 3,4). Most bacterial treatments performed better in a direct-seeded crop than in a transplanted crop. This is evident in the WS 1988 experiment (Table 4) which compared direct-seeded and transplanted IR58 crops. For example, strain 33, 1-14-1 and 7-14 treatments had lower Sh-B severities of 6.4, 0.2 and 2.1% respectively in direct-seeded and had higher Sh-B severities of 9.3, 6.0 and 5.0% respectively in transplanted IR58 crops. This may be due to the greater crop canopy in a direct seeded crop and the ideal ecological niche it provides for the epiphytic survival and colonization by bacteria. That this is likely is supported by Mew and Rosales (10) who have shown that bacteria introduced into the rice seed migrate to aerial parts of the rice plant. Whatever the exact reason, this may prove to be more useful as farmers in more and more rice growing areas resort to direct-seeding as the culture practice in intensive rice production.

It is noteworthy that validamycin, the fungicide routinely used for Sh-B control consistently performed better in a transplanted crop than in a direct-seeded crop (Table 3,4). Again, this may provide circumstantial evidence to what has been mentioned above : the increased multiplication of bacteria in the dense crop niche of a direct-seeded crop cannot be achieved by a chemical.

The work on the isolation of antiblast substances suggests that these substances are quite different from the phenazine carboxylic acid characterized as the antibiotic produced by P. fluorescens strain 2-79 by Gurusiddaiah et al (21) and shown to have a essential role in the biological control of Gaeumannomyces graminis var. graminis and other root pathogens of wheat (15). It is also different from pyoluteorin and pyrrolnitrin antibiotics (6,7) by their spectral characteristics. However, the antiblast properties (inhibition of conidial germination) of these water soluble substances resemble those of blasticidin and ediphenphos used routinely in Japan for blast control (26).

These first set of field evaluations on biological control of rice blast and Sh-B by bacterial treatments offer promising results and also leave questions to be answered by future investigations. However, these results can be exploited to improve rice disease management. In addition if genes for antiblast (and antisheath blight) activities can be identified, cloned and incorporated into rice genotypes, biological control can be further strengthened. This will help the resource poor rice farmers of the tropics, particularly those in the upland rice growing areas, in their efforts to sustain rice yields.

REFERENCES

1. Strategy Report., "IRRI Towards 2000 and Beyond", International Rice Research Institute, 1989.

2. Ou, S.H., Rice diseases. 2nd Ed. Commonwealth Mycological Institute, Kew, Surrey, England, 1985, 380 pp.

3. Sugiyama, M., Rice sheath blight and chemical control in Japan. Japan Pesticide Information, 1988, **52**, 9-12.

4. Research Briefings 1987., Report of the Research Briefing Panel on Biological Control in Managed Ecosystems, National Academy Press, Washington, D.C., 1987.

5. Gnanamanickam, S.S. and Mew, T.W., Biological Control of rice blast with antagonistic bacteria. International Rice Res. Newsl. 1989, **14**(2), 34 - 35.

6. Howell, C.R. and Stipanovic, R.D., Control of _Rhizoctonia solani_ on cotton seedlings with _Pseudomonas fluorescens_ and with an antibiotic produced by the bacterium. Phytopathology, 1979, **69**, 480-482.

7. Howell, C.R. and Stipanovic, R.D., Suppression of _Pythium ultimum_ - induced damping-off of cotton seedlings by _Pseudomonas fluorescens_ and its antibiotic, pyoluteorin. Phytopathology 1980, **70**, 712 - 715.

8. Kloepper, J.W., Effect of seed piece inoculation with plant growth-promoting rhizobacteria on populations of _Erwinia carotovora_ on potato roots and daughter tubers. Phytopathology 1983, **73**, 217 - 219.

9. Kloepper, J.W., Leong, J., Teintz, M. and Schroth, M.N., Enhanced plant growth by siderophores produced by plant growth-promoting rhizobacteria. Nature 1980, **286**, 885-886.

10. Mew, T.W., and Rosales, A.M., Bacterization of rice plants for control of sheath blight caused by _Rhizoctonia solani_. Phytopathology, 1986, **76**, 1260 - 1264.

11. Sakthivel, N. and Gnanamanickam, S.S., Evaluation of _Pseudomonas fluorescens_ for suppression of sheath rot disease and for enhancement of grain yields in rice, _Oryza sativa_, L. Appl. Environ. Microbiol. 1987, **53**, 2056-2059.

12. Scher, F.M. and Baker, R., Effect of _Pseudomonas putida_ and a synthetic iron chelator on induction of soil suppressiveness to _Fusarium_ wilt pathogens. Phytopathology, 1982, **72**, 1577 - 1573.

13. Stutz, E.W., Defago, G. and Kern, H., Naturally occuring _Pseudomonas fluorescens_ involved in suppression of black root rot of tobacco. Phytopathology, 1986, **76**, 184-185.

14. Schippers, B., Lugtenberg, B. and Weisbeek, P.J., Plant growth control by fluorescent Pseudomonas. In Innovative Approaches to Plant Disease Control ed, I.Chet, ed). John Wiley and Sons, New York., 1987, pp. 19-39.

15. Thomashow, L.S. and Weller, D.M., Role of a phenazine antibiotic from Pseudomonas fluorescens in biological control Gaeumannomyces graminis var. tritici. J. Bacteriol., 1988, **170**, 3499 - 3508.

16. Weller, D.M., Biological control of soilborne plant pathogens in the rhizosphere with bacteria. Annu. Rev. Phytopathol., 1988, **26**, 319 - 407.

17. Weller, D.M., and Cook, R.J., In Iron, Siderophores, and Plant Disease, ed. T.R. Swineburne, Plenum Publ. Corp. New York. 1985, pp. 99-107.

18. Sakthivel, N and Gnanamanickam, S.S., Incidence of different biovars of Pseudomonas fluorescens in flooded rice rhizospheres in India. Agriculture, Ecosystem and Environment, 1989, **25**, 287 - 298.

19. Palleroni, N.J., Family I. Pseudomonadaceae Wilson, Broadhurst, Buchanan, Krumwide, Rogers and Smith, 1917. In Bergey's Manual of Systematic Bacteriology, Vol. 1. eds, N.R. Krieg and J.A. Hott, Williams & Wilkins, Baltimore, MD, pp. 143-213.

20. Logan, N.A. and Berkeley, R.C.W., Identification of Bacillus strains using the API system. J. Gen. Microbiol. 1984, **130**, 1871 - 1882.

21. Gurusiddaiah, S., Weller, D.M., Sarkar, A. and Cook, R.J., Characterization of an antibiotic produced by a strain of Pseudomonas fluorescens inhibitory to Gaeumannomyces graminis var. tritici and Pythium spp. Antimicrob. Agents Chemother, 1986, **29**, 488 - 495.

22. Mackill, A.O. and Bonman, J.M., New hosts of Pyricularia oryzae. Plant Disease, 1986, **70**, 125 - 127.

23. Misaghi, I.J., Olsen, M.W., Cotty, P.J. and Dondellinger, C.R., Fluorescent siderophore-mediated iron deprivation - a contingent biological control mechanism. Soil Biol. Biochem., 1988, **20**, 573 - 574.

24. Garrote, B.P., Mercado, A and Garrity, D.P., Soil fertility management in acid upland environment. Philipp. J. Crop. Sci., 1986, **11**, 113 - 123.

25. Gnanamanickam, S.S., Candole, B.L. and Mew, T.W., Influence of soil factors and culture practice on biological control of sheath blight of rice with antagonistic bacteria. Progress Report, Department of Plant Pathology, International Rice Research Institute, Philippines, June 1989.

26. Yoshino, R., Present status of occurrence and control of blast disease in Japan. Japan Pesticide Information, 1988, **52**, 3-8.

FUNGICIDE USE FOR THE CONTROL OF MAJOR RICE DISEASES IN JAPAN

YUKIO MIYAGI
Biological Research Center
Nihon Nohyaku Co.Ltd
4-31 Honda-cho Kawachi-Nagano Osaka 586, Japan

ABSTRACT

About 40 per cent(2.1 Mha) of the arable land in Japan was devoted to rice production in 1988 and which in 1989 represented 42 per cent of the total fungicide market($314M). The most important diseases are blast and sheath blight caused by Pyricularia oryzae and Rhizoctonia solani, respectively. Fungicides for the control of these two diseases were valued at $280 M, 89 per cent of rice fungicides. The leading blast-fungicides are isoprothiolane, probenazole,tricyclazole, pyroquilon, iprobenfos, edifenphos, phthalide, kasugamycin and blasticidin-S. Flutolanil,pencycuron, diclomezine,validamycin, mepronil, organoarsenates and polyoxin are for sheath blight control. Other diseases like soil-borne seedling blight caused by Pythium, Fusarium, Rhizopus, Rhizoctonia and Trichoderma are the main problems in nurseries. These are controlled by hymexazole, hymexazole-metalaxyl and chlorothalonil. Among bacterial diseases, bacterial grain rot caused by Pseudomonas glumae is more common than bacterial leaf blight caused by Xanthomonas campestrips pv. oryzae. Most rice seeds are treated with TMTD-benomyl or TMTD-thiophanate methyl to control seed-borne diseases such as bakanae disease, blast and brown spot caused by Gibberella fujikuroi, P. oryzae and Cochliobolus miyabeanus, respectively. Nine per cent ($27M) of rice fungicides were used for the control of these seed-borne dieases. Though many diseases are effectively controlled by the current fungicides, fungicide resistance to antibiotics or organophosphorus compounds and to carbendazim generators has occured with P.oryzae and G.fujikuroi. New fungicides with different modes of action are needed to cope with them. Japanese rice culture is facing serious problems such as the decrease in rice consumption, the decrease in younger farmers and the impacts to the free trade of rice. The area under rice decrease to some extent. These problems need an urgent solution because rice culture is important being the staple food of Japan. Rice production necessarily needs fungicides as the climate in Japan favors the outbreak of many diseases.

INTRODUCTION

The Japanese islands provides a mild and humid climate favorable for the outbreak of many diseases on many crops. Rice, the most crop is frequently damaged by many diseases. Rice fungicide turnover was $314M, 42 per cent of the total fungicide turnover($743M) in 1987. This showes the importance of rice culture and rice disease control in Japan. However, Japanese rice culture has encountered serious problems. These are the decrease in rice consumption, the decrease in younger farmers and the impact to the free trade of agricultural products including rice. These problems has brought about the argumentation on the future of Japanese agriculture and rice production.

Agriculture in Japan has changed since Araki reported in 1986 the outline of chemical control of plant diseases (1). This paper presents the recent status of rice production, major diseases and their control by fungicides in Japan. Current problems and the future are discussed.

CURRENT STATUS OF RICE PRODUCTION IN JAPAN

Up to a few decades ago, Japan suffered from a shortage of rice, the staple food crop. This was overcome by improving culture methods. Surplus rice production became the problem in 1960's. The increased supply of various foods reduced rice consumption. The mean rice consumption per person was 118.3 kg in 1963 but 71.9 kg in 1987. Rice area has been politically reduced since 1965 (TABLE 1). Rice area was 3.1 M hectare in 1960 but 2.1 M hectare in 1988. Rice production was concomitantly decreased to 9.9 M ton in 1988 from 12.5 M ton in 1960 (TABLE 1). The average yield per hectare has been

TABLE 1
Rice production in Japan

Year		1960	1965	1970	1975	1980	1985	1986	1987	1988
Area	(Mha)	3.1	3.1	2.8	2.7	2.4	2.3	2.3	2.1	2.1
Production	(Mt)	12.5	12.4	12.5	13.1	9.7	11.6	11.6	10.6	9.9
Yield/ha	(t)	4.0	3.9	4.4	4.8	4.1	5.0	5.1	5.0	4.7

According to Ministry of Agriculture, Forestry and Fishery

going up and recently reached about 5 ton,   though it was lower in 1980 and 1988 due to severe damage by cold injury and panicle blast (TABLE 1 and Figure 1). As people favor rice with good taste and high quality, farmers are willing to grow a few popular cultivar to bring more income. These cultivars are not resistant to stresses such as cold injury and blast. The shift to these cultivars is thought to be one of the reasons for the recent fluctuation in the yield per hectare.

The average size of each farm is very small(0.6 ha) and 96 per cent of rice farms are less than 2 hectare.

MAJOR RICE DISEASES IN JAPAN

Major rice diseases are listed in TABLE 2 together with fungicide treated area and treatment frequency in 1989. Blast, sheath blight, bakanae disease and seedling blight were broadly controlled followed by brown spot, stem rot, false smut, bacterial grain rot and bacterial leaf blight.

TABLE 2
Major rice diseases, fungicide treated area and treatment frequency in 1989

Disease	Pathogen	Treated area(Mha)	Frequency
Bakanae disease	Gibberella fujikuroi	1.73	1.0
Seedling blight	Pythium, Rhizoctonia,etc.	1.56	1.3
Seedling blast	Pyricularia oryzae	0.39	1.0
Leaf blast	Pyricularia oryzae	1.29	1.4
Panicle blast	Pyricularia oryzae	1.74	1.8
Sheath blight	Rhizoctonia solani	1.36	1.3
Brown spot	Cochliobolus miyabeanus	0.18	1.2
Stem rot	Magnaporthe salvinii	0.09	1.1
False smut	Claviceps virens	0.09	1.3
Bacterial grain rot	Pseudomonas glumae	0.09	1.1
Bacterial leaf blight	Xanthomonas campestris pv.oryzae	0.03	1.4

According to Ministry of Agriculture, Fishery and Forestry

Seed-borne and soil-borne fungal diseases(1-4)
Rice seedlings are prepared in the upland nursery tray, usually 60cm x 30cm(3 cm in depth) in size, for mechanical transplanting. About 200 g of seeds are sown in a tray and 150 to 200 trays are used for one hectare. As seedlings are densely grown, seed-borne and soil-borne diseases tend to occur due to high humidity among seedlings. Seed-borne fungal diseases are bakanae disease, blast and brown spot. Diseased

seedlings are the main inoculum sources in the field when
these are transplanted. Soil-borne seedling blight is caused
by <u>Pythium</u>, <u>Fusarium</u>, <u>Rhizoctonia</u>, <u>Rhizopus</u> and <u>Trichoderma</u>.
In addition to these fungi, <u>Pseudomonas</u> spp. often causes bac-
terial seedling rot.

Blast(2,5,6)
Blast is the most serious disease because of damage it causes.
An outbreak of blast depends on the climatic conditions and
disease occurrence varies year by year. The affected area,
however, has been decreasing since 1976 (Figure 1). The
reasons for this decrease are thought to be 1)annual application of excellent fungicides reduced the inoculum potentials,
2)climate was not favorable for disease outbreak, 3)field
sanitation was thorough among rice growers and 4)less application of nitrogen fertilizers reduced the susceptibility of
rice plants to blast.

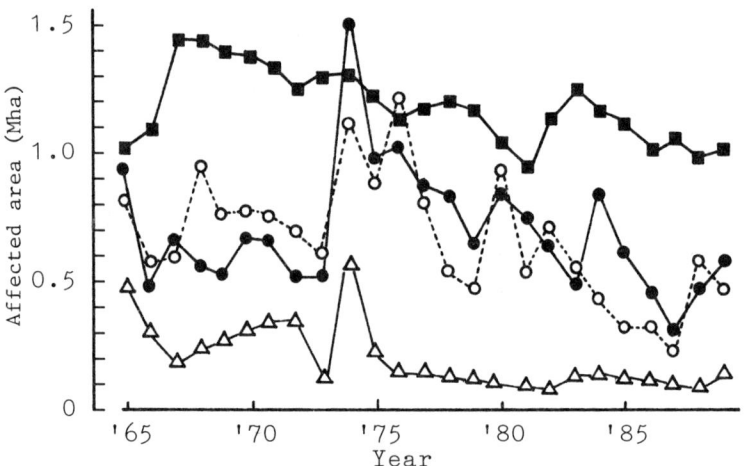

Figure 1. Affected rice area by major diseases
■;sheath blight,   ●;leaf blast,   o;panicle blast
△;bacterial grain rot and bacterial leaf blight

Sheath blight(2,7,8)
Sheath blight is the second most important rice disease
following blast. Though the disease was limited to south-west
Japan before 1950, it now occured widely following the
introduction of new rice cultivars, dense planting for
mechanization, heavier fertilization and early season culture.
Under these culture conditions, humidity among hills is high
and favorable for disease development. The area affected

by sheath blight has decreased in relation to the decrease in rice cultivation since 1967, but it still affects about a hal of rice fields(TABLE 1 and Figure 1).

Bacterial diseases(2,5,9)

Bacterial leaf blight has decreased markedly since 1975 because of the introduction of highly resistant cultivars, improved irrigation systems and seedling culture in upland nursery trays. Bacterial grain rot has occured more frequently and widely, though it was limited to south-west Japan. This increase is thought to be due to the use of infested seeds for seedling culture. Area affected area by these two bacterial diseases are shown in Figure 1. P. glumae as well as other Pseudomonas spp. causes seedling rot, which is not included in Figure 1.

FUNGICIDES USE FOR THE CONTROL OF MAJOR RICE DISEASES

Control of blast(2,5,6)

Many compounds have been developed for the control of blast (TABLE 3). Organomercury compounds and pentachlorobenzyl alcohol have been withdrawn from the market. The other compounds have been extensively used in various formulations. Iprobenfos and the fungicides developed after 1975 such as isoprothiolane, probenazole, tricyclazole and pyroquilon are easily translocated to the upper parts of rice plants through roots. They are formulated into granules and one application either to irrigation water of paddy fields or to nursery trays can control leaf blast. Panicle blast can be controlled by another treatment to irrigation water. Thus, two applications can cover the whole cropping season. This is labor saving compared with more than four times of foliar sprays or dusts. These systemic fungicides have additional activities other than disease control. Isoprothiolane enhances the growth of rice plants and grain maturation,leading to increase in yield. Iprobenfos shortens the stem length of rice and prevents the rice plant from lodging. These activities are advantageous for farmers to get more yield and income. Fungicide treatments for leaf blast control was 1.29 M hectares at a frequency of 1.4 times in 1989 giving a total treatment frequency area of 1.81 M hectare. For panicle blast control, they were 1.74 M hectare, 1.8 times and 3.13 M hectare, respectively (Table 2). Fungicide turnover for blast control was valued $208M in 1987 (TABLE 4), 66 per cent of total rice fungicides ($314M) and 28 per cent of total fungicide turnover($743M).

Control of sheath blight(2,7,8)

Historical aspects of fungicides for sheath blight control are shown in TABLE 3. These seven active ingredients are now used. Organoarsenates, polyoxin and validamycin have played the principal roles for a long time. Mepronil, flutolanil,

pencycuron and diclomezine have been put into the market in
these 10 years. These fungicides were used on 1.36 M hectare,
1.3 times, to give a total treatment frequency area of 1.77 M
hectare (TABLE 2). Fungicide turnover for sheath blight control was valued $71M (TABLE 4), 23 per cent of total rice
fungicides turnover. Granule formulation of flutolanil, a
systemic fungicide, was introduced firstly for sheath blight
control in 1988 and about 1,000 ton of granules were applied
to 26,000 hectare.

TABLE 3
Development of fungicides for blast and sheath blight control

For blast control		For sheath blight control	
Year launched	Fungicides	Year launched	Fungicides
1950's	Organomercuries	1950's	Organoarsenates
1961	Blasticidin-S		.
1965	Kasugamycin		
1966	Iprobenfos, PCBA		
1967	Edifenphos	1967	Polyoxin
1971	Phthalide	1972	Validamycin
1975	Isoprothiolane	1982	Mepronil
	Probenazole	1986	Flutolanil
1983	Tricyclazole		Pencycuron
1986	Pyroquilon	1988	Diclomezine

PCBA = Pentachlorobenzyl alcohol

Control of Seed-borne and soil-borne fungal diseases(2-5,9)
TMTD-benomyl or TMTD-thiophanate methyl can effectively control seed-borne diseases such as bakanae disease, blast and
brown spot and the fungicides used was valued $27M in 1987.
Rice seeds for 1.73 M hectare were treated once prior to
sowing (TABLE 2). Bakanae disease has been gradually
increased because of the frequent occurrence of the pathogen
resistant to these carbendazim-generators. Imidazole compounds
such as triflumizole and pefurazoate have been recently
registered for seed treatment and they may have some positions
in this field.
        Seedling blight caused by Pythium, Fusarium, Rhizoctonia,
Rhizopus and Trichoderma is controlled by hymexazole,
metalaxyl and chlorothalonil. Fungicides are incorporated
into soil prior to sowing or drenched just after sowing.
Seedlings for 1.56 M hectare were treated 1.3 times on average
(TABLE 2). Seedlings often wilt suddenly in the nurseries due
to imbalance between absorption of water through roots and
transpiration under unfavorable conditions. Isoprothiolane,
hymexazole and hymexazole-metalaxyl can protect this
physiological disorder, called "murenae".

Control of bacterial and virus diseases(2,5,9)
Bacterial grain rot has become the major bacterial rice disease and is not well controlled because the chemicals such as
nickel compound and kasugamycin are not fully effective.
Probenazole and pyroquilon, the blast-fungicides, were
registered for this disease but they only show some degree of
efficacy. Seedling rot caused by P. glumae was partially
controlled by seed treatment of kasugamycin and seeds for 0.05
M hectare were treated in 1989. Oxolinic acid, a new
quinolone compound, was registered for seed treatment in 1989
and is expected to give better control.

Control of other diseases(1)
Rice stripe and rice dwarf disease are the major virus diseases. They can be controlled only by treatment of insecticides controlling planthoppers as vectors. The affected
area has been gradually decreased and was 0.2 M hectare in
1989. However, these virus diseases may increase because of
the recent increase in planted area of wheat and barley on
which vectors are overwintering.
    Brown spot, stem rot and false smut have occurred locally
and fungicide treated area was not broad (TABLE 2). A few
blast-fungicides are somewhat effective. Organophosphorus
fungicides control former two and isoprothiolane controls stem
rot. Inorganic copper fungicides can control false smut.
Rhizoctonia spp. and Sclerotium spp. cause stem diseases whose
symptom are quite similar to that of sheath blight. Some
fungicides for sheath blight control act effectively on them.

PROBLEMS AND THE FUTURE

Rice production
Japanese agriculture is facing serious problems. These
are decrease in rice consumption, the decrease in younger
farmers and the impact of free trade in agricultural products
including rice.
    Rice area has been politically reduced to balance rice
production with rice consumption. This decrease in rice
consumption will continue for a while because different types
of food will be produced or imported and an increase in
population is not expected. Rice cultivation, which was 2.1 M
hectare in 1988, will probably reduced to some extent.
However, rice must be the biggest of all crops even in the
future because rice culture is very important not only for
self-supply of the staple food but also its function to
prevent flood by keeping a lot of rainwater in paddy fields.
    The size of each farm for rice production is quite small
and farmers cannot make their living from agriculture. Thus,
part-time farmers who work in small farms just on weekends
have increased and were nearly 90 per cent of all farmers in
1988. The number of farmers is also decreasing, 10.25 million
in 1960 and 6.18 million in 1987. Farmers younger than sixty

were 73 and 53 per cent in 1960 and 1987, respectively. This means a decrease in manpower and lower rice production. However, it is certainly a good opportunity to enlarge the size of each farm for rice production. Part-time farmers cannot pay much attention on the incidence of diseases or insect pests and are apt to miss adequate timing of pesticide treatments. However, they will take a major part in rice production for a while and fungicides or fungicide application methods should be changed to favor them.

Pressure towards free trade of agricultural products has threatened Japanese agriculture, especially rice culture. Japan. is now the biggest food-importing country and Japanese self-supply of food was 49 per cent on a calorie base, much lower than U.S.A and most EC countries in 1987. This will reduce within a few years when all agricultural products except thirteen (rice etc.) are exposed to the free trade. The Japanese Government fears that lower priced imported rice may destroy Japanese rice culture and worsen self-supply of this staple food in terms of national security. Price of agricultural products in Japan are high because of high land and labor cost of small scale production. This is enhanced by the exchange rate of yen to foreign currencies. In fact, most Japanese do not feel that the price of rice is high because only a few per cent of our income is spent on rice. The Government has planned to make agriculture more attractive and competitive with foreign agriculture by increasing professional farmers holding large farms. The Governmental financial support to agriculture is thereby suppressing the price of rice. Japanese agriculture is now at the turning point. These changes should be considered in developing new fungicides or new methods of disease control.

Rice disease control

Many fungal diseases can be controlled by current fungicides but there are some problems to be solved; lack of effective eradicants; lack of effective compounds for bacterial and virus diseases; resistance to fungicides; unfavorable effects by fungicides on the environment. Though the fungicide market in Japan has peaked(1), new fungicides should still be developed.

Several blast-fungicides listed in TABLE 2 have some eradicant activity by foliar application but are not fully effective. Protectants are necessarily treated within a short period, just before but not after disease appearance. Eradicants, however, can be applied after disease appearance thus providing greater flexibility in application timing.

None of the bactericides is fully effective and the kinds of bactericide are quite a few, in contrast to those for fungal disease control. As bacteria can develop resistance more rapidly than fungi, new effective bactericides are needed. As for virus disease control, insecticides play the principal role by reducing the population of their insect vectors.   Systemic inhibitors of virus multiplication must be of great value.

Fungicide resistance is one of the most important problems in the world. P. oryzae causing rice blast developed resistance to antibiotics (9,10) and organophosphorus compounds (11,12). Though laboratory-selected mutants with high resistance to iprobenfos were reported to be also resistant to isoprothiolane (13), none of the field isolates were highly resistant. Some field isolates moderately resistant to iprobenfos were not resistant to isoprothiolane(14). It is suggested that laboratory selected mutants are less competitive than field isolates and their population hardly increase in the field(15). This suggests that resistant problems should be considered through intensive ecological studies in the field. G. fujikuroi causing bakanae disease developed resistance to benomyl and thiophanate methyl and lead to unsuccessful control(3,4). Resistant isolates of G. fujikuroi to triflumizole were of less pathogenicity and problems in practice have not been reported. As any fungicide cannot escape from the risk of resistance, compounds with different modes of action should be used rotationally or in binary mixtures as it is now recommended.

TABLE 4
Fungicide formulations used for blast and sheath blight control in 1987

		Blast		Sheath blight	
	Dust	Granule	WP & EC	Dust	WP & EC
Volume (1,000t)	70.5	21.4	3.1	43.6	2.2
Value (M$)	80.0	87.9	40.0	46.4	25.0
Treated area(Mha)	2.1	0.9	1.8	1.2	1.0

Treated area are roughly calculated.
WP and EC are treated either by ground-liquid spray or by aircraft.

Application of dust formulations is quite popular among rice growers in Japan (TABLE 4) because it is of low cost and is labor saving. More than 80 per cent of dusts were mixtures of fungicides and insecticides for the control of concomitantly occurring diseases and insects. Mixture dusts of fungicides for blast and sheath blight control were included in each statistics in TABLE 4. They are usually applied by a knapsack type power duster with boom-type blow head and two persons are enough to cover a broad area in a short time. Application of dusts, however, tends to drift and cause some unfavorable effects to the environment. Though dust have been improved to DL type dust(drift-less) with bigger particle sizes, the problem was not completely solved. Use of dust is now decreasing on that account.

Wettable powder(WP) and emulsifiable concentrate(EC) are applied either by ground-liquid spray or by aerial spray. These formulations are more frequently applied by the latter method. Aerial application is quite labor saving and 29 per cent of rice area(0.6 Mha) was treated 2.9 times for controlling blast, sheath blight and insects in 1988 (16). But EC's and WP's also drift from the target area, causing a degree of environmental concern. Remote control aerial spray system suitable for small target areas where drift may cause problems is under development(16). Ground-liquid spray has less risk of drift but is labor intensive because of difficulties of handling a long hose. An airblast sprayer is on trial. It can spray from footpaths between rice fields and is labor saving(17).

Granule formulations have no drift problems and the treatment frequency can be minimized. About 0.9 M hectare were treated by granules for blast control in 1987 (TABLE 4). Part-time farmers and those whose rice fields border other crop fields prefer to use granules because they are labor saving and have no risk of drift. Application of granules to irrigation water in paddy fields require, however, higher doses than that for foliar application and it tends to cause residue problems in water. Recommended dose of isoprothiolane to control leaf blast, for a example, is 0.4 kg per hectare for foliar spray but more than 3.6 kg for granule application. The dose is higher for granule application even though its efficacy is comparable to more than twice foliar sprays. Adsorption of the fungicides to soil is one of major factors for this dose difference. Systemic fungicides, however, have many advantages in their application. Treatment to seeds, nursery tray and paddy water would be possible to control diseases in the field. These treatments certainly have little risk of environmental pollution. If these fungicides are effective at low doses, They are clearly the most promising treatments for the future.

ACKNOWLEDGEMENTS
The author thanks Dr. Fujio Araki, the manager of Biological Research Center, for his kind suggestion to this work.

REFERENCES
1. Araki,F.,Chemical control of plant diseases in Japan. Chemistry and Industry, 1986, 20 January, 54-60.

2. Ohata,K., Change in the outbreak of rice diseases in mechanized transplant culture. Japan Pestic. Inform., 1981, 38, 9-12.

3. Ogawa,K., Damage by "Bakanae" disease and its chemical control. Japan Pestic. Inform., 1988, 52, 13-15.

4. Yoshino,R., Current status of bakanae disease occurrence and its control.Shokubutsu boueki(Plant Protection), 1988, **7**, 321-325.

5. Yamaguchi,T., Nursery-tray application of fungicides for the control of rice diseases. Japan Pestic.Inform., 1986,**49**,10-14.

6. Yoshino,R., Present status of occurrence and control of blast disease. Japan Pestic.Inform., 1988, **52**, 3-8.

7. Hori,M., Present status of occurrence and chemical control of sheath blight. Japan Pestic.Inform., 1984, **44**, 6-10.

8. Sugiyama,M., Rice sheath blight and chemical control in Japan. Japan Pestic.Inform., 1988, **52**, 9-12.

9. Horino,O., Epidemiology and control of bacterial leaf blight and other bacterial diseases of rice. Japan Pestic.Inform., 1986, **49**, 3-6.

10. Miura,H.,Katagiri,M.,Yamaguchi,T.,Uesugi,Y. and Ito,H., Mode of occurrence of kasugamycin resistant rice blast fungus. Ann.Phytopathol.Soc.Japan, 1976, **42**, 117-123.

11. Uesugi,Y., Resistance of phytopathogenic fungi to fungicides. Japan Pestic.Inform., 1978, **35**, 5-9.

12. Katagiri,M. and Uesugi,Y., Development to organophosphorus fungicides in Pyricularia oryzae in the field. Nippon Nohyaku Gakkaishi(J.Pestic.Sci.), 1980, **5**, 417-421.

13. Katagiri,M. and Uesugi,Y., Similarities between the fungicidal action of isoprothiolane and organophosphorus thiol fungicides.Phytopathology, 1977,**67**, 1415-1417.

14. Miyagi,Y.,Hirooka,T. and Araki,F., An approach for grouping field isolates of Pyricularia oryzae Cav.on the basis of their sensitivities to fungicides. Nippon Nohyaku Gakkaishi(J. Pestic. Sci.), 1983, **8**, 81-86.

15. Miyagi,Y.,Hirooka,T. and Araki,F.,Relative parasitic fitness of isolates of Pyricularia oryzae Cav.with different sensitivities to fungicides. Pestic.Sci., 1986, **17**, 653-658.

16. R.Ichikawa, Utilization of remote control helicopter in aerial application of pesticides. Shokubutsu boueki(Plant Protection), 1989, **43**, 319-324.

17. Fukazawa.H.,Performance efficiency of airblast sprayer for paddy field. Shokubutsu boueki(Plant Protection), 1990, **44**, 2-6.

TRICYCLAZOLE FOR CONTROL OF <u>PYRICULARIA</u> <u>ORYZAE</u>
ON RICE: THE RELATIONSHIP OF THE MODE OF ACTION AND DISEASE
OCCURRENCE AND DEVELOPMENT

LANCE G. PETERSON
Development Manager/Asian-Pacific Area
DowElanco Pacific, Ltd.
40/F Sung Hung Kai Center
30 Harbour Road
Hong Kong

ABSTRACT

Tricyclazole (5-methyl-1,2,4-triazolo[3,4-b]benzothiazole) is
a unique fungicide for control of <u>Pyricularia</u> <u>oryzae</u> on rice.
Tricyclazole is systemic in rice and will control rice blast
disease in any stage of plant development by a variety of
application methods. Tricyclazole protects plants from infection
by <u>P</u>. <u>oryzae</u> by preventing penetration of the epidermis by the
fungus. The compound acts by inhibiting melanization within the
appressorium, thus causing a lack of rigidity in the appressorial
wall. Tricyclazole has no apparent effect on spore germination
although sporulation is reduced. Tricyclazole is not curative
but is protective in its activity.

INTRODUCTION

Rice is cultivated in more than 100 countries. In total area
planted, rice is second only to wheat. However, more people depend
on rice as a food crop than on wheat. The occurrence of rice blast
disease has been reported in almost every rice growing area. Rice
blast is caused by the parasitic fungi <u>Pyricularia</u> <u>oryzae</u> and is
the most serious disease of rice in the world. It reduces rice
yield and grain quality resulting in economic losses.

When rice blast disease and the proper timing of tricyclazole (5-methyl-1,2,4-triazolo[3,4-b]benzothiazole, BEAM) applications for blast control are discussed, specific stages of rice growth are mentioned. It is important to be familiar with the names of these stages and to recognize these stages in the field.

The rice blast fungus can infect rice in all stages of growth, if environmental conditions are favorable. Thus, one can expect to see blast in the vegetative stages of growth, such as seedlings and at tillering. Blast also appears in the reproductive stage on ligules, stems, panicles and grain. Disease symptoms are different on the various plant parts and are given different names.

RICE BLAST DISEASE

Leaf blast symptoms result from <u>Pyricularia oryzae</u> infection of leaves, which is more common during the vegetative than the reproductive stage of rice growth. Leaf spots are typically elliptical with pointed ends. The centers of mature spots are usually gray or white and the margins are usually brown or reddish brown. Surrounding the brown margins are yellow, chlorotic areas. Both the size and shape of the spots can vary, depending on age of infection, environmental conditions, and degree of susceptibility of the rice variety. Under favorable environmental conditions, leaf blast can become so severe that it kills rice plants.

Fungus infection of the stem causes disease symptoms called stem blast, culm blast, neck blast, or rotten neck. This is probably the most frequently encountered rice blast symptom. Also, it is probably the most damaging because diseased stems do not allow the proper flow of nutrients to the panicle. The grain does not fully mature, and weakened stems are susceptible to breakage and total loss of the grain.

Disease symptoms are generally seen on the stem just below the junction, or node, of the stem and panicle. The diseased area is characterized by brown coloration and is readily observed when the remainder of the stem is still green. As the plant matures and the stem loses its green color, stem blast is more difficult to see. However, close observation can distinguish between diseased and healthy tissues, even on fully mature plants, which are tan or buff colored.

Panicle blast looks similar to stem blast. The major difference is the location. Panicle blast occurs within the panicle, above the junction of the stem and panicle. When the site of infection is in the basal portion of the panicle, severe damage to the total panicle can result, similar to stem blast. On the other hand, infection higher in the panicle affects only that portion above the infection site. Thus, panicle blast can have a large or small effect on grain quality and total yield depending on the site of infection.

When the infection occurs on the ligule, disease symptoms appear at the junction of the leaf blade and the sheath. Often, the result of this infection is death of the flag leaf. Sometimes ligule blast can lead to stem or panicle blast.

Node blast results from infection of the uppermost node of the stem. This node either is fully exposed or is enclosed by the sheath of the flag leaf. Node blast results in breakage of the stem and loss of the panicle.

The rice blast fungus can infect the rice grain. Diseased grain is gray or dark colored. Grain diseases caused by other fungi can result in similar symptoms. Other diseases of rice frequently are encountered in the field. Some of these can be confused with rice blast symptoms.

For example, brown spot disease, caused by fungus Helminthosporium oryzae, infects leaf blades, sheaths, and maturing grain. Spots are small, oval shaped, dark colored, and sometimes have gray centers. Brown spot symptoms are similar to leaf blast on a blast-resistant rice variety.

Narrow brown spot disease is caused by the fungus Cercospora oryzae. Symptoms are characterized by long narrow dark spots sometimes surrounded by yellow areas.

Tricyclazole is specific for Pyricularia oryzae and will not control other rice diseases (3).

Rice blast has been reported in almost every country which cultivates rice. In general, the most severe rice blast occurs in temperate areas of the world. Rice grown closer to the equator has less blast. There are notable exceptions. For example, blast can be severe in mountainous areas of Colombia and India. Conditions favorable for disease occur locally in the Philippines and Guyana. A number of factors affect when and to what degree blast appears on rice. These factors will be discussed by first looking at the disease cycle.

The rice blast spore is disseminated from diseased to healthy plant tissue by wind. After germination of the spore, a round structure called an appressorium is formed through the epidermis and into the interior of the plant where an infection site is established. From that site, fungus hyphae penetrate adjacent plant cells and kill plant tissues. The result of this parasitism is visible disease symptoms we call leaf blast, stem blast, and panicle blast. From these diseased plant tissues the fungus produces new spores which are carried by wind to re-infect rice. With repetition of this cycle and after numerous infections, severe rice blast damage results.

Several factors favor rice blast occurrence. These include a susceptible rice variety and a virulent strain of the fungus. Weather conditions favorable for disease are moderate temperatures and abundant moisture. High plant population results in dense foliage which holds moisture and favors disease. High nitrogen fertilization and irregular irrigation make rice more susceptible to disease. There are other factors that encourage rice blast disease, but these are the most important ones.

As mentioned, moderate temperature and abundant moisture are necessary for rice blast occurrence. Abundant moisture is the more critical factor for infection by the fungus. Without moisture on the plant surface, infection will not take place.

In general, if the temperature is between 20° and 30°C, with high moisture content in the air (relative humidity) and water droplets on the plant surface, the entire fungus life cycle can be completed in five to seven days. These conditions are also favorable for good rice growth. The rice blast fungus can infect rice in every growth stage if weather conditions are favorable.

TRICYCLAZOLE APPLICATION

Proper timing of tricyclazole applications in relation to rice blast occurrence assures good product performance and minimum number of applications (1,8).

Tricyclazole may be applied as a seed treatment for either drill seeding or aerial broadcast seeding. A rate of 0.75 to 2 grams ai/kg seed will protect seedling rice from leaf blast for 30 to 60 days.

Tricyclazole may also be applied to rice seedlings just prior to transplanting as another means to protect the plant from leaf blast. The grower may use a root soak method of bare rooted plants at a rate of 3 grams ai/liter of water. The seedlings can be hand transplanted after 24 hours.

Where mechanical transplanting equipment is used, tricyclazole may be applied as a granule to the transplant flat at a rate of 2.5 grams ai per standard transplant flat just before planting. Either method of transplant application will provide 30-70 days protection against leaf blast in the field.

A tricyclazole application of foliar spray or dust at first appearance of leaf blast symptoms in established rice, will provide 18-20 days of protection from later leaf infections. This would be sufficiently long protection to get the crop through its susceptible leaf blast period. Spores that infect stems and panicles can be stopped by a foliar tricyclazole application at late booting to 5 percent heading. Foliar application rates range from 225-400 grams ai/hectare.

Stem and panicle blast incidence may be severe and require two tricyclazole applications. In addition to the application at late booting to 5 percent heading, treatment is repeated in 10-14 days when 75-100 percent of the panicles have emerged.

Variable plant maturation can occur in a large rice field because of different planting depths, non-uniform irrigation, non-level soil, and other factors. This variable growth makes timing of tricyclazole application difficult. If the application is delayed until the majority of the field is in the booting stage, the advanced plants may be in full heading with severe disease, while the majority of the field is protected.

Timing of the tricyclazole application in a field with variable plant maturation is slightly different from the normal recommendations. Two tricyclazole applications are necessary to provide good disease control. Timing of the first application is when the most advanced rice is in the very early heading stage of growth. The second application is made 10-14 days later. At this time the majority of the field is in the proper stage for a tricyclazole application.

There are several important items to consider when applying tricyclazole for rice blast control. a) Tricyclazole is a protectant fungicide that must be applied before major disease occurrence. b) Identify when rice blast appears during the growing season. This pattern can be different for areas just a few kilometers apart because of different environments. c) Apply tricyclazole at the first appearance of leaf blast, if leaf blast is expected to be severe in that area. d) Always apply tricyclazole at late booting to 5 percent heading. If no other application is made, make this one. e) If stem and panicle blast are expected to be severe due to continued favorable weather, apply a second tricyclazole spray at 75-100 percent panicle emergence or 10-14 days after the first application. f) If rice growth is variable, time the first tricyclazole application according to growth of the more advanced plants. It is better to apply tricyclazole too early than too late.

TRICYCLAZOLE MODE OF ACTION

Tricyclazole is an excellent fungicide for the protection of rice from blast disease. One of the reasons for its outstanding performance is that tricyclazole acts systemically in the rice plant (2). This means that tricyclazole is absorbed by the rice plant and moved inside, or translocated, to different parts of the plant. Fungicides which are not systemic remain on the plant surface.

There are several advantages to the use of a systemic fungicide for plant disease control. Two that are important to tricyclazole for rice blast control are: 1) Tricyclazole moves into and protects untreated portions of the plant, and

2) Tricyclazole is not susceptible to removal by rain. These advantages mean that even though the spray application does not completely cover the rice plant, tricyclazole will still protect untreated plant parts. Also, tricyclazole does not have to be reapplied if rainfall occurs as soon as one hour after application.

Tricyclazole is compatible when tank-mixed with other fungicides and insecticides applied to rice. Tricyclazole can be applied in as little as 20 liters per hectare or up to as much as 1000 liters per hectare. The key is the accuracy of application. Tricyclazole can be foliar applied to rice using any equipment normally used to apply fungicides and insecticides to rice. It is important with any sprayer to obtain uniform coverage of all plant surfaces for maximum protection by tricyclazole.

The chemistry of tricyclazole is unique. There are no other triazolobenzothiazole fungicides. Its different chemistry means a unique mode of action or mechanism of controlling disease.

A second unique feature of tricyclazole is its narrow spectrum. Tricyclazole controls rice blast caused by the fungus Pyricularia oryzae, and diseases caused by the fungus Colletotrichum, such as occur on cucurbits and soybeans. However, control of rice blast is more effective than control of Colletotrichum diseases. For all practical purposes, tricyclazole is considered at this time to be a fungicide for the control of Pyricularia oryzae (3).

The rice blast disease cycle is as follows: an airborne spore of the rice blast fungus lands on the surface of the rice plant. During favorable weather, the spore germinates, an appressorium is formed, and within 8-18 hours the fungus penetrates the plant surface to establish an infection site. From the infection site the fungus invades and kills adjacent plant cells, resulting in disease symptoms we recognize as blast. The fungus sporulates from disease symptoms and releases spores to the air to repeat the disease cycle. The complete cycle may occur in 5 to 7 days under favorable weather conditions.

Tricyclazole has essentially no effect on spore germination at recommended rates of application (7). Appressorium formation is inhibited to a slight degree. Maximum inhibition occurs when the fungus attempts to penetrate the plant surface. Tricyclazole prevents establishment of the fungus inside the leaf.

When tricyclazole is applied as a curative treatment, after establishment of the infection site inside the leaf, rice blast disease symptoms are not prevented. However, tricyclazole does reduce sporulation and those spores that do form are less virulent (4). Tricyclazole is not considered a curative fungicide. Tricyclazole applied as a curative treatment will not stop infections in progress but will prevent future infections from starting.

The inhibition of sporulation by tricyclazole may be significant in disease control. In a greenhouse study, a selection of rice blast fungicides were applied to diseased rice plants on day 0. On days one through six, sporulation was measured for each treatment. Tricyclazole provided moderate reduction of sporulation on day one but provided highly significant reduction by day six (4). Sporulation inhibition from the other treatments was strongest on day one and reduced significantly by day six. The reduction in the spread of disease may be more significant for tricyclazole than other blasticides due to its long-term inhibition of Pyricularia sporulation.

Although tricyclazole does not inhibit growth of Pyricularia oyrzae at low rates in vitro, tricyclazole does inhibit the formation of a dark pigment in the fungus called melanin (5,6). A tricyclazole concentration as low as 0.1 ppm will completely inhibit melanin production in Pyricularia oryzae. Studies in the United States and Japan have shown that melanin is necessary for the rice blast fungus to parasitize the rice plant. Tricyclazole prevents this parasitism by blocking melanin biosynthesis in the fungus (6). Under favorable growth conditions, the fungus penetrates the epidermal cell wall of the plant and produces infection hyphae which parasitize underlying cells.

The processes of melanization in the appressorium and penetration of the host surface are closely correlated. It is hypothesized that melanin provides strength and rigidity to the appressorial wall, except for that portion in contact with the plant surface. A hydrostatic force builds up inside the appressorium after it is cemented to the plant surface. The unmelanized appressorial wall is forced against the plant surface with increasing pressure until the surface is sheared and the fungus flows through. Penetration appears to be by mechanical force only. There is no evidence that enzymes dissolve a path through the plant surface (7).

A tricyclazole-treated appressorium that lacks melanin also lacks a rigid wall. The hydrostatic force which forms inside this abnormal appressorium cannot be concentrated against the plant surface. Rather, the force is dissipated to the entire appressorial wall and the appressorium swells in size. The end result is that the treated appressorium fails to form a penetration hypha and eventually dies without entering the rice plant. Tricyclazole interferes with normal appressorial function by inhibiting melanin biosynthesis and formation of penetration hyphae.

CONCLUSIONS

Tricyclazole may be applied, as a preventative treatment for the control of rice blast, to any stage of growth that is susceptible to attack. Tricyclazole is equally effective when applied through a variety of methods.

Tricyclazole is a systemic compound that is rapidly taken up and translocated throughout the rice plant.

Tricyclazole is essentially selective for activity against Pyricularia oryzae.

Tricyclazole inhibits the formation of a dark pigment, called melanin, in Pyricularia oryzae. This occurs at rates lower than required to inhibit mycelial growth or spore germination in vitro. It is suspected that melanin provides strength and rigidity to the appressorium.

By preventing melanin formation, the tricyclazole-treated appressorium fails to produce a penetration hypha and does not enter the plant. Therefore, tricyclazole interferes with appressorial function by inhibiting melanin biosynthesis and host penetration.

There may be other mechanisms of inhibition not yet identified.

REFERENCES

1. Froyd, J.D., Guse, L.R., and Kushiro, Y., Methods of applying tricyclazole for control of Pyricularia oryzae on rice. Phytopath., 1978, 68, 818-22.

2   Froyd, J.D., Paget, C.J., Guse, LR., Dreikorn, B.A., and Pafford, J.L. Tricyclazole: A new systemic fungicide for control of Piricularia oryzae on rice. Phytopath., 1976, 66, 1135-39.

3   Lilly Research laboratories. A Systemic Fungicide for the Control of Rice Blast (Pyricularia oryzae). Technical Report on BEAM. Lilly Research Laboratories. A Division of Eli Lilly and Company. Indianapolis, IN, April, 1981.

4   Mukelar, A. and Yamaguchi, T., Action of EL-291 on rice blast fungus, Pyricularia oryzae Gav. JICA Technical Training Report, National Institute of Agricultural Sciences, Tokyo, Japan, March-September, 1975.

5   Tokousbalides, M.C. and Sisler, H.D., Site of inhibition by tricyclazole in the melanin biosysthetic pathway of Verticillium dahliae. Pest. Biochem. and Physiol., 1979, 11, 64-73.

6   Woloshuk, C.P., Sisler, H.D., Tokousbalides, M.C., and Dutky, S.R. Melanin biosynthesis in Pyricularia oryzae: Site of tricyclazole inhibition and pathogenicity of melanin-deficient mutants. Pest. Biochem. and Physiol., 1980, 14, 256-64.

7   Woloshuk, C.P., Sisler, H.D., and Vigil, E.L.   Action of the antipenetrant, tricyclazole, on appressoria of Pyricularia oryzae.   Physiol. Pl. Path., 1983, 22, 245-59.

8   Yamaguchi, T.,   Nursery-tray application of fungicides for the control of rice diseases.   Japan Pesticide Information, 1986, 49, 10-14.

THE IMPORTANCE OF CROP GROWTH STAGES FOR DETERMINING THE APPLICATION TIMING OF DISEASE CONTROL AGENTS ON RICE

K.M. Chin (1) & A. Bhandhufalck (2)

(1)Ciba-Geigy Agricultural Expt. Station, Beg Berkunci
71309 Rembau, NS, West Malaysia; (2) Ciba-Geigy
(Thailand) Ltd., P.O. Box 747, Bangkok 10501, Thailand

ABSTRACT

Effective timing for the application of chemicals in the control of rice diseases requires a clear understanding of the relationships between crop growth stage, disease development and yield. Recent studies on the importance of growth stage as a decision aid in the field management of blast and sheath blight diseases are discussed. Because such studies demand a precise definition of crop growth stage, adoption of a recent modification of the 'Zadoks' decimal code for cereals is proposed.

INTRODUCTION

Blast and sheath blight are major problems of rice in most areas of the world where the crop is grown. High yield losses occur on susceptible cultivars under conditions favourable to disease development. For example, yield losses attributable to blast range from 43% in the tropics to 90% in temperate areas (1,2) when disease is severe. Sheath blight is also known to cause losses of up to 30-44% when infection extends above the 4th leaf sheath (3,4).

Control of blast and sheath blight is now possible with a number of effective compounds (2,4,5,6). Modern systemic compounds active against Pyricularia oryzae include IBP, isoprothiolane, probenazole, tricyclazole, pyroquilon and ferimzone. Unlike the other products, probenazole is reported to have the property of inducing the production of antifungal substances by the host plant.

Strong activity against Thanatephorus cucumeris (Rhizoctonia solani) has been reported with validamycin, pencycuron, flutolanil, diclomezine and propiconazole. Apart from propiconazole which has a broad spectrum of activity against a range of fungi, the other compounds are specific to the basidiomycetes.

The degree of disease control achieved, yield and economic return from the application of each fungicide is, however, greatly influenced by application timing. Protective fungicides, for example, have to be applied at least one incubation period ahead of the time when disease levels are expected to exceed economic damage thresholds (7).

The exact definition of growth stages of a crop is important for at least three reasons. Firstly, they affect the degree to which disease severity is related to yield, so that damage thresholds vary depending on the growth stage of the crop at which disease is measured. Secondly, when damage is restricted to certain crop stages, monitoring of disease levels and application timings may be carried out at these stages. Finally, precise description of growth stages at which applications are made enable an analysis of product performance at the end of crop growth.

GROWTH STAGES OF THE RICE CROP

The 'Zadoks' decimal code system (8) is currently used widely to describe growth stages (GS) in small grain cereals. The first digit of the two digit code describes major growth stages and the second, details of each stage (Figure 1).

. Except for growth stages 21-29, the scale is largely concerned with the development of a single tiller and is, therefore, most appropriate for describing growth of synchronously developing cereal crops; since in these cases the development of the entire crop is the same as that of a single tiller. Unlike other cereals, rice is mainly transplanted and most cultivars are bred for a high tillering capacity. Because it is multiple tillering (up to 30 tillers per plant), development in a crop of rice is asynchronous, with considerable differences in growth stage between tillers.

For example, a hill of rice with the primary tiller at anthesis, is likely to have tillers at growth stages ranging from mid-booting to full emergence. To describe these populations differences in crop development, Chin et al. (9) proposed a third digit to measure the percentage (in tens) of the population at the most common stage of development plus those beyond that stage (Figure 2). If it is important to know the exact proportions of tillers at different stages, the frequencies of individual growth stages may be recorded. In the following discussion, growth stages are described using the modified 'Zadoks' scale.

APPLICATION TIMING FOR LEAF BLAST CONTROL

Leaf blast is an explosive disease capable of an apparent infection rate (10) 'r' of 0.66 per unit disease per day (K.M. Chin, unpublished). In comparison, potato blight, in a blight year may achieve an 'r' rate of only 0.46 (11).

GROWTH STAGE

Codes

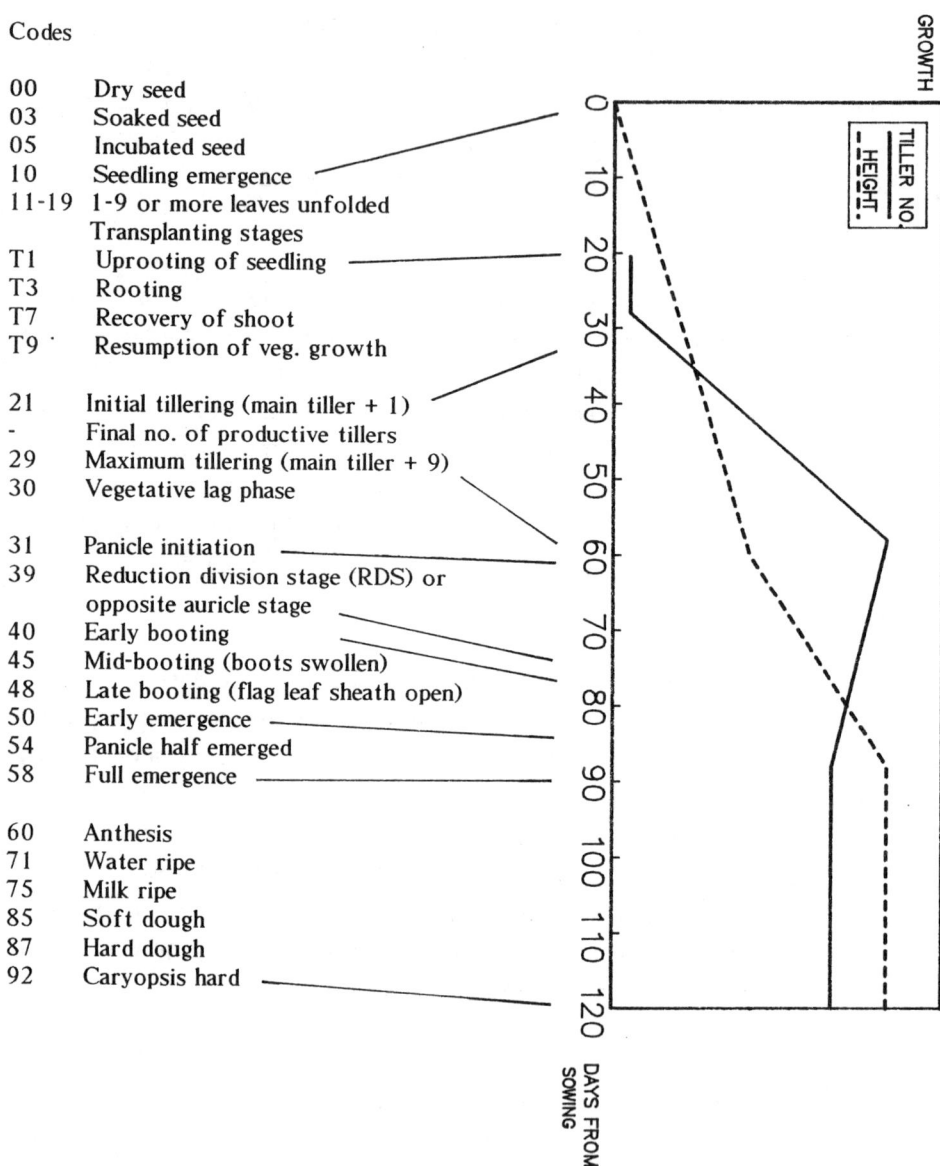

00	Dry seed
03	Soaked seed
05	Incubated seed
10	Seedling emergence
11-19	1-9 or more leaves unfolded
	Transplanting stages
T1	Uprooting of seedling
T3	Rooting
T7	Recovery of shoot
T9 ·	Resumption of veg. growth
21	Initial tillering (main tiller + 1)
-	Final no. of productive tillers
29	Maximum tillering (main tiller + 9)
30	Vegetative lag phase
31	Panicle initiation
39	Reduction division stage (RDS) or
	opposite auricle stage
40	Early booting
45	Mid-booting (boots swollen)
48	Late booting (flag leaf sheath open)
50	Early emergence
54	Panicle half emerged
58	Full emergence
60	Anthesis
71	Water ripe
75	Milk ripe
85	Soft dough
87	Hard dough
92	Caryopsis hard

FIG 1. Growth stages of a 120-day, non-photoperiod sensitive rice cultivar grown in the tropics by transplanting. Adapted from Sugimoto (1971) and Zadoks, Chang and Konzak (1974).

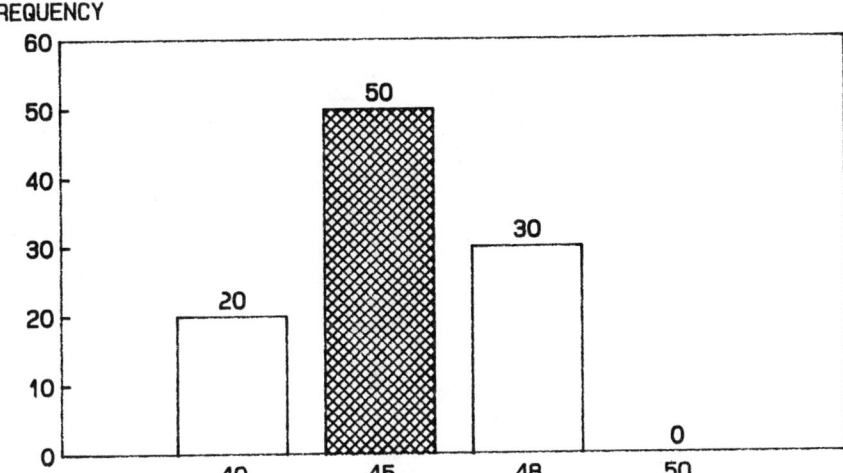

Fig 2: Frequencies of growth stages in a plot of rice cv.
Cimandiri. The growth stage (GS) is 458 since 80%
of the tillers are at (the most common) stage 45 and
beyond.

One new leaf is produced every 3-4 days during the blast susceptible,
early vegetative growth stage of the rice plant, so that non-systemic
compounds have to be applied very frequently to maintain control over a
rapidly expanding leaf area. Systemic compounds are advantageous but need
a long residual activity. Because the majority of the currently available
compounds are mainly non-curative in activity, protective applications
before or at early infection are needed.

For example the systemic compound, pyroquilon (CORATOP 2.7G), was
broadcast once, at 4 rates (250, 375, 500 and 750 g ai ha^{-1}) and 3
timing dates (ten and five days before infection [dbi], and five days
after infection [dai]) in three lowland rice trials at Chachoengsao in
Thailand in 1987.

The results (Figure 3) demonstrate the interaction between application
rate and timing. Untreated plots had a mean disease severity of 73% at 67

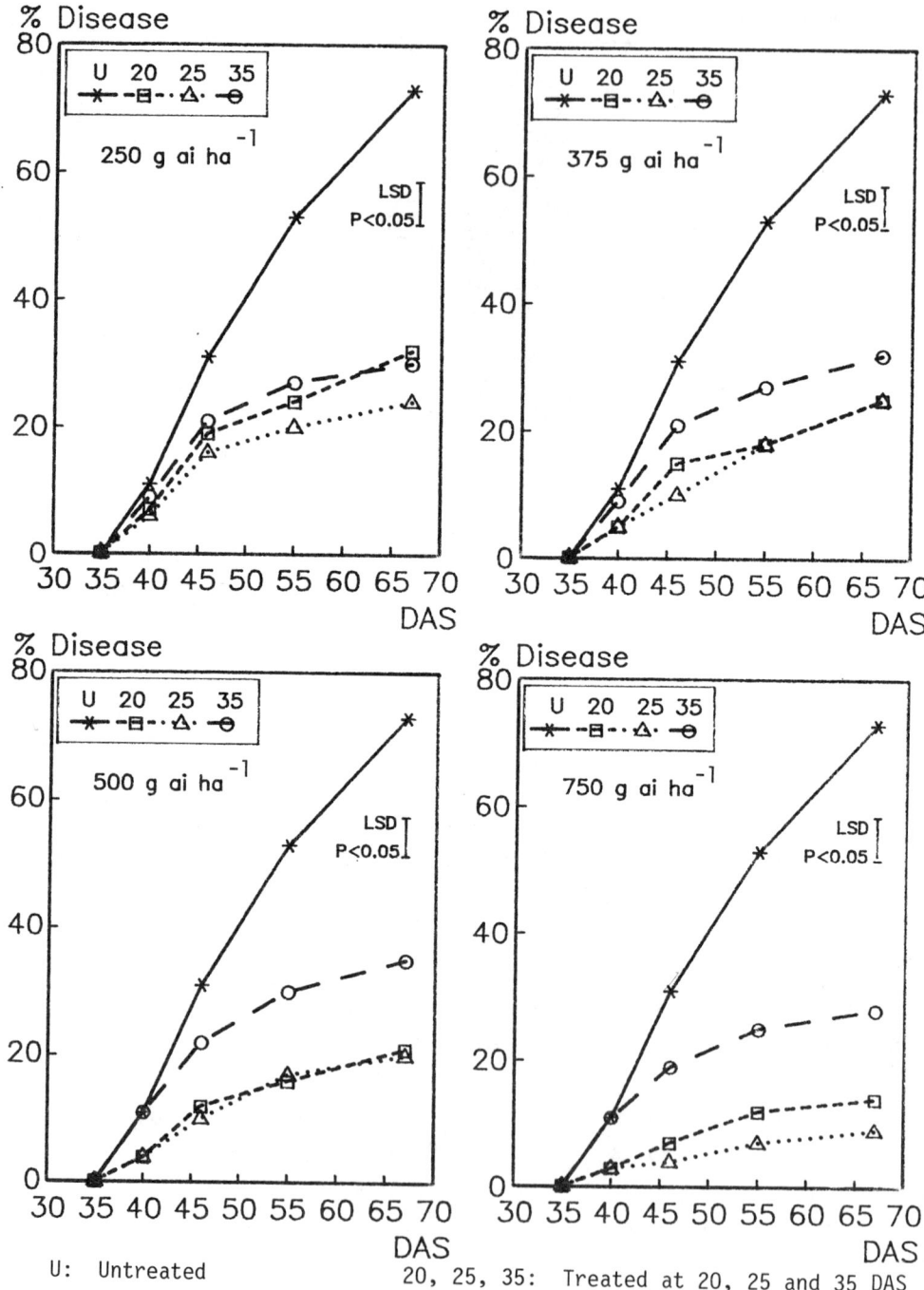

U: Untreated   20, 25, 35: Treated at 20, 25 and 35 DAS

Figure 3: Effect of application rate and timing of pyroquilon on leaf
blast infection (Mean of 3 trials)

days after sowing (GS 315). Plants treated shortly before infection (5
dbi at GS 22-23) gave the best control, even at the lowest rate of 250 g
ai ha^{-1}. Too early application (10 dbi) generally gave less
satisfactory control at low rates. High application rates improved
control, compensating for the earliness in application timing. Late
application (5 dai) after symptom appearance gave the least satisfactory
control which was little improved by increasing application rates.

The above interactions are likely to be more pronounced with compounds
having residual activity inferior to that of pyroquilon; so that it seems
clear that application against leaf blast must be done either before or
immediately after first symptom appearance, to achieve the most efficient
level of disease control.

In locations like the Chachoengsao area of Thailand where
environmental conditions favour disease development regularly at the same
stage of crop development (GS 22-23) each year, application can be made
according to growth stage if environmental conditions are conducive and
susceptible cultivars are grown. Alternatively, disease monitoring can be
intensified during the period, so that applications can be made as soon as
symptoms first become apparent.

Resistance of the plant to foliar infection increases with age almost
irrespective of host cultivar (12); so that if infection occurs too late
in the season (beyond maximum tillering or GS 29), the leaf phase of the
disease is unlikely to be damaging and control is unnecessary. However,
residual infection on leaves can be an important source of infection on
panicles (see application timing against neck blast).

The time of infection also influences the method of fungicide
application. In southern Japan, seed boxes may be treated
prophylactically against subsequent leaf blast in the field if infection
is expected to occur shortly after transplanting. In the north, where
infection appears later, seed box treatment may have insufficient residual
activity and into-water applications with systemic compounds are made in
the field, 7-10 days before the expected disease occurence (2).

In Latin America where blast is often the biggest threat to upland rice cultivation, foliar application has not been successful for the control of the disease. Only prophylactic seed treatment with pyroquilon has offered any hope of controlling the disease on susceptible cultivars (14).

APPLICATION TIMING AGAINST NECK AND PANICLE BLAST

The timing of fungicidal application against neck and panicle blast is equally critical. Because of the shortness of the economically significant part of the epidemic (from ear emergence to dough grain or about 2 weeks), and the extreme vulnerability of panicle development to infection at the neck region, prophylactic treatment is usually needed whenever a susceptible cultivar is grown under environmental conditions which favour disease. The damage is already largely done by the time first symptoms are apparent in the crop.

The results of a trial in Malaysia show the sensitivity of disease control to slight differences in application timing (Figure 4). Application of tricyclazole at GS 545 (50% of tillers at stage where panicles have half emerged from their flag leaf sheaths), reduced disease to 26% of that in untreated plots. A two day delay in application timing resulted in a substantial loss of control, with disease increasing to 38% of that in untreated plants.

Clearly application of control agents either as into-water applications or as foliar sprays must be timed to ensure presence of the active ingredient at the panicle raches and rachillae before inoculum arrival at the infection court. Precise definition of growth stage for application is important. The term 'heading' is insufficient; neither is '50% heading', which could mean 50% of heads emerged or 50% of the length of a panicle emerged from the flag leaf sheath. Application at GS 585 (50% of panicles with full emergence) may be too late for most protective compounds.

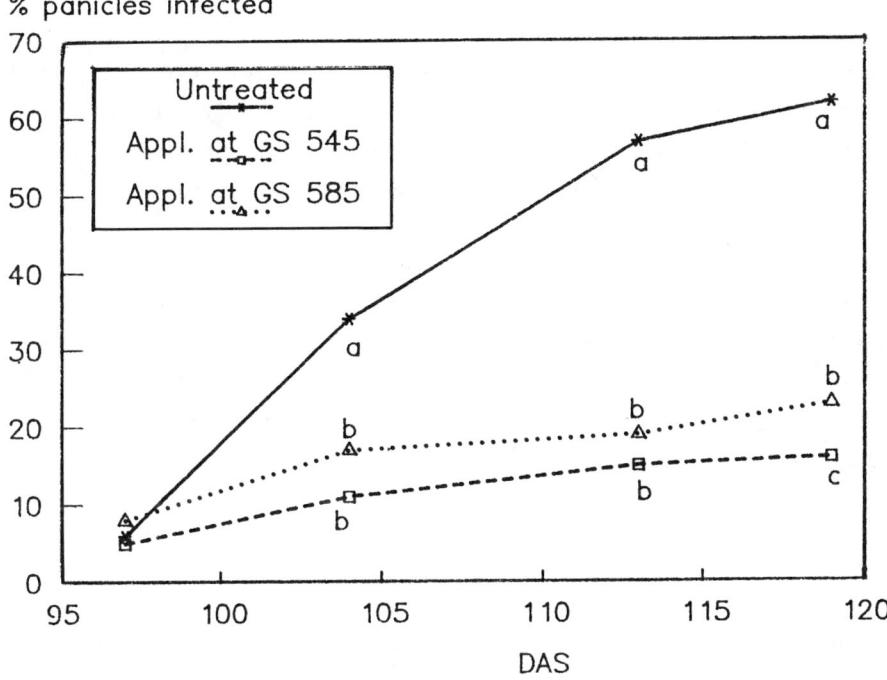

% panicles infected

Fig. 4 : Effect of application timing on control of neck
blast with tricyclazole (means at any one recording time
followed by the same alphabet are not significantly
different at P< 0.05)

Equally important, the reasons for unsatisfactory control may be best
examined when precise information is available of crop growth at the time
of application.

When leaf infection is common in the same crop, an additional
application at booting (GS 455 to 485) is often recommended as a sanitary
measure to reduce 'initial inoculum'. Antisporulant properties are
obviously needed for compounds used in this manner. Because they are
often applied 1-2 weeks before heading to allow time for the product to
reach the desired site of action, into-water applications serve both as
sanitary eradicative (on leaves) and protective (on panicles) measures.

APPLICATION TIMING AGAINST SHEATH BLIGHT

When inoculum is present and temperatures are sufficiently high (28°C), the appearance of sheath blight in the field is largely determined by growth stage. First symptoms usually appear at maximum tillering (GS 29) as the canopy begins to close over and humidity within the crop increases.

Significant yield losses are expected when the disease extends beyond the fourth leaf sheath. Damage on the third, second and flag leaf sheaths are estimated to cause yield losses of 15, 20 and 30% respectively(4). Therefore, fungicides should be applied at panicle initiation (GS 315) and booting (GS 455) (5), or once at GS 455 (4), if infection is progressing from the fourth to the third leaf sheath.

CONCLUSIONS

The efficient use of disease control agents requires that they be applied in a manner that brings about the maximum economic return to the farmer. This implies the construction of forecasting systems to predict on the basis of epidemiological parameters, likely disease curves; and the establishment of economic damage thresholds and control thresholds.

A number of forecasting systems have been developed for the blast disease (13,15,16,17), although none are at present widely used (7). Damage thresholds on the other hand are dependent, amongst other factors, upon cultivars grown, crop growth stages and price indices, and are therefore often quite location and time specific.

As discussed earlier in this paper, crop growth stages are an important factor in determining the time of application of disease control agents. Effective timing of fungicidal applications for the control of rice diseases requires a clear understanding of the relationships between crop growth stage, disease development and yield.

Precise definition of the growth stages of rice, as described for example in the modified 'Zadoks' decimal code for cereals (9) are a necessary prerequisite of studies to optimize application timing.

REFERENCES

1. Chin, K.M., Fungicidal control of the rice blast disease. Malaysian Agricultural Journal, 1975, 50, 221-228.

2. Yoshino, R., Present status of occurence and control of blast disease in Japan. Japan Pesticide Information, 1988, 52, 3-8.

3. Chin, K.M., Chemical control of sheath blight disease of rice caused by Thanatephorus cucumeris (Frank) Donk. Malaysian Agricultural Journal, 1977, 51, 238-243.

4. Sugiyama, M., Rice sheath blight and chemical control in Japan. Japan Pesticide information, 1988, 52, 9-12.

5. Jones, R.K. & Jeger, M.J., Evaluation of benomyl and propiconazole for controlling sheath blight of rice caused by Rhizoctonia solani. Plant disease, 1987, 71, 222-225.

6. Okuno, T., Furasawa, I., Matsura, K. & Shishiyama, J., Mode of action of ferimzone, a novel systemic fungicide for rice diseases: biological properties against Pyricularia oryzae in vitro. Phytopathology, 1989, 79, 827-832.

7. Kranz, J., Epidemiological information as an aid in pest management. In Progress in Upland Rice Research, IRRI, Los Banos, Philippines, 1985, pp. 355-362.

8. Zadoks, J.C., Chang, T.T. & Konzak, C.F., A decimal code for the growth stages of cereals. Weed Research, 1974, 14, 415-421.

9. Chin, K.M., Sozzi, D., Lohken, A. & Williams, R.J., A modified Zadoks decimal code for the growth stages of rice. In preparation 1990.

10. Vanderplank, J.E., Plant Diseases: Epidemics and Control. Academic Press, New York, 1963.

11. Wheeler, B.E.J., Diseases in crops. Edward Arnold, London, 1976.

12. Koh, Y.J., Hwang, B.K., & Chung, H.S., Adult-plant resistance of rice to leaf blast. Phytopathology, 1987, 77, 232-236.

13. Suzuki, H., Studies on the behaviour of rice blast fungus spore and application to outbreak forecast of rice blast disease. Bulletin Hokuriku Agricultural Experiment Station, 1969, 10, 1-118.

14. Prabhu, A.S. & Morais, O.P., Blast disease management in upland rice in Brazil. In Progress in Upland Rice Research, IRRI, Los Banos, Philippines, 1986, pp. 383-392.

15. Ono, K., Principles, methods and organization of blast disease forecasting. In The rice blast disease. Proceedings of the symposium at IRRI, July, 1963, John Hopkins Press, Baltimore, Maryland, 1965, pp. 173-194.

16. Sasaki, T. & Kato, H., A statistical method of predicting outbreaks of rice panicle blast. Phytopathology, 1972, 62, 1126-1132.

17. Chang, K.K., Improved methods for rice blast forecasting. Korean Journal of Plant Protection, 1982, 21, 19-22.

18. Sugitomo, K., Plant-water relationships of indica rice in Malaysia. Tropical Agricultural Research Center Technical Bulletin, 1971, 1, 1-80.

VARIATION IN PEST AND NATURAL ENEMY POPULATIONS – RELEVANCE TO BROWN PLANTHOPPER CONTROL STRATEGIES

M.F. Claridge
School of Pure and Applied Biology, University of Wales, Cardiff CF1 3TL

ABSTRACT

The brown planthopper, Nilaparvata lugens (Stal), shows complex variation in its ability to attack and damage different cultivars of rice. Populations with distinct patterns of virulence have been termed "biotypes". Such biotypes represent adaptive responses of the species to local host plant variation and care should be taken when identifying the "same" biotype from different regions. Generalisation from laboratory biotype cultures to field populations should be treated with caution.

Studies on hybridisation and acoustic mate recognition signals show clear genetic differentiation of populations in different parts of Asia and Australasia. A supposed "non–virulent biotype" of N.lugens associated with the weed, Leersia hexandra, is a separate, but morphologically indistinguishable, biological species. The term biotype has thus been used for a variety of biological phenomena and is not helpful in determining appropriate control strategies.

Resistance breeding programmes for N.lugens may be expected to be of long term advantage only in combination with other control strategies, particularly biological control. In the absence of pesticide use, natural enemies of N.lugens generally maintain good levels of control. Effective pest management in tropical countries will depend on an integration of resistance breeding and biological control.

INTRODUCTION

The idea that evolutionary biology might have any real relevance to practical control problems of crop pests would probably have been doubted by most agriculturalists, at least until very recently. This is surprising since inherited variation in crop plants has been exploited for the production of desirable cultivars since the earliest days of agriculture. Indeed very successful control strategies have been developed for a diverse array of crops by the incorporation of pest and disease resistant factors by plant breeders (1,2). Very often in such breeding programmes, variation is found in the responses of pests to different host cultivars. Generally much less attention has been paid to the nature of this variation than to that in the crop plants themselves. Often the designation biotype has been used as a general term by which to refer to both pest and disease organisms with different patterns of virulence to their hosts (3,4). In few examples is any detailed information available on the genetic nature of such biotypes. Many parthenogenetic insects develop clones with distinctive patterns of virulence, and within which variation may be minimal. Because of the nature of sexual reproduction such genetically uniform populations are very unlikely to be found in biparental species. An exceptional example is that of the hessian fly of wheat, Mayetiola destructor (Say), in north America. A series of biotypes characterised by genes for virulence which correspond to genes for resistance in different wheat cultivars has been described (5). In recent years it has been assumed, with very little evidence, that such simple correspondences of single genes in parasites and hosts are commonplace for sexually reproducing insects. In fact the hessian fly is the only example in which such a system has been described.

Here I shall discuss the value of studies on variation in the rice brown planthopper and some of its natural enemies for developing a control strategy in tropical countries.

THE RICE BROWN PLANTHOPPER

The Brown Planthopper, Nilaparvata lugens (Stal), is one of the most important pests of irrigated rice in Asia (IRRI, 6 1979). In tropical regions it has only become a serious problem in the past twenty years or so with widespread expansion in the cultivation of high yielding and rapidly maturing cultivars. The major strategy for control, developed to great effect at the International Rice Research Institute (IRRI), Philippines, has been the breeding of resistant varieties. This is still currently a successful technique in many parts of south east Asia, but has been less successful in India and Sri Lanka. Insects in these latter countries, or at least in certain areas of these countries, seem to be generally more virulent to a wider array of cultivars (8,9).

IRRI biotypes

The introduction of IR8, the first of the high yielding varieties released by IRRI, immediately preceded the emergence of N.lugens as a pest in the Philippines and other south east Asian countries. A massive screening programme identified a number of traditional varieties, with resistant properties to feeding by the insect, most of which derived originally from southern India and Sri Lanka. Among these, groups of varieties were identified with one or two distinct major genes for resistance (7). Populations of the insects were reared on those varieties at IRRI and were shown to have clear cut and distinct patterns of virulence in mass screening tests. On a basis of these patterns populations were identified as distinct biotypes and numbered 1, 2 and 3. Biotype 1 was characterised by its ability to damage only varieties that lack genes for resistance (eg. TN1 and IR8). In addition biotype 2 could also attack varieties that incorporate the gene Bph 1 (eg. Mudgo and IR26) and biotype 3 could attack varieties that include the gene bph 2, but not Bph 1 (eg. ASD7 and IR36).

These biotype populations have been maintained continuously in culture at IRRI, and have been used widely in screening trials for identifying further sources of resistance (eg. 8). Similar populations have been reared under similar conditions in other parts of Asia. At least two further major genes for virulence have been identified at IRRI, Bph. 3 and bph. 4, and incorporated into improved varieties (10).

With the widespread cultivation of improved varieties, populations of brown planthopper with similar patterns of virulence to the laboratory reared biotypes have been reported widely in Asia and have often been identified by use of the same numbering system (6). Because of the implicit assumption that these biotypes resemble those of the hessian fly in being an example of a gene – for – gene relationship, the genetic identity, or at least close similarity, of the same biotypes in different regions has also been widely assumed.

Although the IRRI biotype cultures show very distinct differences in mass screening trials, studies on variation in virulence within each biotype have revealed extreme variation (3). The insects show wide variation in their abilities to feed on standard cultivars. For example in our experiments a few individuals from biotype 1 showed the ability to feed on Mudgo or ASD7 and thus resembled biotypes 2 and 3 respectively. Similar aberrant individuals were found also in the biotype 2 and 3 cultures despite many generations of inbreeding (Figure 1) (3). Crosses between the biotypes showed no clear segregation patterns for virulence. Contrasting with the gene – for – gene hypothesis, the results suggested that virulence is inherited by a system of polygenes (11).

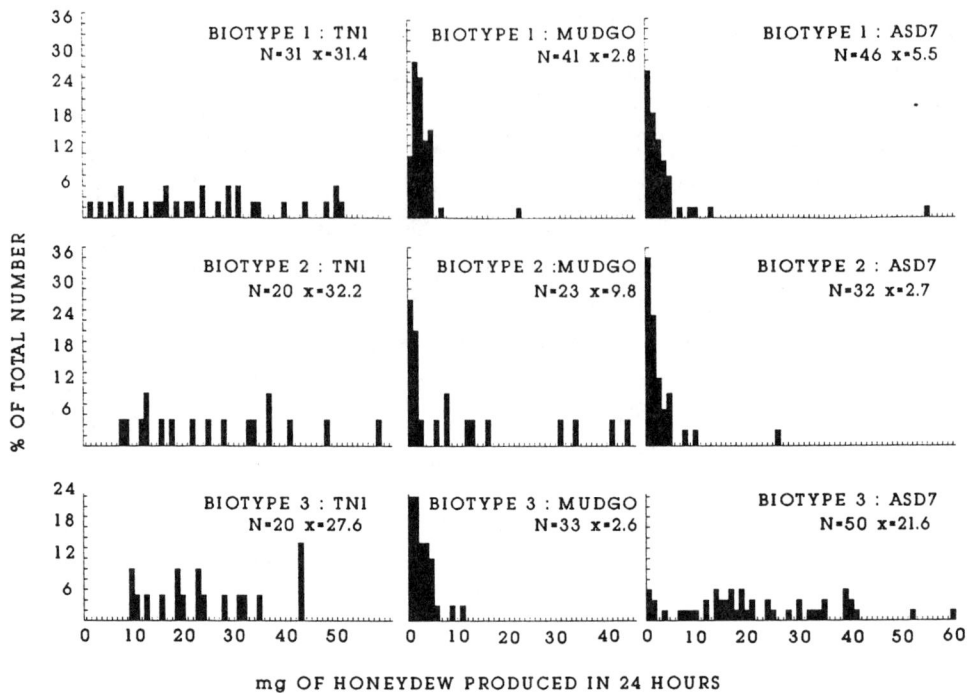

Figure 1. Honeydew produced by individuals of biotypes 1, 2 and 3 from IRRI on rice cultivars TN1, Mudgo and ASD7. After Claridge and Den Hollander (3).

Thus even the long inbred IRRI biotype cultures show considerable variation in virulence. Selection experiments have also demonstrated that after no more than about ten generations of rearing on appropriate cultivars, the biotypes can be effectively changed in virulence pattern (12).

Because of the clear cut patterns of virulence obtained in mass screening trials, many workers have attempted to find some other characters by which the different biotypes of N.lugens may be recognised. Particularly Saxena and co-workers at IRRI have claimed recently that morphometric analyses allow complete identification of biotypes 1, 2 and 3 from IRRI and that the data may be used to identify such biotypes in field samples (13,14). We have followed up these studies with similar morphometric analyses of the same biotype cultures (15). We confirmed that such techniques do indeed allow the statistical separation of the three populations when reared on their normal culture host plants (Figure 2). However, these differences are only of value if they can be shown to be inherited characteristics of the populations. In order simply to test this, we reared each biotype for one generation only on the same susceptible

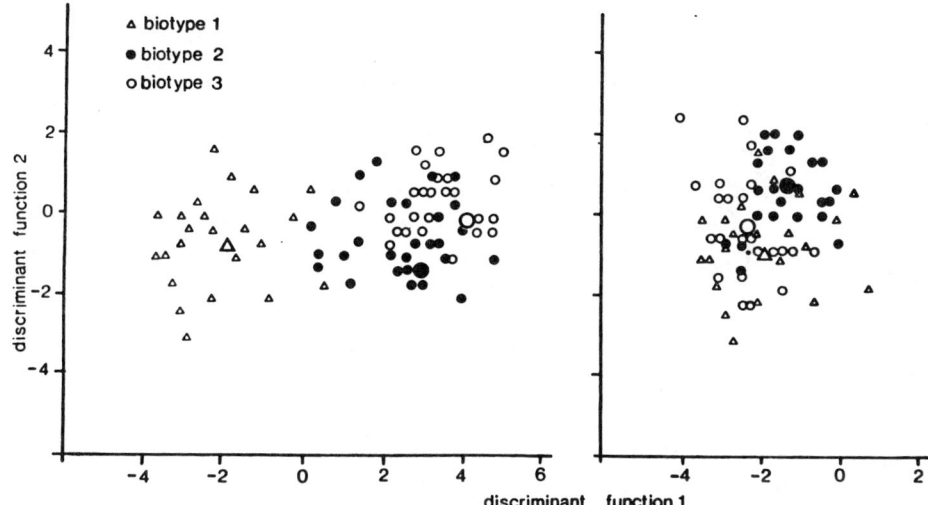

Figure 2. Plots of discriminant function 1 against discriminant function 2 for canonical analyses of samples of the three IRRI biotypes on cultivars TN1, Mudgo and ASD7 (left), and all on TN1 (right). After Claridge, Den Hollander and Haslam (15).

variety, TN1, and then took the same series of measurements as previously on samples from each. The morphometric differences were dramatically reduced (Figure 2) suggesting that most of the previously demonstrated differentiation was due to differences in nutritional quality, or related characteristics, of the three cultivars. Unfortunately Saxena and Rueda (13) reported no such experiments, so that their differentiation of the three biotypes must be assumed also to be due to environmentally induced effects. Saxena and Barrion (14), though briefly citing our study, made no attempt to confront our arguments.

Saxena and his colleagues have several times suggested that the morphometric differences reported by them were correlated with the important virulence differences between the biotypes, and thus that such techniques could be used to identify biotypes in the field. To date they have presented no evidence to validate this suggestion. Our own results suggest that this is very unlikely.

. Thus the IRRI biotypes are populations selected from an original field population in Luzon and inbred in culture cages for more than ten years. It must therefore be dangerous to generalise from them to field populations and to use them in green house screening tests (eg. 8).

Virulence patterns of field populations

IR26 was the first BPH resistant variety to be developed and was released in the Philippines in 1973 (16). By 1976 damage to IR26 was reported in parts of the country and when the insects were tested in the laboratory they killed Mudgo and were therefore regarded as biotype 2. More recently populations were reported in Mindanao, southern Philippines, which damaged IR36. These were then assumed to be biotype 3. However, when tested in the standard way, the insects were found not to repond as IRRI biotype 3 (17). Instead they showed some characteristics of both biotypes 2 and 3. It is clear then that the laboratory cultures may have little significance in explaining, and much less in predicting, the development of field virulence patterns.

All authors have agreed that over the wide area of distribution covered by N.lugens, different regions are characterised by major differences in virulence or biotype pattern (6,7,8). For example, populations in India and Sri Lanka show generally higher levels of virulence than do those from South East Asia. Indeed in parts of southern India and Sri Lanka there are virtually no varieties which may be said to be truly resistant. Considerable and confusing variability in responses of insects in different parts of India has often been noted (7,8,18).

In Cardiff we have been able to import live samples of brown planthoppers from many localities in Asia and Australasia (19). When tested for virulence characters some startling results were obtained. Most remarkably, when populations were first imported from Australia, the supposed universally susceptible variety, TN1, which is said to lack any gene for resistance, was found to be strongly resistant (20). The insects were then cultured on the japonica varieties – Delta from France and Towada from Japan. However selection experiments over about ten generations with the same insects produced a laboratory population virulent to TN1. It thus then resembled IRRI biotype 1 in gross virulence pattern.

In order to test the variation in virulence that might occur in the field over a relatively small geographical area, we studied populations collected from five different cultivars, including improved and traditional varieties, and a wild rice, Oryza rufipogon, from different localities in Sri Lanka. None of the localities were separated from the others by more than 200 kilometres. Insects were collected and brought back to Cardiff where a new generation was reared from each locality on the same variety from which it was collected. Each population was then tested for individual weight change and honeydew excretion on each of the six host plants and also on the standards, TN1, Mudgo and ASD7 (21). Each population was found to have a distinct pattern of virulence, different to each of the others. It might perhaps then be suggested that each population should be regarded as a different biotype! This would hardly be practicable and certainly would not be helpful in understanding the development of virulence in the field.

Thus, it may be concluded that the extreme variability in virulence demonstrated in laboratory populations is also characteristic of field populations. Local populations will adapt to local conditions and to local cultivars throughout the species range. Similar patterns of virulence (or biotype) will be expected to evolve independently on many occasions. If the term biotype is used in this context then it should only be in a very general sense to denote populations with similar virulence patterns. Suggestions that the same biotype may have other distinguishing features, whether morphological or otherwise, are likely to be misleading and to exacerbate the sorts of problems discussed above with insects virulent to IR36 in Mindanao. Similar difficulties in trying to force field populations into a predetermined biotype system have also been found in Taiwan (22).

Geographical variation

It is clear that differences in patterns of virulence show considerable geographical variation, but as shown above, these are highly labile and susceptible to relatively rapid change. We have studied populations of N.lugens from more than 20 localities in Asia. No consistent differences in morphology were found between any of them. Some electrophoretic differences were discovered, but no diagnostic alleles (23). In hybridisation experiments varying degrees of success were achieved between insects from geographically separated populations. In all crosses where F1 hybrids were obtained, they were viable and fertile and produced viable F2 insects with apparently normal sex ratios. It was thus suggested that difficulties in hybridisation were due to some premating barriers and not to general genetic incompatibilities.

In most leafhoppers and planthoppers substrate transmitted acoustic signals are of major importance in mate recognition (24). In N.lugens both sexes produce characteristic signals. Analyses of pulse repetition frequencies of male calls from different populations showed the greatest differences for those populations between which it was most difficult to obtain hybrids. For example, all populations from Australia differed very significantly in this character from any from Asia investigated by us (18). Thus, perhaps surprisingly, characteristics of courtship calls provide the most sensitive means we have yet found for identifying geographically discrete populations of N.lugens.

Weed – feeding populations

N.lugens has usually been regarded as a specific feeder only on rices, both wild species of Oryza and cultivars of O.sativa (25).   In 1982 Medrano and Heinrichs first briefly reported populations of N.lugens feeding and reproducing on the weed grass, Leersia hexandra, in the Philippines.   Saxena and colleagues (14,26) considered these insects to be "a non virulent biotype of N.lugens".   However, when we tested them in mate choice experiments with sympatric rice – feeding forms, they showed complete preference for mating with their own type (27). In addition we found significant differences in pulse repetition frequencies of both male and particularly female acoustic signals.   Laboratory reared female hybrids between the two populations produced intermediate signals, significantly different from either parent.   Though the populations are widely sympatric in the Philippines we found no evidence of field hybridisation between them.   We therefore concluded that these two populations represent distinct biological species.   More recently we have collected both species also from several different localities in Sri Lanka and India where they show very similar acoustic characteristics to those from the Philippines (28).

The nature of N.lugens biotypes

Diehl and Bush (4) reviewed the biotype concept as applied to insects and concluded "that future application of this term (biotype) be restricted to use as a temporary and provisional designation for cases where biological differences have been observed between organisms but where the genetic basis and evolutionary status of the differences have yet to be ascertained."   I agree with this suggestion.   It is clear that the use of the term for N.lugens has meant much more than this to many authors.   As outlined above, it is clear that biotype has been used on the one hand for laboratory selected populations that show particular virulence patterns with respect to rice cultivars with known major genes for resistance, as in the IRRI biotypes 1,2 and 3.   On the other it has also been used for a distinct biological species associated with the weed, Leersia hexandra.   In addition and more importantly it has been used of field populations in different areas with apparently similar patterns of virulence (7,8). No term that embraces so many different concepts can be helpful in our specific understanding of the phenomena concerned.   It is particularly undesirable to develop a biotype nomenclature or numbering system that forces the natural diversity and plasticity of these insects into an unrealistic and predetermined system of classification.   It would be more helpful to undertake detailed studies of virulence and genetic differentiation widely throughout the species range, if we are to understand the nature of variation in these remarkable insects.

CONCLUSIONS AND PEST MANAGEMENT

The use of varietal resistance as a control strategy for brown planthopper has been startlingly effective, especially in parts of south east Asia. In India and Sri Lanka fewer resistant varieties are available and results are distinctly variable from place to place. However, despite massive screening programmes (most recently summarised in 29), there are only a limited number of genes for resistance known to this insect. Thus, even in south east Asia and much more in south Asia, it would be very unwise to rely solely on a resistance breeding programme as a strategy for control.

In recent years detailed studies in the field on the population dynamics of brown planthopper have demonstrated a significant role for natural enemies. Kenmore et al. (30) showed that several groups of predators, but particularly the assemblage of spiders, in rice fields in the Philippines may respond numerically to planthopper density. They presented powerful evidence to show that these predators play a key role in reducing the pest populations. Clearly also effects of more specialised parasitoids play an important part in the overall field mortality of planthoppers, but much detailed research is necessary before management strategies can be developed to use them. Kenmore et al. showed that in their experimental field most of the mortality during the brown planthopper life cycle "occurred before the first nymphal instar (65 – 91%)". Among the most important factors involved were egg predation and parasitism, both of which are extremely difficult to study in the field. Egg parasitism particularly may be very high and many species of Mymaridae and Trichogrammatidae have been reared. Unfortunately host records are confused because usually several different hosts are included in any plant sample. Our own sampling in the Philippines, Sri Lanka, India and Australia, shows that some parasites show greater host specificity than is usually suspected (31). Also differences in species occur in Sri Lanka than in the Philippines, and completely different species are involved in Australia to those in Asia. Thus, variability in these natural enemies provides the possibility not only for managing indigenous parasite populations, but also for possible introductions of species or variant populations to new areas.

Many studies suggest that the development of N.lugens as a major pest of tropical rice was largely due to the results of increased and indescriminate use of pesticides in the 1970's following the introduction of high yielding cultivars. Well documented evidence is provided by Kenmore et al. (30). It is clear then that if reliable pest management strategies are to be developed for brown planthopper in Asia, some combination of varietal resistance and biological control must provide the solution. Complete reliance on single resistant varieties is a very dangerous strategy and must be expected to lead to the evolution of new patterns of virulence in pests. The maintenance of some diversity of crop genotypes, both seasonally and spatially, is an important way of reducing the evolution of pest virulence and of maintaining natural enemy diversity.

ACKNOWLEDGMENTS

I thank my colleagues of the rice research group in Cardiff for their help and assistance over the years in our studies on N.lugens which have been supported by successive contracts from the Centre for Overseas Pest Research, Tropical Development and Research Institute and the Overseas Development and Natural Resources Institute, all scientific units of the U.K. Government Overseas Development Administration, and also from the European Economic Community. Live insects are maintained in Cardiff under the terms of Welsh Office Agriculture Department licence.

REFERENCES

1. Harris, M.K. (ed.), Biology and breeding for resistance to arthropods and pathogens in agricultural crops. Texas A & M University, 1980, pp. 1–605.

2. Maxwell, F.G. and Jennings, P.R. (eds.), Breeding Plants Resistant to Insects, John Wiley, New York, 1980, pp. 1–683.

3. Claridge, M.F. and Den Hollander, J., The 'biotypes' of the rice brown planthopper Nilaparvata lugens. Entomologia experimentalis et applicata, 1980, 27, 23–30.

4. Diehl, S.R. and Bush, G.L., An evolutionary and applied perspective of insect biotypes. Annual Review of Entomology, 1984, 29, 471–504.

5. Everson, E.H. and Gallun, R.L., Breeding approaches in wheat. In Breeding Plants Resistant to Insects, eds. Maxwell, F.G. and Jennings, P.R., John Wiley, New York, 1980, pp. 513–533.

6. IRRI, Brown planthopper: Threat to rice production in Asia. IRRI, Los Banos, Philippines, 1979, pp. 1 – 369.

7. Pathak, M.D. and Khush, G.S., Studies of varietal resistance in rice to the brown planthopper at the International Rice Research Institute. In Brown Planthopper: Threat to Rice Production in Asia, IRRI, Philippines, 1979, 285 – 301.

8. Seshu, D.V. and Kauffman, H.E., Differential response of rice varieties to the brown planthopper in international screening tests. IRRI Research Paper Series, 1980, 52, 1 – 13.

9. Heinrichs, E.A. and Khush, G.S., Levels of resistance of rice varieties to biotypes of the Brown Planthopper, Nilaparvata lugens, in south and southwest Asia. IRRI Research Paper Series, 1982, 72, 1 – 14.

10. IRRI, Highlights 1985 Accomplishments and Challenges. IRRI, Los Banos, Philippines, 1986, pp. 1 – 101.

11. Den Hollander, J. and Pathak, P.K., The genetics of the 'biotypes' of the rice brown planthopper, Nilaparvata lugens. Entomologia experimentalis et applicata, 1981, 29, 76 – 86.

12. Claridge, M.F. and Den Hollander, J., The biotype concept and its application to insect pests of agriculture. Crop Protection, 1983, 2, 85 – 95.

13. Saxena, R.C. and Rueda, L.M., Morphological variations among three biotypes of the brown planthopper Nilaparvata lugens in the Philippines. Insect Science and its Application, 1982, 3, 193 – 210.

14. Saxena, R.C. and Barrion, A.A., Biotypes of the brown planthopper Nilaparvata lugens (Stal) and strategies in deployment of host plant resistance. Insect Science and its Application, 1985, 6, 271 – 289.

15. Claridge, M.F., Den Hollander, J. and Haslam, D., The significance of morphometric and fecundity differences between the 'biotypes' of the Brown Planthopper, Nilaparvata lugens. Entomologia experimentalis et applicata, 1984, 36, 107 – 114.

16. Khush, G.S., Genetics and breeding for resistance to the brown planthopper. In Rice Brown Planthopper: Threat to Rice Production in Asia, IRRI, Philippines, 1979, pp. 321 – 332.

17. Medrano, F.G. and Heinrichs, E.A., Response of resistant rices to brown planthopper (BPH) collected in Mindanao, Philippines. International Rice Research Newsletter, 1985, 10 (6), 14 – 15.

18. Kalode, M.B. and Khrishna, T.S., Varietal resistance to brown planthopper in India. In Rice Brown Planthopper: Threat to Rice Production in Asia, IRRI, Philippines, 1979, pp. 187 – 199.

19. Claridge, M.F., Den Hollander, J., and Morgan, J.C., Variation in courtship signals and hybridization between geographically definable populations of the rice brown planthopper, Nilaparvata lugens (Stal). Biological Journal of the Linnean Society, 1985, 24, 35 – 49.

20. Claridge, M.F. and Den Hollander, J., Virulence to rice cultivars and selection for virulence in populations of the brown planthopper Nilaparvata lugens. Entomologia experimentalis et applicata, 1982, 32, 213 – 221.

21. Claridge, M.F., Den Hollander, J., and Furet, I., Adaptations of brown planthopper (Nilaparvata lugens) populations to rice varieties in Sri Lanka. Entomologia experimentalis et applicata, 1982, 32, 222 – 226.

22. Cheng, C.H. and Chang, W.L., Studies on varietal resistance to the brown planthopper in Taiwan. In Brown Planthopper: Threat to Rice Production in Asia, IRRI, Philippines, 1979, pp.251 – 271.

23. Den Hollander, J., Electrophoretic studies on planthoppers and leafhoppers of economic importance. In Electrophoretic Studies on Agricultural Pests, eds. Loxdale, H.D., and Den Hollander, J., Clarendon Press, Oxford, 1989, pp. 297 – 315.

24. Claridge, M.F., Acoustic behaviour of Leafhoppers and Planthoppers: species problems and speciation. In The Leafhoppers and Planthoppers, eds Nault, L.R. and Rodriguez, J.G., John Wiley, New York, 1985, pp. 103 – 125.

25. Mochida, O. and Okada, T., Taxonomy and biology of Nilaparvata lugens (Hom., Delphacidae). In Rice Brown Planthopper: Threat to Rice Production in Asia, IRRI, Philippines, 1979, pp.21 – 43.

26. Saxena, R.C., Velasco, M.V. and Barrion, A.A., Morphological variations between brown planthopper biotypes on Leersia hexandra and rice in the Philippines. International Rice Research Newsletter, 1983, 18 (3), 3.

27. Claridge, M.F., Den Hollander, J., and Morgan, J.C., The status of weed – associated populations of the brown planthopper, Nilaparvata lugens (Stal) – host race or biological species? Zoological Journal of the Linnean Society, 1985, 84, 77 – 90.

28. Claridge, M.F., Den Hollander, J. and Morgan, J.C., Variation in hostplant relations and courtship signals of weed – associated populations of the brown planthopper, Nilaparvata lugens (Stal), from Australia and Asia: a test of the recognition species concept, Biological Journal of the Linnean Society, 1988, 35, 79 – 93.

29. Heinrichs, E.A., Medrano, F.G. and Rapusas, H.R., Genetic evaluation for insect resistance in rice. IRRI, Los Banos, Philippines, 1985, pp. 1 – 356.

30. Kenmore, P.E., Carino, F.O., Perez, C.A., Dyck, V.A. and Gutierrez, A.P., Population regulation of the rice brown planthopper (Nilaparvata lugens Stal) within rice fields in the Philippines. Journal of Plant protection in the Tropics, 1984, 1, 19 – 37.

31. Claridge, M.F., Claridge, L.C. and Morgan, J.C., Anagrus egg parasitoids of rice – feeding planthoppers. Proceedings of 6th Auchenorrhycha Meeting, Turin, Italy, 1988, 617 – 621.

RESISTANCE MANAGEMENT OF BROWN PLANTHOPPER, <u>NILAPARVATA LUGENS</u>, IN INDONESIA

MARK E. WHALON, HUGO E. VAN DE BAAN, AND KASUMBOGO UNTUNG[1]

Department of Entomology and Pesticide Research Center, Michigan State University, East Lansing, MI 48823, U.S.A., and [1]Department of Entomology, Gadjah Mada University, Yogyakarta 55581B, Indonesia

ABSTRACT

In the 1970's and early 1980's, during rice production intensification in Indonesia, the brown planthopper, <u>Nilaparvata lugens</u> Stal, became a major pest of rice and seriously threatened Indonesia's rice self-sufficiency. Factors that contributed to the increasing problems of brown planthopper were: injudicious use of pesticides which caused pest resurgence, the elimination of natural enemies and the development of resistance; breakdown of host plant resistance, and; lack of integration of different pest management tactics. In 1986, because of the increasing problems with brown planthopper, the Indonesian government declared Integrated Pest Management (IPM) the national rice pest management strategy and banned 57 pesticides for their use on rice based on expert advice. Although this IPM program is highly effective, brown planthopper will continue to adapt to pesticides and resistant rice varieties used in the current IPM program. Therefore, in order to develop a sustainable rice IPM program, pesticide and host plant resistance management strategies need to be implemented.

THE BROWN PLANTHOPPER PROBLEM

In the 1970's and early 1980's one of the major goals of rice production in Indonesia was to reach self-sufficiency [1]. Through a rice production intensification program Indonesia became rice self-sufficient by 1983. This was primarily achieved by combining high pesticide and fertilizer input with the use of high yielding rice varieties that were resistant to insect pests [1,2].

The intensification of rice production, however, caused increasing pest problems [1,2,3]. During this period, the brown planthopper, <u>Nilaparvata lugens</u> Stal, became a major pest of rice in Indonesia. This insect causes both direct damage to rice by feeding on the rice plants, causing 'hopperburn', and indirect damage by transmitting grassy stunt, a mycoplasma [4]. The brown planthopper was first reported as a rice pest in Indonesia in 1969, and in subsequent years there was a dramatic increase in brown planthopper populations and losses in rice yield due to damage caused by this insect [2, see Fig. 1]. From the period 1977 to 1979 alone, over 2 million hectares of rice were lost due to brown planthopper damage [1,2]. Since 1979 damage caused by brown planthopper decreased, however, brown planthopper outbreaks in 1984 and 1986 reduced rice yields nation-wide [2,3].

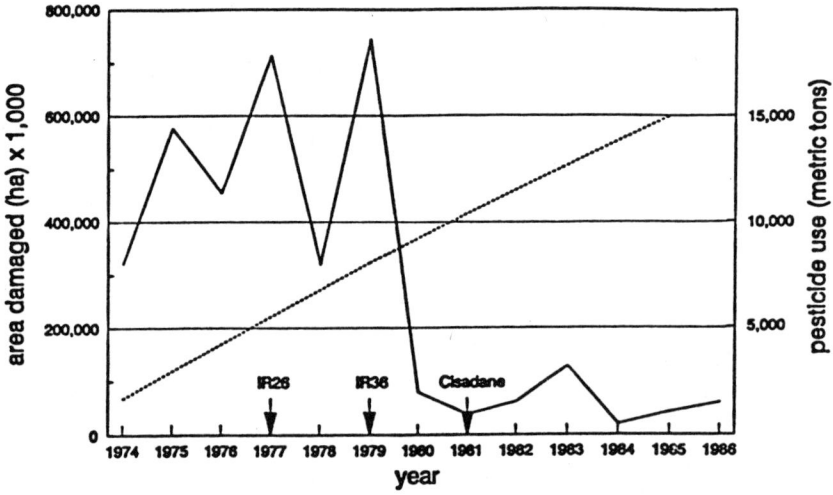

Figure 1. Economic damage caused by brown planthopper, N. lugens, (—) and the use of pesticides (--) and release of resistant rice varieties (IR26, IR36, Cisadane) for its control in Indonesia.

Strategies for controlling brown planthopper in the 1970's and early 1980's depended mainly on the use of brown planthopper resistant rice varieties and insecticides, although the use of some cultural practices were suggested to overcome the brown planthopper problem [2,5, see Fig. 1]. However, the unilateral dependence on resistant rice varieties and insecticides, to which brown planthopper was able to adapt, and the lack of integration of different pest management strategies caused brown planthopper to become a major pest in this period. The following factors contributed to the increasing brown planthopper problem;

1) Inappropriate cultural practices: the lack of crop rotation and staggered planting provided a continuous food source for brown planthopper [2,5]. Nitrogen fertilizers trigger ovipositional response in brown planthopper, and increased use of fertilizers

during the rice production intensification program led to dramatic population increases [5]. Although resistant rice varieties were available, many farmers planted old, brown planthopper-susceptible, varieties because of better taste.

2) Pesticides: three major factors contributing to the failure of chemical control have been the resurgence of brown planthopper after insecticide applications, the elimination of natural enemies of brown planthopper due to broad spectrum chemicals, and the development of insecticide resistance in brown planthopper. Resurgence, a significant increase in brown planthopper populations after insecticide treatment, was observed in Indonesia since 1979 as well as elsewhere in Southeast Asia [1,2,3,6,7]. Studies on the effect of insecticides on populations of brown planthopper on central Java indicated that all the major groups of insecticides (carbamates, organophosphates, and pyrethroids) caused brown planthopper resurgence [3,8]. These studies also indicated that the use of broad-spectrum insecticides eliminated natural enemies of brown planthopper, allowing the pest to reach damaging levels. Brown planthoppers were effectively controlled by natural enemies if no disruptive insecticides were used. The excessive use of insecticides during the rice production intensification program caused high selection pressure on brown planthopper populations, resulting in the development of insecticide resistance. Populations of brown planthopper from Java were reported to be resistant to organophosphates, carbamates, and pyrethroids [9,10], and resistance levels were related to patterns of insecticide use [10].

3) Breakdown of host plant resistance: in order to overcome the increasing problems with brown planthopper, more effort was put into the propagation and distribution of brown planthopper-resistant rice varieties in Indonesia. As early as 1967, varieties resistant to brown planthopper had been identified at the International Rice Research Institute, Los Banos, Phillipines [11]. In Indonesia, IR26 and other resistant varieties containing the Bph 1 gene, were introduced in 1977-1978 [3, see Fig. 1]. However, host plant resistance was easily broken down by brown planthopper, and resistance to these varieties did not last for more than 2 cropping seasons (less than one year), after their introduction [1]. After 1979, brown planthopper outbreaks could be controlled by the introduction of varieties containing the bph 2 gene, such as IR36 and Cisadane [3]. However, this narrow base of host plant resistance made the rice production system vulnerable to brown planthopper outbreaks. The acreage of rice fields damaged by brown planthopper increased from 19,000 ha in 1984 to 60,000 ha in 1986 [3, see Fig. 1]. The ability to breakdown host plant resistance has been related to the development of 'biotypes', based on the observations that laboratory cultures of brown planthopper obtained from field populations and selected on host plants containing different resistant genes led to the development of strains capable of

surviving host plant resistance [12]. Although some authors speculate that the occurence of biotypes in the field may even eventually lead to sympatric speciation [13], other workers conclude that biotypes are simple genetic variants rapidly selected with no mating barriers [14,15]. The ability of brown planthopper to quickly breakdown host plant resistance as well as to develop resistance to insecticides indicates that this insect is highly adaptive to selection pressure exerted through different means on the population. Researchers start to realize that control of brown planthopper solely based on the use of pesticides or resistant varieties will not be effective in the long term.

INTEGRATED PEST MANAGEMENT OF BROWN PLANTHOPPER

Based on the findings that natural enemies can effectively control brown planthopper if no disruptive insecticides are used, pilot studies were conducted in the early 1980's to implement pest control based on the conservation of natural enemies as part of a new Integrated Pest Management approach [2,3,8]. Integrated Pest Management (IPM) is a philosophy of pest control that utilizes the "best set" of management strategies, tactics and tools to limit pests below an economic threshold with mimimum environmental and socioeconomic impacts [16]. Various tools, tactics, and strategies were developed and evaluated in order to implement a more sustainable rice production system (see Fig. 2). The rice IPM strategy emphasized the use of insecticides only when needed and the use of locally acceptable resistant rice varieties. An important aspect of this program was the training of extension personnel and farmers to diagnose and monitor pest problems in the field and to make decisions accordingly. Results of these pilot studies demonstrated the feasibility of the IPM approach for larger areas of rice production in Indonesia.

Because of the increase in damage caused by brown planthopper in the mid 1980's (see Fig. 1) and the availibility of an IPM alternative, the Indonesian Government declared on November 5, 1986, by Presidential Decree 3 (Inpress 3/1986) IPM the national pest control strategy for rice [1,2]. The Indonesian legislation was based on expert advice from the Indonesian (Gadjah Mada University, Yogyakarta, and Central Research Institute for Agriculture, Bogor) and international (International Rice Research Institute and Food and Agricultural Organization, Phillipines) rice research community. The Presidential Decree emphasized that insecticides should only be used when control thresholds in effect were reached (5 brown planthoppers/tiller), thus mandating a monitoring strategy. The decree banned the use of 57 insecticides on rice because of their implication on brown planthopper resurgence. Only four compounds

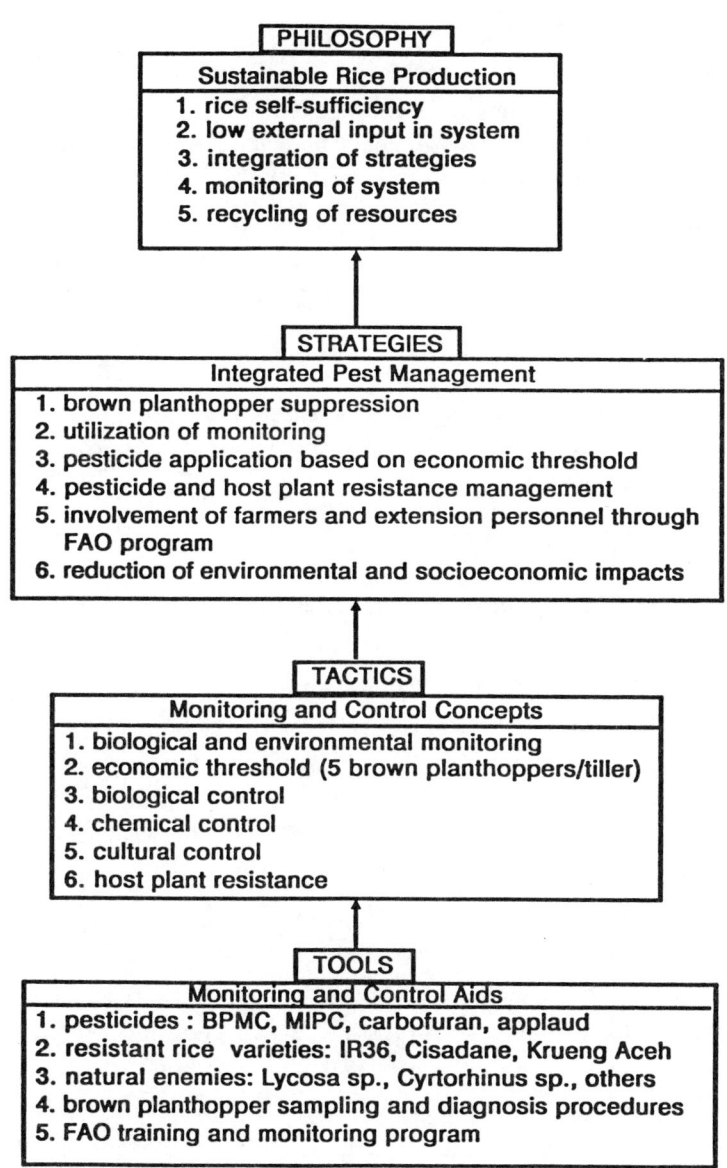

Figure 2. Tools, tactics, strategies, and philosophy of Integrated Pest Management of brown planthopper, N. lugens, in Indonesia.

were allowed for rice pest management, the carbamates MIPC, BPMC, and carbofuran and the insect growth regulator buprofuzin (Applaud^R). The use of brown planthopper resistant rice varieties was also required. An appropriate cropping system, including synchronous planting, rotation with non-rice crops, and a considerable free rice crop period was recommended. In order to implement this program, the Presidential Decree stated that extension personnel and farmers should be trained to conduct IPM of brown planthopper.

Presidential Decree 3 resulted in effectively controlling brown planthopper and improving rice pest management in Indonesia. The number of insecticide applications dropped from 4.5/ha in 1986 to 0.5/ha in 1988 [1]. This resulted in a reduction of insecticide costs to the farmers from 7,500 rupiah/ha in 1986 to only 2,200 rupiah/ha in 1988, even though insecticides were more expensive due to a reduction in government subsidy compared with 1986. The rice yield increased from 6 tons/ha in 1986 to 7.5 tons/ha in 1988. Throughout the introduction of IPM in Indonesia, an extensive training program of field personnel, extension workers and farmers was put in place.

RESISTANCE MANAGEMENT OF BROWN PLANTHOPPER

Although the current rice IPM program in Indonesia emphasizes the integration of various tools for pest control, resistant rice varieties as well as insecticides remain important components of the overall IPM strategy. The availibility of only 4 insecticides (MIPC, BPMC, carbofuran, and buprofezin) and 3 major brown planthopper-resistant rice varieties (IR36, Krueng Aceh, and Cisadane) for brown planthopper control, imposes a substantial selection pressure on populations of this pest. Because of the propensity of brown planthopper to quickly adapt to selection pressure exerted on populations by resistant host plants or insecticides, as indicated by strains of brown planthopper able to breakdown host plant resistance or to develop insecticide resistance, the application of resistance management is necessary. We define resistance management as a strategy within an IPM system that seeks to limit the selection for resistance alleles to major population suppression strategies such as host plant resistance and insecticides. Through resistance management one seeks to prolong the life of a pesticide or resistant host plant variety by preventing, delaying or reverting resistance development to the pesticide or ability to breakdown host plant resistance by the pest. The goal of resistance management is to implement a sustainable IPM system which allows for long term control of a pest or pest complex. Because of the importance of managing resistance in brown planthopper for overall pest control in rice, a strategy of resistance management of brown planthopper has been developed as an initial step to implement a resistance management program in Indonesia. Figure 3 shows the different tools, tactics, and strategies of such a resistance management program.

Pesticide resistance management

An important requirement for a resistance management program is the ability to detect resistance in a population at a sufficiently early stage to reduce selection pressure. Through resistance monitoring one attempts to measure changes in the frequency or degree of resistance in time and space in a pest species. Resistance monitoring is therefore essential for the evaluation of strategies, validation of tactics, and implementation of an ongoing IPM program.

Insecticide resistance in brown planthopper in Indonesia has been mainly observed through field failure of insecticides [10]. Because of the importance of detecting resistance at an early stage, resistance monitoring techniques have been currently developed as initial steps in resistance management of brown planthopper. Toxicity bioassays, including topical application of insects, dipping of insects, and exposing insects to insecticide residues seem appropriate assays for evaluating the efficacy of pesticides. However, disadvantages of such toxicity bioassays are that only one insecticide can be tested per insect, relatively large numbers of insects are needed, and results are only known after 24 or 48 hrs. More recently biochemical assays have been developed for various insects such as aphids and mosquitoes in which the activity of detoxification enzymes can be measured in individual insects which is an indication of resistance to a certain insecticide [17,18]. Advantages of such biochemical tests are that they provide information about resistance frequencies within populations, require fewer insects, and are more sensitive and less time consuming than toxicity bioassays [19]. Studies on the biochemistry of resistance in brown planthopper from Java showed that esterases are important in conferring resistance to organophosphates, carbamates, and pyrethroids [9,10]. Therefore, biochemical assays were developed for the detection of esterase activity in individual brown planthoppers, using either a microtitre plate assay and an ELISA reader or a portable photometer set-up. The latter allows for the detection of resistance levels in populations of brown planthopper in the field. These biochemical assays are simple and easily transferred to relatively intrained field personnel and seem useful tools for resistance monitoring.

The availability of resistance monitoring techniques may allow for effective resistance management of brown planthopper in Indonesia in the future. The following strategies supported by laboratory data and/or field experience have been generally considered useful in managing resistance in arthropods [20] and will be evaluated for brown planthopper during the development and implementation of a

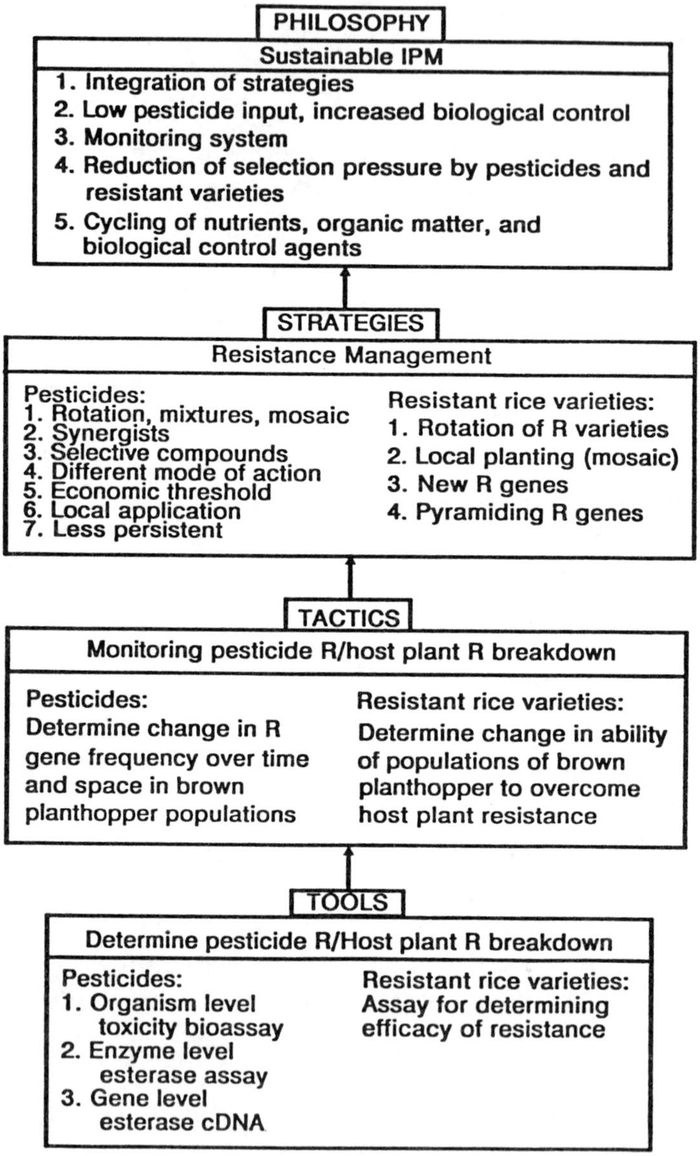

PHILOSOPHY		
Sustainable IPM		
1. Integration of strategies		
2. Low pesticide input, increased biological control		
3. Monitoring system		
4. Reduction of selection pressure by pesticides and resistant varieties		
5. Cycling of nutrients, organic matter, and biological control agents		

STRATEGIES

Resistance Management

Pesticides:
1. Rotation, mixtures, mosaic
2. Synergists
3. Selective compounds
4. Different mode of action
5. Economic threshold
6. Local application
7. Less persistent

Resistant rice varieties:
1. Rotation of R varieties
2. Local planting (mosaic)
3. New R genes
4. Pyramiding R genes

TACTICS

Monitoring pesticide R/host plant R breakdown

Pesticides:
Determine change in R gene frequency over time and space in brown planthopper populations

Resistant rice varieties:
Determine change in ability of populations of brown planthopper to overcome host plant resistance

TOOLS

Determine pesticide R/Host plant R breakdown

Pesticides:
1. Organism level toxicity bioassay
2. Enzyme level esterase assay
3. Gene level esterase cDNA

Resistant rice varieties:
Assay for determining efficacy of resistance

Figure 3. Tools, tactics, strategies, and philosophy of Resistance Management of brown planthopper, N. lugens, in Indonesia.

resistance management program: local rather than areawide insecticide applications; treatments only when the economic threshold is reached; use of less persistent insecticides; mixtures, rotations or mosaics of applications of the carbamates MIPC, BPMC, carbofuran and the insect growth regulator buprofezin; use of synergists; use of selective compounds to protect natural enemies; use of compounds with different mode of action.

Host plant resistance management

Regarding host plant resistance management, monitoring for the ability to breakdown the resistances of different resistant varieties in field populations of brown planthopper under standardized conditions will be an essential component. Strategies that may reduce the speed at which brown planthopper will break down host plant resistance are the following: rotation of resistant varieties, i.e. rotation of different resistant genes; more local than areawide planting of a certain resistant variety, i.e. host plant mosaic; pyramiding existing resistant genes from different varieties into a new variety; introduction of new convential selected resistant genes (laboratory biotypes of brown planthopper may be useful for genetic screening of new resistant varieties). Biotechnology also offers the possibility of creating new resistant varieties, but there is no reason a priori to assume that these exotic genes could not be overcome by brown planthopper biotypes.

Future perspectives

Because of the great demand for rice, rice production in Indonesia will continue to be based on high-input intensive farming. As part of the current IPM program, the use of insecticides as well as resistant rice varieties will therefore continue, thus selection pressure will continue and brown planthopper populations will eventually adapt. However, it is hoped that the integration of insecticide and host plant resistance management strategies will result in stable control of brown planthopper in Indonesia. Because, brown planthopper is a highly adaptive species, we believe that continuous monitoring and strategy alteration will be necessary for brown planthopper management. In our view, resistance management is therefore an essential strategy within a sustainable IPM approach to brown planthopper management.

ACKNOWLEDGEMENT

This research was funded, in part, by grant no. (8.395) 936-5542 from USAID.

REFERENCES

1. FAO, Integrated pest management in rice in Indonesia, Jakarta, Indonesia, 1988, 13 pp.

2. Anonymous, History and present status of brown planthopper as rice pest in Indonesia. Indonesian country report, presented at the International Workshop on Brown Planthopper, Yogyakarta, Indonesia. December 1986.

3. Untung, K., The study on the effect of insecticides on the populations of Nilaparvata lugens (Stal) and its natural enemies. Paper presented at GIFAP meeting on IPM on rice, Jakarta, Indonesia, April 1988.

4. Sogawa, K., The rice brown planthopper: feeding physiology and host plant interactions. Ann. Rev. Entomol., 1982, 27, 49-73.

5. Oka, N., Cultural control of the brown planthopper. In Brown planthopper: threat to rice production in Asia. IRRI, Los Banos, Philippines, 1979, pp. 357-69.

6. Chui, S.-C, Biological control of the brown planthopper. In Brown planthopper: threat to rice production in Asia. IRRI, Los Banos, Philippines, 1979, pp. 335-56.

7. Heinrichs, E.A. and Mochida, O., From secondary to major pest status: the case of insecticide-induced rice brown planthopper, Nilaparvata lugens, resurgence. Prot. Ecol., 1984, 7, 201-18.

8. Rahardja, U., The influence of the change in the three rice hopper species as compared to the population and prey preference of Lycosa spp. M.S. thesis (in Indonesian, with English abstract), Gadjah Mada University, Yogyakarta, Indonesia, 1982.

9. Chang, C.K. and Whalon, M.E., Substrate specificities and multiple forms of esterases in the brown planthopper, Nilaparvata lugens (Stal). Pest. Biochem. Physiol., 1987, 27, 30-5.

10. Sustrino, Study on brown planthopper, Nilaparvata lugens (Stal), resistance to organophosphate and carbamate insecticides. Ph.D. Dissertation (in Indonesian, with English translation), Gadjah Mada University, Yogyakarta, Indonesia, 1989.

11. Pathak, M.D., Cheng, C.H. and Fortuno, M.E., Resistance to Nephotettix impicticeps and Nilaparvata lugens in varieties of rice. Nature, 1969, 223, 502-4.

12. Pathak, M.D. and Saxena, R.C., Breeding approaches in rice. In Breeding plants resistant to insects, eds. F.G. Maxwell and P.E. Jennings, Wiley, New York, 1980, pp. 421-55.

13. Saxena, R.C. and Barrion, A.A., Biotypes of the brown planthopper Nilaparvata lugens (Stal) and strategies in deployment of host plant resistance. Insect Sci. Applic., 1985, 6, 271-89.

14. Claridge, M.F. and Den Hollander, J., The biotype concept and its application to insect pests of agriculture. Crop Protection, 1983, 2, 85-95.

15. Gallagher, K.D., Effects of host plant resistance on the microevolution of the rice brown planthopper, Nilaparvata lugens (Stal) (Homoptera: Delphacidae). Ph.D. Dissertation, University of California, Berkeley, 1988.

16. Whalon, M.E. and Weddle, P., Implementing IPM strategies and tactics in apple: an evaluation of the impact of CIPM on apple IPM. In Integrated pest management on major agricultural systems, eds. R.E. Frisbie and P.L. Adkisson, Texas Agricultural Experiment Station MP-1616, 1985, pp. 619-37.

17. Sawicki, R.M., Devonshire, A.L., Rice, A.D., Moores, G.D., Petzing, S.M. and Cameron, A., The detection and distribution of organophosphorus and carbamate insecticide-resistant Myzus persicae (Sulz.) in Britain in 1976. Pestic. Sci., 1976, 9, 189-201.

18. Brogdon, W.G., Beach, R.F., Stewart, J.M., Castanaza, L., Microplate assay analysis of the distribution of organophosphate and carbamate resistance in Guatemalan Anopheles albimanus. Bull. World Health Org., 1988, 66, 339-46.

19. Brown, T.M. and Brogdon, W.G., Improved detection of insecticide resistance through convential and molecular techniques. Ann. Rev. Entomol., 1987, 32, 145-62.

20. National Academy of Sciences, Pesticide resistance: strategies and tactics for management. National Academy Press, Washington, 1986, 471 pp.

THE ROLE OF MIGRATION IN THE PEST STATUS OF BROWN PLANTHOPPER IN TEMPERATE AND TROPICAL AREAS

J.R. RILEY, J. HOLT[1] and D.R. REYNOLDS
Overseas Development Natural Resources Institute (ODNRI), Radar
Entomology Unit, R.S.R.E., Leigh Sinton Road, Malvern, Worcs. WR14 1LL
[1] ODNRI, Central Avenue, Chatham Maritime, Chatham, Kent ME4 4TB, U.K.

ABSTRACT

A radar study of the migration of brown planthopper (*Nilaparvata lugens* Stal) during the autumn in east central China revealed that the flight duration was strikingly different to that previously found for the same species during the dry season in the Philippines. Flights were largely confined to periods of about 30 minutes at dusk and dawn in the Philippines, with minimal activity at other times, whereas in China, migrants which took off in the period from late afternoon to dusk continued in downwind migratory flight for several hours and some may have flown all night. These different durations of migration may be characteristic of populations from the humid tropics and from the temperate zone, respectively. The relative importance in the ecology of *N. lugens* is correspondingly different in the two zones. In temperate areas, the initial long-distance migrants arrive in localities where rapid and sustained population increase is possible because of the ineffectiveness of regulatory factors, particularly natural enemies. In contrast, in the tropics, mortality due to natural enemies will usually prevent the build-up of damaging populations which might otherwise follow any immigration from a distant source area. However, migration could still be a major determinant of outbreaks in the tropics in certain situations, for example where short-range mass movements occur between asynchronously-planted contiguous rice cultivations. Migration of brown planthopper may also have an impact on its pest status through the spread of virus diseases, or of *N. lugens* genotypes which carry resistance to insecticide or which are able to overcome the varietal resistance of rice plants. The effect of the newly-arrived genotypes is often difficult to assess, however, and in some cases immigration may be beneficial, hindering the local evolution of virulent genotypes.

EFFECTS OF NITROGENOUS FERTILIZER, INSECTICIDES AND PLANT SPACING ON INSECT PESTS AND YIELDS OF FLOODED RICE IN EASTERN INDIA

CHAKRABORTY, D.P.[1], MASLEN, N.R.[2] & HOLT, J.[2]

[1] Hindustan Fertilizer Corporation Ltd., 52A Shakespeare Sarani, Calcutta-700 017, India

[2] Overseas Development Natural Resources Institute, Central Avenue, Chatham Maritime, Chatham, Kent ME4 4TB, U.K.

ABSTRACT

Field trials performed in flooded rice at nine locations in West Bengal, Orissa and Bihar during the 1986 wet (kharif) and 1987 (boro) dry season showed that rice yields as well as densities of major rice pests were higher in the high nitrogen treatments. Scheduled insecticide applications reduced densities of all pests but only by 20 - 30%. Insecticide did not increase rice yields but pest densities, with the exception of yellow stemborer, were below application thresholds which are currently recommended. At most sites brown planthopper was more abundant than green leafhopper during the boro season and *vice versa* during the kharif. Stemborer densities exceeded the application threshold (1 moth /m²) at most sites during the boro only. Even in 'pest endemic areas', rice pest problems in the trials were few, suggesting that insecticide use was rarely necessary over the region as a whole during these two seasons. This emphasizes the importance of need-based insecticide use rather than prophylaxis.

INTRODUCTION

Measures to increase food grain production in eastern India have included the introduction of short duration high yielding rice varieties, the use of nitrogenous fertilizers, and where possible, increased cropping frequency. In common with most of South and Southeast Asia, this has led to an increase in rice pest problems [1] [2]. In much of the southern half of West Bengal and coastal Orissa, two or three crops can be grown per year and major pests such as brown planthopper, yellow stemborer and leaffolder are found throughout the year. In contrast, a single rainfed crop is grown in the drier areas of Bihar and pest problems are generally infrequent and less severe [3].

For reasons of availability and cost, rice farmers in eastern India are largely restricted to using wide spectrum insecticides, which kill natural enemies as well as pests. Despite this, farmers who can afford them may apply insecticides on a schedule basis at set times during the season, this being easier than need-based application. The timing of insecticide application is critically important for effective brown planthopper control [4]. Scheduled spray application times of 35 and 55 days after transplanting (DAT) have performed reasonably well in fertilizer demonstration plots (D.P. Chakraborty, unpublished), although this practice is increasingly being replaced by need-based control. While scheduled insecticide inputs have provided satisfactory pest control in local fertilizer trials (e.g. [5]), need-based control has given better returns when pest densities were low [6]. Although insecticide induced resurgence of brown planthopper does occur (N.R. Maslen and S.V. Fowler, pers comm.), the risk of resurgence appears to be less than in parts of South east Asia [7].

These trials were performed to help evaluate basic rice cultivation practices, used with high yielding varieties, as a starting point for the development of a more need-based pest management strategy. This paper describes an investigation of the effects of the scheduled insecticide applications, nitrogen fertilizer and plant spacing on pest population densities and crop yields at a range of sites and using a number of different rice varieties.

METHODS

Field trials were carried out over three states in eastern India during the 1986 wet season (kharif) and the 1987 dry season (boro). Eight trial sites were used in the kharif and nine in the boro. Most of the sites were located in West Bengal with two in Orissa and one in Bihar.

At each site, three replicates of each of 18 treatment combinations were laid out in a split-plot design, there being 3 spacing treatments (Narrow 10x10 cm, Medium 15x15 cm, Narrow with 2 rows left unplanted after every 6th row [corridor]), 3 insecticide treatments (quinalphos, endosulfan, No pesticide) and 2 nitrogen treatments (50 kg/ha and 100 kg/ha). The 27 main plots measured 10x5 m, each being split into two 10x2.5 m sub-plots for the two nitrogen treatments. Insecticides were applied routinely at 35 and 55 days after transplanting (DAT), quinalphos as Ekalux 25% EC and endosulfan as Thiodan 35% EC, at rates of 280g and 525g a.i./ha (0.0375% and 0.07%), respectively. Nitrogen, in the form of urea, was applied as follows: 50% at transplanting, 25% at 35 DAT and 25% at 55 DAT. In addition to nitrogen, basal applications of 50kg P_2O_5/ha (single superphosphate) and 50kg K_2O/ha (muriate of potash) were applied at transplanting.

The effects of these treatments on both rice yield and the densities of insect pests were examined. Pest populations were sampled using a D-vac suction sampler. Each sample comprised the total catch from five randomly selected points in the sub-plot. The abundant insect species were identified and counted; pest densities were estimated, the area of crop sampled being the D-vac nozzle area x the number of sample points, 0.071 x 5 = 0.354 m². In the kharif season, insect samples were taken at three crop stages: primordial, booting and milk ripe. In the boro season samples were taken at the latter two stages. The first two samples were made 1 to 5 days following the insecticide applications, depending on the trial (Table 1).

TABLE 1
Details of sites and number of days after transplanting (DAT) of pest sampling and harvest times

Site	State	Variety (DAT)	Sample times (DAT)			Harvest
			1st	2nd	3rd	
Kharif season 1986						
Barrabazar	W.Bengal	IR36	41	60	80	90
Joypur	W.Bengal	IR36	40	60	80	94
Galsi-II	W.Bengal	CR190	41	61	80	89
Bolpur	W.Bengal	IR36	36	61	81	88
Durgapur	W.Bengal	Ratna	44	60	70	90
Pipli	Orissa	CR1030	41	61	81	112
Satyabadi	Orissa	CR190	46	66	86	95
Sindri	Bihar	Ratna	42	62	82	91
Boro season 1987						
Barrabazar	W.Bengal	IR36	-	56	73	93
Joypur	W.Bengal	IR36	-	56	76	94
Galsi-II	W.Bengal	IR50	-	55	76	86
Khanakul	W.Bengal	Parijat	-	57	77	88
Jagatballavpur	W.Bengal	Ratna	-	55	75	92
Kharagpur	W.Bengal	IR36	-	55	75	84
Balianta	Orissa	Pratap	-	56	76	87
Satyabadi	Orissa	CR190	-	55	74	98
Sindri	Bihar	Ratna	-	53	72	87

Yield and pest density differences were analysed separately for the two seasons. For those pest species present in all samples, the basic 3 x 3 x 2 factorial analysis of variance (with split plots) had three replicates, 2 or 3 sampling occasions and 8 or 9 sites. A square root (x+1) transformation was performed on the insect counts.

As this paper is primarily concerned with results of general significance, independent of location, treatment effects were compared against variation between sites, revealing effects which show up despite site to site variation. Patterns associated with sites and seasons are also examined briefly. If consistent, such patterns may reflect differences in the risk of pest problems in different locations and seasons. Comparison with other years is needed to assess the consistency of such variation.

RESULTS AND DISCUSSION

General treatment effects over all sites

In the 1987 boro season, three pest species were relatively abundant, and present in most samples: brown planthopper, BPH (*Nilaparvata lugens* (Stal)); green leafhopper, GLH (*Nephotettix* spp.) and Yellow stemborer, YSB (*Scirpophaga incertulas* (Walker)). In the 1986 kharif season, the main species were green leafhopper, brown planthopper and whitebacked planthopper, WBPH (*Sogatella furcifera* (Horvath)). In an analysis of variance including all sites, treatment effects were examined for these species. Results were very consistent between pest species and there was strong evidence in both seasons that, a. insecticides decreased pest densities and, b. higher nitrogen increased pest densities (Table 2, Fig.1). The other piece of strong evidence was that higher nitrogen, but *not* insecticides, increased yields (Table 2, Fig.2). Figures 1 and 2 illustrate the boro season; treatment means show an identical pattern in the kharif.

TABLE 2

The significance of plant spacing, insecticide and nitrogen treatment effects, judged against between-site variation (treatment*site interaction mean square used as denominator to calculate F values)

Season	boro				kharif			
Pest	BPH	GLH	YSB	Yield	BPH	GLH	WBPH	Yield
Factor								
Plant spacing, S	ns	ns	*	ns	ns	ns	ns	ns
Insecticide, P	***	***	***	ns	***	***	**	ns
Nitrogen, N	***	***	**	***	***	***	***	**
Sample occasion, T	ns	*	ns	–	ns	ns	ns	–
S*P	ns	ns	ns	ns	ns	ns	ns	ns
S*N	*	ns	ns	ns	ns	ns	ns	ns
P*N	ns	ns	ns	ns	ns	ns	ns	ns
S*T	ns	ns	ns	–	ns	ns	ns	–
P*T	ns	ns	ns	–	**	**	*	–
N*T	ns	ns	ns	–	ns	ns	ns	–

Significance of F: ns P>.05, * P<=.05, ** P<=.01, *** P<= .001, – not applicable. No significant three-way interactions were found.

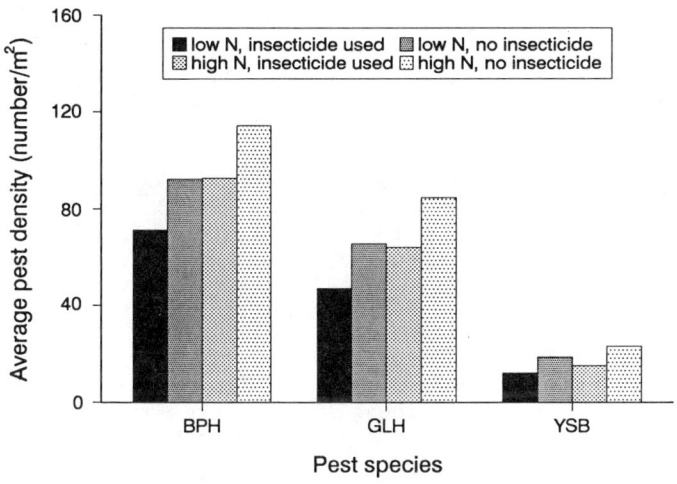

Fig. 1. Responses of insect pests to nitrogen and insecticide
treatments, average of all sites for the boro season

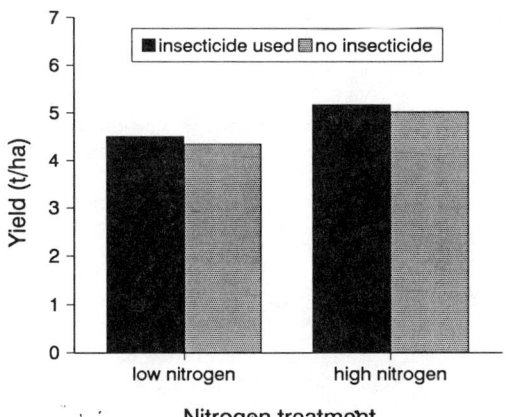

Fig. 2. Rice yield response to nitrogen and insecticide
treatments, average for all sites for the boro season

With the exception of yellow stemborer, pest densities (Fig. 1), remained below the insecticide application thresholds which are thought to be appropriate for this area: BPH, WBPH - 5 to 10 hoppers/hill (equivalent to 220-440/m² at 15x15 cm hill spacing) ; GLH - 20 hoppers/hill (880/m²); YSB - 1 moth/m² [8]. In general insecticides were probably unnecessary and therefore yields could not be improved by their use. The reduction in pest densities observed as a result of insecticide use was typically about 20 % (Fig. 1). Higher efficacies would have been unlikely to result in yield improvements, however, as pest densities were generally too low.

A number of other points emerged from the trials. There was no conclusive evidence that either of the two insecticides was more effective in reducing pest (BPH, GLH, YSB & WBPH) populations, although there was some trend in favour of quinalphos at some sites. Insecticide effects on yield were significant (P<=.05) at some individual sites during the boro season but the effect was not consistent.

The effects of plant spacing were relatively minor. There were significantly more yellow stemborer caught in the corridor treatment during the boro and at some individual sites, there were significantly more BPH in this treatment also. YSB and BPH densities were very similar in the other two spacing treatments. In the boro season, the corridor treatment resulted in lower yields in most sites; there was a significant effect of spacing at five of the nine sites. Yields from the other two spacing treatments were very similar.

Although pest population dynamics were impossible to assess from two or three sampling occasions during a season, there were no general trends of population change between sampling occasions which showed up over site to site variation, except for GLH during the boro where density increased from Sample 2 to 3 (Table 2). The interaction between insecticides and time, for BPH, GLH and WBPH in the kharif (Table 2), suggests greater insecticide efficacy for the application at booting stage (55 DAT) than at the primordial stage (35 DAT). Treated-untreated differences at booting and primordial stages were 42% and 9%, 57% and 12%, 66% and 15%, for BPH, GLH and WBPH, respectively.

The only other significant interaction, between spacing and nitrogen, for BPH in the boro (Table 2), was very small compared to the main nitrogen effect. With low nitrogen, BPH densities were similar at narrow and medium spacings, but with high nitrogen there were slightly more BPH at the narrower spacing.

Patterns associated with season and location

Yields in the boro season were generally higher than those in the kharif because the boro season is longer and growing conditions are better. A range of factors, especially variety and local growing conditions contribute to the variability between sites. Brown planthopper was the most abundant pest species during the boro season, especially in the West

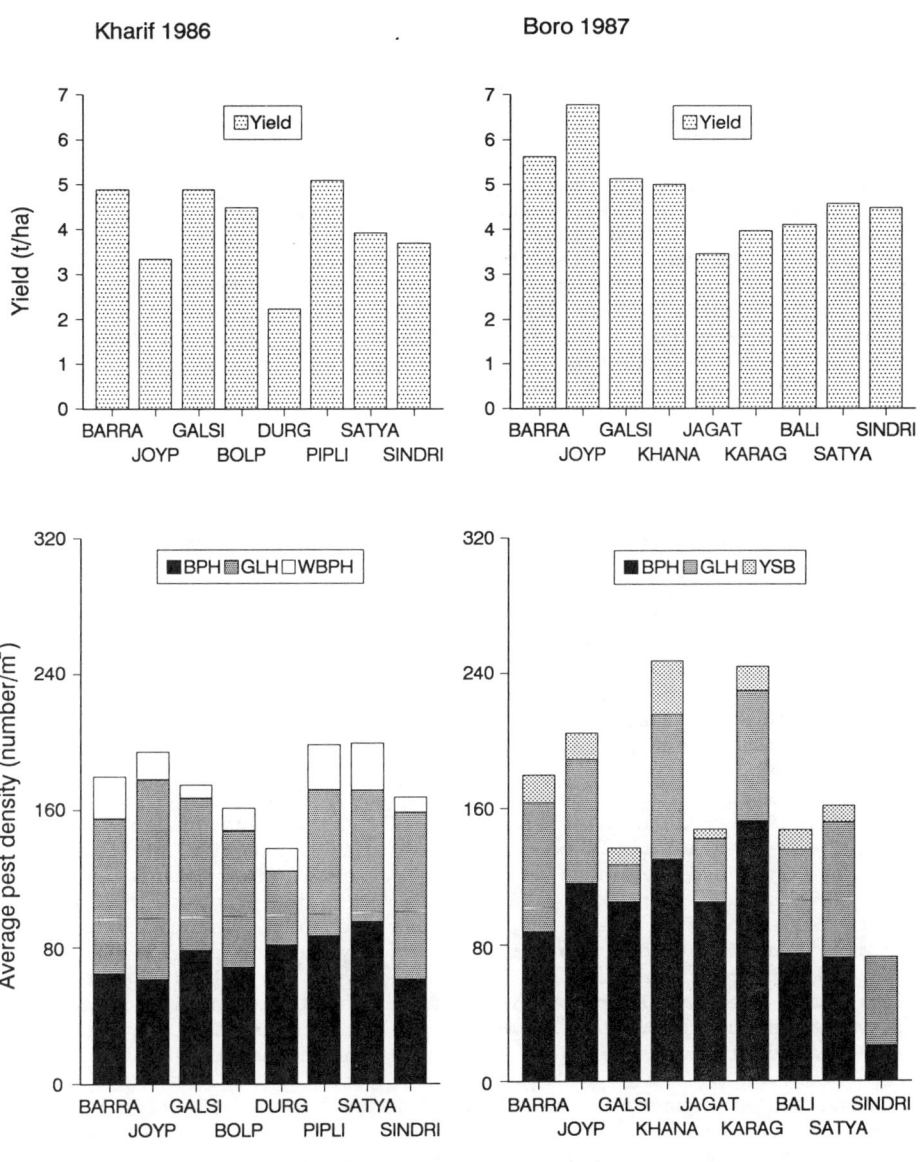

Fig. 3. Rice yields and average densities of most abundant pest species at each trial site during the 1986 wet (kharif) season and the 1987 dry (boro) season

Bengal sites (Fig. 3), whereas during the kharif season, green leafhopper was typically more abundant (Fig.3). The relatively low abundance of pests at Sindri during the boro season is likely to be associated with the limited area of rice cultivation which may serve as a source of immigrant pests in this part of Bihar.

There were significant positive effects of insecticides on yield at a number of individual sites: Joypur, Khanakul, Jagatballavpur and Sindri (boro season) and Pipli and Sindri (kharif season). Such effects did not appear to be associated with the abundance of particular pests at these sites. During the boro, when stemborer densities exceeded insecticide application thresholds at some locations, significant insecticide effects on yield were observed at sites with a wide range of stemborer densities (Fig. 3). It is possible that yield increases were due to the direct effect of insecticides on the crop.

CONCLUSIONS

In common with other studies (e.g. [9],[10]), all pest species were generally more abundant in the treatments with high fertilizer inputs. This effect was seen for all rice varieties in the trial. Even in the high nitrogen treatments, however, pest densities at most sites in both seasons remained below treatment thresholds. Yellow stemborer, at some sites in the boro season, was the exception but there was no evidence of reduced yields associated with higher stemborer numbers. The suction sampling technique may have over-estimated stemborer adult densities, by also taking those at canopy level from adjacent plants.

Pest mortality achieved by insecticide application at the rates used in the trials appeared to be very low. Applications causing pest mortality less than 30% would be unlikely to give successful control even if pest densities had warranted treatment. Mortality of natural enemies may also have been low, possibly explaining the lack of any resurgence in these trials, although in general, predators are more susceptible to pesticides than are BPH and sub-lethal doses may lead to resurgence [11]. Movement of both pests and natural enemies between the relatively small plots may have been partly responsible for the apparent ineffectiveness of the insecticide treatments. However, insecticide application rates were also rather low compared with rates often recommended [12].

In the kharif season, the spray at 55 DAT was more effective for brown planthopper control than that at 35 DAT. This is consistent with the results of a simulation analysis of BPH management in the Philippines [13], from which it is possible to predict spray times which are likely to be most effective, if required. Clearly, insecticides are only required if pest density thresholds are exceeded. The low pest densities at all sites suggests that insecticide use was rarely necessary over the region as a whole during these two seasons, and certainly highlights the importance of a need-based approach to pest control.

ACKNOWLEDGEMENTS

Mr. G.C. Ghose, Mr. P.C. Srivastava and other staff at the Hindustan
Fertilizer Corporation (HFC), Sindri Research Station, carried out the
field sampling and sample sorting. Dr. C. Gay provided statistical
advice and assistance. The work was carried out as part of the Indo-
British Fertilizer Education Project (IBFEP) supported by HFC and the UK
Government Overseas Development Administration. We are grateful to Dr.
P.K. Das, Chief Agricultural Scientist, and Dr. M.H. Ali, IBFEP Project
Manager, for their encouragement throughout.

REFERENCES

1. Bhandyopadhyay, A.K., Das, M.K., Ghosh, J.K. and Pawar, A.D.,
Feasibility of biological control in the integrated management of brown
planthopper, *Nilaparvata lugens* (Stal) in Burdwan, West Bengal.
Unpublished report, Central Biological Control Station, Burdwan-713 103,
1987, 7pp.

2. Litsinger, J.A., Second generation insect pest problems on high
yielding rices. *Tropical Pest Management*, 1989, 35, 235-242.

3. Chakraborty, D.P., Srivastava, P.C., Ghose, G.C., Maslen, N.R., Holt,
J. and Fowler, S.V., Contrasting rice pest abundance in Bihar and Orissa.
International Rice Research Newsletter, 1990, in press.

4. Cheng, J.A., Norton, G.A. and Holt, J., A systems analysis approach to
brown planthopper control in Zhejiang Province, China. II. Investigation
of control strategies, 1990, 27, 100-112.

5. Mathew, G., Chakraborty. D.P., Ghosh, G.C. and Dhua, S.P., Efficacy of
different fertilizer-pesticide mixtures as granular application in rice
(var. Jaya). *Pesticides*, 1978, 12, 23-25.

6. Gangwar, S.K. and Dasgupta, M.K., Efficacy and economics of pest
management in evolving an appropriate integrated pest management system
of rice in West Bengal. *Oryza*, 1984, 21, 188-200.

7. Kenmore, P.E., Carino, F.O., Perez, C.A., Dyck, V.A. and Gutierrez,
A.P., Population regulation of the rice brown planthopper (*Nilaparvata
lugens* Stal) within rice fields in the Philippines. *Journal of Plant
Protection in the Tropics*, 1984, 1, 19-37.

8. *Project Manual on Pest Management*, Indo-British Fertilizer Education
Project, Hindustan Fertilizer Corporation Ltd, Calcutta, 162pp.

9. Heinrichs, E.A. and Medrano, F.G., Influence of N fertilizer on the
population development of brown planthopper. *International Rice Research
Newsletter*, 1985, 10, 20-21.

10. Gopalakrishna Pillai, K., Kalode, M.B. and Rao, A.V., Effects of nitrogen levels, plant spacings and row orientation on the incidence of the brown planthopper of rice. *Indian Journal of Agricultural Science*, 1979, **49**, 125-129.

11. Yi, S.H. and Choi, S.Y., Effects of sublethal doses of some pesticides on the biotic potential and population density in brown planthopper, *Nilaparvata lugens* Stal. *Korean Journal of Plant Protection*, 1986, **25**, 139-149.

12. Waibel, H, *The economics of integrated pest control in irrigated rice*, Springer-Verlag, Berlin, 1986, 196pp.

13. Wareing, D.R., Holt, J. and Cheng, J.A., Use of computer tools for the design of pest management strategies, this volume, 1990.

THE POTENTIAL OF FUNGI FOR CONTROL OF RICE PESTS

ADRIAN GILLESPIE[1] and JAIME JIMENEZ[2]
[1] Chr. Hansen's Bio Systems A/S, Bøge Alle 10-12,
DK-Hørsholm, Denmark
[2] CAB International Institute of Biological Control,
Silwood Park, Ascot, Berks.

ABSTRACT

Rice provides a favourable environment for the exploitation of fungi as mycoinsecticides, but this potential has yet to be realised. This short paper provides an outline for a rational programme of strain selection, using examples from studies on Beauveria bassiana and Metarhizium anisopliae as pathogens of Nephotettix virescens and Nilaparvata lugens.

INTRODUCTION

The tropical rice crop, growing in flooded soil at moderate temperatures, appears to provide an uniquely favourable environment for the exploitation of fungi as mycoinsecticides. The high humidity occurring within the crop canopy, at least during the night time, can provide the necessary relative humidity for germination of fungal conidia, and the temperatures from 20 -35°C are favourable for rapid conidial germination and growth. There are currently, however, no mycoinsecticides registered for use against rice pests. This paper will examine the biology of entomogenous fungi with particular emphasis on their requirements for moisture and response to temperature. Examples from studies on Metarhizium and Beauveria as pathogens of Nilaparvata lugens (Delphacidae) and Nephotettix virescens (Cicadellidae) will be used to describe the complexities of developing fungi as mycoinsecticides.

Water relations

Entomogenous deuteromycetes will not germinate at relative humidities below 93%, and require higher humidities to germinate optimally. Generally, fungi kill insects slowly and

require prolonged high humidity periods for infection of
insects. Thus, <u>N. virescens</u> treated with fungi and maintained
at 100% r.h. and 25°C, only die after 3-5 days.

Temperature

Temperature is less crucial to entomogenous fungi than
relative humidity as most grow rapidly from 25-30°C, temperatures commonly encountered in tropical rice. Nevertheless,
intra-specific variation is considerable and isolates of <u>M.
anisopliae</u> can have optimum temperatures for growth ranging
from 23-30°C. This factor should be considered when selecting
isolates for detailed study.

Selection of strains - virulence

The virulence of fungi is difficult to predict, for example
<u>Verticillium lecanii</u> isolates are highly virulent to aphids,
whitefly, thrips and mites, and might also be expected to be
effective against hoppers. In fact, both <u>N. lugens</u> and <u>N.
virescens</u> are almost totally resistant to <u>V. lecanii</u>.

Strains of both <u>M. anisopliae</u> and <u>B. bassiana</u> vary greatly in
their virulence to both <u>N. virescens</u> and <u>N. lugens</u>. For
example LC 50 values of <u>B. bassiana</u> against <u>N. virescens</u>
ranged from 2 - 148 x 10^6 spores per ml under suboptimal r.h.
conditions and from 0.3 - 27.6 x 10^6 at constant high humidity. Assays should preferably be conducted under similar
conditions to those occurring in the rice paddy. This is often
difficult or impossible to achieve.

Virulence of single spore isolates

Single spore cultures taken from a multispore isolate can vary
greatly in virulence to insects. Thus, when 100 <u>B. bassiana</u>
isolates were examined for virulence to <u>N. virescens</u>, mortalities after 4 days ranged from 0 to 100%. Interestingly,
the most virulent single spore isolates failed to sporulate
on dead insects, presumably because death was at least in part
due to toxins produced before the fungus had colonized the
insect.

Stability of virulence

The stability of a virulence trait is an important factor to
consider when selecting candidate mycoinsecticides. When 2
single spore isolates and their multispore parent were
passaged through insects, or subcultured on agar, virulence
altered dramatically. Thus, insect passage increased virulence
initially, but continued passages resulted in a decline until
after 4-6 passages, virulence had returned to the initial

level. In contrast, all isolates slowly lost virulence after subculturing on agar until the sixth ·transfer, when one single spore isolated started to lose virulence rapidly. By the 9th subculture this isolate was virtually avirulent with a LT 50 of 155 hours. This data strongly illustrates the dangers of repeated subculturing, even of the reputedly more stable single spore isolates. It is recommended that fungal strains are stored in liquid nitrogen as soon as possible after isolation.

Production

To be used as a mycoinsecticide, a fungus must be amenable to mass production, and this is often cited as a reason for the limited commercial interest in fungi. However, some fungi such as V. lecanii can be easily produced in submerged culture using conventional stirred tank reactors. To date, it has proved impossible to produce spores of M. anisopliae economically using such technology, though Bayer are developing a product based on mycelia. Some isolates of B.bassiana can be grown in submerged culture and depending on the media, produce either blastospores or submerged conidia. Furthermore, these spores can be considerably more virulent than aerial conidia. Conidia can also be produced by semi-solid fermentation, and this may well prove to be the most suitable technology for production of entomogenous fungi, as these spores can survive for long periods.

Storage

In view of the foregoing illustrations of intra-specific variation, both fungal species and strains vary greatly in their survival after drying. One isolate of B. bassiana maintained conidial viability for 200 days at $20^{\circ}C$, while other Beauveria strains lost 90% viability after 100 days. Since storage is so important for commercial development of mycoinsecticides, spore survival should be studied early in the development programme.

Persistence

The persistence of fungi can be considerable in field conditions. However, after application, spores adhere strongly to leaves and are not efficiently picked up by insects. Thus, in practice, only insects receiving a lethal spore dose during spraying become diseased and subsequent insect generations do not become infected, unless dead insects remained attached to the host plant.

The result is, of course, poor control of the pest, particularly with rapidly reproducing insects such as N. lugens.

In practice, most dead insects fall into the water and do not adhere to the plant. The ability of fungi to adhere dead insects to the plant is quite rare in <u>B. bassiana</u> and <u>M. anisopliae</u>. However, where this feature does occur, it can result in very efficient disease transmission to nymphs.

Conclusion

In this paper I have tried to suggest an appropriate research programme for mycoinsecticide development, which is applicable to both rice and other agroecosystems. I hope to have persuaded companies on the fringes of biological control research, that the development of a microbial control agent is a complicated business requiring significant investment in specialized personnel. If these investments are made, the chances of mycoinsecticide use, particularly in rice, look bright.

INSECT PEST MANAGEMENT IN RICE
IN THE UNITED STATES

M. O. Way
Texas Agricultural Experiment Station
Route 7, Box 999
Beaumont, Texas   77713

ABSTRACT

Rice is grown in the United States in varied locations where local environmental conditions impact cultural practices and insect diversity and abundance. Each location harbors a unique complex of insect pests which attack rice from seeding to harvest. However, the key insect pests are the rice water weevil (*Lissorhoptrus oryzophilus*), distributed throughout the rice growing areas of the United States, and the rice stink bug (*Oebalus pugax*), found only in the southern rice producing states. Insecticides are the main tactics of control but research is underway to develop alternative methods. Economic thresholds have been developed for these and other pests but research continues to refine the damage/insect density relationships in light of new high yielding varieties, associated cultural practices, and complex government subsidy programs. Entomologists must cooperate with researchers in other disciplines to solve these complicated problems and make the most of limited research dollars.

Introduction

The rice producing states in the U.S. are Arkansas, California, Florida, Louisiana, Mississippi, Missouri, and Texas with a combined area of production in 1988 of 1,171,590 ha [1]. In all states, rice farming is a highly technological industry which produces some of the highest yields in the world. For instance, in 1988 the average U.S. rice yield was 6177 kg/ha [1]. Only 17 man-hours are required to produce one hectare of rice in the U.S. while more than 740 man-hours are needed in Asia and Africa [2]. In the U.S. land is tilled with large tractors equipped with plows, discs, and harrows; divided into basins surrounded by levees to impound water and facilitate irrigation; and planted by air or large tractor drawn seeders. Rice is fertilized and treated by air with pesticides and harvested with self propelled combines. Active research and extension programs exist in all the rice producing states. These programs are funded by producer, industry, state, and federal contributions. Research is aimed at increasing yields, improving quality, decreasing production inputs, minimizing environmental problems associated with rice production, and improving marketing of the commodity.

Growing conditions are different for each state (and often within a state) which results in diverse production practices. Thus, rice in each state is attacked by a unique complex of aquatic, semi-aquatic, and terrestrial insect pests. For instance, rice seed midges (*Chironomus* spp., *Tanytarsus* spp., and *Cricotopus* spp.) and the rice leaf miner (*Hydrellia griseola*) are pests of rice under a continuous or pinpoint flood which is practiced in California and parts of Texas and Louisiana [3, 4, 5, 6, 7]. In a continuous flood, rice is water seeded and the flood is maintained until rice nears maturity [8]. Rice is also water seeded in a pinpoint flood but within a day or two after seeding, the flood is removed for three to five days to encourage root growth which helps prevent seedling drift [8]. After the drain period, another flood is applied and maintained until rice nears maturity. The fall armyworm (*Spodoptera frugiperda*) can defoliate young rice and chinch bugs (*Blissus leucopterus leucopterus*) can attack water-stressed rice under a prolonged drainage which is practiced extensively in the southern rice producing states [9]. Here rice is dry seeded by air or ground. If sufficient soil moisture is present, seeds germinate and rice seedlings emerge through the soil; however, when soil is too dry at planting, a temporary flood is applied and drained which is called a flush irrigation. Occasionally, rainfall occurs at the proper time and substitutes for a flush. Additional flushes are applied as needed until rice is actively tillering, approximately four weeks after seeding. At this time a flood is applied and maintained until rice nears maturity. These irrigation methods have a profound influence on population dynamics of pest insects, plant response to damage, and type and timing of control tactics.

Regardless of irrigation method, the rice water weevil (*Lissorhoptrus oryzophilus*) is a key pest of all rice producing states [10]. The adult emerges from overwintering sites (perennial grasses) in early spring and invades flooded rice where eggs are laid in rice stems underwater. Eggs hatch and larvae move to rice roots where feeding can cause an increase in weed competition, delay in maturity and reduction in yield.

Larvae pupate in mud cells attached to roots. Upon completion of the pupal stage adults emerge and seek an overwintering site or reinfest rice. Generally, only a single generation occurs in California where all weevils are females and reproduce parthenogenetically [11]. In all other rice producing states, two generations and both sexes occur. Rice water weevil damage appears to be more severe when rice is exposed to a continuous or pinpoint rather than a prolonged drainage flood which is probably attributable to the presence of earlier infestations when rice is smaller and more vulnerable to attack.

The aster leafhopper (*Macrosteles fascifrons*) is a minor pest of rice in California. Infestations tend to increase on broadleaf weeds and sedges and have reduced rice yields in small experimental plots [12]. Generally, producers can avoid aster leafhopper problems by controlling aquatic weeds.

Another key insect pest, which occurs only in the southern rice producing states, is the rice stink bug (*Oebalus pugnax*) [10]. Adults overwinter in ground litter and infest graminaceous weed hosts for one or two generations during spring and early summer. When rice begins to head, adults move to rice. The insect inserts its mouth parts into the rice kernel and extracts the contents [13, 14]. Microorganisms such as *Curvularia lunata, Bipolaris oryzae, Cercospora oryzae, Trichonis caudata, Fusarium oxysporum, Alternaria* spp. and *Nematospora coryli* are associated with this feeding activity which can result in a discolored area on the grain causing a quality problem called peck [13, 15, 16]. If severe enough, peck can lower the grade of rice and reduce the percentage of whole grains since pecky rice breaks more easily during milling than undamaged rice. Price discounts are levied against pecky and broken rice. Rice stink bug populations and consequent damage to rice are affected by proximity and stage of growth of nearby rice, sorghum, and pasture.

The rice stalk borer (*Chilo plejadellus*) and the sugarcane borer (*Diatraea saccharalis*) occur in the south where populations rarely require treatment [17]. The larvae bore into the stem and weaken it which can result in lodging. More severe damage causes partial or total blanking of the panicle. Generally, rice on levees and in areas where the stand is thin is more heavily infested. Another late season pest, the armyworm (*Pseudaletia unipuncta*), occasionally defoliates foliage and damages panicles in California [18]. Defoliation must exceed 25% before yield is affected and infestations are easily controlled with carbaryl.

The remainder of the paper will deal with the key insect pests of rice in the U.S.--the rice water weevil and the rice stink bug. Emphasis will be placed on describing and analyzing the current methods of chemical control in relation to integrated pest management and the future of rice insect pest control in the U.S.

Rice Water Weevil

With the exception of Florida, the recommended method of control of the rice water weevil is application of granular carbofuran. This is the only insecticide registered by the Environmental Protection Agency for control of the weevil in the U.S. Currently, granular carbofuran is under special review and may be banned by this agency due to the insecticide's high, acute toxicity to avian species which are abundant in and around the rice agroecosystem. To date, rice water weevil resistance to carbofuran has not been documented. However, resistance to aldrin was reported as early as 1965 in Arkansas [19]. Since only female rice water weevils occur in California, resistance, if it develops, would be expected to occur sooner in the southern rice producing states which harbor more highly, genetically variable populations composed of males and females. On the other hand, the low number of weevil generations produced annually (one in California and Arkansas and two in the remaining rice producing states), the practice of applying one treatment of carbofuran per year, the relatively small amount of total rice acreage treated with carbofuran, and the wide host range of the insect should discourage development of resistance [20, 21]. Other methods of control have been investigated with little success. For instance, draining basins kills weevils but soil must crack from drying before mortality occurs [22]. In addition, basins must be reflooded which means more labor and higher water costs. Draining also increases loss of nitrogen and weed-growth so additional fertilizer and herbicide must be applied. Consequently, draining is not cost effective except in Florida where rice is grown on soils extremely high in organic matter. Producers routinely drain these soils which releases nitrogen to the rice plants (D.B. Jones, 1989, personal communication). Thus, the primary benefit of draining in Florida is to increase nitrogen availability and a fortuitous benefit is control of the rice water weevil.

California, Louisiana and Texas have participated in host plant resistance research but progress has been slow. California released a genotype as breeder material with WC1403 as the source of tolerance (A.A. Grigarick, 1990, personal communication). However, levels of tolerance are not expected to preclude insecticide use. Emphasis in Louisiana is on antixenosis; however, again, levels of resistance are not expected to be high enough to replace carbofuran. Texas routinely screens genotypes targeted for release for susceptibility to the weevil but is not involved in developing a resistant variety.

Some research has been conducted on natural biological control; however, this tactic also appears less than promising. The fungus *Beauveria* sp. has been isolated from overwintering rice water weevils but rates of infection are quite low (A.A. Grigarick, unpublished data). A mermithid nematode was found parasitizing adults in Arkansas but again levels of parasitism were not high enough to influence population dynamics of the weevil [23]. However, more effort should be aimed at describing and analyzing the flooded rhizosphere where natural mortality of the immature stages occurs so that control tactics can be aimed at manipulating the properties of this environment which affect rice water weevil survival.

Economic thresholds exist for producer use in all states. These guidelines are based on adult feeding scar and/or immature densities. In all cases, granular carbofuran is the recommended treatment. In California, the weevil concentrates its activity near the levees and margins of basins, thus, farmers are advised to apply 5G Furadan only to those areas [24]. Generally, treatments are prophylactic, based on past field history, and applied preplant to dry soil. Producers have the option of sampling for adult feeding scars after plants emerge through the water. If scar density exceeds the threshold, basins can be drained and carbofuran applied followed by reflooding. In the remaining states, weevil activity is more evenly distributed throughout the basins, thus, entire basins must be treated with 5G or 3G Furadan between seven and 14 days after onset of the permanent flood [9].

Rice Stink Bug

Little research has been conducted on developing non-chemical controls for the rice stink bug. Reasons for this lack of effort include a short period of host plant vulnerability (from heading to harvest ca. 30 days for most varieties), the insect is highly mobile, economic threshold densities are very low, and chemical controls are relatively inexpensive. Recently, entomologists and agricultural economists in Texas revised the economic thresholds for the rice stink bug [25]. Biological data were collected for three years utilizing large field plots, natural infestations of the rice stink bug, insecticide exclusion techniques applied at various stages of grain maturation (heading, milk and dough), and current high yielding, semidwarf varieties. Data were subjected to dynamic programming analysis to generate the revised, dynamic thresholds which include projected yield and price, planting date, cost of insecticide application and stage of grain maturation. Sampling is based on the number of adult rice stink bugs collected in 10 consecutive sweeps of an insect sweep net. Additional research in Texas has emphasized the evaluation of the residual activity of labeled insecticides for rice stink bug control [26]. Methyl parathion is used by most producers due to its low cost but has little residual activity whereas encapsulated methyl parathion and carbaryl have longer activity but are more expensive. The choice of insecticide is crucial to producers, particularly if adult rice stink bugs reinfest treated rice. Drees and Plapp reported no evidence of resistance to carbaryl or methyl parathion in two geographically separated populations of rice stink bugs in Texas [27].

Future of Integrated Pest Management in Rice in the U.S.

U.S. rice production is growing increasingly complex. Intricate government subsidy programs, the advent of high yielding varieties which require more inputs and high levels of management, rising production costs, low profit margins, and increased environmental concerns profoundly influence integrated insect pest management. Alternative methods of pest control, including chemical, must be developed. Presently, U.S. rice entomologists are evaluating novel insecticides (such as entomogenous nematodes and insect growth regulators) possessing minimal environmental effects in response to the possible ban of carbofuran. Scientists with training in tissue culture and genetic engineering are beginning to work with conventional plant breeders to develop new rice varieties with special qualities. These efforts are and will markedly reduce the time required to develop a cultivar for commercial use. A new variety will have a unique set of optimal production practices and phenological and morphological properties which will impact insect pest biology and damage. Thus, insect pest management guidelines, ideally based on economic thresholds, should be tailor made for each new variety targeted for release. To keep pace with these demands, rice integrated pest management research must receive more funding and support; however, at the present time, competition for research dollars is extremely keen. Thus, the prognosis for rice integrated pest management in the U.S. is somewhat unclear. A definite challenge exists and much work needs to be done but adequate support may not be available to accomplish all goals. A possible partial solution is to better coordinate research activities among states to minimize duplication and establish research priorities. Currently, U.S. rice entomologists are working towards that goal under the auspices of a Cooperative State Research Service Regional Project. Also, entomologists should cooperate in more interdisciplinary efforts to utilize existing support from other disciplines.

With increasing complexity in rice farming, decision making becomes more critical and difficult. Although production guidelines, insecticide control recommendations and economic thresholds for key pests exist, producers often do not have the time or expertise to properly sample for insect pest populations. Extension scientists will play an increasingly important role in educating producers in integrated pest management. Private consultants, hired by producers to monitor insect populations, will also take on more responsibility for implementing research and extension recommendations.

REFERENCES

1. Texas Agricultural Statistics Service. 1989. Texas Agricultural Facts. Jan 25.

2. Rutger, J.N. and D.M. Brandon. 1981. California rice culture. Scientific American. Feb. 42-51.

3. Way, M.O. and R.G. Wallace. 1989. First record of midge damage to rice in Texas. Southwestern Entomologist. 14(1):27-33.

4. Clement, S.L., A.A. Grigarick, and M.O. Way. 1977. Conditions associated with rice plant injury by Chironomid midges in California. Environ. Entomol. 6:91-96.

5. Smith, C.M., J.L. Bryant, S.D. Linscombe, and J.F. Robinson. 1986. Insect pests of rice in Louisiana. Louis. Agric. Exp. Sta. Bull. 774.

6. Grigarick, A.A. 1959. Bionomics of the rice leaf miner, *Hydrellia griseola* (Fallen) in California. Hilgardia. 29:1-80.

7. Way, M.O., F.T. Turner, and J.K. Clark. 1983. The rice leaf miner, *Hydrellia griseola* (Fallen), a potential pest of rice in Texas. Southwestern Entomologist. 8(3):186-189.

8. Louisiana State University Agricultural Center. 1987. Rice Production Handbook. Publication 2321.

9. Texas Agricultural Extension Service. 1989. Rice Production Guidelines. D-1263.

10. Bowling, C.C. 1980. Insect pests of the rice plant. In Rice: Production and Utilization. Ed. B.S. Luh. AVI Publishing Company Inc. Westport, Connecticut.

11. Grigarick, A.A. and G.W. Beards. 1965. Ovipositional habits of the rice water weevil in California as related to a greenhouse evaluation of seed treatments. J. Econ. Entomol. 58:1053-1056.

12. Way, M.O., A.A. Grigarick, and S.E. Mahr. 1984. The aster leafhopper (Homoptera:Cicadellidae) in California rice:herbicide treatment affects population density and induced infestations reduce grain yield. J. Econ. Entomol. 77:936-942.

13. Douglas, W.A. and E.C. Tullis. 1950. Insects and fungi as causes of pecky rice. U.S. Dept. Agric. Tech. Bull. 1015.

14. Swanson, M.C. and L.D. Newsome. 1962. Effect of infestation by the rice stink bug, *Oebalus pugnax*, on yield and quality in rice. J. Econ. Entomol. 55:877-879.

15. Hollay, M.E., C.M. Smith, and J.F. Robinson. 1987. Structure and formation of feeding sheaths of rice stink bug (Heteroptera:Pentatomidae) on rice grains and their association with fungi. Ann. Entomol. Soc. Am. 80:212-216.

16. Marchetti, M.A. 1984. The role of *Bipolaris oryzae* in floral abortion and kernel discoloration in rice. Plant Disease. 68(4):288-291.

17. Browning, H.W., M.O. Way, and B.M. Drees. 1989. Managing the Mexican rice borer in Texas. Tex. Agric. Ext. Ser./Exp. Sta. B-1620.

18. Rice, S.E., A.A. Grigarick, and M.O. Way. 1982. Effect of leaf and panicle feeding by armyworm (Lepidoptera:Noctuidae) larvae on rice grain yield. J. Econ. Entomol. 75(4):593-595.

19. Rolston, L.H., R. Mayes and Y.H. Bang. 1965. Aldrin resistance in the rice water weevil. Arkansas Farm Research. Nov.-Dec.

20. Grigarick, A.A. 1984. General problems with rice invertebrate pests and their control in the United States. Protection Ecology. 7:105-114.

21. Lauck, J.E. 1972. Host relationships and impact of herbicide treatments on population fluctuations of the rice water weevil. Proc. Rice Tech. Working Group. 14:51.

22. Cooperative Extension Service, University of Arkansas. 1987. Control of Insects Attacking Rice. EL 330.

23. Bunyarat, M., N.P. Tugwell, and R.D. Riggs. 1977. Seasonal incidence and effect of a mermithid nematode parasite on the mortality and egg production of the rice water weevil, *Lissorhoptrus oryzophilus*. Environ. Entomol. 6:712-714.

24. University of California Statewide Integrated Pest Management Project. 1983. Integrated Pest Management for Rice. Publication 3280.

25. Harper, J.K. 1988. Developing economic thresholds for rice stink bug management in Texas using dynamic programming. Ph.D. diss. Texas A&M University, College Station.

26. Way, M.O. and R.G. Wallace. 1990. Residual activity of selected insecticides for control of rice stink bug (Hemiptera:Pentatomidae). J. Econ. Entomol. (in press).

27. Drees, B.M. and F.W. Plapp. 1986. Toxicity of carbaryl and methyl parathion to populations of rice stink bugs, *Oebalus pugnax* (Fabricius). Tex. Agric. Exp. Sta. PR-4415.

Synthesis and properties of the Silaneophane HOE 084498 - a promising new insecticide

H.H.SCHUBERT, G.SALBECK, W.KNAUF, R.SCHAUB, A.WALTERSDORFER, K.H.LEIST,
U.SCHOLLMEIER, R.FISCHER and G.GÖRLITZ
Hoechst AG, R & D Agrochemicals Department
P.O. Box 800320, D-6230 Frankfurt am Main 80
Federal Republic of Germany

ABSTRACT

The paper presents a summary of different synthetic approaches to the Silaneophane Hoe 084498, which is a new Silicon based insecticide with promising efficiency in the protection of important crops, especially rice. In addition, some biological, ecological and toxicological data are given.

INTRODUCTION

Since the discovery of the ethofenprox type insecticides by Mitsui Toatsu Chem. Inc. at the beginning of the eighties (1,2), considerable efforts have been made to further improve the insecticidal and ecological properties of such products by chemical modification. Some of this work led to the discovery of a new class of insecticidal compounds, which all contain the so called Silaneophylradical as a characteristic structural moiety. We therefore named these chemicals "Silaneophanes".

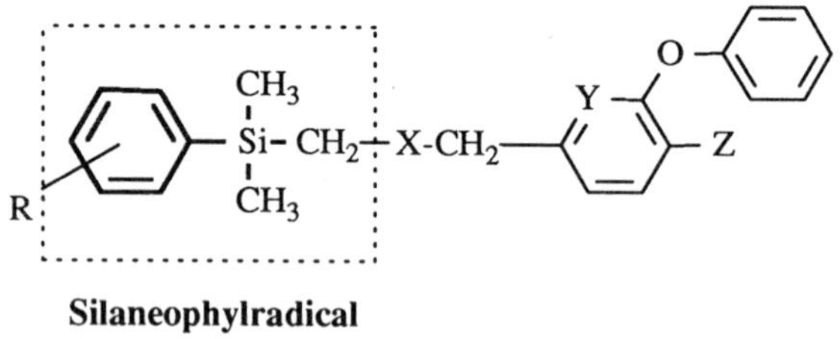

Silaneophylradical

Figure 1. Insecticidal Silaneophane Derivatives

Silaneophanes, which have been discovered at roughly the same time by research groups at seven different companies (3-12), are the first known insecticides with a silicon atom in an essential position of their chemical structure.

In the case of Hoe 084498, the Silaneophyl moiety is connected with the rest of the molecule by a methylene group, thus giving rise to a substituted Aryl-dimethyl-propyl-silane. Table 1 summarizes some general data of Hoe 084498, which is a colourless, odourless and distillable liquid.

TABLE 1

General data of the Silaneophane Hoe 084498

Chemical structure:

Chemical name:	(4-Ethoxyphenyl)-(dimethyl)-<3-(4-fluoro-3-phenoxyphenyl)-propyl>-silane
Empirical formula:	$C_{25}H_{29}FO_2Si$
Molecular Weight:	408.59
Physical state:	liquid
Colour:	colourless
Odour:	odourless
Boiling point:	$200°C/0.02$ mbar
Vapour pressure:	5.5×10^{-8} mbar
Density:	1.066 g/cm^3
Solubility:	soluble in most organic solvents water: 0.001 mg/l

In the following paragraphs the chemical, biological, ecological and toxicological aspects of Hoe 084498 will be described.

SYNTHESIS

In general, the synthesis of Hoe 084498 is effected by a combination of metalorganic procedures, Ullmann biarylether condensations and Hydrosilylation reactions, using commercially available aromatics, aliphatics and silicon precursors. All different approaches to the target molecule can be classified in two groups:

- Figure 1 shows three ultimate synthetic steps, which all result in the introduction of the diphenylether moiety of Hoe 084498.

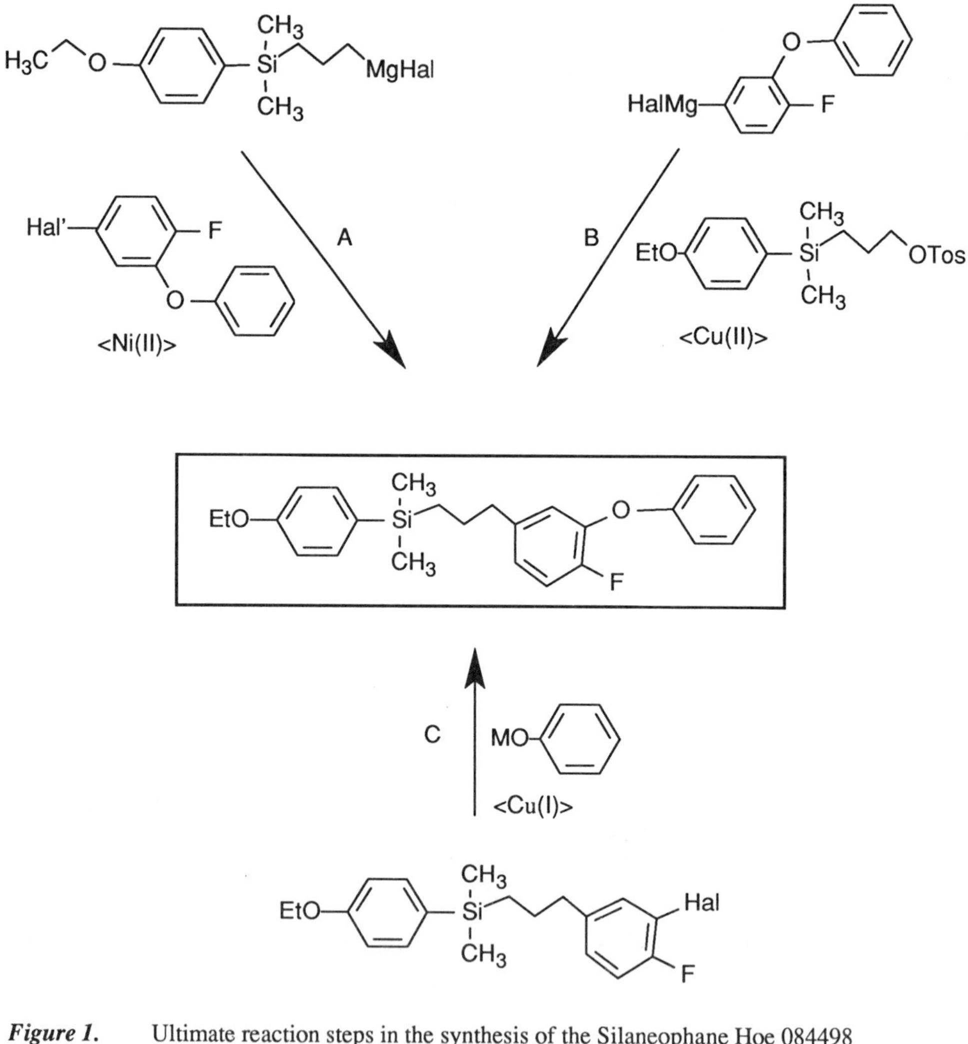

Figure 1. Ultimate reaction steps in the synthesis of the Silaneophane Hoe 084498
- Introduction of the diphenylether moiet -

Figure A is based on the Nickel catalyzed coupling of alkylgrignard reagents with halogenåted aromatics (13). The second possibility (Method B) is a copper catalyzed reaction of the corresponding metallated diphenylether with the appropriate alkyl tosylate (12). Approach C uses the wellknown Ullmann diarylether synthesis, which also is catalyzed by copper ions (14). So far, all these methods gave only moderate yields of the desired Silaneophane.

- Due to its chemical structure as a silane, ultimate reaction steps which result in the formation of one or more of the carbon/silicon bonds of Hoe 084498 are preferred from the synthetic point of view. Figure 2 gives a selection of reasonable processes.

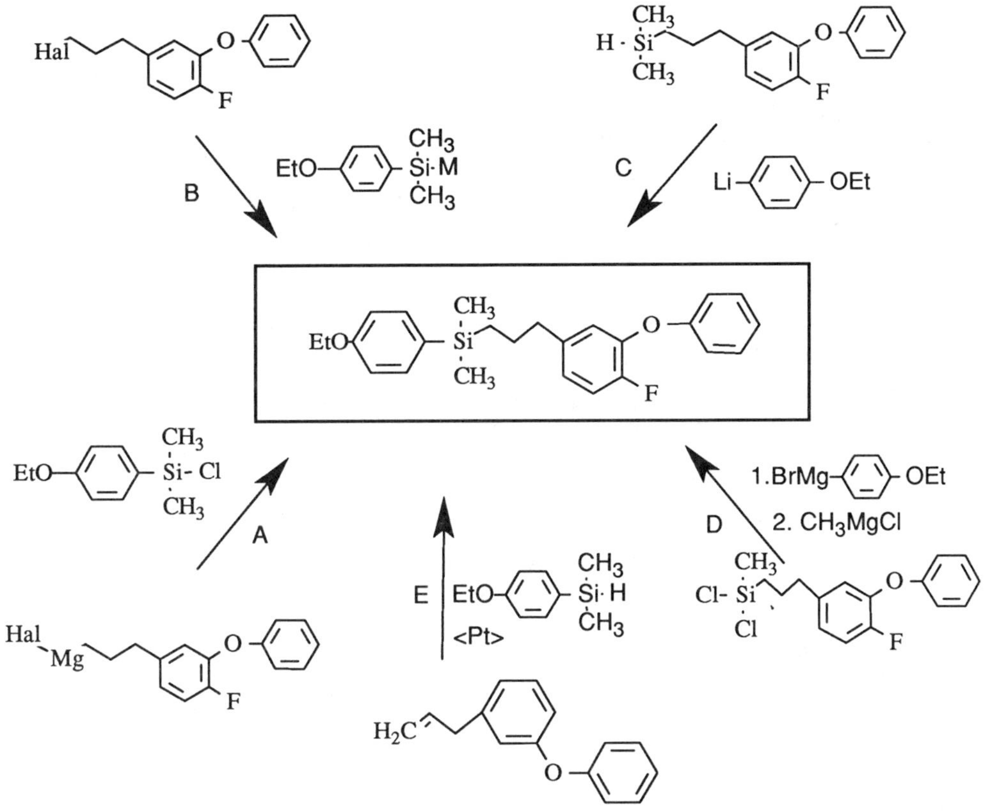

Figure 2.   Ultimate reaction steps in the synthesis of the Silaneophane Hoe 084498
      - Formation of a Carbon/Silicon bond -

Routes A (7,9,10), C (12) and D(15) proceed via the substitution of nucleofugic leaving groups at the positively polarized silicon atom. In sharp contrast to these variants, route B (4, 11) uses a negatively polarized silicon synthon as the key intermediate. Last not least, route E (7) yields the target molecule Hoe 084498 as result of a platinum catalyzed addition reaction between (4-Ethoxyphenyl)-dimethylsilan and the appropriate olefin.

Figure 3 summarizes some of the synthetic pathways to (4-Ethoxyphenyl) silanes, which are prepared starting from dichloro-dimethyl-silane, diethoxy-dimethyl-silane, chloro-dimethyl-silane or dichloro-methyl-silane.

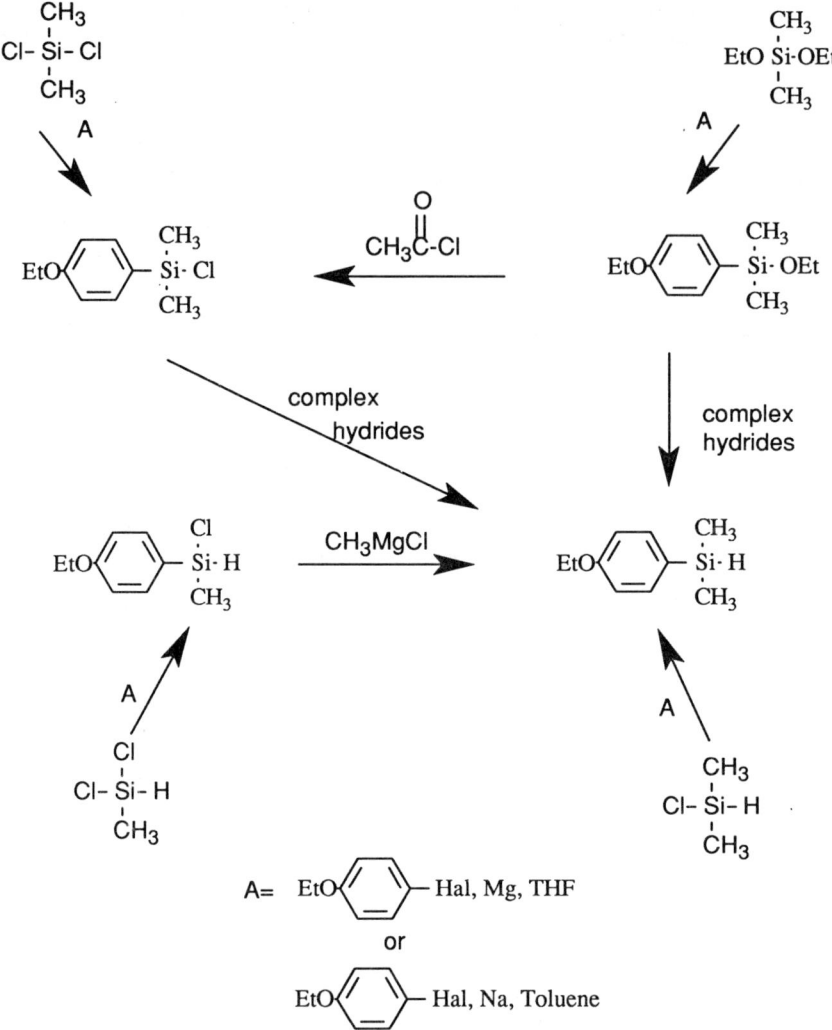

Figure 3. (4-Ethoxyphenyl)-silanes for the synthesis of the Silaneophane Hoe 084498

All these educts come directly or indirectly from the technical Rochow process and are commercially available. Coupling with p-metallated phenetoles produces the corresponding (4-Ethoxyphenyl)-silanes, which can be converted into each other by standard procedures of reduction, halogenation or alkylation.

Figure 4 gives an impression of the synthetic possibilities to produce 2-Fluoro-5-(3-halopropyl)-diphenylethers 5.

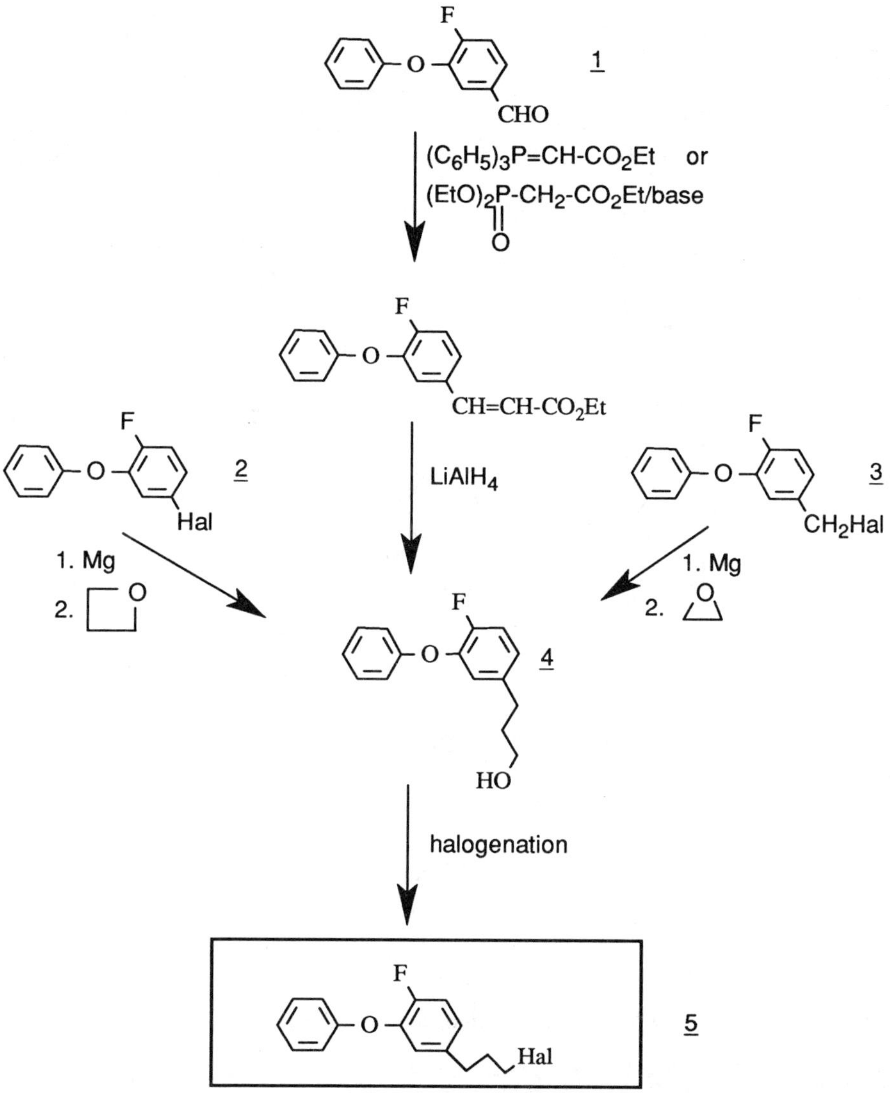

Figure 4. The preparation of 2-Fluoro-5-(3-halopropyl)-diphenylether (5)

Starting from the later described Fluoro-halo-diphenylether 2 (11), the literature known benzaldehyde 1 (7,10) or the corresponding benzylchloride 3 (12), 2-Fluoro-5-(3-hydroxypropyl)-diphenylether (4) is prepared as the common intermediate, which is then transformed to the desired halopropyl derivative 5 by standard techniques.

Another valuable intermediate in the preparation of the Silaneophane Hoe 084498 is 2-Fluoro-5-(propen-3-yl)-diphenylether (7), the synthesis of which is described in figure 5.

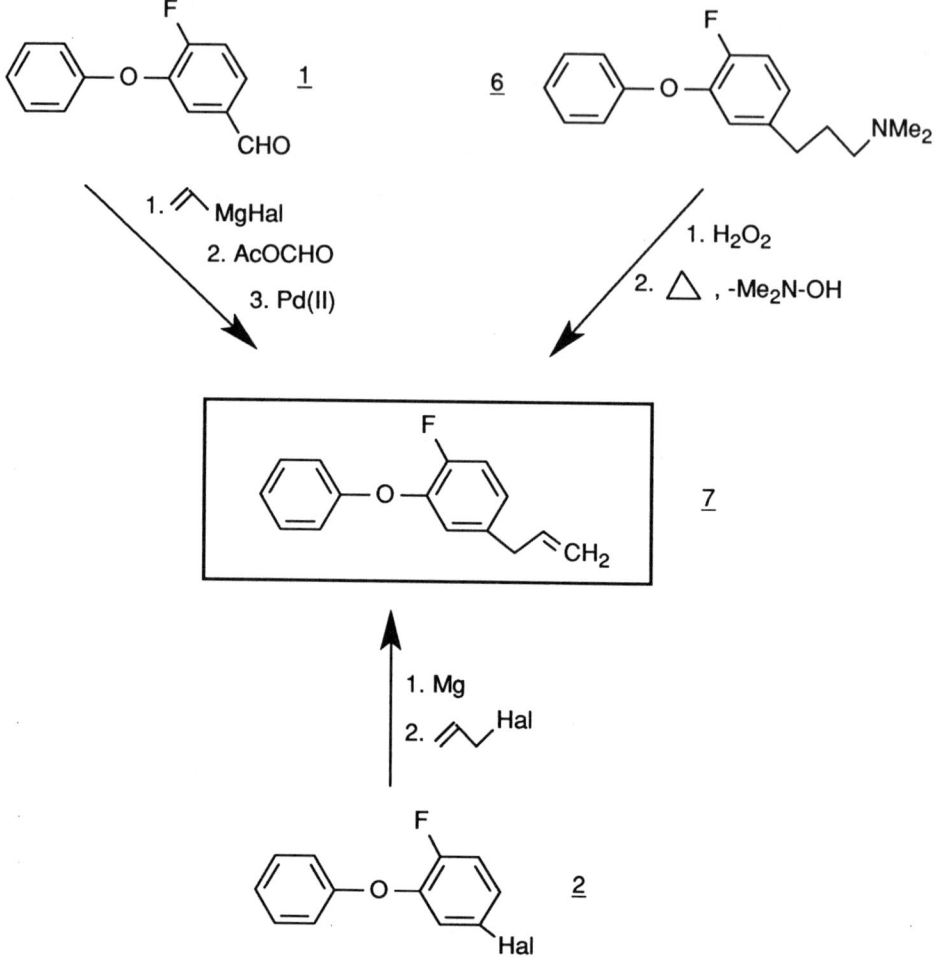

Figure 5.   The preparation of 2-Fluoro-5-(propen-3-yl)-diphenylether (7)

The already mentioned benzaldehyde 1 yields the target olefin 7 as the result of a three step process (10). Alternatively, 7 can be prepared by metallation of the Halodiphenylether 2, followed by allylation of the intermediate grignard reagent (7). Last not least, a Cope elimination reaction of the oxydized dimethylamin 6 also gives the propenylbenzene 7 (16).

The synthesis of the precursor 6 is shown in figure 6.

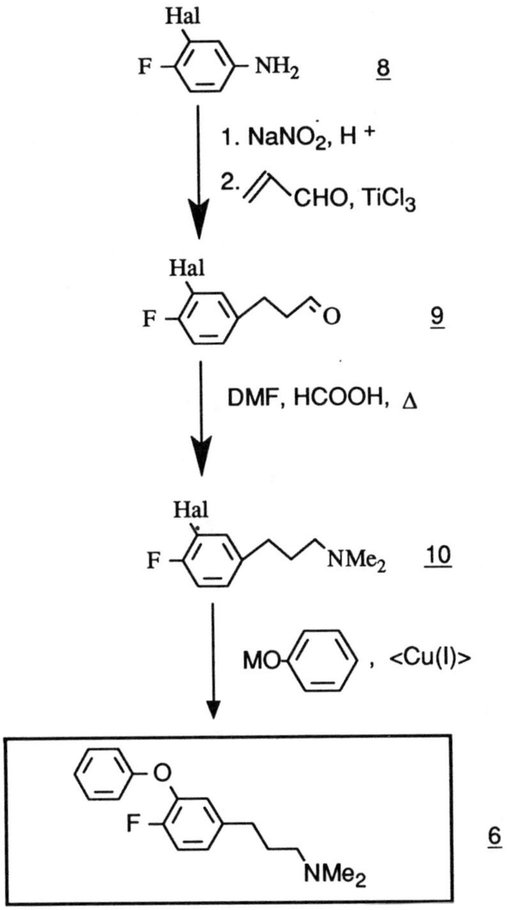

Figure 6. The preparation of 5-(3-N,N-Dimethylamino-propyl)-2-fluoro-diphenylether

Key step of this reaction sequence is a reductive arylation of acrolein, which is conducted by using the system aryldiazonium salt/TiCl₃ (16). Reductive amination of the resulting aldehyde, followed by an Ullmann diarylether synthesis finally gives the desired product 6.

One of the mostly used reagents in the chemistry of Hoe 084498 is 2-Fluoro-5-halodiphenylether (2). We therefore intensively studied the aspects of its synthesis. Some of the results are summarized in figure 7.

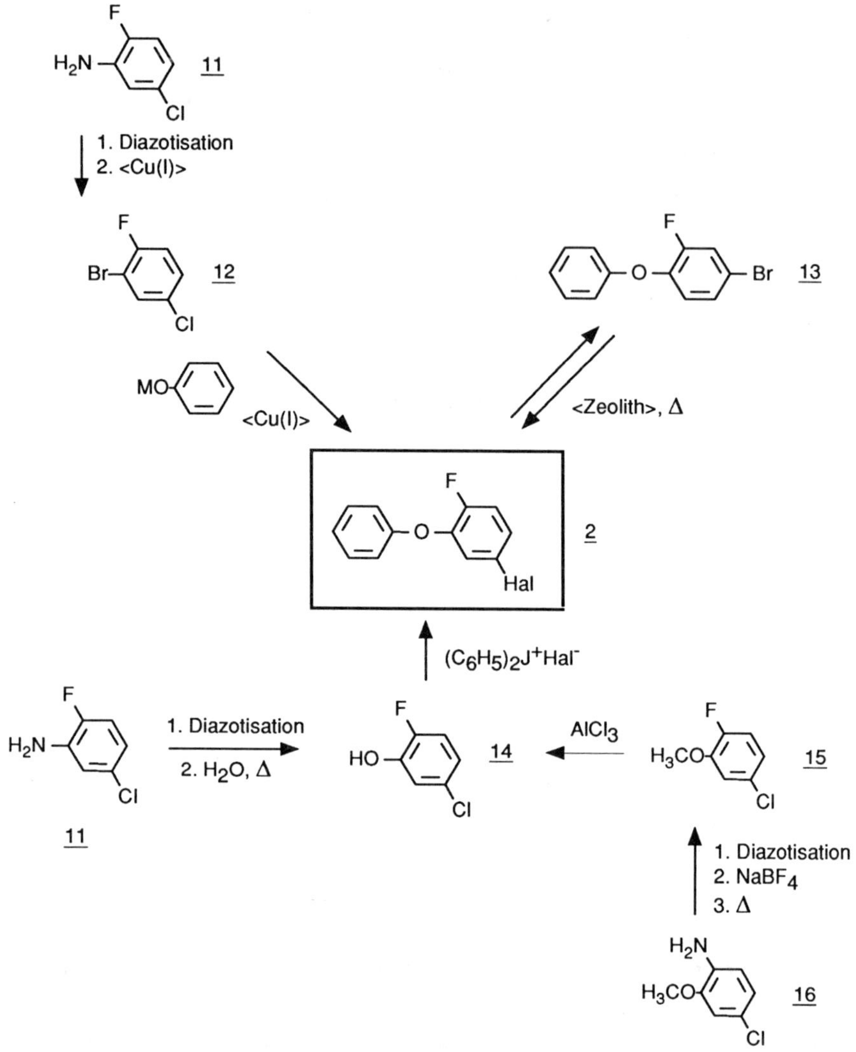

Figure 7. The preparation of 2-Fluoro-5-halodiphenylether (2)

The literature known aniline 11 can be converted into the synthon 2 in two different ways. Bromination of 11 yields the trihalobenzene 12, which is then reacted with phenolate in the presence of copper ions to give the target molecule 2. On the other hand, 11 can be transformed into the phenol 14, which is then arylated with for example a diphenyl iodonium halide. Intermediate 14 is also accessible from the aniline 16 via anisole 15. Another - rather exotic - access for 2-Fluoro-5-halodiphenylether 2 is the zeolith catalyzed rearrangement of 4-Bromo-2-fluoro-diphenylether.

INSECTICIDAL PROPERTIES

Silaneophane Hoe 084498 was tested around the world against main pests of numerous crops. In all cases the compound showed good to excellent performance. As listed in table 2, main indications of Hoe 084498 are pests in rice, topfruit, vine, several field crops (maize, tobacco and e.g. cotton) and vegetables as well, where the recommended dose to have adequate pest control is within the range of 50 - 200 g a.i./ha. Because of its safety to men and environment Silaneophane Hoe 084498 should also be useful as pest controling agent in other crops where intensive field testing is under way. Further results will be published elsewhere.

Table 2

Silaneophane Hoe 084498 - Insecticidal properties -

Main Indications for Hoe 084498

Indication	Pest	Recommendation g a.i./ha
Rice	Cicadina Coleoptera Heteroptera Lepidoptera	100 - 200
Top fruit Vine	Lepidoptera Cicadina	100 - 200 100
Field Crops (maize, tobacco, cotton etc.)	Lepidoptera Coleoptera Cicadina	100 - 200 100 - 200 50 - 100
Vegetables	Lepidoptera Coleoptera	50 - 200 100 - 200

TOXICOLOGY

As can be seen from table 3, the technical material Hoe 084498 exhibited no toxic properties in various species following acute treatment using different ways of application.

After inhalation for 4 hours to the highest technically applicable concentration up to and including 6.61 mg/l, Hoe 084498 was tolerated without any signs of intoxication.

Table 3

Silaneophane Hoe 084498 - Toxicological data -

Acive ingredient:		
Acute oral LD_{50}	Rat (male/female)	> 5000 mg/kg
Acute dermal LD_{50}	Rat (male/female)	> 5000 mg/kg
Acute oral LD_{50}	Mouse (male/female)	> 5000 mg/kg
Acute skin	Rabbit	non irritant
Acute inhalation		no effects in highest doses tested
Ames test		negativ
Bird toxicity:		
Acute oral LD_{50}	Quail	> 2000 mg/kg
Acute oral LD_{50}	Mallard duck	> 2000 mg/kg
Fish toxicity:		
Acute LC_{50} (96 hours)	Carp	>1000 mg/l
Acute LC_{50} (96 hours)	Zebrafish	> 50 mg/l
Daphnia magna:		
Acute LC_{50} (3 hours)		> 10 mg/l

In addition, Hoe 084498 proved to be non irritant to the skin and also did not reveal a mutagenic potential in the Ames Test. It has no toxic properties to birds (Mallard duck, Quail) after acute oral administration including a dose of 2000 mg/kg. Hoe 084498 is safe to fish. Its LC_{50} value for water flea was evaluated according to Japanese guide lines.

ABIOTIC DEGRADATION

This section deals with a theme that might have some influence on the ecological performance of the Silaneophane. In contrast to its carbon analogue MTI 800, Hoe 084498 possesses a labile Aryl/Silicon bond, that makes it more likely to be hydrolyzed to non active degradation products (Figure 8).

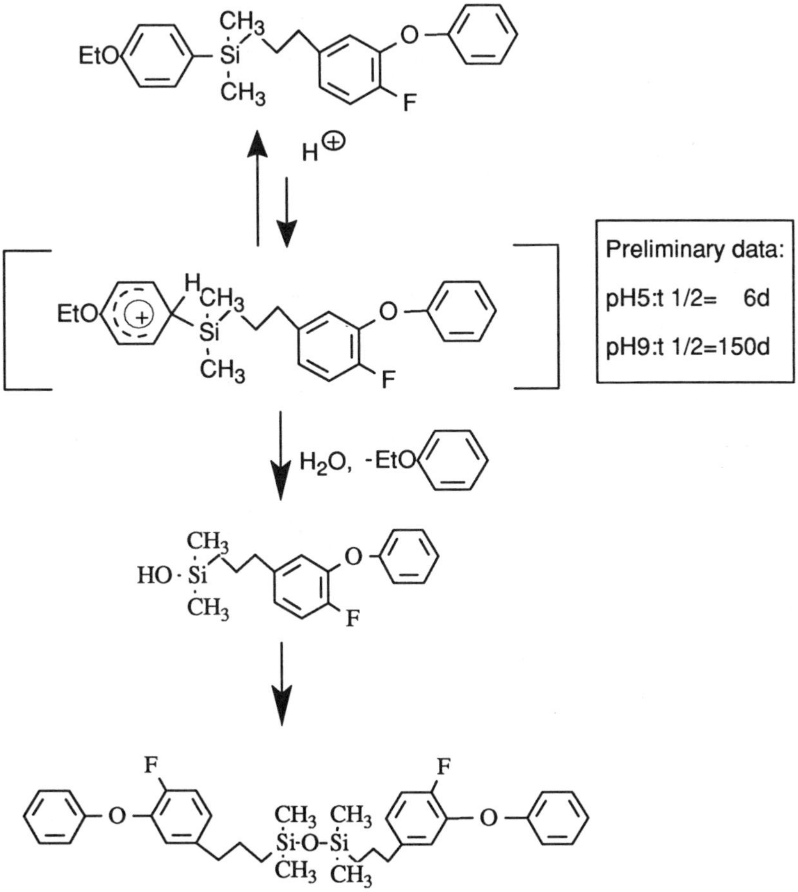

Figure 8.   Silaneophane Hoe 084498 - Abiotic hydrolysis -

As a matter of fact, the compound shows a significant pH dependence of its hydrolysis rate under abiotic conditions. This phenomenon is compatible with the literature known mechanism of such a process, which was published by Eaborn and coworkers in the fifties (17 - 19). The resulting disiloxane can be detected by analytical means.

REFERENCES

1. K. Nakatani et al (Mitsui Toatsu Chem., Inc.), JP 55 057 872 (2.5.1980)

2. K. Nakatani et al (Mitsui Toatsu Chem., Inc.), JP 57 082 473 (18.05.1982)

3. Y. Yamada et al (Sumitomo Chem. Co., Ltd.), JP 60 123 491 (08.12.1983)

4. Y. Katsuda (Dainihon Jochugiku Co., Ltd.), JP 61 087 687 (05.10.1984)

5. Y. Yamada et al (Sumitomo Chem. Co., Ltd.), JP 61 229 883 (04.04.1985)

6. Y. Hayase et al (Shionogi & Co., Ltd.), JP 62 105 465 (16.05.1985)

7. H.H. Schubert et al (Hoechst AG), EP 0 224 024 (26.10.1985)

8. A. Kitajima et al (Mitsui Toatsu Chem., Inc.), JP 62 108 885 (08.11.1985)

9. H. Franke et al (Schering AG), DP 3604781 (13.02.1986)

10. S.M. Sieburth (FMC Corp.), USP 4709068 (02.06.1986)

11. Y. Katsuda et al (Dainihon Jochugiku Co., Ltd.), WO 88/01271 (19.08.1986)

12. Y. Yamada et al, J. Pesticide Sci., 1987, 12, 683

13. K. Tamao et al, J. Amer. Chem.Soc.1972, 4, 4374;
    R.J.P. Corrin et al, J. Chem. Soc., Chem. Commun., 1972, 144

14. H.H. Schubert (Hoechst AG), EP 0308788 (19.09.1987)

15. H.H. Schubert (Hoechst AG), patent application

16. H.H. Schubert (Hoechst AG), EP 0308808 (19.09.1987)

17. C. Eaborn, J. Chem. Soc. 1953, 3149

18. C. Eaborn, J. Chem. Soc. 1956, 4858

19. F. B. Deans und C. Eaborn, J. Chem. Soc. 1959, 2299

BIOLOGICAL PROPERTIES OF THE SILANEOPHANE HOE 084498 - A NEW INSECTICIDE FOR THE CONTROL OF RICE PESTS

W. Knauf, M. Kern R. Schaub, A. Waltersdorfer
Biologische Forschung, Hoechst AG; Frankfurt; BRD

SUMMARY

Summarizing the data, Silaneophane Hoe 084498 shows its good activity against the most important rice pests from different entomological groups at low field rates. Together with its low fish toxicity and safety to vertebrates, Silaneophane combines the most important properties required for new rice insecticides in a very favourable way. It is thus also especially suitable for rice cropping under restricted conditions as possible environmental side effects are concerned. Silaneophane will also open up new strategies in pest control in crops other than rice and may be considered as an alternative to conventional insecticides.

INTRODUCTION

Rice as a major crop, especially in the Far and the Near East, Africa and Central and South America, is liable to infestation by a number of pests which have to be controlled since pest management and modern agriculture permit the grower to optimize yields.

For use in rice-cropping, a compound should offer specific properties with regard to biological activity. This activity should cover Lepidoptera and Homoptera, some Coleoptera (e.g. black rice water weevil), and also Hepteroptera. Above all, there should be no adverse environmental side effects which might lead to restrictions on applications in rice-cropping areas.

In particular, side effects on fish could appreciably restrict the use of insecticides. This was one of the reasons why e.g. modern pyrethroids have been unable to enter a substantial market segment in rice.

During the 80's, Hoechst AG decided to increase its efforts to develop modern insecticides for this specific market, with a view to the problems referred to above. Within this strategy, screening was focussed, principally on high activity against hoppers and Lepidoptera and on minimum fish toxicity.

Fortunately, it proved possible to find a compound which fitted well into modern rice-cropping philosophy.

The compound Silaneophane (Hoe 084498) belongs to a specific class of organic silicone compounds, new in the insecticidal area and has been carefully tested in recent years in the glasshouse and in the field throughout the world.

CHEMICAL STRUCTURE

Silaneophane ((4-Ethoxyphenyl)[3-(4-fluoro-3-phenoxy-phenyl)-pro-pyl]-(dimethyl) silane (IUPAC)) Hoe 084498 is described in its chemistry recently [1]. Patents IP 61087687 (Katsuda), 1986) and EP 224024 (Hoechst AG, 1987) are pending.

FORMULATION

The following formulations were tested successfully against the different pest species:
oil in water emulsion, emulsifiable concentrate, dust, wettable powder, floating granule and soil granule.

The formulations mentioned are adapted to the specific market segments where the product is to be used.

MODE OF ACTION

The mode of action of Silaneophane Hoe 084498 has been discussed else-where [2]. It could be demonstrated that this compound showed remarkable differences from other insecticides such as p.esters, carbamates and pyrethroids. Unlike those products it has no knock-down effect, which is evident also from a number of physiological phenomena such as CO_2 production or transpiration rate and activity. In addition to its contact toxicity, it also acts as a strong stomach poison and does not accumulate in the fat body of the insects tested (Blaberus craniifer). Unlike, for example pyrethroids, enzymatic studies revealed no changes in important enzymatic activity levels (e.g. creatin kinase, glutamate dehydrogenase, lactate dehydrogenase or mono-amine oxidase and others).

Despite this, it interferes with the nervous system and disturbs specifically the homoeostasis of the insect blood-brain barrier.

TOXICITY TO NON-TARGET ORGANISMS

The active substance and the product are of a low order of acute oral and dermal toxicity to higher animals [1].

The very important fact of practically no fish toxicity opens a wide potential in the water rice market with no risk of possible fish hazards.

Trials regarding side effects on beneficials are in progress.

BIOLOGICAL ACTIVITY AGAINST TARGET PESTS

In general the insecticidal properties of Silaneophane cover a fairly wide range of entomological groups:

- effective against Lepidoptera, Coleoptera, Cicadina, Heteroptera, Thysanoptera and Orthoptera

- moderately effective against Diptera

- slightly effective against Aphidina and Aleyrodina.

Despite this activity, neither acaricidal nor nematicidal effects could be found in laboratory or glasshouse tests.

LABORATORY TRIALS

- **Lepidoptera:**
  The activity of the product is remarkably high in important species from all types of crops. Table 1 shows the effect against P. litura, H. virescens and P. maculipenis. The slow manifestation of intoxication symptoms, despite an early feeding stop, clearly distinguishes the compound from other standard insecticides.

TABLE 1
Toxicity of the Silaneophane Hoe 084498 to 5 lepidopteran species
(L$_3$, sprayed on artifical diet, 5 days)
expressed as % mortality (lab trial)

ppm a.i.	250	125	63	31	16	8	4	2
P. litura	100	100	100	100	90	30	0	0
H. virescens	100	100	100	80	40	0	0	0
H. armigera	100	100	100	100	20	0	0	0
P. maculipanis	100	100	100	100	70	30	0	0
A. reticulana	100	100	100	100	100	100	100	60

- Coleoptera:
Early laboratory data showed the outstanding efficacy against
some Coleoptera, as demonstrated in table 2 with for example
E. varivestis. On the other hand some species were unaffected
when treated in the same trial. This shows that Silaneophane
is not as broad-acting as for example pyrethroids despite the
high potency especially to important rice pests.

TABLE 2
Effect of different insecticides to E. varivestis (larvae)
on Phaseolus vulgaris (5 days), expressed as % mortality

ppm a.i.	31	16	8	4	2	1	0,5	0,25
Hoe 084498	100	100	100	100	100	90	70	20
Dimecron	100	70	60	60	30	0	0	0
Lannate	100	80	80	70	50	20	0	0
Mipsin	100	10	0	0	0	0	0	0
Applaud	0	0	0	0	0	0	0	0
Bassa	100	100	80	80	35			

- Cicadinae:
Leaf and plant hoppers are among the main insect pests in
rice, which not only damage the plant by sucking but also by
transferring pathogens such as viruses.

Therefore adequate hopper control is a "must" for modern pest
management in all major rice cropping areas throughout the world.
All hopper species tested are very sensitive to the product.
The discovery of this property during screening was one of the
main reasons for forcing development at a very early stage.

Table 3 shows the activity against N. lugens in comparison with
other important rival products.

Not only larval stages but also adults can be effectively controlled in very low doses.

TABLE 3
Effect of Hoe 084498 to N. lugens
(test with dipped plants; larvae; 5 days),
expressed as % mortality

ppm a.i.	250	63	16	4
Hoe 084498	100	100	40	0
Bassa	100	70	40	0
Dimecron	100	100	30	0
Lannate	100	70	10	0
Mipsin	100	70	30	0

SPECIFIC BIOLOGICAL PROPERTIES

- **Time course of intoxication:**
  As has been stated already, the Silaneophane is not a knock-down insecticide. The intoxication process increases slowly when an individual incorporates the Silaneophane; this phenomenon is to a remarkable extent independent of the level of active ingredient available to the organism (Table 4).

TABLE 4
Time course of intoxication of Hoe 084498
tested with Prodenia litura (L$_4$, artifical diet, sprayed),
expressed as % mortality

time (h)	0.5	1	3	6	24	48	72
ppm (a.i.)							
63	0	0	60	100	100	100	100
31	0	0	40	70	80	100	100
16	0	0	0	0	20	60	90
8	0	0	0	0	0	0	20

- **Intoxication pathway:**
  To check different routes of intoxication, specific tests were carried out to differentiate the toxicity caused by contact and by ingestion.

  It can be seen that, in addition to the contact toxicity, this Silaneophane showed a remarkable effect when ingested and absorbed via the gut.

INTOXICATION IN THE GASEOUS PHASE

When tested in a closed environment, Silaneophane remains active by diffusion over a distance from the location where it was applied. Under conditions within the range of the situation tested, it was indicated that additional interaction with the molecule is possible via the tracheola system of the insects.

TRANSPORT IN PLANTS

The Silaneophane showed no systemic action in plant tissues either through the green parts of the plant or through the roots. This is an advantage as far as the possible residue situation in the harvested crop is concerned.

FIELD TRIALS

It is well documented that the performance of the Silaneophane in the laboratory correlates well with other species in the field [3]. Good activity could be found against the following important rice lepidoptera

- Chilo suppressalis
- Cnaphalocrocis medinalis
- Marasmia patnalis
- Parnara guttata
- Pseudaletia separata

In addition, a large number of other Lepidoptera species are effectively controlled in other crops.

In general, the different trials showed that 100 g - 200 g active ingredient per hectare will ensure successful control of rice lepidoptera.

Up to now the following Homoptera species are well controlled in all stages

- Nephotettix cincticeps
- Sogatella furcifera
- Nilaparvata lugens
- Laodelphax striatellus
- Oxya yezsensis

Good control was demonstrated in all trials over a number of years in Japan, the Philippines and other countries in Eastern Asia.

A field rate of 100 - 200 g a.i./ha is also recommended against hoppers.

With regard to Coleoptera, it turned out that Silaneophane worked well against the rice water weevil L. oryzaephilus, one of the dangerous new rice pests. This beetle can be controlled with application rates of 150 - 300 g a.i./ha.

USAGE

As demonstrated in the sections above all trials in rice up to the present indicate that the recommended doses for Lepidoptera, Homoptera and Coleoptera are all in the same range. This means that a user can apply one product in one dose to be sure of an effect on important rice pests from different entomological groups.

It is also certain that the recommended rate of Silaneophane will not cause fish toxicity and is also acceptably to the user is concerned.

REFERENCES

Schubert, H.H., Salbeck, G., Knauf, W., Schaub, R., Waltersdorfer, A., Leist, K.H., Schollmeier, U., Fischer, R. and Görlitz, G.: Synthesis and properties of the Silaneophane Hoe 084498 - a promising new insecticide

SCI Symposium: Pest Management in Rice, London, June 4-7, 1990

Kern, M., Basaller, W., Grötsch, H., Kellner, H.-M., Knauf, W. and Schacht, U.:

Insecticidal Mode of Action of the Silaneophane Hoe 084498 - A new Type of Insecticide

The 15th Japan Pesticide Science Conference, Tokyo, Mar 27-29, 1990

Stübler, H.; Takagaki, T.; Salbeck, G.; Schaub, R. and Kern, M.:

The Silaneophane Hoe 498, a novel Silico containing Insecticide II. A Suitable Insecticide for Rice Pest Control

The First Asia-Pacific Conference of Entomology (APCE), Chiangmai, Thailand, Nov. 8-13, 1989

Buprofezin; A Reliable IGR for the Control of Rice Pests

TAKAMICHI KONNO
Biological Research Center
Nihon Nohyaku Co., Ltd
4-31 Honda-cho, Kawachi-Nagano, Osaka 586, JAPAN

ABSTRACT

Buprofezin (APPLAUD®) is the first insect growth regulator (IGR) registered in the world for the control of rice pests. It shows high activity on homopterous pest insects such as rice planthoppers and leafhoppers without any adverse effects on their predators and parasitoids. Since the chemical inhibits larval molting, egg-laying and/or induces oviposition of unhatchable eggs, it suppresses the population density of hoppers even in the progeny of the treated generation with long lasting activity in the paddy field. Buprofezin causes no resurgence of hoppers by itself nor with the combination of other insecticides, working rather preventive. The safe properties on non-target organisms and the stable control effects on pest insects have led this compound to a prominent IGR for integrated pest managements in rice and other crops.

INTRODUCTION

Homopterous insects such as brown planthopper (BPH), *Nilaparvata lugens* Stål, white-backed planthopper (WBPH), *Sogatella furcifera* Horvath, smaller brown planthopper (SBPH), *Laodelphax striatellus* Fallen, and green rice leafhopper (GRLH), *Nephotettix cincticeps* Uhler, are important pests of rice in Southeast and Far East Asia. Among them, BPH is probably the most serious pest, because it causes direct damage by sucking plant sap which often results in the complete withering of the plants known as hopperburn and because it transmits viral diseases (1,2). In order to control BPH, several resistant varieties of rice were released in some areas, but their success was limited by the appearance of biotypes of BPH (3). Thus, insecticides have been mainly used for the control of hoppers in rice production. What makes it more difficult to control them is that they have developed resistance to organophosphates, carbamates (4-7) and even to synthetic pyrethroids (8). Some of these insecticides are commonly used to control

lepidopterous pests and bugs in rice, but they cause resurgence of planthoppers after field application in South and
Southeast Asia, mainly due to their toxicity to natural
enemies and their stimulative effect on reproduction of BPH at
sub-lethal doses (9,10). Therefore, new selective insecticides
have been desired to be developed in rice.

Buprofezin (APPLAUD®) developed by Nihon Nohyaku Co Ltd
(11) is a new type of insect growth regulator (IGR) and has
shown long-term control effect against plant- and leafhoppers
(11-14), whiteflies (15-17), and scales (18). It is now registered in more than 25 countries worldwide (as of July, 1989)
and is used in effective integrated pest managements.

In this paper, properties of biological activity of
buprofezin are summarized to present how it works for the
control of homopterous pest insects in paddy fields.

BIOLOGICAL ACTIVITY

1. Activity on hoppers

1-1. Larvae
Buprofezin prevents the molting process of larvae by disruption of the regulation of ecdysteroids (19) followed by inhibition of chitin biosynthesis (20,21). Therefore, the chemical
reveals larvicidal activity at the end of the instar as shown
in FIGURE 1 (12). When larvae were treated with chemicals for
a limited period of developing stage (TABLE 1), the insecticidal activity of buprofezin changed dramatically with treatment timing even in an instar, whereas those of conventional
insecticides changed little (22). These findings are enough to
explain that buprofezin is a typical insect growth regulator
rather than an insecticide.

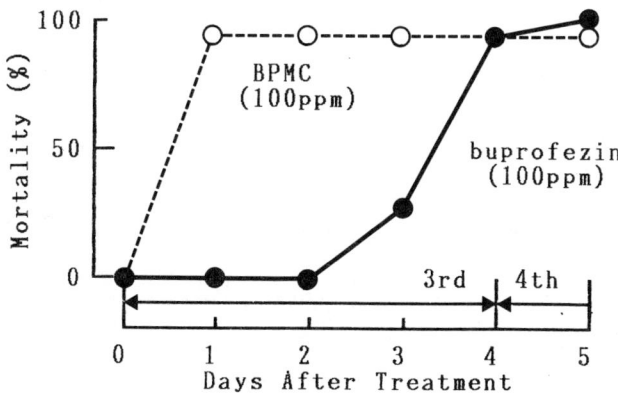

FIGURE 1. Cumulative mortality of 3rd instar larvae of BPH
  treated with buprofezin and BPMC (after Kajihara *et al*.).[a]

[a] The newly ecdysed 3rd instar larvae of BPH were released
  on rice plants treated with the chemical solution.
  Assessment was made each day after treatment.

TABLE 1

Insecticidal activity of buprofezin related to the
treatment period in the 4th instar larvae of BPH
(after Kuriyama *et al.*).[a]

Period of treatment (hr after molting)	LC$_{50}$ value (ppm)		
	buprofezin	BPMC	diazinon
0 - 6	570	11	8.5
12 - 18	210	-	-
24 - 30	85	16	8.8
42 - 48	0.47	-	-
48 - 54	0.71	20	12
54 - 60	120	-	-
54 - E[b]	150	-	-

[a] The newly ecdysed 4th instar larvae were chosen and maintained on rice seedlings until used. After being kept on
rice plants treated with each chemical for indicated period,
insects were transferred to untreated plants and the assessment was made in 3 days. LC$_{50}$ values were calculated, based
on the mortality.
[b] E ; Ecdysis from 4th to 5th instar larvae.

TABLE 2 shows the toxicity of buprofezin to 3rd instar
larvae of 4 different species of hoppers. Buprofezin showed
excellent larvicidal activity against all species tested, and
its potency was 10 to 100 times higher than those of conventional insecticides (12).

TABLE 2

Toxicity of buprofezin to 3rd instar larvae of 3 species
of planthoppers and a leafhopper (after Kajihara *et al.*).[a]

Chemical	LC$_{50}$ value (ppm)			
	BPH	WBPH	SBPH	GRLH
Buprofezin	0.31 (0.28-0.35)[b]	0.05 (0.05-0.06)	0.06 (0.05-0.06)	0.41 (0.36-0.46)
BPMC	18.64 (15.11-22.92)	2.53 (2.06-3.25)	5.10 (4.15-6.22)	57.76 (44.65-75.99)
Propaphos	4.33 (3.52-5.33)	0.92 (0.81-1.08)	1.21 (1.04-1.45)	4.73 (3.73-5.79)

[a] Rice plants were treated with chemical solution, and 3rd
instar larvae of each species were released separately
after air-dried. Assessment was made 5 days after treatment
and LC$_{50}$ values were calculated, based on the mortality.
[b] Values in parentheses represent 95 % confidence limits.

1-2. Adults

Buprofezin does not have lethal activity against adults, but affects reproductive system. Some aspects are shown in TABLES 3 and 4. It is clear that all the activities observed are age-dependent. When immature adults (24hr-old) were treated with buprofezin at a concentration of 250 ppm, the oviposition was almost completely inhibited, while mature adults (more than 24hr-old) laid eggs as many as the untreated control with almost no hatchability (23). Such effects were still observed at a concentration of 10 ppm, but was not below 2 ppm (24). It should be emphasized that buprofezin has no adverse effect such as the stimulation of reproduction at sub-effective doses.

The shortening of the longevity of immature females was also observed with a bunch of eggs in their ovaries (23,25). This finding led to study on the mode of action in relation to prostaglandins known to stimulate oviposition in other insect species (26,27) and recently, it has been suggested that the suppression of oviposition by buprofezin was attributable to the inhibition of biosynthesis of prostaglandin E_2 (25). The study on the mechanisms of the inhibition of hatching is ongoing and related to ecdysteroid metabolism mentioned above.

TABLE 3
Effects of buprofezin against reproduction of BPH
(after Asai *et al.*).[a]

Age (Hours after emerging)	Buprofezin concentration (ppm)	No. of eggs laid / female		Hatchability (%)
0 - 24	250	0.4	(0)[b]	0.0
	0	171.6	(100)	17.4
24 - 48	250	121.3	(117)	0.8
	0	103.8	(100)	87.2
72 - 96	250	175.0	(72)	1.0
	0	243.5	(100)	75.6
120 - 144	250	194.3	(97)	4.3
	0	201.4	(100)	65.6
168 - 192	250	202.7	(72)	1.4
	0	283.0	(100)	27.7

[a] Female and male adults of indicated hours-old were released onto rice plants treated with buprofezin, and the number of larvae hatched and eggs unhatched was counted respectively by dissecting rice stems 50 days after release.
[b] Values in parentheses represent percentage compared to control.

TABLE 4
Effects of buprofezin against oviposition of BPH at
sub-effective concentrations (after Kanaoka *et al.*)[a] .

Age (Hours after emergence)	Buprofezin concentration (ppm)	No. of eggs laid / female[a]			Hatch- ability (%)
0 - 24	50	33	± 43	a	0.0
	10	118	± 21	b	47.5
	2	125	± 31	bc	83.7
	0.4	136	± 24	bc	88.5
	0	150	± 22	c	90.7
72 - 96	250	219	± 48	d	0.0
	50	248	± 24	d	6.5
	10	241	± 14	d	64.7
	2	252	± 27	d	86.9
	0	258	± 22	d	82.6

[a] Mean ± standard deviation of 5 to 9 replication with
5 pair of adults in each experiment. Values followed by
a common letter are not siginificantly different at 5%
level by Duncan's Multiple Range Tests (DMRT).

1-3. Eggs
Although buprofezin shows unique effects on eggs through
adults, it does not have strong ovicidal activity and is not
practicaly useful, either (28).
     TABLE 5 summarizes the level of effects of buprofezin
against several developmental stages of BPH (29). EC_{50} and
EC_{95} values are represented as the concentration of buprofezin in rice plants. The activity is extremely high against
larvae followed by adults and eggs.

TABLE 5
Effects of buprofezin against several stages of BPH
(after Asai *et al.*).[a]

Stage	Age	(ppm in rice plants) EC_{50}	EC_{95}
Egg	0 - 24hr old	53	680
Larva	1st instar	0.04	0.16
	3rd instar	0.06	0.56
Adult	0 - 24hr old	3.7	11
	120 - 144hr old	4.3	25

[a] Rice plants in a pot were treated with buprofezin
and insects were released and assayed as described
above. The amount of buprofezin in rice plants was
determined by gas chromatography (30).

2. Cross resistance

As shown in TABLE 6, buprofezin revealed almost the same activity against a susceptible strain and field strains of BPH
resistant to organophosphorus insecticides and carbamates (31).

TABLE 6
Susceptibility of field strains of BPH to buprofezin
(after Ikeda et al.)[a] .

| Chemical | LD$_{50}$ (μ g/g) | | | Resistance factor (R/S) | |
| | Field strain[b] | | Susceptible strain[c] | | |
	1978	1982		1978	1982
Buprofezin	3.90	2.76	5.63	0.69	0.49
BPMC	5.98	19.98	0.85	7.0	23.5
MTMC	6.81	12.81	1.60	4.3	8.0
Diazinon	62.26	117.6	7.26	8.6	16.2
Fenitrothion	133.1	222.6	3.32	40.1	67.0
Malathion	155.3	1117	3.13	49.6	356.9

[a] Each chemical was applied topically on the dorsal surface
of 4th instar larvae of BPH. Assessment was made 5 days
after treatment.
[b] Field strains were collected at Saga Prefecture in Japan
and maintained in the laboratory.
[c] The susceptible strain was collected at Osaka Prefecture
in 1970 and maintained in the laboratory.

3. Influence on natural enemies of hoppers

Most experiments have been carried out at the International
Rice Research Institute in Philippines and at several prefectural Agricultural Experimental Stations in Japan. Some of
results are shown in TABLES 7-9. It is clear that buprofezin
is a quite safe chemical for predaceous spiders (32), bugs (13,
33) and parasitic wasps of hoppers (34).

TABLE 7
Influence of buprofezin on *Cyrtorhinus lividipennis* and
Microvelia atrolineata (after IRRI Annual Report for 1981)[a]

| Chemical | Stage tested | Mortality (%) | | | |
| | | *C. lividipennis* | | *M. atrolineata* | |
		1DAT	6DAT	1DAT	6DAT
Buprofezin	Adult	1 b	9 b	1 b	4 bc
	Nymph	0 b	10 b	1 b	10 b
Carbofuran	Nymph	99 a	99 a	14 a	34 a
Control	Adult	1 b	8 b	0 b	3 c
	Nymph	1 b	9 b	1 b	9 b

[a] Insects were treated with each chemical at a dosage of 750
gai/ha in potter's spray tower. Means followed by a common
letter are not significantly different at 5% level of DMRT.

TABLE 8

Influence of buprofezin on larvae of *Lycosa pseudoannulate*
(after IRRI Annual Report for 1981)[a] .

Chemical	Mortality (%)				Molting (%)			
	1DAT	3DAT	6DAT	9DAT	1DAT	3DAT	6DAT	9DAT
Buprofezin	0	0	0	0	0	7	47	100
Carbofuran	93	93	93	93	0	0	3	7
Control	0	0	0	0	13	20	43	100

[a] Larvae of spiders were treated with each chemical at a
dosage of 750 gai/ha in potter's spray tower.

TABLE 9
Influence of buprofezin on *Paracentrobia andoi*
(after Kajihara *et al.*)[a] .

Experimental condition	No. of eggs of GRLH used	No. of wasps emerged	% of parasitised
Treated plants in treated field	331	62	18.7
Untreated plants in untreated field	433	52	12.0

[a] Rice plants treated with buprofezin at a concentration of
250ppm were oviposited by GRLH in a pot and placed in the
field previously treated with the chemical. Untreated plants
and fields were prepared as control. After parasitation of
P. andoi for 5 days, rice sheaths with eggs of GRLH were cut
and put into a test tube with some water. The number of
adult *P. andoi* was counted after emergence and the number of
eggs left was also counted with rice sheaths dissected.

4. Control effects on hoppers with practical formulations

4-1. Residual activity
The most important property of buprofezin in a practical use
is a much longer residual activity than conventional insecticides. As shown in FIGURE 2, buprofezin is effective on larvae
of BPH even at 5 weeks after application on rice plants in
pots at the recommended dosages of each type of formulation,
that is, 1.5% Dust, 25% Wettable Powder and 2% Granule at the
dosage of 600, 375 and 800 gai/ha, respectively (29). The
reason why buprofezin has such a long persistence of effect is
not because   it is stable in rice plants but because its
EC_{95} value in rice plants against larvae is extremely low
(TABLE 5). The degradation of buprofezin in rice plants treated with 1.5% Dust at a dosage of 450 gai/ha is shown in FIGURE
3 (29). Buprofezin degraded logarithmically in rice plants and
reached the concentration equivalent to EC_{95} on immature adult
in 10 days and on 1st instar larvae in as many as 50 days
after application.

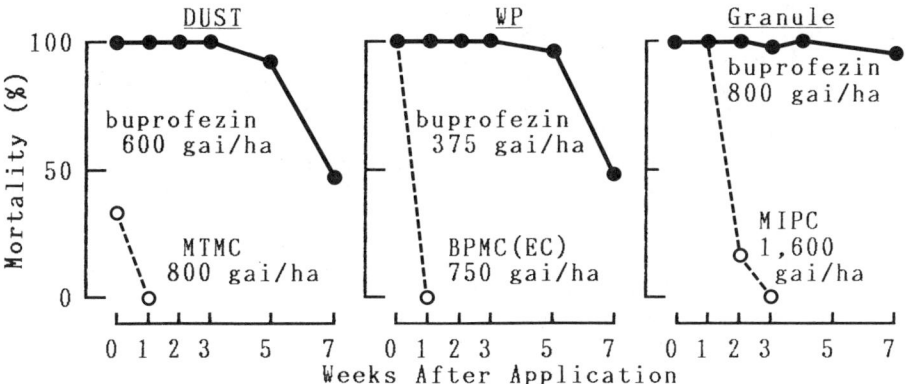

FIGURE 2. Residual activity of buprofezin on 3rd instar larvae of BPH with three types of formulation (after Asai *et al.*).[a]

[a] Each chemical was applied on rice plants in a pot and insects were released at one week interval. Assessment was made 7 days after each **release.**

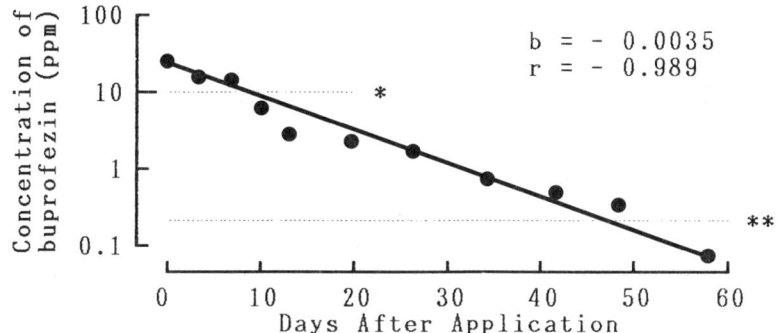

FIGURE 3. Degradation of buprofezin in rice plants treated with 1.5% Dust (after Asai *et al.*).[a]

[a] The dust formulation was applied on rice plants in a pot at a dosage of 450 gai/ha. The amount of buprofezin was determined as described previously.
* EC$_{95}$ value against immature adults of BPH (from TABLE 5).
** EC$_{95}$ value against 1st instar larvae of BPH (from TABLE 5).

4-2. Vapour and systemic activity
Another advantage of buprofezin in its practical use is that the chemical has vapour activity. It was observed when untreated rice plants in a pot were put in the paddy field treated with buprofezin and assayed with larvae of BPH (35). Although the volatility of the chemical persists only for a few days at the recommended doses in the paddy field, such property is somewhat useful for the practical control of insects because regular applications of dust formulation or foliar sprays do

not always cover whole plants with chemicals. The relatively
high vapour pressure of the chemical (9.4 × 10⁻⁶ mmHg at 25°C)
brings about this bonus effect.

   On the other hand, the systemic activity of buprofezin
through roots is very weak and is not expected for practical·
effect. However, a floating type of granule formulation has
been developed. The mechanism of performance effect is elaborate, that is, the active ingredient in granules floats on
the surface of the water after applied into paddy field, gets
to rice stems and travels to the habitat of insects with capi-
llarity of the water (36). FIGURE 4 shows the vertical distribution of the active ingredient of buprofezin in rice plants
in a pot treated with two types of the granule formulation.
The amount of buprofezin on the base of rice stems was much
higher in the floating granule than in the submerging one,
indicating the better availability of the former.

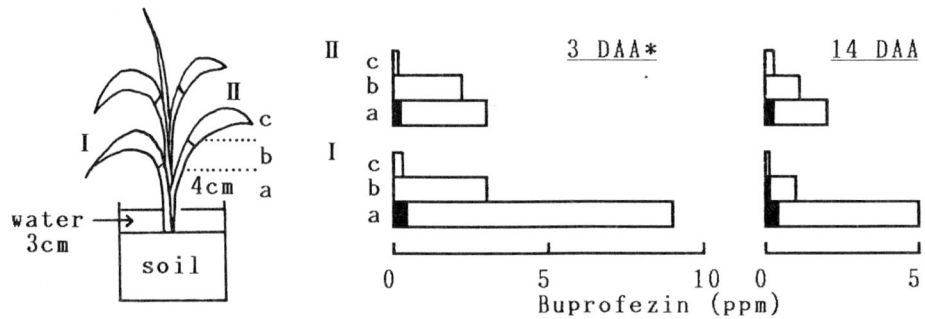

FIGURE 4. Vertical distribution of buprofezin in rice plants
   treated with two types of the granule formulation in pots
   (after Asai *et al.*)ᵃ .

   ⬜ : Floating granule,   ⬛ : Submerging granule.
ᵃ Rice plants in a pot were treated with 2% floating and 4%
submerging granules at a dosage of 40kg/ha, respectively.
The first (I) and the second (II) sheath and leaf were cut
to 3 pieces (a,b,c, from the base) and the amount of buprofezin was determined by gas chromatography.
* Days after application.

5. Control performance on hoppers in the field
The typical result of field trial is shown in TABLE 10, where
buprofezin alone and its mixture with insecticides such as
organophosphates, carbamates and synthetic pyrethroids worked
very well to suppress the population-density of BPH, while
common insecticides alone caused tremendous resurgence after
application in paddy field in Philippines (33). It is clear
that the selectivity of buprofezin on natural enemies (TABLES
8-10), non-adverse effect on adults of hoppers at sub-effective dosages (TABLE 4) and long persistence effects on larvae
(FIGURE 2) contributed to the good control performance. The
reason why buprofezin prevented other insecticides from
causing resurgence in their combination is probably due to the
longer persistence of control effect of buprofezin on hoppers

than those of adverse effects of insecticides on natural
enemies, or maybe due to the cancelling effect of buprofezin
on the stimulation of reproduction by other insecticides.

TABLE 10. Field evaluation of effect of buprofezin mixed with
commonly used insecticides for BPH control
(after IRRI Annual Report for 1982)[a] .

| Chemical[b] | BA[c] | Average BPH larvae / hill | | | |
| | | DAFA [d] | | DASA [e] | |
		5	15	5	15
buprofezin	23.57	7.5 e-h	1.76 h	2.73 f	0.25c
azinphos ethyl(AZ)	31.91	19.3 bc	19.2 b-d	47.75 c-e	749.37a
fenitrothion(FE)	44.00	17.8 c	46.15 ab	69.10 bc	1709.00a
phenthoate(PH)	52.33	18.0 bc	42.95 ab	64.50 ab	553.12a
BPMC	43.12	8.4 d-h	14.37 b-f	16.86 de	55.06b
MTMC	25.97	10.7 d-h	25.45 b-e	14.23 e	49.02b
deltamethrin(DE)	57.43	20.1 bc	137.9 a	160.8 a	2848.00a
buprofezin + AZ	43.31	6.4 f-h	4.48 d-h	2.62 f	1.25c
buprofezin + FE	52.62	6.2 f-h	7.38 c-h	3.17 f	0.50c
buprofezin + PH	30.55	6.5 f-h	3.87 e-h	2.42 f	4.75c
buprofezin + BPMC	51.75	5.3 gh	2.54 gh	2.81 f	1.65c
buprofezin + MTMC	48.93	5.3 gh	2.71 f-h	3.35 f	0.67c
buprofezin + DE	38.02	7.6 d-h	157.80 ab	4.38 f	1.58c
Control	34.89	38.6 a	33.70 ab	33.97 b-d	56.35b

[a] The experiment was carried out at Calauan, Laguna,
Philippines in 1982. In a column, means followed by a common
letter are not significantly different at 5% level by DMRT.
[b] Chemicals were applied at the dosage of 750 gai/ha, except
for buprofezin and deltamethrin. They were applied at the
dosage of 125 and 12.5 gai/ha, respectively.
[c] Before application.
[d] Days after first application.
[e] Days after second application.

CONCLUSIONS

In Japanese history, there have been several food famines
caused by planthoppers. The most serious one occurred in 1732
and it is said that more than 10,000 people died because of
the starvation, according to the literature. Nowadays, such
sad incidents never happen owing to many good insecticides.
However, lacking persistence of efficacy in the field,
most of them need frequent applications, resulting in the
resistance problem. Besides, the resurgence of planthoppers
caused by non-selective insecticides has become another problem in Southeast Asia. As presented in this paper, a prominent
IGR, buprofezin, solves all these problems and surely contributes to the stable rice production and the human welfare.

ACKNOWLEDGEMENTS

The author thanks Dr. Fujio Araki, the manager of Biological Research Center, for reading the manuscript and members of the Laboratory of Entomology for helping work.

REFERENCES

1. Dyck, U.A. and Thoma, B., The brown planthopper problems. In *The brown plantho*pper :Threat to Rice Production in Asia, IRRI, Los Banos, Laguna, 1979, pp. 3-17.

2. Brader, L., Recent trends of insect control in the tropics. *Entomol. Exp. Appl.*, 1982, **31**, 111-20.

3. Khush, G.S., Disease and insect resistance in rice. *Adv. Agron.*, 1977, **29**, 265-341.

4. Nagata, T., Masuda, T. and Moriya, S., Development of insecticide resistance in the brown planthopper. *Appl. Ent. Zool.*, 1979, **14**, 264-69.

5. Lin, Y.H., Sun, C.N. and Feng, H.T., Resistance of *Nilaparvata lugens* to MIPC and MTMC in Taiwan. *J. Econ. Entomol.*, 1979, **72**, 901-3.

6. Kilin, D., Nagata, T. and Masuda, T., Development of carbamate resistance in the brown planthopper, *Nilaparvata lugens. Appl. Ent. Zool.*, 1981, **16**, 1-6.

7. Chung, T.C., Sun, C.N. and Hung, C.Y., Resistance of *Nilaparvata lugens* to six insecticides in Taiwan. *J. Econ. Entomol.*, 1982, **75**, 199-200.

8. Dai, S.M. and Sun, C.N., Pyrethroid resistance and synergism in *Nilaparvata lugens* in Taiwan. Ibid.,1984, **77**, 891-7.

9. Heinrichs, E.A., Aquino, G.B., Chelliah, S., Valencia, S.L. and Reissig, W.H., Resurgence of *Nilaparvata lugens* populations as influenced by method and timing of insecticide applications in lowland rice. *Environ. Entomol.*, 1982, **11**, 78-84.

10. Heinrichs, E.A. Reissig, W.H., Valencia, S. and Chelliah, S., Rates and effect of resurgence inducing insecticides on populations of *Nilaparvata lugens* and its predators. Ibid., 1982, **11**, 1269-73.

11. Kanno, H., Ikeda, K., Asai, T. and Maekawa, S., 2-*tert*-Butylimino 3-isopropyl-5-phenylperhydro-1,3,5-thiadiazin-4-one (NNI-750), a new insecticide. *Pro. 1981 Br. Crop Prot. Conf. Pests and Diseases*, pp. 59-66.

12. Kajihara, O., Asai, T,, Ikeda, K. and Lim, S.S., Bupro-fezin, a new insecticide for control of brown planthopper, *Nilaparvata lugens. Pro. Intr. Conf. Plant Protection in*

the Tropics., 1982, pp. 1-7.

13. Heinrichs, E.A., Basilio, R.P. and Valencia, S.L., Buprofezín, a selective insecticide for the management of rice planthoppers, Homoptera Delphacidae, and leafhoppers,Homoptera Cicadellidae. *Environ. Entomol.*, 1984, **13**, 515-21.

14. Nagata, T., Timing of buprofezin application for the brown planthopper, *Nilaparvata lugens. Appl. Entomol. Zool.*, 1986, **21**, 357-62.

15. Naba, K., Nakazawa, K. and Hayashi, H., Long-term effects of buprofezin spray in controlling the greenhouse whitefly, *Trialeurodes vaporariorum*, in vinylhouse tomatoes. Ibid., 1983, **18**, 284-6.

16. Ishaaya, I., Mendelson, Z. and Melamed-Madjar, V., Effect of buprofezin on embryogenesis and progeny formation of sweetpotato whitefly, Homoptera aleyrodidae. *J. Econ. Entomol.*, 1988, **81**, 781-4.

17. Wilson, D. and Anema, B.P., Development of buprofezin for control of whitefly *Trialeurodes vaporariorum* and *Bemesia tabaci* on glasshouse crops in the Netherlands and the UK. *Pro. 1988 Br. Crop Prot. Conf. Pests and Diseases,* pp. 175-80.

18. Yarom, I., Blumberg, D. and Ishaaya, I., Effects of buprofezin on California red scale, Homoptera Diaspididae, and Mediterranean black scale, Homoptera Coccidae. *J. Econ. Entomol.*, 1988, **81**, 1581-5.

19. Kobayashi, M., Uchida, M. and Kuriyama, K., Elevation of 20-hydroxyecdysone level by buprofezin in *Nilaparvata lugens. Pestic. Biochem. Physiol.*, 1989, **34**, 9-16.

20. Izawa, Y., Uchida, M., Sugimoto, T. and Asai, T., Inhibition of chitin biosynthesis by buprofezin analogs in relation to their activity controlling *Nilaparvata lugens* Stål. Ibid., 1985, **24**, 343-7.

21. Uchida, M., Asai, T. and Sugimoto, T., Inhibition of cuticle deposition and chitin biosynthesis by a new insect growth regulator, buprofezin, in *Nilaparvata lugens* Stål. *Agri. Biol. Chem.,* 1985, **49**, 1233-4.

22. Kuriyama, K., Kimura, M. and Kajihara, O., The properties of the activity of buprofezin on larvae of the rice brown planthopper. *Pro. 1987 Ann. Conf. Jap. Soc. Appl. Ent. Zool.,* Tsukuba, p.141 (in Japanese).

23. Asai, T., Kajihara, O., Fukada, M. and Maekawa, S., Studies on the mode of action of buprofezin II. Effects on reproduction of the brown planthopper, *Nilaparvata lugens* Stål. *Appl. Ent. Zool.,* 1985, **20**, 111-7.

24. Kanaoka, A., Kuriyama, K. and Kajihara, O. (Unpublished.)

25. Uchida, M., Izawa, Y. and Sugimoto, T., Inhibition of prostaglandin biosynthesis and oviposition by an insect growth regulator, buprofezin, in *Nilaparvata lugens* Stål. *Pest. Biochem. Physiol.*, 1987, **27**, 71-5.

26. Destephano, D.B. and Brady, U.E., Prostaglandin and prostaglandin synthetase in the cricket, *Acheta domesticus*. *J. Insect Physiol.*, 1977, **23**, 905-12.

27. Loher, W., Ganjian, I., Kubo, I., Stanley-Samuelson, D. and Tobe, S.S., Prostaglandins: Their role in egg-laying of the cricket, *Teleogryllus commodus. Proc. Natl. Acad. Sci.* USA, 1981, 7835-8.

28. Asai, T., Fukada, M., Maekawa, S., Ikeda, K. and Kanno, H., Studies on the mode of action of buprofezin I. Nymphcidal and ovicidall activities on the brown rice planthopper, *Nilaparvata lugens* Stål. *Appl. Ent. Zool.*, 1983, **18**, 550-2.

29. Asai, T., Kajihara, O. and Maekawa, S., Effectiveness of various formulations of buprofezin on rice planthopper in irrigated rice. *Pro. 1984 Br. Crop Prot. Conf. Pests and Diseases.*, pp.1039-44.

30. Uchida, M., Nishizawa, H. and Suzuki, T., Hydrophobicity of buprofezin and flutolanil in relation to their soil adsorption and mobility in rice plants. *J. Pestic. Sci.*, 1982, **7**, 397-400.

31. Ikeda, K., Yasui, M., Kanno, H. and Maekawa, S., Development of a new insecticide, buprofezin. Ibid., 1986, **11**, 287-95.

32. International Rice Research Institute, Annual Report for 1981, IRRI, Los Banos, Laguna, Philippines, 1983.

33. International Rice Research Institute, Annual Report for 1982, IRRI, Los Banos, Laguna, Philippines, 1984.

34. Kajihara, O., Konno, T. and Fukada, M. (Unpublished).

35. Kajihara, O. and Fukada, M. (Unpublished).

36. Asai, T., Higashimura, M., Fukada, M. and Maekawa, S., The properties of the granule formulation of buprofezin. *Pro. 1984 Ann. Conf. Jap. Soc. Appl. Ent. Zool.*, Utsuno-miya, p.72 (in Japanese).

HEXAFLUMURON (DOWCO* 473 Pesticide)
A POTENTIAL NEW RICE INSECTICIDE IN JAPAN

J. KATO, N. KONDO and C. D. FORGIE
DowElanco Japan Ltd.
1-6-12, Toranomon, Minato-ku
Tokyo, Japan

ABSTRACT

Hexaflumuron (proposed common name), a promising
benzoylphenyl urea (BPU) compound, is being developed for
upland crop and rice pests in Japan. Dust formulations are
most preferred in Japan for application of rice insecticides.
Since formulation to provide ultrafine particle size of BPU
provides better efficacy generally, a new type of dust
formulation was developed for hexaflumuron, which was 5 to
10 times more efficacious than the conventional type of dust
formulation. Several combination dust formulations were
developed using this new concept in order to provide a wide
insecticidal spectrum. Official trials demonstrated that
those combinations including hexaflumuron showed excellent
efficacy against important rice chewing insects, such as
grass leafroller Cnaphalocrocis medinalis and rice stem borer
Chilo suppressalis.

INTRODUCTION

Several benzoylphenyl urea (BPU) compounds have been
developed as new generation insecticides. BPUs are highly
effective to insects by ingestion and act by means of
inhibiting the insect molting process. Because of novel
chemistry and unique mode of action, BPUs are very effective
to control conventional insecticide-resistant pests, e.g.
organophosphate/pyrethroid resistant diamond-back moth,
Plutellla xylostella, and pyrethroid resistant cotton
leafworm, Spodoptera littoralis. Hexaflumuron, a BPU
compound provides excellent control of Lepidoptera,
Coleoptera and Diptera pests (1), (2). In this paper, we
elaborate on the effectiveness of hexaflumuron against
important rice pests.

*Trademark of DowElanco

DEVELOPMENT OF DUST FORMULATION

The biological efficacy of BPUs is enhanced by the particle
size of the active ingredient: i.e., the formulation
including the smaller particle size of the product provides
better efficacy (3). Laboratory screening tests demonstrated
5 - 10 times the application dosage of wettable powder (WP)
and 1.5 - 3 times the dosage of suspension concentrate (SC)
was required to provide the same efficacy as the emulsifiable
concentrate (EC). Therefore EC and SC were mainly developed
for upland crop use as the preferable formulation of BPU. On
the other hand, dust (DP) is the most popular formulation as
conventional insecticide use on rice in Japan. This meant
that the development of an efficacious dust formulation was
essential to provide a cost-effective product for the rice
market. At first, a conventional type of dust (dry blended
dust) was produced, and its efficacy was compared with
hexaflumuron EC for control of rice leaf beetle (RLB), Oulema
oryzae and grass leafroller (GLR) in a rice field trial (Figs
1 and 2). Results indicated that the conventional dust of
hexaflumuron provided poor efficacy on both RLB and GLR in
comparison with the EC formulation. This meant the
conventional dust of hexaflumuron would not be acceptable
practically in the field.

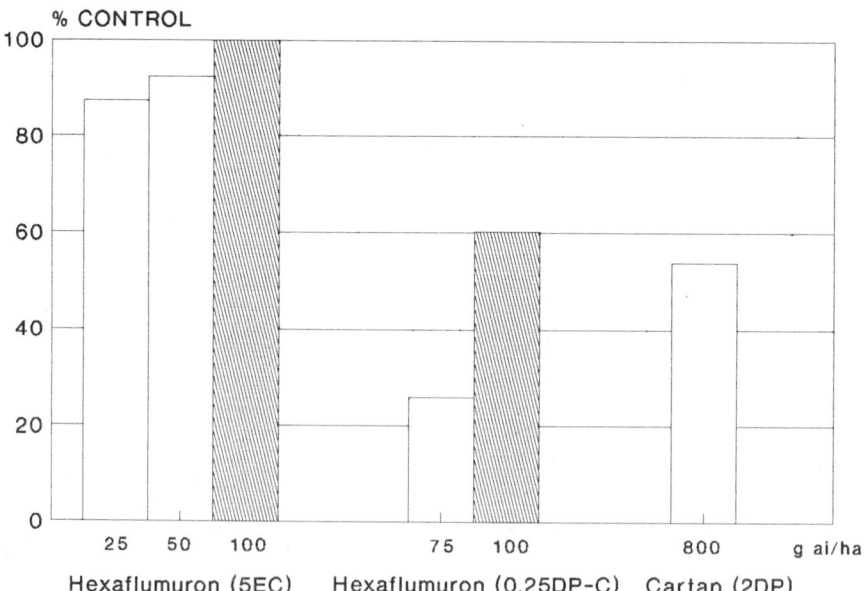

Figure 1. Hexaflumuron EC and Conventional Dust Formulation
Activity on Rice Leaf Beetle (18 DAT) Field Trial, 1987

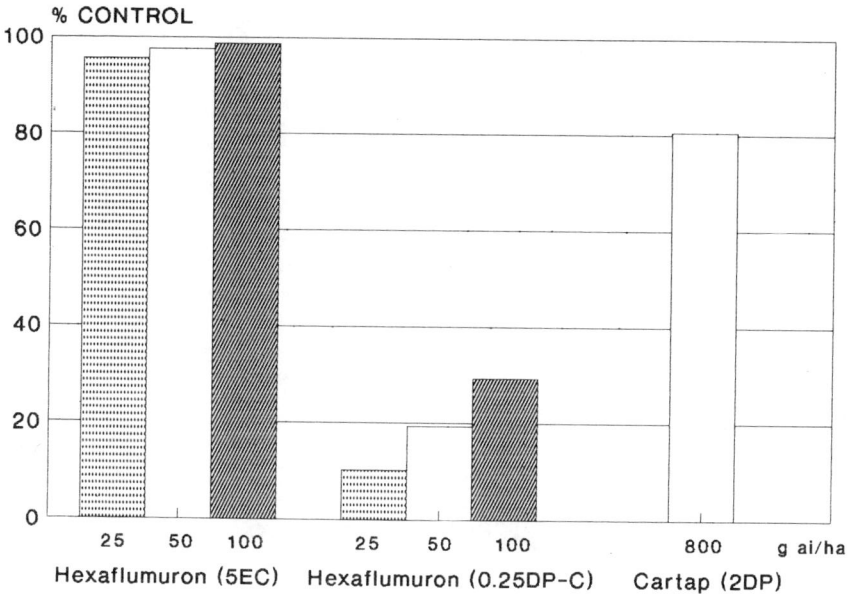

Figure 2.   Hexaflumuron EC and Conventional Dust Formulation
         Activity on Grass Leafroller (30DAT) Field Trial, 1987

We then developed a new type of dust formulation using
non-conventional formualtion technology. The new dust
formulation was evaluated against rice stem borer (RSB), and
compared with the conventional dust in the laboratory using
the Bell-Jar Dust Sprayer (Fig 3). LD values of two dusts
against RSB were as follows:

	LD values (g ai/ha)	
	50	90
Conventional dust	98.4	169
New dust	9.8	30.5

The new dust prototype was 10 and 5 times more efficacious
than the conventional blended dust with respect to LD50 and
90 values. Dose response curves for the two formulations are
provided in Figure 3.

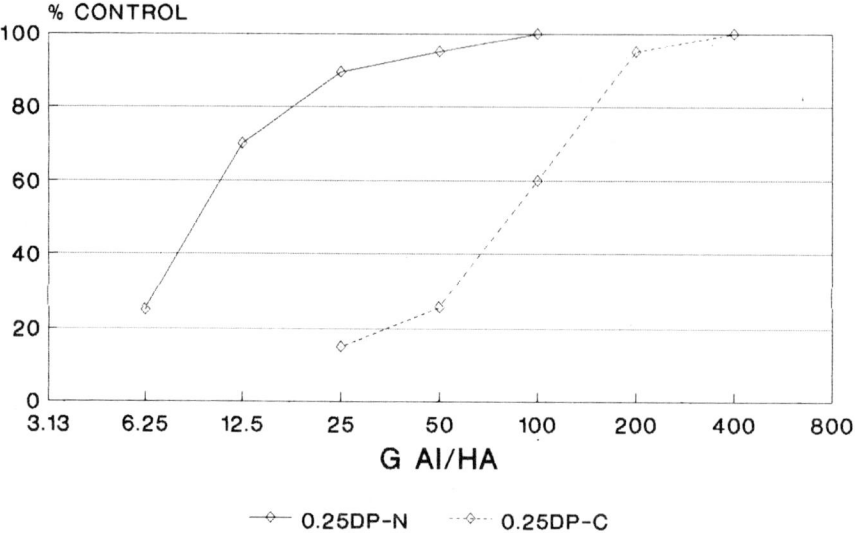

Figure 3.   Hexaflumuron Conventional Dust and New Dust
        Fomulation Activity on Rice Stem Borer (5DAT) Lab
                        in 1988

As the next step, a commercialy acceptable dust formulation
was developed from several candidate formulations and field
tested for control of grass leafroller in comparison with EC
formulation (Fig 4).   Results indicated that:

a.   Although the new dust formulation (DP-N) was slightly
      inferior to EC formulation, at greater than 40 g ai/ha
      it provided commercially acceptable control of GLR.

b.   The new hexaflumuron dust formulation at 40 and 100 g
      ai/ha was superior to cartap 2% DP at 800 g ai/ha, a
      commercial standard, against GLR.

Based on the above results, this new hexaflumuron 0.25% DP
was further evaluated in official trials.

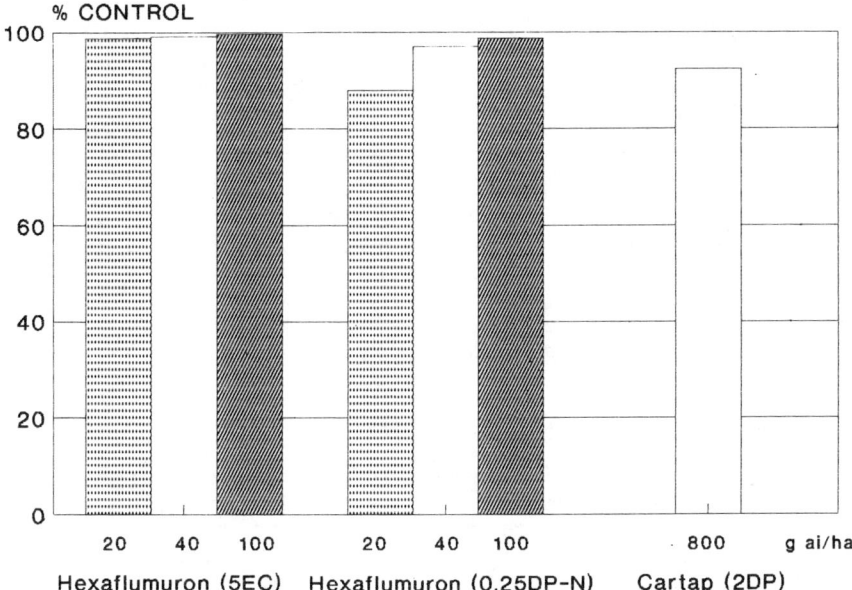

Figure 4.   Hexaflumuron EC and New Dust Formulation Activity
            on Grass Leafroller (28DAT) Field Trial, 1988

OFFICIAL TRIAL OF HEXAFLUMURON DUST

Among rice pests, the serious lepidopteran insects are grass
leafroller (GLR) and rice stem borer (RSB). These two
insects would be the major target pests of hexaflumuron. The
evaluation of hexaflumuron 0.25% dust against GLR and RSB was
locally conducted at agricultural experimental stations in
official trials for registration purpose.

Sucking insects such as rice hoppers: brown planthopper
(BPH), Nilaparvata lugens, and green leafhopper (GLH),
Nephotettix cincticeps, are also serious rice pests. Because
hexaflumuron is not effective at economic dosage on these
insects, combination dust formulations with rice hopper
control agents were developed as follows:

        DWI-101
              Hexaflumuron   0.125%
              Chlorpyrifos-methyl   1.5%

        DWI-102
              Hexaflumuron   0.25%
              Ethofenprox   0.5%

```
MKS-289
     Hexaflumuron              0.25%
     XMC                       3.0%

NNI-8902
     Hexaflumuron              0.25%
     Buprofezin                1.5%

NNIF-8903
     Hexaflumuron              0.25%
     Buprofezin                1.5%
     Flutolanil                1.5%
```

Efficacy of these combination dusts was also evaluated against GLR and RSB in local agricultural experimental stations.

Efficacy against grass leafroller (GLR)

Results from 10 locations "a" through "j" are summarised in Figure 5. Hexaflumuron single dust provided excellent activity against GLR in official trials confirming data from the in-house trial. Similarly the five combination dusts demonstrated excellent activity against GLR.

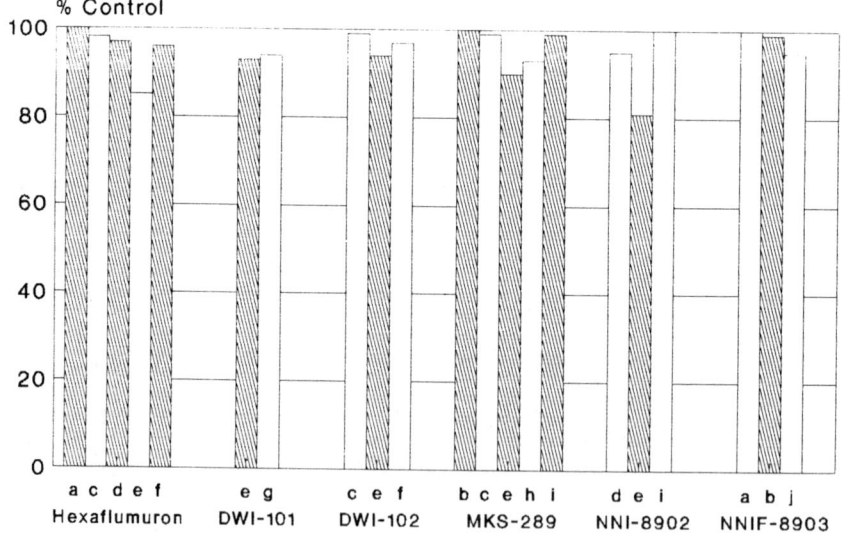

Figure 5. Efficacy of Hexaflumuron and Combination Dust Formulations on Grass Leafroller (40 kg formulation/ha)

Efficacy against rice stem borer (RSB)

Results from 5 locations "a" through "e" are summarized in
Figure 6. Although hexaflumuron as a single or combination
dust was relatively less active against RSB compared with
GLR, the efficacy of the products is still considered
commercially acceptable.

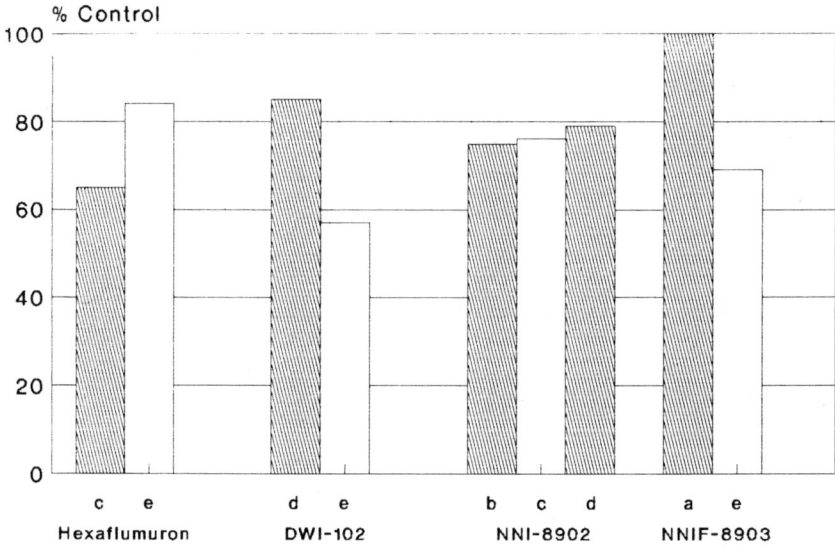

Figure 6.   Efficacy of Hexaflumuron and Combination Dusts
            on Rice Stem Borer (Hexaflumuron 100 g ai/ha)

Efficacy against BPH and GLH

Results in official trials of hexaflumuron alone and
combination dust formulations are summarized in Table 1.
Available data on the combination dust formulations against
green leafhopper (GLH) and planthoppers (PH) are limited, but
the combination partners, ethofenprox, buprofezin and XMC,
are already established commercialy as excellent rice hopper
control agents and the combination dusts would therefore be
recognized as excellent rice hopper cotnrol agents. In
conclusion, these combination dusts can be expected to
perform as broad spectrum rice insecticides controlling both
serious chewing and sucking insects.

TABLE 1

Efficacy of hexaflumuron 0.25 dust and five combination dusts against major rice pests (40 kg formulation/ha)

Product	Average percent control			
	GLR	RSB	GLH	PH
Hexaflumuron	95.2	74.5	–	–
DWI-101	93.5	–	–	–
DWI-102	96.7	71.0	99.0	100.0
MKS-289	96.2	–	56.5	77.2
NNI-8902	92.0	76.7	–	–
NNIF-8903	98.0	84.5	–	–

CONCLUSION

An improved dust formulation has been developed for hexaflumuron which provides efficacy against lepidopteran pests of rice similar to emulsifiable concentrate formulation. This same formulation concept can be applied to combinations of hexaflumuron with rice hopper control agents to produce dusts with broadspectrum activity against key rice pests in Japan.

REFERENCES

1.   Komblas, K.N. and Hunter, R.C., A benzoylphenyl urea for rational control of pests on fruit and vegetables. British Crop Protection Conference - Pests and Diseases, 1986, Proceedings Volume 3, PP.907 - 914.

2.   Kawamura, N., Kato, J., Liao, P.H. and Mendoza, N.S., XRD-473 A promising new insect chitin inhibitor with possible application towards some important pests of rice, tea and crucifers in Asian countries. Eleventh International Congress of Plant Protection, 1987, Proceedings

3.   Munakata, N., Kohyama, Y. and Shigematsu, T., Optimization of teflubenzuron EC formulation. Fourteenth Annual Meeting of Pesticide Science Society in Japan, 1989.

USE OF COMPUTER TOOLS FOR THE DESIGN
OF PEST MANAGEMENT STRATEGIES

DAVID R. WAREING
Imperial College, Silwood Park, Ascot, Berkshire, SL5 7PY.
JOHN HOLT
Overseas Development Natural Resources Institute,
Chatham Maritime, Kent, ME4 4TB.
JIA AN CHENG
Zhejiang Agricultural University, Hangzhou, Zhejiang, China.
GEOFF A. NORTON
Imperial College, Silwood Park, Ascot, Berkshire, S15 7PY.

ABSTRACT

The use of systems analysis and computer modelling is
described in connection with developing management
strategies for the control of brown planthopper in temperate
and tropical rice. The model developed for rice in Zhejiang
Province, China, has been used to determine the best times
to make one or two applications of insecticide. In either
case an initial application at 30 DAT is a remarkably robust
strategy. If required, a second application may be made 10
days later. This model has also been used to determine the
necessity for insecticide application and a series of
treatment thresholds has been developed for a range of field
situations.
     A model developed for rice in the Philippines has
highlighted the importance of the aggregative behaviour of
natural enemies in regulating brown planthopper populations
and also the importance of early season pest population
processes within the crop. This model has been used to help
identify those situations which lead to a high risk of a
brown planthopper outbreak, whether they arise through
immigration from adjacent crops or the interaction between
this and the direct use of insecticides, either against
brown planthopper or another insect pest. If natural
enemies are unaffected then the situation is similar to that
seen in China, where spraying at 30 DAT can achieve good
control of brown planthopper. However, when insecticide use
affects natural enemies, which are central to the regulation
of the population, then sprays at this time can *cause* a
brown planthopper outbreak. Resurgence risk is much lower
when applications are made before 20 or after 40 DAT.

INTRODUCTION

This paper describes two examples of the use of systems analysis and computer modelling in the development of management strategies for the control of brown planthopper (*Nilaparvata lugens* Stal), in irrigated rice in Zhejiang Province, China, and in the Philippines. The main aim is to illustrate the value of these tools in helping us understand the ecological processes involved, the situations where outbreaks of brown planthopper are likely to occur and how effective various control strategies are likely to be under different circumstances and, in particular, the likelihood of causing brown planthopper resurgence through the use of insecticides against either brown planthopper or another pest.

MODEL DEVELOPMENT

To simulate the development of a brown planthopper population the key components and processes need to be expressed in a mathematical form that allow incorporation into a computer model. This raises the question - What are the key components? One systems analysis technique which helps determine these components is the interaction matrix. The matrix used in the development of the simulation model for the Philippines is illustrated in Figure 1. The matrix aims to identify the important components of the system that need to be considered in selecting the key components to be included in the model itself. A dot or black square indicates, respectively, that a minor or major direct relationship is thought or known to occur; the row component affecting the column component.

The matrix is not only useful in helping to determine the components to include in the model, but also highlights where knowledge is lacking and helps to support decisions on field and laboratory experiments aimed at improving our understanding of brown planthopper ecology. A detailed

Figure 1   Interaction matrix constructed during development
of the Philippines model

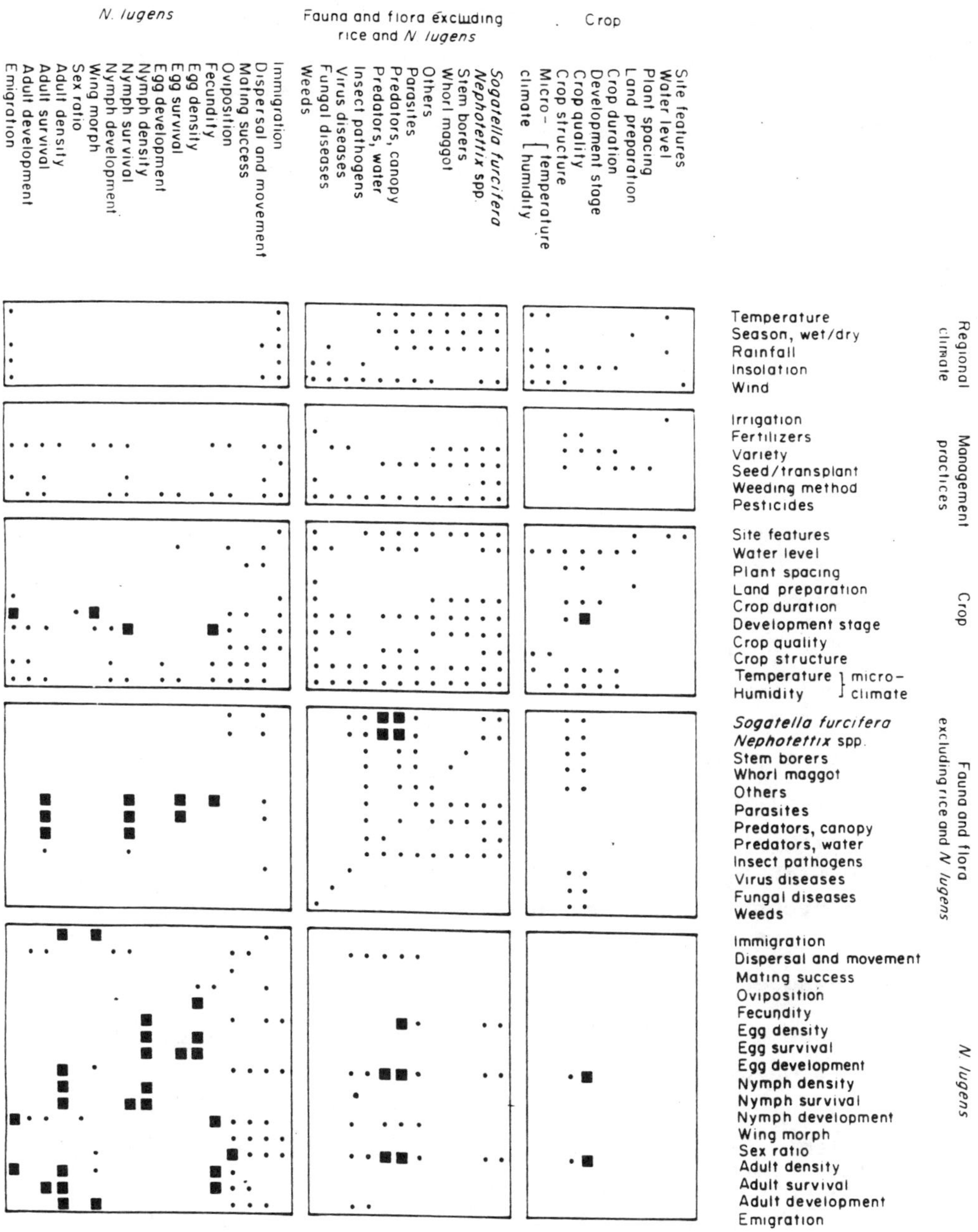

description of the development of the models may be found elsewhere [1,2,3]. The first goal with both simulation models was a reasonably accurate prediction of observed population trends in the field. Through this process of validation, the models were tested using independent data and found to give reasonably accurate simulations. The models were then used to address questions relating to brown planthopper outbreaks and how they might be caused through immigration from resurgence in nearby crops or directly through the use of insecticides.

RESULTS

Before looking at the results of the modelling work, we first need to consider the differences between the population dynamics of brown planthopper in temperate and tropical rice systems.

In tropical areas, such as the Philippines, brown planthopper normally remains at a low density [4]; indeed, insecticides which kill natural enemies commonly induce brown planthopper resurgence [5,6].

By contrast, in temperate rice, natural enemy regulatory mechanisms do not maintain the stability of brown planthopper populations and it is possible to predict outbreaks by monitoring the size of the immigrant population [2]. Such prediction is not possible in the tropics; the difficulties hinge around the complexity of the interaction between brown planthopper and its natural enemies [1,9].

Identifying outbreak conditions

Turning our attention to tropical rice, we can first address the question of the impact of immigrants on the occurrence of brown planthopper outbreaks. Previously, no clear relationship was apparent between immigration and outbreaks of brown planthopper in the tropics. Thus, one clear objective was to quantify the circumstances under which immigration might lead to an outbreak, which here we define as a peak density of brown planthopper adults and nymphs greater than 100 per hill. A range of immigration levels and times was simulated and this identified those combinations which lead to a brown planthopper outbreak, as illustrated in Figure 2. The results show that if brown planthopper immigration begins before 20 DAT then a relatively low immigration level might cause an outbreak. This is particularly important since it is very difficult to monitor such low levels. Nervertheless, very early immigration of a magnitude necessary to cause outbreaks is rare. After 20 DAT a very high level of immigration is needed.

Figure 2   Effect of immigration, lines join points of equal brown planthopper peak density.

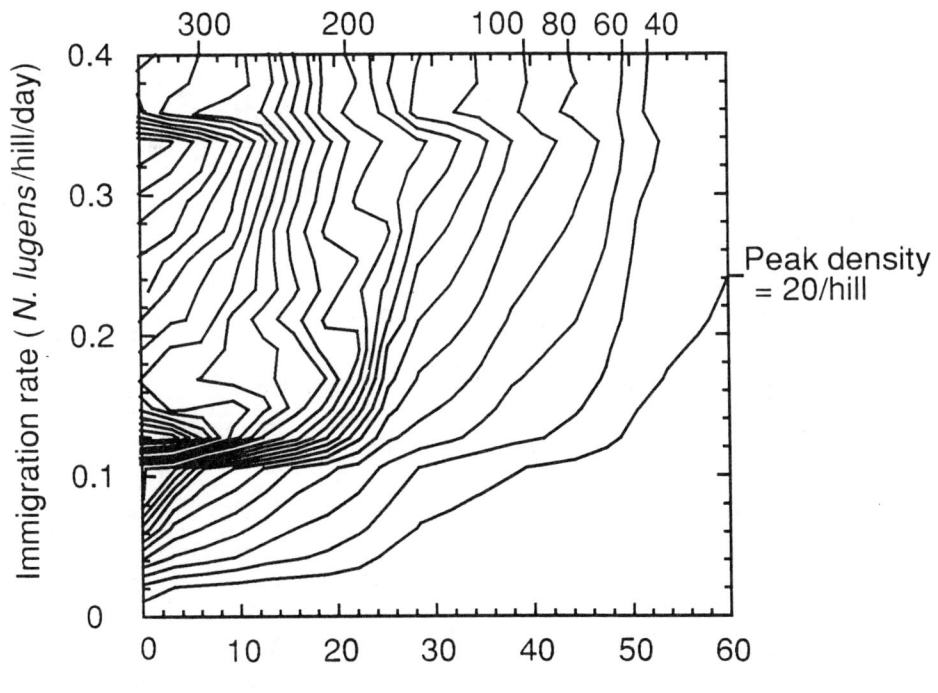

Developing management strategies.

Tropical rice

Having identified those conditions which can lead to an outbreak in tropical rice we can now focus on how the model can help in determining a satisfactory control strategy against rice pests when brown planthopper is present in the crop. Specifically, we can address the question - What is the best time to apply insecticide to control either brown planthopper or another pest without causing resurgence of brown planthopper?

Figure 3   Risk of causing brown planthopper resurgence by the application of three types of insecticide at different times.

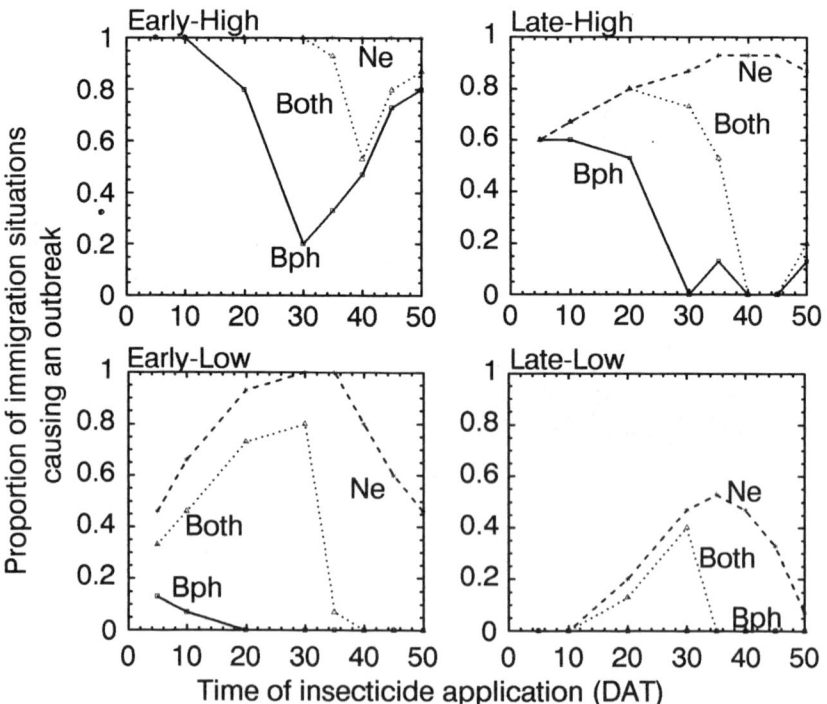

To answer this question the model was used to simulate
insecticide applications at a range of times in each of the
immigration scenarios described above. In each case this
resulted in a change in the peak density. Three different
insecticide types were simulated, which killed either brown
planthopper, natural enemies or both. The impact of these
insecticide applications is illustrated in Figure 3.
Application of an insecticide at 30 DAT can be the most
effective time to control brown planthopper, but it is
important to be careful that the spray does not effect
natural enemies since at this time there is also the highest
risk of resurgence. In general, delay to 40 DAT may be a
safer, but still effective option.

Temperate rice

In general, there are two central questions associated with
pesticide use against rice pests: the necessity for
application and the timing of application. A simulation
model can help to determine the most effective time by
simulating all possible times at which to spray for a range
of situations, and recording the outcome. This can be
expressed either in terms of a reduction in pest density or,
by the use of a damage relationship, in terms of increased
crop yield.

To illustrate, we turn our attention to the simulation
model of brown planthopper on rice in China and ask two
questions - Given that insecticide is to be applied, when
is the best time to apply it and, if necessary, when should
a second application be made? The model has been used to
simulate a single application of MTMC insecticide to the
rice crop at different times: 25, 30, 35, 40 45 and 50 DAT
[3]. Figure 4 shows that the lowest peak density of brown
planthopper results when an application of insecticide is
made at 30 DAT. Since there is a linear relationship
between density and yield loss, the most effective time to
spray MTMC is 30 DAT. This result agrees with findings of
field trials in Zhejiang Province where single applications
of MTMC were made at different times. Because brown

Figure 4   Best time for a single application of insecticide
in Zhejiang Province.

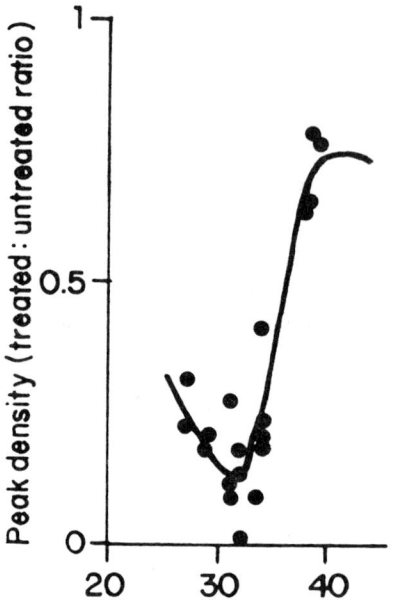

planthopper immigration into the crop nearly always starts
at 1 DAT, by 30 DAT egg density is almost at a minimum.
Consequently, most of the population is vulnerable,
insecticides having little effect against brown planthopper
eggs.

To assess whether spraying at 30 DAT is a robust strategy
the model has been used to study other parameters that are
likely to vary, particularly other management practices,
such as transplanting time.   Figure 5 shows the result of a
series of simulations where the effect on brown planthopper
of both insecticide application time and transplanting time
is determined.   In most cases, a single application of MTMC
at 30 DAT is both adequate (defined here as a peak density
of brown planthopper below 8 adults and nymphs per hill) and
most effective.   However, when the crop is planted early, a
single application of MTMC is no longer adequate and in this
case a second application is necessary.   To evaluate the
best times for two applications of insecticides, numerous

Figure 5 Effect of transplanting time of the efficacy of a single insecticide application.

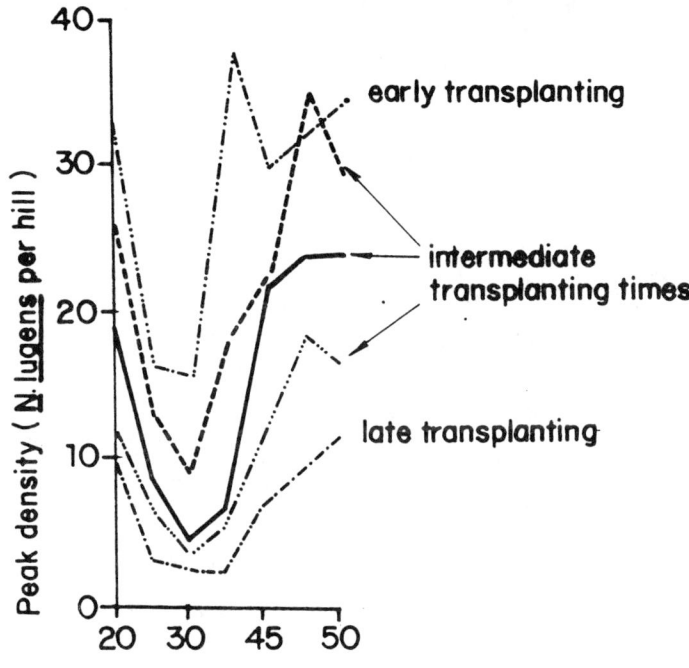

simulations were carried out where an initial insecticide application was made between 20 and 60 DAT and a second from 5 to 40 days after the first. The results (Figure 6) show that the minimum peak density of BPH is obtained by an application at 25 DAT and a second one 10 days later. However, making the initial application at 30 DAT allows much greater flexibility in the timing of the second application while still achieving acceptable control.

Thus, in Zhejiang Province, whether a single or two insecticide applications are to be made, the general rule is the same - apply the first treatment at 30 DAT. However, this begs the question of whether it is worth applying insecticide. To investigate this, we examined action thresholds for insecticide application.

In Zhejiang Province, from 0 up to 4 pesticide applications are made against brown planthopper in rice crops in a year. Local extension officers give advice each

Figure 6 Best times for two insecticide applications, lines join points of equal brown planthopper peak density.

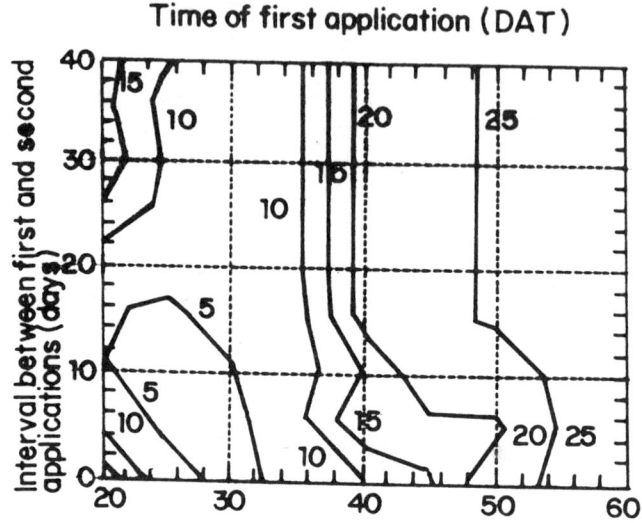

Time of first application (DAT)

year on the basis of monitored information on brown planthopper. Insecticide application is recommended to achieve the practical objective of limiting yield loss caused by brown planthopper to less than 5% [2]. To determine the brown planthopper density at which treatment should be recommended (treatment or action thresholds), the model was used to determine the density of the first generation which led to a second generation level associated with a 5% yield loss. For a set of standard values used in the model and a single application of MTMC, this was 2.6 nymphs and adults per hill.

To determine the treatment threshold for two applications, the model was again used to determine the first generation density associated with a 5% yield loss but this time after an initial application 30 DAT. Clearly the determination of treatment thresholds is complicated by variations in temperature and transplanting time, which affect the rate of brown planthopper development and therefore the level of crop damage. To account for this, nine scenarios incorporating three transplanting times (27 July, 1 and 6 August) and three temperatures (favourable, average, unfavourable) were considered. This resulted in a range of treatment thresholds, as shown in Table 1.

TABLE 1

Treatment thresholds for a single and two applications of MTMC as derived from the simulation model.

Transplanting time	Temperature	Daily average temperature (51-70 DAT)	Peak density per hill of 2nd generation that will cause 5% yield loss	Treatment (No. per 100 hills in first generation)	
				1 SPRAY	2 SPRAYS
	Favourable	21.9	3.4	3	13
July 27th	Average	20.9	5.8	6	26
	Unfavourable	19.9	10.2	13	62
	Favourable	21.1	5.1	6	24
August 1st	Average	20.1	8.9	17	100
	Unfavourable	19.1	15.5	39	187
	Favourable	20.4	7.7	13	94
August 6th	Average	19.4	13.3	45	343
	Unfavourable	18.4	23.1	122	777

DISCUSSION

As illustrated by the examples described, simulation modelling techniques can be of value in a variety of ways. During the development of the model, more qualitative systems analysis techniques can help target research and development towards the key questions which arise from highlighting gaps in our knowledge. Second, the modelling exercise synthesises data, knowledge and experience, allowing the development of improvements in management strategies.

Finally, it should be emphasised that these techniques should not be carried out at the expense of laboratory studies, field trials and other research and extension activities, but rather should complement these activities. Good models cannot be built without good data, knowledge and experience of the pest, but also, our understanding of the pest may be incomplete without models which can help us to test hypotheses rigorously and to the full.

REFERENCES

1. Holt, J., Cook, A. G., Perfect, T. J. and Norton, G. A. Simulation analysis of Brown Planthopper population dynamics in tropical rice: a simulation analysis. *Journal of Applied Ecology*, 1987, **24**, 87-102.

2. Cheng, J. A. and Holt, J. A systems analysis approach to Brown Planthopper control on rice in Zhejiang Province, China. I. Simulation of outbreaks. *Journal of Applied Ecology*. 1990 (in press).

3. Cheng, J. A., Norton, G. A. and Holy, J. (1990) A systems analysis approach to Brown Planthopper control on rice in Zhejiang Province, China. II. Investigation of control strategies. *Journal of Applied Ecology*. 1990 (in press).

4. Cook, A. G. and Perfect, T. J. The population characteristics of the brown planthopper, *Nilaparvata lugens*, in the Philippines. *Ecological Entomology*, 1989, **14**, 1-9.

5. Kenmore, P. E., Carino, F. O., Perez, C. A., Dyck, V. A. and Gutierrez, A. P. Population regulation of the rice brown planthopper (*Nilaparvata lugens* Stal) within rice fields in the Philippines. *Journal of Plant Protection in the Tropics*, 1984, **1**, 19-37.

6. Heinrichs, E. A. and Mochida, O. From secondary to major pest status: The case of insecticide-induced rice brown planthopper, *Nilaparvata lugens*, resurgence. *Protection Ecology*, **7**, 201-218.

7. Holt, J., Wareing, D. R., Norton, G. A. and Cook, A. G. The impact of immigration on brown planthopper population dynamics: a simulation analysis. *Journal of Plant Protection in the Tropics*, 1990 (in press).

IPM FOR TROPICAL ASIAN RICE FARMERS - A BLUEPRINT FOR INDUSTRY PARTICIPATION

WILLIAM T. VORLEY

Agricultural Division
CIBA-GEIGY Ltd.
CH-4002 Basel, Switzerland

ABSTRACT

Responsible members of the agrochemical industry are becoming increasingly aware that their commitment to IPM must extend beyond the search for selective 'IPM' compounds, to the voluntary modification of marketing techniques and customer service for established products. Messages of support and highly publicised token gestures are not enough to bring companies into line with the demands of IPM-friendly marketing. Changes in product use recommendations on labels and other literature, changes in sales language, and support for IPM research and extension are concrete steps that each agrochemical company can take towards harmonisation with IPM principles.

The needs of the IPM end user, the farmer, are best met through collaboration between industry and national authorities, rather than enforcing 'IPM' with restrictive legislation.

INTRODUCTION

Traditionally, industry has viewed its major role in integrated pest management (IPM) as the contribution of safe, selective chemicals [e.g. 1], and has so far made little contribution to IPM extension. This reluctance of industry to assist in the 'downstream' implementation of IPM has been attributed to concern that IPM will reduce market size (especially for the older, broad-spectrum compounds) and that increased extension costs will reduce the competitiveness of its products. In fact, the marketing activities of the agrochemical industry have been cited as a significant constraint to IPM implementation [2, 3, 4], which has further entrenched 'green' attitudes towards industry.

The divide between industry and pro-IPM groups is aggravated by confusion on both sides about the objectives and benefits of IPM. This paper attempts to bridge that divide, using the example of insect pest control in tropical rice, and challenges agrochemical companies, governments and environmental pressure groups. For companies which are interested in improving the quality of use and stewardship of their products, and maintaining the long-term security of their markets, the challenge is for them to put their full weight behind appropriate IPM research and extension [5,6]. Because many companies are unclear on how to react to the IPM challenge, a blueprint model of an 'IPM-responsive company' is presented. For governments and research organisations which are interested in the farmer acceptability and sustained implementation of IPM, the challenge is for them to work in collaboration with industry [7], rather than enforcing 'IPM' on them by restrictive legislation. Pressure groups are challenged to avoid presenting IPM as a universal panacea, because current IPM technologies for rice will only be widely adopted by farmers when they are less complex and offer better improvements in yield and profit than those presently achievable [8].

Why should industry contribute to the IPM research and extension process?

There are quality, policy and product stewardship reasons for industry to assist IPM research and extension:

Quality of use. IPM-trained farmers get the most benefit and the least risk (to themselves and to the environment) from each insecticide application. IPM-trained farmers are quality-conscious farmers.

Policy commitments. Through GIFAP, industry is committed to supporting training in safety and IPM by its endorsement of the FAO Code-of-Conduct [9] (notably Article 3.8) and this may be reinforced by internal policy statements.

Product Stewardship. There is a growing realisation that some broad spectrum insecticides can be fully compatible with IPM, both in terms of selectivity and benefit/costs, provided that farmers are made fully aware of methods for their correct usage. Thus, it is in the interests of farmers that companies should not wait for the development of highly selective products before promoting IPM - such products are anyway likely to be priced out of reach of most tropical rice farmers.

IPM is also seen as an effective resistance management strategy for prolonging the life of useful chemistry [10].

Why should governments and research organisations collaborate with industry for IPM research and extension?

The commercial sector can benefit IPM extension programmes through its access to different extension channels, and through its field experience:

Access to different extension channels. The commercial pesticide distribution channel (salesmen, dealers/retailers) is an important source of farmers' pest control advice [11], which until now has not been properly exploited for the dissemination of information on the effective and safe use of the products which they sell. The agrochemical industry has much stronger links to this channel than extension services.

Experience. Agrochemical companies have experience of education in the safe and effective use of insecticides, and in the promotion of ideas through advertising and training. Company trialists have a broad understanding of pest control tactics, and are experienced in conducting field experiments of a high standard.

PROFILE OF AN IPM-RESPONSIVE COMPANY

1. Awareness of the 'IPM-fit' of products

Pesticides are often sold and used in inappropriate market segments - either they are relatively ineffective against the stated target pest, or are highly toxic to the pest's key natural enemies. An IPM-responsive company would review the activity and selectivity of their product range through quality field trials involving creditable national institutions, for each crop/pest combination which appears on the product label or promotional literature. Recommended products should have proven economic benefits and minimum environmental risks/costs.

Case Study 1: IPM-fit of enolphosphate insecticides in tropical rice

The enolphosphate insecticides including monocrotophos and phosphamidon are highly active against a range of rice pests. Monocrotophos in particular gives excellent control of foliar Lepidoptera and bugs, and good control of stemborer and hoppers. They are popular with farmers because of this spectrum of activity and their low cost. However, they have a reputation (mainly derived from lab screening) of being broad spectrum, non-selective compounds, and this image led to their being included amongst the organophosphates banned in Indonesian rice in 1986 for reasons of supposed IPM incompatibility.

Two useful indices of IPM-fit in flooded rice are (a) toxicity to the key brown planthopper (BPH) predators, spiders, and (b) induction of BPH resurgence [e.g. 12].

Monocrotophos (Nuvacron®) and phosphamidon (Dimecron®) were screened for these characteristics in the 1987/8 and 1988/9 wet seasons in Cikampek and Purwokerto, Java, respectively. Relative to the pyrethroid 'resurgence standard', both products proved to have remarkably low toxicity to spiders, and did not cause BPH resurgence (Figure 1). This finding is supported by IRRI data [13,14]. Thus for this particular ecosystem, monocrotophos and phosphamidon may be classified as IPM compatible.

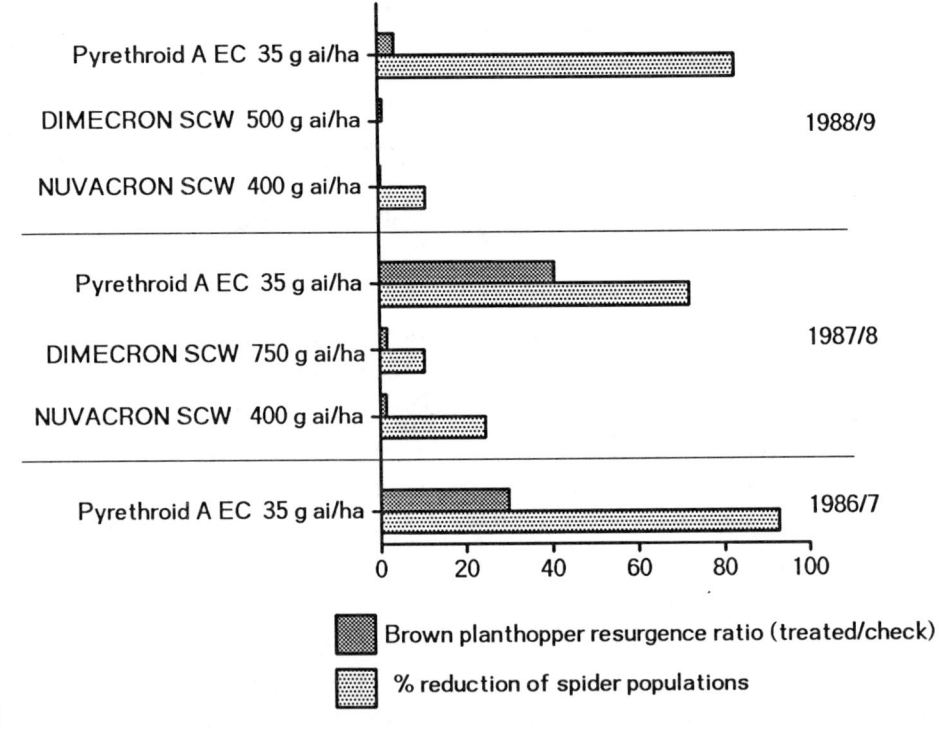

Figure 1. IPM-fit of Dimecron®, Nuvacron® and pyrethroid standard in rice: spider toxicity and brown planthopper resurgence. Java, Indonesia 1986-9

2. Product Use Recommendations (labels etc.)

A typical label for an insecticide for use in rice will give recommendations for calendar applications, such as: 'Apply 500 g ai/ha every 14 days from 15 days after transplanting'. Should an IPM-responsive company replace calendar-based recommendations with thresholds, and is the recommended dose economically justified?

Thresholds. Economic thresholds (ETs) are the cornerstone of present recommendations for the judicious needs-based use of pesticides in IPM programmes. Therefore it follows that an IPM-responsive company should state on labels and product literature that the product should be used only when nationally or regionally recommended ETs are exceeded.

Several arguments are put forward by companies to justify non-adherence to ET principles:

a) *"Calendar spraying gives higher yields than ETs."* The preference for calendar spraying over threshold-based applications can be very logical. Early-season preventative applications timed to coincide with peaks in populations of endemic pests may effectively control the pest at low doses, because the pest is present in young, sensitive instars. Early applications will also give little disruption to the complex of beneficial arthropods. Furthermore, the farmer is spared the labour of weekly sampling. Application in response to ETs, on the other hand, may require a higher dose to get good control, especially if the threshold is based on damage symptoms (such as 'deadhearts' for rice stemborer), which may not be apparent until the pest is well established in the late larval or pupal stages. If recommendation for calendar spraying is based on sound biological information, and consistently out-profits national ET-based recommendations and farmers' practice, and gives better conservation of beneficials, then it can justifiably be termed 'IPM'. But its superiority must first be proven, in collaboration with a reputable national institution, over a wide range of seasons and pest infestations.

b) *"ETs are set too high."* Industry people often complain that national agricultural authorities set ETs too high to give reasonable yield protection. There often seems to be some justification to the claim that some thresholds have little more scientific rationale than an appearance in IRRI literature, and have no proof of profitability under local farm conditions. Insecticide use in response to low ETs may give excellent pest control and yield protection, but because untreated infestations passing low ETs do not always go on to exceed the economic injury level (EIL), the 'economic' aim of the ET may not be fulfilled. Treatment at high ETs, on the other hand, may not give good pest control because the infestation has become too well established, or because the damage has already been done. As thresholds move from low to high ET values, the benefit to the farmer changes from yield protection through telling the farmer when best to control his pests, to cost saving through telling the farmer when not to control. It is the task of all concerned, including industry, to support research to determine where between these two extremes ETs for each pest should be set.

c) *"ETs are too complex and/or too laborious for farmers to use"*. Thresholds are often expressed in units unchanged from their research station prototypes; i.e. units/m2, percentage damage etc. [e.g. 15]. Such units may be too complex even for the extension workers whose task it is to disseminate IPM techniques. Improving the accuracy of ETs may make matters worse - IRRI ETs for stemborer now require scouting for egg masses, and for whorl maggot, searching for minute eggs in neighboring fields [16,17]. Criticisms of excessive labour requirements for sampling may also be justified, especially with the trend towards 'part-time' rice farming and absentee owners.

Doses. The price of insecticides is high in tropical countries relative to a low value crop such as rice [16]. Dose recommendations should reflect this, and be set to a minimum based on optimum returns, not necessarily maximum kill and highest yield.

To summarise, prophylaxis and calendar spraying can be proposed by a company as an alternative to ETs only if it consistently out-profits national ET-based recommendations and farmers' practice, and gives better conservation of beneficials. Criticisms of ETs as "too high" or "too complicated" should not be an excuse to abandon the concept, but a challenge to support research on profitable, farmer-friendly thresholds.

Case study 2: Research into pesticide use recommendations in Indonesia

In six cropping seasons from the 1986/87 wet season to the 1989 dry season, six different insecticide use regimes were compared at the Ciba-Geigy R&D station near Cikampek, West Java, Indonesia. Details of methodology can be found in [5]. These regimes comprised one high-input and one low-input prophylactic regime ('maximum protection' and 'minimum protection'), two ET-based regimes with thresholds set at low or high levels ('low ETs' and 'high ETs'), an average farmer's actions based on weekly surveys in a neighbouring village ('farmers' practice') and an untreated check. Comparisons were made between each regime regarding number of applications, pesticide cost, yield, profit and effect on beneficials.

Predictably, 'maximum protection' gave the highest yield, averaging 12% above untreated for BPH-resistant varieties and 47% for BPH-susceptible varieties. The different regimes varied greatly in terms of profitability. In resistant varieties, the threshold-based treatments and farmers' practice (especially 'high ETs') out-profited the prophylactic regimes, and in the exceptionally pest-free 1988-89 dry season, 'no control' was the most profitable (Figure 2). The use of low ETs and high ETs achieved only 53% and 37% respectively of the maximum protection yield increase (Figure 3).

GAIN FROM INSECTICIDE (US$/HA)

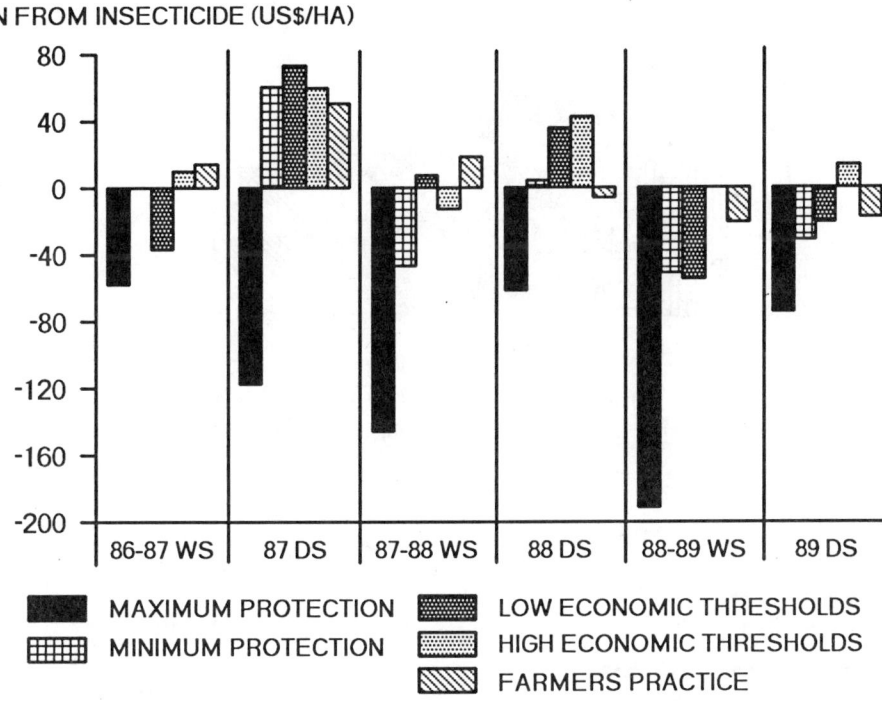

Figure 2. Profitability of different rice insect control regimes, BPH-resistant varieties. 1986/7 wet season (WS) to 1989 dry season (DS). Cikampek, W. Java.

These results clearly show that, under these conditions in W. Java, it would be inappropriate to recommend calendar applications of the products tested. Even populations of endemic pests such as stemborer were not predictable enough to warrant calendar applications of granular products. In this region, farmer intuition and experimentation (together with advice from the extension services) has led to a kind of 'IPM', with low-cost preventative applications of granular insecticides being followed by one to two foliar sprays in response to pest sightings or actions of neighbours. IPM extension should build on this farmers' good sense, rather than attempting to replace it with a completely new system.

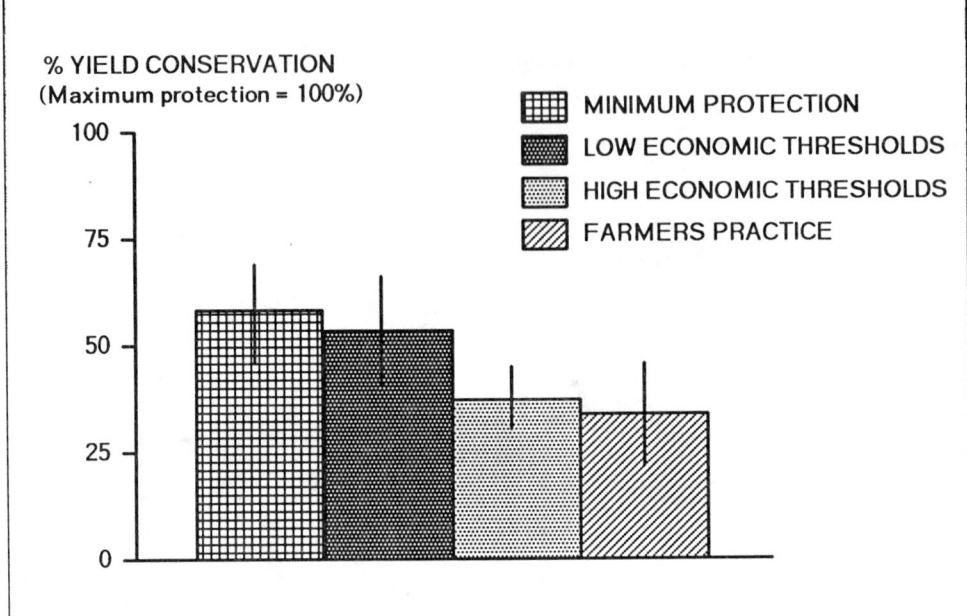

Figure 3. Yield conservation with economic thresholds. Average of 6 seasons (± S.E.), BPH-resistant varieties. Cikampek, W. Java, 1986-9.

3. Sales language.

There is some justification to accusations [3,18] that farmers are sometimes encouraged to waste money on unnecessary pesticide applications by advertisements which increase farmers' fear of the unseen pest. To be in line with the principles of IPM, sales messages and advertisements must move away from promoting 'spray to be sure' and 'spray for a clean field' messages, towards the message of profit optimisation from rational pesticide use. Agrochemicals should not be marketed as the only solutions to pest problems - cultural control methods (resistant varieties, synchronous planting etc.) are part of all 'integrated' programmes, and rice pest control is no exception.

4. Support for national IPM research

Evaluating the bioefficacy of new agrochemicals requires relatively simple, albeit very costly research, involving the assessment of toxicological effects (usually % mortality) of chemical biocides. The development of IPM systems, on the other hand, is a much more demanding task, involving aspects not only of multi-pest population dynamics, crop loss assessment and economics in the research stage, but also sociology and anthropology in the extension phase. IPM recommendations have to be fine-tuned to respond to local variations in pest abundance, crop yield and value and availability of insecticides, and must also be expressed in terms easily understood by target farmers. It is misleading to present IPM as a ripe technology just waiting for extension. Stronger efforts are needed to make IPM more attractive to farmers, and to give improvements in yield over current practices. Not surprisingly, there is much scope for industry to assist research in rice IPM systems, to improve the profitability and acceptability of IPM recommendations and then to actively promote the extension of those recommendations to the farming community.

IPM for yield increase. Data are widely available which confirm the findings in Figure 3, showing that when ETs are used in rice, only a portion (usually less than 50%) of potential yield loss due to insect pests is conserved, and no significant yield improvements over current farmer practices are achieved [16,19,20]. The paramount IPM objective of optimising grower profitability should not lead us to consider these yield losses as inevitable. Asian countries with rapidly growing populations will need to minimise yield losses for intensified rice production. More research is needed to improve ETs regarding yield protection, to allow a more precise and effective chemical intervention once thresholds are passed, preferably excluding ETs based on damage. It is dangerous to root our IPM principles for the tropics in the Western environment of crop surpluses and set-aside [see also 21].

Low-cost pest control technologies. Many new selective compounds are too expensive to be economically viable in tropical rice farming. The yield conservation required to recoup the investment is unrealistically high, especially considering that losses to all pests average only ca. 20% in most tropical conditions [22,23]. Industry should give more support to the development of low cost alternatives, such as locally produced botanical extracts.

User-friendly sampling aids. Once spray decision criteria are established, they should be made available to farmers in the form of simple tools to show if pests are above or below economic thresholds. The peg-board first used for cotton scouting [24] could be a model from which to develop such sampling aids.

IPM in rice-based cropping systems (RBCS). Pest problems and pesticide misuse are often more serious in the vegetable crops grown in rotation with rice. Research on IPM systems in RBCS crops should aim to increase the profitability and reduce the resistance and residue problems associated with present farmers' practice.

5. Training and Extension

Farmer training. Even if considerable research effort has been expended* on the development of IPM systems which offer farmers a profitable and user-friendly alternative to their present practices, companies should not set themselves up as independent IPM specialists. Instead, they should assist national extension authorities to get the IPM message to as many farmers as possible. A concrete and uncontroversial first step would be the sponsorship of pest identification displays in village centres and extension offices. Safety training should go hand in hand with IPM training, and the excellent GIFAP protocols for farmer and dealer training [11,25] are a valuable resource for this purpose. Participation of farmer groups in the needs-identification, planning and implementation phases of IPM projects will ensure that meeting farmers' needs remains the top priority [e.g. 26,27].

Training for dealers and sales representatives. There is no benefit in establishing the IPM-fit of products and recommendations for their use in IPM systems, if sales staff and retailers are not trained and motivated to pass this information onto the customer (see Article 11.1 in [9]). For this to succeed, incentives must extend beyond the usual rewards for meeting sales targets. Good extension must be rewarded, and a challenge for industry's participation in IPM programmes is to make salesmen and retailers more service-oriented through competitions or other incentive schemes.

Case study 3: A multidisciplinary rice IPM project underway in the Philippines

In order to support national IPM policies and enhance its commitment to effective and safe pest control, Ciba-Geigy (Philippines) Inc. has embarked on a comprehensive · three-pronged rice IPM research and extension programme. The three targets are:

a) Supporting basic IPM research at a national research institute, including work on farmer-friendly sampling aids, IPM in rice-based cropping systems, and selectivity of rice insecticides.

b) Carrying out in-house comparisons of the 'IPM-fit' of existing marketing recommendations with national ET-based recommendations and farmers' practices.

c) Supporting farmer education in effective and safe pest control, around a training centre in Nueva Ecija.

HOW GOVERNMENTS CAN ENCOURAGE OR DISCOURAGE THE PARTICIPATION OF INDUSTRY IN IPM PROJECTS

The degree of support or reservation on the part of industry towards national IPM projects is determined to a large extent by the legislation which the government adopts to support their IPM policies. No move can be more certain to antagonise industry than legislation which restricts the availability of useful and popular products without apparent consideration of the end users' real needs.

Indonesia is the first rice-growing country to ban the use of certain products on the grounds of IPM-incompatibility. From November 1986, 57 registered products, mainly organophosphates (OPs), have been banned from use in rice because they 'can bring about resurgence, resistance and other harmful impacts'. It is commendable that governments should support IPM implementation and encourage the rational use of pesticides, for a host of reasons including grower profitability, conservation of foreign exchange and environmental protection. However, the blanket ban of OPs in Indonesia overlooked several crucial factors. Firstly, OPs vary enormously in their toxicity to key beneficials - as mentioned, the enolphosphates monocrotophos and phosphamidon, for example, are relatively selective to spiders in the tropical rice ecosystem. Secondly, the restriction in the variety of insecticide groups down to only the phenylcarbamates isoprocarb and fenobucarb used over wide areas is certainly hastening the onset of carbamate resistance in the brown planthopper, which is now threatening rice security in large areas of central and west Java, in the same way as restriction of plant variety in Indonesia led to Asia's worst BPH biotype problem. Thirdly, the ban has left rice farmers with no foliar chemicals to control the endemic stem pests of rice, especially stemborers and gallmidge, and epidemic leaf feeders such as leaffolders.

In addition to IPM-compatibility, national authorities are reviewing registered insecticides for their toxicological hazard to the applicator, and industry is responding by developing safer formulations, packaging and application techniques. Despite the sensitivity of the subject of applicator hazard, it is again valid to advise against sweeping bans without analysis of real risks and benefits. For knapsack applicators with the normal minimal protective clothing, potential dermal exposure is much greater than oral or respiratory exposure [e.g. 28]. Therefore from a scientific point of view, it is difficult to follow a risk/benefit assessment which leads to a ban on two compounds that have dermal LD_{50}s as different as 10 and 350 mg/kg, because their oral LD_{50}s have put them into the same hazard class.

CONCLUSIONS

The message is clear: Responsible agrochemical companies should not wait for the perfect 'IPM insecticide' before putting IPM principles into their marketing practices, because IPM principles can be developed for currently established products. A proactive rather than a reactive approach to IPM will provide companies with the data and experience to assist in the development of national IPM policies, and therefore reduce the risk of politically motivated legislation based on spurious data, which is not in the farmers' interests. Only a few companies may be prepared (or be able) to go to the lengths of the 'model' IPM company described above, but some action is essential. Helping farmers to correctly identify the pests and beneficials in their fields is a relatively cheap and uncontroversial first move in the IPM direction.

Many of the features of an IPM-responsive company outlined here are inexpensive to implement, but there is no question that full compliance to these principles will increase costs and erode profitability, unless the costs are passed on to the farmer in higher pesticide prices. Efforts must be made by manufacturers associations to ensure that local formulators as well as the large multinationals bear some of the burden of farmer education.

Before translating 'IPM' into law, governments and pressure groups should ensure that the farmer's needs remain a top priority. Current IPM recommendations in tropical rice answer neither of the farmer's aspirations to raise yields and increase profits. The concept of IPM for more accurate pest management and higher yield (rather than IPM for less insecticide input) has hardly been touched, and warrants an intensive research effort.

Governments should be very careful not to legislate farmers out of their most useful and economic insect control weapons - the outcome may be costlier, riskier and less effective pest control, in countries which need to achieve sustainable high yields with minimised insect losses. The alternative, of working in collaboration with industry to improve the effective and safe use of existing products (Article 11.3 in [9]), should be intensively explored by both governments and agrochemical companies.

ACKNOWLEDGMENTS

The contributions of the staff of Ciba-Geigy Indonesia, Philippines and Basle to the development of the concepts expressed in this paper are greatly appreciated.

REFERENCES

1. GIFAP, Integrated Pest Management - crop protection and provision of food and fibre. Groupement International des Associations Nationales de Fabricants de Produits Agrochemiques, Brussels, 1985, 10pp.

2. Van Lenteren, J.C., Environmental manipulation advantageous to natural enemies of pests. In Integrated Pest Management - Quo Vadis? proc. Symp. 'Parasitis 86', Geneva, 9-11 December 1986. Parasitis, Geneva, Switzerland, 1987, pp. 123-162.

3. Bull, D., A growing problem - Pesticides and the Third World Poor. OXFAM, UK, 1982, 192 pp.

4. Van den Bosch, R., The Pesticide Conspiracy. Doubleday, NY, 1978, 226 pp.

5. Vorley, W.T., IPM in rice - an extension role for the agrochemical industry. In Pesticide Management and Integrated Pest Management in Southeast Asia. Consortium for International Crop Protection, Maryland USA, 1988, pp. 213-220.

6. Statement of conclusions from the conference 'Pest Management and the Asian Farmer', Nairobi, Kenya, 22-26 May 1989.

7. Geach, N.G.E., Agrochemical industry needs stewardship. Farm Chemicals International, November 1989, 74-5

8. Wearing, C.H., Evaluating the IPM implementation process. Ann. Rev. Entomol. 1988, 33, 17-38.

9. FAO, International Code of Conduct on the Distribution and Use of Pesticides, FAO, Rome, 1986.

10. Voss, G., Insecticide/acaricide resistance: Industry's efforts and plans to cope. Pestic. Sci. 1988, 23, 149-156.

11. Lowe, J.C., The GIFAP education and training programme. Proc. Brighton Crop Protection Conference - Pests and Diseases-1988, pp. 1068-1076.

12. Vorley, W.T., Spider mortality implicated in insecticide-induced resurgence of white-backed planthopper (WBPH) and brown planthopper (BPH) in Kedah, Malaysia. Int. Rice Res. Newsl., 1985, 10:5, 19.

13. Fabellar, L.T. and Heinrichs, E.A., Relative toxicity of insecticides to rice planthoppers and leafhoppers and their predators. Crop Protection, 1986, 5, 254-258.

14. Reissig, W.H., Heinrichs, E.A. and Valencia, S.L., Effects of insecticides on Nilapavarta lugens and its predators: spiders, Microvelia atrolineata, and Cyrtorhinus lividipennis. Environ. Entomol., 1982, 11, 193-199.

15. Lim, G.S. and Heong, K.L., The role of insecticides in rice integrated pest management. In Judicious and Efficient Use of Insecticides on Rice; Proc. FAO/IRRI Workshop. IRRI, Los Banos, Philippines, 1984, pp.19-39.

16. Smith, J., Litsinger, J.A., Bandong, J.P., Lumaban, M.D. and DelaCruz, C.G., Economic thresholds for insecticide application in rice: profitability and rissk analysis to Filipino farmers. J. Pl. Prot. Tropics, 1989, 6, 67-87.

17. Bandong, J.P. and Litsinger, J.A., Development of action control thresholds for major rice pests. In Pesticide Management and Integrated Pest Management in Southeast Asia. Consortium for International Crop Protection, Maryland USA, 1988, pp. 95-102.

18. Kenmore, P.E., Crop loss assessment in a practical integrated pest control program for tropical Asian rice. In Crop Loss Assessment and Pest Management (ed. P.S. Teng). APS Press, St. Paul, Minnesota, pp. 225-241.

19. IRRI, Insecticide Evaluation for 1981-1986. International Rice Research Institute, Los Banos, Philippines, 1982-1988.

20. Waibel, H., The Economics of Integrated Pest Control in Irrigated Rice. Springer-Verlag, Berlin, 1986, 196 pp.,

21. Ruttan, V.W., Sustainability is not enough. Americal Journal of Alternative Agriculture, 3, 128-130.

22. Litsinger, J.A., Canapi, B.L., Bandong, J.P., Dela Cruz, C.G., Apostol, R.F., Pantua, P.C., Lumaban, M.D., Alviola III, A.L., Raymundo, F., Libetario, E.M., Loevinsohn, M.E., and Joshi, R.C., Rice crop loss from insect pests in wetland and dryland environments of Asia with emphasis on the Philippines. Insect Sci. Applic., 1987, 8, 677-692.

23. Waibel, H., Possibilities of economics as a decision tool for plant protection in developing countries. Paper presented at the International Symposium on Integrated Pest Management in Tropical and Subtropical Cropping systems, Feb 8-15 1989, Bad Duerkheim, W. Germany (in press).

24. Beeden, P., The pegboard - an aid to cotton pest scouting. PANS, 1972, 18, 43-45.

25. GIFAP, Trainers Manual. Groupment International des Associations Nationales de Fabricantes de Produits Agrochimiques, Brussels, 1988, 153 pp.

26. Farrrington, J., Farmer participation in agricultural research. Food Policy, May 1989, 97-100.

27. Putter, C.A.J., Potential opportunities for rational pest control in developing countries. In Rational Pesticide Use (eds. K.J. Brent and R.K. Atkin). Cambridge University Press, 1987, pp. 255-267.

28. GIFAP, Monitoring Studies in the Assessment of Field Worker Exposure. Technical monograph, Groupement International des Associations Nationales de Fabricants de Produits Agrochemiques, Brussels, 1990 (in press).

16. Smith, J., Litsinger, J.A., Bandong, J.P., Lumaban, M.D. and DelaCruz, C.G., Economic thresholds for insecticide application in rice: profitability and rissk analysis to Filipino farmers. J. Pl. Prot. Tropics, 1989, 6, 67-87.

17. Bandong, J.P. and Litsinger, J.A., Development of action control thresholds for major rice pests. In Pesticide Management and Integrated Pest Management in Southeast Asia. Consortium for International Crop Protection, Maryland USA, 1988, pp. 95-102.

18. Kenmore, P.E., Crop loss assessment in a practical integrated pest control program for tropical Asian rice. In Crop Loss Assessment and Pest Management (ed. P.S. Teng). APS Press, St. Paul, Minnesota, pp. 225-241.

19. IRRI, Insecticide Evaluation for 1981-1986. International Rice Research Institute, Los Banos, Philippines, 1982-1988.

20. Waibel, H., The Economics of Integrated Pest Control in Irrigated Rice. Springer-Verlag, Berlin, 1986, 196 pp.

21. Ruttan, V.W., Sustainability is not enough. Americal Journal of Alternative Agriculture, 3, 128-130.

22. Litsinger, J.A., Canapi, B.L., Bandong, J.P., Dela Cruz, C.G., Apostol, R.F., Pantua, P.C., Lumaban, M.D., Alviola III, A.L., Raymundo, F., Libetario, E.M., Loevinsohn, M.E., and Joshi, R.C., Rice crop loss from insect pests in wetland and dryland environments of Asia with emphasis on the Philippines. Insect Sci. Applic., 1987, 8, 677-692.

23. Waibel, H., Possibilities of economics as a decision tool for plant protection in developing countries. Paper presented at the International Symposium on Integrated Pest Management in Tropical and Subtropical Cropping systems, Feb 8-15 1989, Bad Duerkheim, W. Germany (in press).

24. Beeden, P., The pegboard - an aid to cotton pest scouting. PANS, 1972, 18, 43-45.

25. GIFAP, Trainers Manual. Groupment International des Associations Nationales de Fabricantes de Produits Agrochimiques, Brussels, 1988, 153 pp.

26. Farrrington, J., Farmer participation in agricultural research. Food Policy, May 1989, 97-100.

27. Putter, C.A.J., Potential opportunities for rational pest control in developing countries. In Rational Pesticide Use (eds. K.J. Brent and R.K. Atkin). Cambridge University Press, 1987, pp. 255-267.

28. GIFAP, Monitoring Studies in the Assessment of Field Worker Exposure. Technical monograph, Groupement International des Associations Nationales de Fabricants de Produits Agrochemiques, Brussels, 1990 (in press).

INTEGRATED PEST MANAGEMENT IN RICE: PRESENT STATUS AND FUTURE PROSPECTS IN SOUTHEAST ASIA

B. MERLE SHEPARD
Resident Director
Clemson University
Coastal Research and Education Center
2865 Savannah Highway
Charleston, SC 29414

ABSTRACT

Integrated Pest Management (IPM) in rice is in its formulation stages in South and Southeast Asia, although several programs have been initiated. Resistant rice varieties and chemical pesticides have been the major tactics with almost no attention given to the value of indigenous natural enemies (predators, parasites and entomopathogens). Effective IPM programs must be sociologically sensitive and adapted to local farmers' needs. Future activities should focus on 1.) identification of pests and beneficial species, 2.) assessing pest damage, and determining action thresholds 3.) developing sampling/monitoring methods 4.) selecting appropriate control tactics and 5.) integration. Farmer training is the key for successful IPM implementation but education of policy makers to IPM principles also is critical.

INTRODUCTION

The philosophy and principles of integrated pest management (IPM) have been known for over 25 years and although IPM programs are operational for many crops in developed countries, it is still more of an aspiration than a reality in less developed ones [1]. Unfortunately the concept is often misunderstood, especially by nonspecialists, who think of IPM as a new tactic (such as a new resistant variety) or a neat package of tactics and strategies that can be readily applied to any given situation. Further, some have the mistaken impression that IPM will replace all existing pest control technologies.

It is unfortunate that there are several, but actually closely related, definitions of IPM. Kenmore et al (1985) [2] provided a workable definition by suggesting that IPM is the farmers' "best mix" of control tactics based on crop yield, profit, and safety. IPM is actually a philosophy--a way in which crop protection is viewed--not a fixed set of procedures or activities. K.L. Heong (Personal Communication) has summarized past, present, and future IPM approaches in table 1.

IPM grew out of a recognition that total reliance on chemicals for insect control is unsound. The concept originated within the discipline of entomology, evolving from an earlier term "integrated control," which originally meant the integration of biological with chemical means for controlling insects [3]. The IPM concept now embraces all pest control disciplines but insect pests and plant pathogens are the major subjects of most IPM programs in effect today. The integration of weeds and vertebrate pests into IPM programs is evolving slowly as more information is obtained.

There are those who argue that IPM is not tailored to fit the needs of developed countries [1], because, among other reasons, mechanism for technology transfer is seriously limited. It must be recognized, however, that IPM programs need not be complex ones requiring technologically advanced delivery systems. Given the current status of pest control in many rice-growing areas of South and Southeast Asia, there is ample room for IPM development and implementation even in its simplest form.

Resistant rice varieties and chemicals are now considered by most to be the major pest control tactics. Some cultural practices are advocated but these, in general, play a minor role in most rice pest control programs.

As Kenmore et al (1987) [4] pointed out, Philippine rice farmers have used more and more insecticides during the last 30 years. Interestingly, insecticide use on rice began before the new rice varieties were developed. This is contrary to the belief by some that the modern varieties required more pesticides. During the early 1950s insecticide use was very low in the Philippines but by 1965, 60% of the farmers in irrigated rice used those chemicals and by the mid 1980s this increased to over 95% [4]. This example is typical of other regions of Asia.

Unfortunately, the importance of indigenous biological control agents has been largely overlooked. These valuable resources cost the grower nothing and maintain insect pest populations below economically important levels most of the time. Further, these rich communities of parasites, predators, and entomopathogens are relatively permanent, and in more than 50% of the fields (n = 300) observed in the Philippines, no insecticide applications were needed because of the action of the beneficial fauna. In another study over 330 farmers' crops were used to compare insecticide treated and untreated fields. Only 50% of the fields showed

TABLE 1

Past, present and future approaches to integrated pest management. (K.L. Heong – The International Rice Research Institute).

APPROACHES TO PEST MANAGEMENT

	Past or Current	Present–IPM	Future–IPM
Goal	Eliminate or reduce pest	Maximize profits	Multiple economic, ecological and social goals
Target	Single pest	Several pests and their natural enemies	Floura and Fauna of cultivated area
Criteria for Intervention	Calendar treatments	Economic threshold	Multiple criteria
Principal Method	Single method, e.g. pesticide or HPR	Reduction of pest by monitoring, timing, plant breeding, etc.	System designed to minimize outbreaks
Research Goal	Improve pesticides or varieties	More kinds of inter-ventions	Minimize need for intervention

a measurable yield loss to insects [5]. Major reasons for the lack of attention to biological control is that (a) pest control technicians and farmers are not trained to recognize the importance of these species; (b) quantitative information about the efficacy of biocontrol agents is generally lacking.

Exploitation of the full range of IPM strategies and tactics for rice will take time to evolve. However, there is ample evidence to show the profitability of IPM in rice and its benefits in terms of production sustainability and environmental compatibility.

Several agencies and institutions are involved in different aspects of IPM development and implementation. In the Philippines, the Department of Agriculture (DA) has fully embraced the IPM philosophy, and the Food and Agriculture Organization (FAO), the Philippine-German Crop Protection Programme, and the International Rice Research Institute (IRRI) are providing support in the form of training and research. Still it will take years for farmers to put IPM into practice on a wide scale.

DEVELOPING EFFECTIVE RICE IPM FOR THE FUTURE

Identification of potential pest(s), major beneficial species and assessing pest damage.
Identification of pests and beneficial species is an obvious first-step in an IPM program. However, farmers, extension workers, and even researchers are often unable to identify pests and, more importantly, to determine if the density of the pest is sufficiently high to warrant control. Training must be provided for surveys and pest identification. Materials to help identify major pests of rice include: (1) Field Problems of Tropical Rice [6], (2) Integrated Pest Management--Rice: Pocket Reference Manual [7] (3) Major Weeds of the Philippines [8].

The importance of indigenous biological control agents (predators, parasites, and pathogens) cannot be overemphasized. These natural enemies must be considered in future IPM programs. Published materials to help farmers and extension workers identify these species include: (1) An Illustrated Guide to Integrated Pest Management in Rice [9], (2) Natural Enemies of Insect Pests of Rice (a poster) [10] and (3) Helpful Insects, Spiders, and Pathogens [11].

Several thousand farmers have been trained to identify pests and beneficial species by the Philippine Department of Agriculture, FAO, and the Philippine-German Crop Protection Programme in the Philippines. In addition, IRRI's Department of Entomology provides a 17.5 week IPM short course primarily for technicians who, hopefully, will pass their knowledge to other technicians and farmers. A short-course in biological control is planned for 1990 in lieu of the IPM course.

Although there are rich communities of natural enemies in rice [12], it is likely that only a few groups actually play a major role in regulating insect pest populations. It is important to identify these and to quantify their effects on pest populations. Information obtained from these studies can be entered into the decision-making process [13].

Orthopterans such as Metioche vittaticollis, Conocephalus longipennis, and Anaxipha sp. feed on eggs and immature stages of several rice pests. M. vittaticollis may consume over 100 eggs of the striped stem borer, Chilo suppressalis, or 5-6 nymphs of the brown planthopper, Nilaparvata lugens, per day [14]. Conocephalus longipennis is a major predator of eggs of the yellow stem borer, Tryporyza incertulas. A single adult can consume as many as 6 egg masses during a 5-day period. Numbers of leaffolder, Marasmia patnalis, eggs consumed by C. longipennis, Anaxipha sp. and M. vittaticollis almost certainly impact on natural populations of leaffolders.

Field studies have demonstrated the importance of predators and parasites as mortality factors for the yellow stem borer in transplanted and direct-seeded rice [15]. Egg parasitism ranged from 20% to over 60%, and predation was over 30% on the IRRI farm. Both predation and parasitism were slightly higher in transplanted than in direct-seeded rice. Defining ways for rice farmers to use this information is a pressing challenge for IPM specialists.

Epizootics of insect pathogens have been noted in many insect pests of rice. In Mindanao, Philippines, outbreaks of the fungus Nomuraea rileyi arrested populations of green hairy caterpillar, Ruvula atimeta [16]. Outbreaks of Entomophthora on the brown planthopper have been observed during the wet season. Artificial manipulation of these pathogens has been effective against brown planthoppers [17] and black bugs [18].

Future research must focus on assessing the impact of major beneficial species on insect pests. The long-term goal of this research should be to develop IPM programs that recognize the importance of predators, parasites, and entomopathogens and to allow adjustments in action thresholds to be made relative to the density and species of natural enemies. This should reduce insecticide use by conserving these natural enemies and result in long-term production sustainability and profitability.

Determining action thresholds.
Determining actions thresholds is central to IPM. Usually the action threshold is expressed as pest density--that pest density which, if left uncontrolled, can reach the economic injury level and cause losses greater than the cost of treatment. In reality it is difficult to determine experimentally because of the dynamic nature of the crop, pest, natural enemy, and environmental interactions. The occurrence of multiple pest species further complicates the problem. However, simple, though imprecise thresholds which

can be easily implemented by farmers allow insecticide
applications to be more properly timed.

Several experimental approaches have been used to
determine actions thresholds. Often, known numbers of
insect pests are introduced into cages in the greenhouse or
field. Resulting yield losses can be obtained after
allowing time for the pests to feed. Natural populations of
pests occur at different densities within a field that
allows sampling from a range of densities and correlations
to be made between pest density and yield loss. Marking and
monitoring individual plants with a range of damage or
insect pest densities provides solid information on yield
loss pest-damage relationships.

Another approach involves allowing the pest population
to build up to different densities in field plots, then
treating with pesticides to keep the populations at or below
these predetermined densities. This procedure is often
difficult to carry out because of differential patterns of
infestation.

A third approach involves simulation of damage levels
by leaf clipping or detillering. This provides valuable
data on the plant's response to feeding damage by insects.
Computer simulation models also can be helpful but a good
understanding of plant physiology is necessary.

Two or more of the above procedures may be needed to
define workable thresholds. These thresholds must then be
tested in fields over a range of pest densities and
locations.

At the IRRI farm, damage by the yellow stem borer,
Tryporyza incertulas, was simulated in IR36 rice variety by
cutting and removing 15 and 30% of the tillers at 50, 69,
and 84 days after seeding [19]. Results from this study
revealed that early-season detillering of IR36 had little or
no effect on resulting yields. However, detillering at high
levels during panicle initiation and grain-fill stages,
significantly reduced yields. Similarly, artificial
defoliation of IR64 at 10 and 25% at 40 days after
transplanting (maximum tillering stage) did not affect
yield. But defoliation at 60 days after transplanting
caused significant yield losses (Arida and Shepard, unpubl.)

While we recognize that simulated damage does not mimic
exactly the actions of insect attack, gross responses by the
plant can be observed and thresholds refined. It is likely
that early season insect pests cause little, if any, yield
loss. Shepard et al. (in press) [20] found that early
season damage (up to 60%) by the rice whorl maggot,
Hydrellia philippina Ferino, caused no discernable yield
loss. However, there is little doubt that occurrence of
these early-season insect pests, even at noneconomic levels,
has triggered needless insecticide applications.

**Sampling and monitoring methods for insect pests and
beneficial species.**

Entering the field and taking samples is the only way to make intelligent decisions about whether or not a pest population is dense enough to require control. Given that training has been provided, the sheer tedium of sampling, often deters farmers from following procedures that require a large number of samples.

In cooperation with the Philippine Department of Agriculture and the FAO, we developed simple and efficient sampling schemes for brown planthoppers (BPH), whitebacked planthoppers [21] and black bugs [22]. In addition, sampling protocols have been developed that incorporate major predator species as well as hopper pests [13]. These sampling techniques are based on the mathematical distribution of the insects, predetermined damage thresholds, and the risk that the grower was willing to assume. This method is called sequential sampling, because samples are taken in sequence; the number of insects found in a sample determines whether the next sample should be taken. Sequential sampling has been used in several crops with significant savings time required to make a decision about whether to apply an insecticide [23, 24].

Results from sampling hoppers and major predators (predaceous beetles, water bugs, and spiders) revealed that when predators were considered savings of insecticide applications were more often realized. Future sampling procedures should be based on pests and beneficial species as information about the impact of natural enemies is obtained.

Selecting the appropriate control tactics.
Chemical pesticides have spread rapidly in developing countries but proper training to insure their safe and effective use has not kept pace [25]. Severe consequences of insecticide overuse were recently experienced in Indonesia in 1986, where more than 50,000 ha were damaged by resurging populations of the brown planthopper. In this case, President Soeharto banned 57 insecticides in order to stabilize rice production by avoiding destruction of natural enemies. The "hopperburned" areas have been significantly reduced after this time.

A major constraint to moving away from heavy dependence on chemicals is lack of quantitive information on the economic value of nonchemical approaches [26]. It is likely that beneficial species will be considered along with pests in future economic analysis. In many areas where the history of pest outbreaks is known, a resistant variety may be available. However, except for plant- and leaf-hoppers, resistance to insect pests of rice is limited. In most cases, within-season management decisions are related to whether or not to apply a chemical pesticide. In these instances, the farmer uses whatever chemical is available. It is not uncommon to find farmers applying fungicides in an attempt to control insects. Other tactics may include adequate land preparation for weed control, destruction of rice stubble after harvest for insect and disease control, or community-wide programs of synchronous plantings and rodent control.

Integration

In most parts of South and Southeast Asia, the total amount of insecticides applied could be reduced by more than 50% without yield loss. In most areas, it is difficult to see a yield response to increasing frequency of insecticide applications. In Laguna province, Philippines, we monitored fields of different growers who applied insecticides from 0 to 5 times, yet obtained approximately the same yields in all fields (Crisostomo and Shepard, unpubl.). Likewise, IPM experiments were carried out in Nueva Ecija, Philippines, during the 1985-86 dry season and 1986 wet season. During each season six farms ranging in size from 0.5 to 2.5 ha were divided into halves. Farmers were asked to continue to manage one half of the farm as per their usual practice. The remaining half was managed by our IPM working group (personnel from IRRI, FAO, the Department of Agriculture, and the University of the Philippines at Los Banos). Insects, diseases, and weeds were monitored and economic analyses carried out on all operations and materials used on IPM- and farmer-managed portions.

Although yields from one of the farmer-managed portions were slightly higher during the 1986 dry season, the benefit to cost ratio was always higher in the IPM portion of the farms. Farmers treated their portions of the fields more frequently with insecticides thus losing profits without increasing yields.

Sumangil (1984) [27] concluded from observations of 442 farmers' crops in the Philippines that "if farmers can use thresholds to make insect control decisions, they can increase profitability--over preventive insecticide applications."

Research and training are key elements in developing effective IPM programs. There is no way of adapting IPM to different geographical locations except by carrying out research at those locations. Shifts in pest importance and differences in abiotic and biotic environments which affect the pests and crop are often location-specific.

IPM demonstration programs, established at several locations in a region, can serve the important function of allowing farmers to observe IPM in action. A major deterrent to successful implementation of IPM is lack of trained extension personnel, lack of support for them, and lack of training for farmers themselves. Research is far ahead of implementation.

Interestingly, outbreaks of brown planthoppers in the Philippines during the 1970s were directly correlated with preventive insecticide treatments. A practice which is no longer recommended. Many similar examples in other parts of Asia have been observed.

The Philippine Department of Agriculture recently reviewed its policy of pesticide use. Presently the Bureau of Plant Industry (BPI) within the Department fully embraces the concepts and philosophy of IPM. Key individuals responsible for the present status of IPM in rice in the Philippines include: J.P. Sumangil (BPI, DA), P.E. Kenmore (FAO), J. A. Litsinger (IRRI), and Candida Adalla (Univ. of the Philippines). IRRI has provided backup research that has been translated into usable thresholds, resistant varieties, surveillance programs, and identification of natural enemies and definition of their importance. Weeds, diseases, and insects are included in the present surveillance program.

CONCLUSION

The activities outlined above should be followed for developing IPM for rice in the future. The level at which a national program may enter this sequence depends upon the existing knowledge base. Clearly, information generated from research in IPM has not been fully disseminated to the farmers in developing countries. However, efforts are under way in many countries in Asia to increase adoption of IPM. Rice production will become sustainable only after ecologically, economically, and sociologically accepted methods of pest control are adopted. This must extend further than breeding for resistance; otherwise, emergence of new pests, resistance of pests to insecticides and resurgence of insect populations will continue to threaten the crop.

REFERENCES

1. Smith, E.H. Integated pest management (IPM). Specific needs of developing countries. Insect Sci. Applic. 1983, 4: 173-177.

2. Kenmore, P.E., Heong, K.L., Putter, C.A. Political, social, and perceptual aspects of integrated pest management programs. 1985, pp. 47-66. Proceedings of the Seminar on Integrates Pest Management in Malaysia. B.S. Lee; W.H. Loke; R.L. Heong, eds.

3. Stern, V.M., Smith R.F., van den Bosch, R., Hagen K.S. The integrated pest control concept. Hilgaria 1959, 29:81-101.

4. Kenmore, P.E., Litsinger J.A., Bandong, J.P., Santiago, A.C., Salac M.M.. Philippine rice farmers and insecticides: Thirty years of growing dependency and new options for change. 1987, pp. 98-108. Management of Pests and Pesticides. J. Tait and B. Napompeth, eds.

5. Litsinger J.A. Assessment of need-based insecticide application for rice. Paper presented at MA-IRRI Technology Transfer Workshop. 1984.

6. Mueller, K. E. Field problems of tropical rice. The International Rice Research Institute. Los Banos, Laguna, Philippines. 1975.

7. Philippine-German Crop protection Program/Bureau of Plant Industry-Ministry of Agriculture and Food. Integrated Pest Management-Rice. Pocket Reference Manual. Manila, Philippines. 1986, 136 pp.

8. Moody, K.I., Monroe, C.E., Lubigan, R.T., Paller, E.C. Jr. Major weeds of the Philippines. Weed Science Socity of the Philippines, University of the Philippines, Los Banos, Philippines. 1984, 328 pp.

9. Reisseg, W.H., Heinrichs, E.A., Litsinger, J.A., Moody, K., Feidler, F., Mew T.W., Barrion, A.T. Illustrated guide to integrated pest management in rice in tropical Asia. 1986. The International Rice Research Institute, Laguna, Los Banos, Philippines. 411 pp.

10. IRRI/FAO. Natural enemies of insect pests of rice. The International Rice Research Institute (1984) Laguna, Philippines. 1985, (a poster).

11. Shepard, B.M., Litsinger, J.A., Barrion, A.T. Helpful insects, spiders and pathogens. The International Rice Research Institute, Los Banos, Laguna, Philippines. 1987, 127 pp.

12. Yasumatsu, K., Torii T. Impact of parasites, predators, and diseases on rice pests. Ann. Rev. Entomol. 1968, 13: 295-323.

13. Shepard, B.M., Ferrer, E.R., Kenmore, P.E. Sequential sampling planthoppers and predators in rice. Journal of Plant Protection in the Tropics. 1988, 5(1): 39-44.

14. International Rice Research Institute in Annual report for 1984. Los Banos, Philippines 1985, pp. 166-167.

15. Shepard, M., Arida, G.S. Parasitism and predation of the yellow stem borer, Scirpophaga incertulas (Walker) (Lepidoptera: Pyralidae) eggs in transplanted and direct seeded rice. J. Agric. Entomol. 1986, 21:26-32.

16. International Rice Research Institute. Annual report for 1983. Los Banos, Philippines. 1984, pp.201-202

17. Rombach, M.C., Aguda, R.M., Shepard, B.M., Roberts, D.W. Infection of the rice brown planthopper, Nilaparvata lugens (Homoptera; Delphacidae) by field application of entomopathognic hyphomycetes (Deuteromycotina). Environ. Entomol. 1986a **15**: 1070-1073.

18. Rombach, M.C., Aguda, R.M., Shepard, B.M. and Roberts, D.W. Entomopathogenic fungi (Deuteromycotina) in control of the black bug, Scotinophara coarctata (Hemiptera: Pentatomidae) of rice. J. Invert. Path. 1986b, **48**: 174-179.

19. Rubia, E.G., Shepard B.M. Simulation of insect damage on rice. Paper presented to Training Workshop on systems Analysis and Simulation for Rice Production, 12-23 Jan 1987, International Rice Research Institute, Los Banos, Philippines. 1987.

20. Shepard, B.M., H.D. Justo, E.G. Rubia and D.B. Estano. Response of the rice plant to damage by the whorl maggot Hydrellia philippina Ferino (Diptera: Ephydridae). Journal of Plant Protection in The Tropics 1990, (In press).

21. Shepard, M., Ferrer, E.R., Kenmore, P.E., Sumangil, J.P. Sequential sampling: planthoppers on rice. Crop Protect. 1986, **5**: 319-322.

22. Ferrer, E.R., Shepard, B.M. Sampling Malayan black bugs. Environ. Entomol. 1987, **16**:259-263.

23. Pieters, E.P. Bibliography of sequential sampling plans for insects. Bull. Entomol. Soc. Amer. 1978, **24**:373-374.

24. Shepard, M. Sequential sampling plans for soyean arthropods. Sampling methods in soybean entomology. M. Kogan; D.C. Herzog, eds., Springer Verlag, New York, NY, USA. 1980, pp. 79-83

25. Bottrell D.G. Government influence on pesticide use in developing countries. Insect Sci. Applic. 1984, **5**:151-155.

26. Bottrell D.G. Applications and problems of integrated pest management in the tropics. In Proceedings of the Second International Conference on Plant Protection in The Tropics, Genting Highlands, Malaysia. 1986, 139.

27. Sumangil, J.P. IPM for cost reduction in the Masagana-99 rice production program. Paper presented at the Agri-Tech Fair, Philcite, Manila, Philippines. 1984.

PEST INTERACTIONS IN RICE IN THE PHILIPPINES

KEITH MOODY
International Rice Research Institute
P.O. Box 933, Manila, Philippines

ABSTRACT

Interactions between weeds, vertebrate pests, mollusks, insect pests, diseases, and nematodes are discussed in this paper. Weed growth is affected by many cultural practices. Weeds also harbor, are hosts or sources of food for insect pests, diseases and other pests, as well as the natural enemies of these pests. Control or failure to control weeds may have positive or negative effects on other rice pests. Some of these interactions are discussed in this paper. Listings of weeds that serve as alternate hosts of insect pests and diseases of rice are given.

RELATIVE IMPORTANCE

Weeds are the most severe and most widespread biological constraint to crop production in the Philippines (66). Farmers in Guimba, Nueva Ecija and Claveria, Misamis Oriental regarded weeds as the most important pest of transplanted and upland rice, respectively (29, 30). They regarded insects as being second most important and diseases were least important. In a study on the decision making process of farmers, Tuettinghoff (1987, cited in 108) noted that herbicides were placed before insecticides and the latter before fertilizers, in terms of their expected contributions to yield increases.

VERTEBRATE PESTS

Rats

Rat damage is widespread in the rice growing regions of the world. Heavy damage in a few fields can mean a large economic loss to the small-scale farmer (32). In Mindanao, Philippines (110), a rice weed control experiment was destroyed by a rat plague which was difficult to control because of the unweeded treatments.

Rats do their damage by cutting tillers and the rice has some power to compensate for the damage caused. An important factor that often affects this potential compensation is weed competition. When the rice is cut back, the regenerating plants face more competition than do recently seeded or transplanted crops.

Drost and Moody (27) reported that a) rat infestation level varies from field to field and from season to season, b) rats will selectively damage weedy plots before attacking weed free plots, c) rat damage is lower in weeded plots and increases as weed weights increase (Table 1), d) rat damage is higher in weeded plots when tiller density is high, and e) rat damage causes an overestimation of the yield loss caused by weeds.

TABLE 1

The effect of weeding treatments on total weed weight, rat damage and yield of rainfed transplanted rice (Adapted from 27)[a]

Treatment	Weed weight (g/m^2)	Rat damage (%)	Yield (t/ha)
Weed free	9.1	6.8	3.3
Weeded twice[b]	12.2	8.5	3.2
Weeded once[c]	98.0	23.9	2.6
No weeding	201.3	26.0[d]	1.3

[a]Average of five fields and three replications per field.
[b]Two and five weeks after transplanting.
[c]Two weeks after transplanting.
[d]Average of three fields and three replications per field.

In addition, a) rat damage is less when rice is transplanted at wider spacing b) rat damage is greater in wet-seeded rice and at higher seeding rates, c) more weeds grow in rat-damaged fields, and d) rat-damaged rice suffers more from weed competition than rice that has not been damaged.

Hoque and Olvida (38) reported that there was significantly more rat activity (41.2% vs. 21.4%) and significantly more damaged tillers (18.5% vs. 1.8%) in weedy transplanted rice plots than in clean plots. Rat damage was directly dependent on weed density. Singh and Moody (unpublished) found that rat damage at harvest and weed weight at 35 days after transplanting were in the order of weedy check > rotary weeded twice > herbicide treated > hand weeded twice. Rat damage increased as weed weight increased. Rat damage also increased as harvest date was delayed.

Cultivation practices that should be recommended as part of an integrated management scheme for rat control are synchronous transplanting, maintenance of low and weed-free bund systems and refraining from planting subsidiary crops on bunds particularly during the intercrop period (16). Large fields with small bunds are those that tend to have the least problem with rats.

Birds

Quisumbing (80) reported that some farmers allow *Echinochloa crus-galli* (L.) P. Beauv. to grow side by side with rice. It is said to drive the birds away because of the long awns of the spikelets and thus protect the rice from bird attack. I observed, however, that *Echinochloa glabrescens* Munro ex Hook.f. in the ripening stage attracted birds to a rice field; the birds fed on the weed seeds. In Hawaii, I observed greater bird damage in sweet corn (*Zea mays* L.) in weed free plots than in weedy plots. The weeds, two of which (*Amaranthus spinosus* L. and *Cenchrus echinatus* L.) had spines, either obscured the sweet corn or repulsed the birds.

MOLLUSKS

The freshwater snail, *Ampullaria luzonica* Reeve, which is used as a substitute for fish in the diet of inhabitants of the region around Laguna de Bay, Luzon, Philippines feeds on the roots, stems and leaves of many aquatic plants such as *Ceratophyllum demersum* L., *Ipomoea aquatica* Forssk., *Pistia stratiotes* L., *Eichhornia crassipes* (Mart.) Solms, filamentous algae and occasionally *Cyperus elatus* L. (67).

Golden apple snails, [*Pomacea canaliculata* (Lamarck)], have become important pests of rice in the Philippines. During the pre-rice planting period, they may be found in irrigation canals and drainage ditches, where they may be feeding on weeds (15). They appear to have a wide host range and to feed on almost all lowland weeds (Table 2). They prefer the soft, young parts of plants (Table 3). I have observed them feeding on weed seedlings in lowland rice fields. They should, therefore, be controlled before conducting weed control trials to prevent erroneous results.

Litsinger *et al.* (57) reported that fentin acetate at rates ranging from 0.6-1.2 kg ai/ha was the most effective molluskicide for the control of golden apple snail infestations in rice fields. However, it is highly toxic to fish (57).

Pablico and Moody (73) found that fentin acetate was toxic to wet-seeded rice (pregerminated-seed sown on puddled soil), water-seeded rice, *P. stratiotes* and *Azolla pinnata* R.Br. They recommended that a) it should be used with caution in wet-seeded rice if it is applied immediately after seeding at rates greater than 0.3 kg ai/ha, b) it should not be used in water-seeded rice, c) it should not be used in weed control experiments involving *P. stratiotes*, and d) it should not be used in areas where azolla is being used as an organic fertilizer.

TABLE 2
Golden apple snail host preference (15)[a]

Plant species	Plants (%) consumed by snails at different times after infestation			
	4 hr	24 hr	48 hr	72 hr
Monochoria vaginalis	50 a	100 a	100 a	100 a
Echinochloa glabrescens	40 a	100 a	100 a	100 a
Cyperus difformis	35 a	90 a	100 a	100 a
Oryza sativa	0 b	100 a	100 a	100 a
Fimbristylis miliacea	0 b	100 a	100 a	100 a
Paspalum distichum	20 ab	55 b	85 ab	90 a
Ipomoea aquatica	0 b	20 c	75 b	85 a
Sphenoclea zeylanica	0 b	48 b	52 c	52 b
Pistia stratiotes	0 b	30 bc	45 c	48 b

[a]Average of ten replications. In a column, means followed by a common letter are not significantly different at the 5% level by DMRT.

TABLE 3
Percentage of young and old weeds consumed by golden apple snail 48 hours after infestation (15)[a]

Weed species	Young	Old
Monochoria vaginalis	93 a	54 b
Paspalum distichum	78 a	15 b
Echinochloa glabrescens	77 a	6 b
Cyperus difformis	68 a	58 b
Fimbristylis miliacea	59 a	27 b
Sphenoclea zeylanica	50 a	32 b
Pistia stratiotes	28 a	24 b
Ipomoea aquatica	6 a	1 b

[a]Average of five replications. In a row, means followed by a common letter are not significantly different at the 5% level by DMRT.

INSECTS

Weedy fields increase the brown planthopper [*Nilaparvata lugens* (Stål)] population (20). IRRI (40) reported that near rice crop maturity, the brown planthopper tends to be more abundant in weedy than in weeded plots, probably because the dense vegetation of weedy fields provides an environment suitable for the insect. After harvest the insect usually transfers to weeds but does not hibernate (78).

Janiya *et al.* (unpublished) found that caseworm [*Nymphula depunctalis* (Guenée)] and leaffolder (*Cnaphalocrosis medinalis* Guenée) populations were highest in unweeded plots while defoliation by green semilooper/green hairy caterpillar [*Naranga aenescens* Moore/*Rivula atimeta* (Swinhoe)] and whorl maggot (*Hydrellia philippina* Ferino) was highest in the hand weeded plots (Table 4). The increase in the number of caseworms and leaffolders in the unweeded plots could be due to entrapment of the insects in the thick weed foliage. The presence of weeds in the unweeded plots made the water inconspicuous to adults of green semilooper/green hairy caterpillar and the whorl maggots and as a result, the insects were not attracted to the rice.

Litsinger *et al.* (55) noted that weeding during the first month after upland rice establishment will force insect pests such as armyworms [*Mythimna separata* (Walker)], which prefer grassy weeds, on to the rice. Soil insects will move to the rice crop and clean culture also may force termites to attack rice. Fields with grassy surroundings and with stubble left after harvest are more susceptible to attack by the yellow stemborer [*Scirpophaga incertulas* (Wlk.)] than clean ones. The moths are active at night. During the day, they spend more time in grassy areas than in the rice field (76, 91).

Barrion and Litsinger (13) reported that parasitization of the swarming caterpillar (*Spodoptera mauritia acronyctoides* Guenée) and the common cutworm [*Spodoptera litura* (Fabricius)] by a tachinid fly [*Peribaea orbata* (Wiedemann)] was higher in unweeded rice fields containing *Rottboellia* and *Amaranthus* than in weeded fields.

Biological control of rice insect pests is an important component of an integrated pest management programme. Members of the subfamily Trigonidiinae are often mentioned as predators of the early development stages of rice insect pests. For example, the cricket, *Metioche vittaticolis* (Stål), preys on the eggs of leaffolders and *Marasmia patnalis* Bradley (9) and striped stemborers (*Chilo suppressalis* Walker), and young nymphal stages of brown planthopper and green leafhopper [*Nephotettix virescens* (Distant)] (92).

Crickets are frequently seen among foliage of tropical understory vegetation (82). Rubia (92) reported that among the weeds associated with the rice plant, *M. vittaticolis* preferred to oviposit on *Fimbristylis miliacea* (L.) Vahl and *Monochoria vaginalis* (Burm.f.) Presl (Table 5). These may serve as reservoirs by providing food, shelter and a reproductive site to maintain a pest population. For example, *M. vaginalis* is an alternate host of some lepidopterous defoliators (92).

TABLE 4

Effect of weed control method on weed weight and insect infestation
in transplanted IR64 rice in Guimba, Nueva Ecija
(Janiya et al., unpublished)[a]

Weed control method	Caseworm (no./hill)	Leaffolders (no./hill)	Naranga/ Rivula (no./hill)	Whorl maggots (% defoliation)	Weed weight (g/m^2)
No weeding	25 a	9 a	10 b	10.8 b	97.0 a
Hand weeding, 21 DAT	8 b	8 ab	12 a	12.6 a	37.8 b
Hand weeding, 21 and 42 DAT	8 b	7 b	12 a	12.8 a	9.4 c

[a]In a column, means followed by a common letter are not significantly
different at the 5% level. DAT = days after transplanting.

TABLE 5

Number of eggs laid by *Metioche vittaticolis* on different plant species
(Adapted from 92)[a]

Plant species	Free choice	No choice
Fimbristylis miliacea	5.0 a	2.0 c
Monochoria vaginalis	4.7 a	5.3 b
Oryza sativa	4.0 a	7.5 a
Sphenoclea zeylanica	3.7 a	2.3 c
Cyperus iria	2.3 ab	0.5 cd
Ludwigia octovalvis	0.3 b	1.0 cd
Leersia hexandra	0.0 b	0.0 d

[a]In a column, means followed by a common letter are not significantly
different at the 5% level by DMRT.

    The brown planthopper population that occurs on *Leersia hexandra*
Sw. is distinct from that on rice because it does not survive on rice and
the rice population does not survive on *L. hexandra*. However, the *L.
hexandra* population is important in the management of brown planthopper
on rice because it is attacked by the same predators, parasites and
pathogens as the rice population (37). Zhang and Saxena (111) reported

that the courtship signals produced by both males and females of biotype 1 of the brown planthopper infesting rice plants significantly differed from those produced by the brown planthopper biotype infesting *L. hexandra* in wave pattern, pulse repetition frequency and sonic spectrum. Claridge *et al*. (23) found that pulse repetition frequencies of female and male calls differed consistently and significantly between rice- and *Leersia*-associated populations. They concluded that the populations from rice and from *L. hexandra* represent two distinct, but very closely allied, sympatric species.

Saxena and Puma (96) reported that the hatchability of eggs of the brown planthopper was reduced when incubated in solutions of trans-aconitic acid, an allelochemic factor in *E. crus-galli*. On the other hand, the eggs of the green leafhopper tolerated trans-aconitic acid levels which were lethal to the brown planthopper eggs. In nature, the green leafhopper thrives on *E. crus-galli* which is unsuitable for the establishment of the brown planthopper.

In rice fields, *Sogatodes pusanus* (Distant) is often confused with whitebacked planthopper, *Sogatella furcifera* (Horvarth). *S. pusanus* is not a rice pest but an Echinochloa-feeding planthopper. Of 15 common rice field plant species, *S. pusanus* completed development on only two species, *Echinochloa. colona* (L.) Link and *Echinochloa crus-galli* (L.) P. Beauv. ssp. *hispidula* (Retz.) Honda (19).

DISEASES

Weedy fields promoted the development of stem rot of rice caused by *Sclerotium oryzae* Catt. (25).

NEMATODES

Weeds, which are always present despite the farmers' weeding practices, sustain nematode populations (17). Some weed species particularly those in the Poaceae and Cyperaceae families which are major limiting factors in rice production are also good hosts of the rice root nematode (*Hirschmanniella spinicaudata* Sch. Stek.) (7). Thus, any programme developed for the control of the rice root nematode must also include weed control. Contrary to what Sankaran and De Datta (95) suggest, the rice root nematode does not have any potential for the biological control of weeds in upland rice because it is a parasitic nematode of rice.

AZOLLA

Azolla is a free-floating fern known for its nitrogen-fixing capacity through its symbiotic association with a green-blue algae. With good water management, the dual culture of rice and azolla provides many benefits such as limited evaporation of water, minimized competition from weeds and yield increases over two rice cycles of as much as 50% (85).

The azolla blanket must be sufficiently developed to suppress weed growth but limited enough to avoid damage to rice seedlings (97).

Viajante and Heinrichs (106) reported that rice plants in plots without azolla had four times more rice whorl maggot eggs than those in inoculated plots. As a result, larval damage was significantly higher in the plots without azolla. Percent damaged leaves in the azolla-inoculated plots was about 50% that of the azolla-free plots. Barrion and Litsinger (11) found that there were more insect predators in fields inoculated with azolla than in fields without azolla.

LAND PREPARATION

Land preparation is an important weed control practice that is frequently overlooked by the farmer. Throughout the tropics, particularly in rainfed areas, farmers often plant in fields that have been hurriedly and poorly prepared. The first and primary reason for land preparation is to provide weed free conditions at planting and the second is to provide favorable conditions for the growth and development of the crop.

In rainfed areas in the tropics, there may be advantages to plowing immediately after the rice crop is harvested than at the beginning of the rainy season. Plowing after the previous crop, to expose the vegetative propagules of perennial weeds to the desiccating effect of the sun during the dry season, is only possible if the vegetative cycle of the rice crop is shorter than that of the rainy season or sufficient residual moisture remains so that the soil is still moist when plowing is carried out (64).

According to Palo (75), plowing fields at the end of the rainy season, will bury and kill many of the sclerotial bodies of R. solani and reduce sources of infection in the following season. Reyes and Palo (84) noted that thorough cultivation and drying of the land at the end of the rainy season, will expose nematodes to desiccation and thus reduce their numbers.

In an upland field in Claveria, Misamis Oriental, Philippines, although the use of the stale-seedbed technique reduced the density of some important weed species, especially Rottboellia cochinchinensis (Lour.) Clayton, delay in the establishment of the rice crop resulted in increased blast, sheath rot, and neck rot (Elliot and Moody, unpublished).

PLANT POPULATION

As the planting distance between hills of transplanted rice is reduced or the seeding rate of wet-seeded rice is increased, i.e., the plant population increases, the crop becomes more competitive against weeds and yield losses due to weeds are reduced. However, excessive plant populations will result in increased intraspecific competition and lower yields may result.

Sanchez (94) reported that the very dense canopy created by heavy tillering of high yielding rice cultivars, heavy fertilization, closer planting distance and continuous flooding of fields have created microclimates that favour certain pests. Custodio *et al.* (24) reported that closer spacing and heavy fertilisation tend to increase disease incidence. Close spacing promotes the development of *R. solani* (75, 102, 105) and *S. oryzae* (25) but Escuro and Aquino (31) found that spacing had no effect on the degree of disease incidence; the most important diseases being *Cercospora* leaf spot, caused by *Cercospora oryzae* Miyake, bacterial blight [*Xanthomonas campestris* pv. oryzae (Uyeda and Ishiyama) Dowson] and *Rhizoctonia* sheath spot, caused by *Rhizoctonia solani* Kuhn.

According to Dyck (28) close plant spacing stimulates the population growth of brown planthopper. There were significantly more brown planthopper per tiller at 10 x 10 cm spacing than at 50 x 50 cm spacing. With closely spaced plants, microenvironments were slightly cooler and more humid. In contrast, Viajante and Heinrichs (106) reported that close planting decreases oviposition and subsequent rice whorl maggot damage. The effect of the rice nymphalid (*Melanitis ledaismene* Cramer) was also more severe at wider plant spacing (89).

Rowar (91) reported that spacing had practically no effect on the degree of infestation of the yellow stemborer but Escuro and Aquino (31) observed that there was a greater incidence of stemborer associated with wider spacing. Litsinger *et al.* (56) noted that low tiller density favors damage by stemborers. Low tiller numbers per unit area allow a higher percentage to be infested from a given population of stemborers. Also, the larger diameter culms under low tiller density are more favorable to stemborer larval development. Higher transplanting densities, also, tolerate stemborer damage better (Litsinger, unpublished).

SEEDLING NUMBER PER HILL

In the Philippines, Kim and Moody (51) reported that weed weight was not affected as the number of rice seedlings planted per hill increased. Elsewhere, other researchers (22, 36) have found that weed weights decreased significantly as the number of rice seedlings planted per hill increased.

Reyes (83) found that increasing plant number per hill resulted in increased stem rot infection caused by *S. oryzae*.

NITROGEN

Higher nitrogen levels often cause proportionately greater weed growth and crop yield reduction (65). Weeds commonly take up added nutrients more rapidly and in larger quantities than do crops.

Otanes (70) stated that fertilization should be practiced whenever necessary because the more vigorous the rice plants, the more they are resistant to the attack of the rice stemborer. Nitrogen helps the plant tolerate insect damage. Litsinger (unpublished) found that rice plants tolerate up to three whiteheads per hill at 90 kg N/ha whereas at 0 kg N/ha even one whitehead per hill caused yield loss. Therefore, rice fields that are well weeded should be able to tolerate insect damage better than those that are not weeded or poorly weeded.

Dyck (28) noted that brown planthopper populations increase at high nitrogen levels. Higher fertilizer levels also increase the incidence of blast (102). According to Litsinger et al. (56) it is well documented that by increasing nitrogen fertilizer levels, higher numbers of a number of insect pests (brown planthopper, stemborers, gall midge, leaffolder) are produced. The cause of the population increase is attributed to several factors: 1) taller plants, 2) larger leaves, 3) softer plant tissues, 4) more tillers per unit area, and 5) more energy in the plants.

Mew et al. (63) reported that lesion length due to bacterial blight infection on cultivars with intermediate (moderate or partial) resistance, such as Pulo and Sailboro 302, increased with increased nitrogen level. But the total lesion length was considerably less than that on cultivars with no functional resistance (IR8 and TN1). When cultivars with adult plant resistance (IR1698 and IR944) were infected at the tillering stage, lesion length did not differ from that on cultivars (IR8) with no resistance to pathotypes 1 and 2. High nitrogen levels did not affect lesion length on cultivars with specific resistance. But lesion length increased if the cultivar-pathotype combination was compatible (i.e., the cultivars' resistance was overcome by pathotypes of different virulence). Thus, when IR20 was infected with pathotype 1 (to which it is resistant), lesion length did not significantly increase. But when infection was caused by pathotype 2 (to which IR20 is susceptible), lesion length increased proportionally with increased nitrogen. When IR1545-339 was infected with both pathotypes, to which it was resistant, lesion length did not increase.

WATER

Good water management is a major factor in good weed control in rice. Keeping the field flooded after planting will kill some weeds and will slow the growth of others. Weed populations decrease as the depth of water increases. Manas et al. (59) reported that the number of weeds/m^2 decreased from 2,940 when the field was not flooded to 100 when the water depth was 12 cm. Deep submergence (15-20 cm) was effective in controlling rice weeds primarily Eleocharis geniculata (L.) Roem. & Schult. and Cyperus difformis L. (6).

The process of puddling and maintaining flooded fields kills many potential soil-borne insect pests such as white grubs, termites, mealy bugs, root aphids which attack dryland rice. However, aquatic insect pests such as the whorl maggot and caseworm are confined to flooded fields and only become problems in flooded rice culture (56).

Pathak and Dyck (79) reported that standing water encouraged the multiplication of the brown planthopper population. Viajante and

Heinrichs (107) reported that there were three to six times more rice whorl maggot eggs in plots flooded 5-cm deep than in saturated plots.

The Malayan black bug [*Scotinophara coarctata* (Fabricius)] lays its eggs at the base of rice plants. When egg masses were not submerged, 967 hatched; none hatched after submergence for 24 hr. When submerged, 6, 12 and 18 hr, only 33, 29 and 3% hatched, respectively (77). Cultural methods involving water management are effective in controlling rice caseworm larvae. Caseworm damage was most severe in plots where the water depth was 20 cm while none was observed when the field was drained and the soil remained moist (6). However, field drainage in rainfed areas is not a practical solution because it favours the growth of weeds, off-setting the insect control effect gained (56).

Reyes and Palo (84) reported that standing water is essential for the production of a rice crop in areas where nematodes are present. They observed that plants growing in standing water grew well whereas those under more arid conditions were badly affected by nematodes. Flooding, even intermittently, is an effective cultural method for controlling plant parasitic nematodes of the genera *Rotylenchulus* and *Meloidogyne* which are important grain legume pests in the Philippines (18). Reyes and Palo (84) recommended that land be flooded immediately after rice harvest and kept. under water for about 2 weeks to reduce the population of parasitic nematodes tentatively identified as a species of *Ditylenchus*.

CROP ROTATION

Rotations among crops having drastically dissimilar life cycles or cultural conditions so as to break the cycle of the pests are among the most effective of all pest control methods. Crops commonly rotated with rice are maize (*Zea mays* L.), soybean [*Glycine max* (L.) Merr.], peanut (*Arachis hypogaea* L.), mungbean [*Vigna radiata* (L.) Wilczek], wheat (*Triticum aestivum* L.), barley (*Hordeum vulgare* L.), and pastures. Crop rotation was recommended for the control of stem rot caused by *S. oryzae* and false smut caused by *Ustilaginoidea virens* (Cke) Tak. (102). However, Palo (75) reported that crop rotation cannot eliminate *R. solani* because it attacks almost all field crops and weeds.

HERBICIDE-DISEASE INTERACTIONS

Lacson (53) found that IR 8 was more susceptible to stem rot disease caused by *S. oryzae* when 2,4-D was applied. In contrast, Manila and Lapis (60) reported that both MCPA and 2,4-D reduced the severity of bacterial blight and had no effect on the severity of blast (*Pyricularia oryzae* Cav.) or the susceptibility of 12 rice cultivars to sheath blight.

HERBICIDE-INSECT INTERACTIONS

Manwan (61) reported that rice plants treated with 2,4-D had significantly more dead hearts due to yellow stemborer than plants to which no 2,4-D had been applied (Table 6). Also the newly-emerged larvae had higher percentage survival and were larger on the plants treated with 2,4-D. According to Manwan (61), application of 2,4-D to the flood water caused the plant to become more favourable for larval development. Ishii and Hirano (46) reported that rice plants treated with 2,4-D had more nitrogenous compounds than untreated plants.

TABLE 6
Effect of 2,4-D treatment on dead heart formation in rice and survival and weight of yellow stemborer (Adapted from 61)[a]

Treatment	Dead hearts (%)	Surviving larvae (%)	Weight of larvae (mg/larva)
2,4-D applied 3 DAT	37.3 a	40.0 a	36.5 a
2,4-D applied 25 DAT	36.2 a	31.1 b	30.9 b
No 2,4-D applied	28.5 b	25.8 b	24.5 c

[a]In a column, means followed by a common letter are not significantly different at the 5% level by DMRT.

WEED-INSECTICIDE INTERACTIONS

IRRI (42) reported that the effect of weed competition on grain yield was more apparent in treatments where there was no weeding and where insect control was initiated when the damage caused by insect pests reached the economic threshold level. IRRI (43) reported that insecticide treatments had no effect on weed weights in wet-seeded rice. However, in a trial conducted in Zaragosa, Nueva Ecija, weed weights were consistently higher when insecticides were applied (average 172 g/m^2 vs 121 g/m^2) indicating that the insecticides were controlling insects that were attacking the weeds (45). In contrast, Heinrichs (unpublished) reported that rice plots that received insecticides had a more dense canopy due to less insect damage and as a result had less weeds due to competition form the rice plants.

SANITATION

Selective elimination of suitable hosts and habitats (sanitation) is one of the cultural methods that has been suggested for the control of a number of rice pests including *Helminthosporium oryzae* Breda de Haan (102), the yellow stemborer (70), the rice bug (*Leptocorisa acuta*

Thunberg) (71), and rice cutworms (*Spodoptera mauritia* Boisd, *S. litura*) (72).

ALTERNATE HOSTS

There are numerous records in the literature which associate rice insect pests and diseases with alternate weed hosts. Examples for the Philippines are given in Tables 7 and 8. These are uncritical listings of literature records and some are probably incorrect. Only through careful observations under controlled conditions can these associations be verified.

BIOLOGICAL CONTROL

The weevil *Cyrtobagous salviniae* Calder and Sands which has given spectacular control of *Salvinia molesta* D.S. Mitchell in Australia (34), Papua New Guinea (90), and India (48) was introduced into the island of Panay in August 1989.

BAITS

According to Philippine folklore, rice bugs, are believed to be attracted to baits consisting of an aquatic weed, *Hydrilla verticillata* (L.f.) Royle. which is wrapped in cloth bags and hung on poles (104).

TABLE 7
Alternate hosts of rice diseases in the Philippines

Alternate host and disease	Reference
Amaranthus spinosus Tungro (bacilliform)[a] + tungro (spherical)[a]	IRRI (44)
Axonopus compressus Ragged stunt Tungro (spherical)[a]	Anjaneyulu *et al.* (5) Anjaneyulu *et al.* (4)
Azolla pinnata *Rhizoctonia solani*	Kroeck and Watanabe (52)
Brachiaria distachya *Pyricularia oryzae*	Mackill and Bonman (58)
Brachiaria mutica *Pyricularia oryzae* Tungro (spherical)[a]	Paje *et al.* (74) Anjaneyulu *et al.* (4)
Chloris sp. *Rhizoctonia solani*	Mew *et al.* (62)
Cleome rutidosperma *Rhizoctonia solani*	Mew *et al.* (62)
Cynodon dactylon Grassy stunt Ragged stunt *Rhizoctonia solani* Tungro (spherical)[a]	Anjaneyulu *et al.* (3) Anjaneyulu *et al.* (5) Mew *et al.* (62) Valdez (105) Anjaneyulu *et al.* (4)
Cyperus brevifolius Tungro (spherical)[a]	Anjaneyulu *et al.* (4)
C. difformis Tungro (spherical)[a]	Anjaneyulu *et al.* (4)
C. rotundus Grassy stunt Ragged stunt *Rhizoctonia solani* Tungro (bacilliform)[a] Tungro (spherical)[a]	Anjaneyulu *et al.* (3) Anjaneyulu *et al.* (5) Valdez (105) Anjaneyulu *et al.* (4) Khan (50) Anjaneyulu *et al.* (4) Khan (50)

TABLE 7. Continued.

Alternate host and disease	Reference

Dactyloctenium aegyptium
    Tungro   Rivera *et al.* (88)

Digitaria ciliaris
    Tungro (spherical)[a]   Anjaneyulu *et al.* (4)

Digitaria sp.
    Rhizoctonia solani   Mew *et al.* (62)

Echinochloa colona
    Grassy stunt   Anjaneyulu *et al.* (4)
    Pyricularia oryzae   Mackill and Bonman (58)
    Rhizoctonia solani   Mew *et al.* (62)
                           Valdez (105)
    Tungro   Wathanakul (109)
    Tungro (bacilliform)[a]   Khan (50)
    Tungro (spherical)[a]   Anjaneyulu *et al.* (4)
                           Khan (50)

E. crus-galli
    Rhizoctonia solani   Valdez (105)
    Tungro   Wathanakul (109)

E. crus-galli ssp. *hispidula*
    Rhizoctonia solani   Mew *et al.* (62)
    Tungro (bacilliform)[a]   Khan (50)
    Tungro (bacilliform)[a] +
        tungro (spherical)[a]   IRRI (44)
    Tungro (spherical)[a]   Khan (50)

E. glabrescens
    Ragged stunt   Salamat *et al.* (93)
    Tungro (bacilliform)[a]   Khan (50)
    Tungro (bacilliform)[a] +
        tungro (spherical)[a]   IRRI (44)
    Tungro (spherical)[a]   Khan (50)

E. stagnina
    Rhizoctonia solani   Valdez (105)

Eleusine indica
    Ragged stunt   Salamat *et al.* (93)
    Rhizoctonia solani   Valdez (105)
    Tungro   Wathanakul (109)
    Tungro (bacilliform)[a]   Anjaneyulu *et al.* (4)
                           Khan (50)
    Tungro (spherical)[a]   Anjaneyulu *et al.* (4)
                           Khan (50)

TABLE 7.   Continued.

Alternate host and disease	Reference
Eragrostis tenella	
Ragged stunt	Anjaneyulu *et al.* (5)
Tungro	Rivera *et al.* (88)
Tungro (spherical)[a]	Anjaneyulu *et al.* (4)
Fimbristylis miliacea	
Ragged stunt	Anjaneyulu *et al.* (5)
Tungro (bacilliform)[a]	Anjaneyulu *et al.* (4)
Tungro (spherical)[a]	Anjaneyulu *et al.* (4)
Imperata cylindrica	
Rhizoctonia solani	Valdez (105)
Tungro (spherical)[a]	Anjaneyulu *et al.* (4)
Ischaemum rugosum	
Rhizoctonia solani	Valdez (105)
Tungro	Rivera *et al.* (88)
Leersia hexandra	
Grassy stunt	Anjaneyulu *et al.* (3)
	IRRI (41)
Pyricularia oryzae	Mackill and Bonman (58)
	Alvenda-Bernardo (1)
Ragged stunt	Anjaneyulu *et al.* (5)
	IRRI (41)
Tungro	Rivera *et al.* (88)
Tungro (bacilliform)[a]	Anjaneyulu *et al.* (4)
	Khan (50)
Tungro (spherical)[a]	Anjaneyulu *et al.* (4)
	Khan (50)
Leptochloa chinensis	
Pyricularia oryzae	Mackill and Bonman (58)
Rhizoctonia solani	Mew *et al.* (62)
Tungro (bacilliform)[a]	Khan (50)
Tungro (spherical)[a]	Khan (50)
Leptochloa filiformis	
Xanthomonas campestris pv. *oryzae*	Dalmacio and Esconde (26)
Monochoria vaginalis	
Grassy stunt	Anjaneyulu *et al.* (3)
Ragged stunt	Salamat *et al.* (93)
Rhizoctonia solani	Mew *et al.* (62)
Tungro (bacilliform)[a]	Anjaneyulu *et al.* (4)
Tungro (spherical)[a]	Anjaneyulu *et al.* (4)

TABLE 7.   Continued.

Alternate host and disease	Reference
Panicum auritum	
Rhizoctonia solani	Valdez (105)
P. repens	
Pyricularia oryzae	Paje *et al.* (74)
Rhizoctonia solani	Valdez (105)
Tungro (bacilliform)[a]	Khan (50)
Tungro (spherical)[a]	Khan (50)
Paspalidium flavidum	
Rhizoctonia solani	Mew *et al.* (62)
Paspalum distichum	
Ragged stunt	Salamat *et al.* (93)
Rhizoctonia solani	Mew *et al.* (62)
P. scrobiculatum	
Rhizoctonia solani	Valdez (105)
Tungro	Rivera *et al.* (88)
Pennisetum glaucum	
Tungro	Rivera *et al.* (88)
Rottboellia cochinchinensis	
Pyricularia oryzae	Mackill and Bonman (58)
Rhizoctonia solani	Mew *et al.* (62)

[a]See Omura *et al.* (68) for a description of rice tungro-associated viruses; Khan (50) concluded that weeds are not major sources of tungro-associated viruses but can act as natural reservoirs and can provide virus sources to rice on rare occasions.

TABLE 8
Alternate hosts of rice insect pests in the Philippines

Alternate host and insect pest	Reference
Brachiaria distachya	
Chaetocnema basalis	Barrion and Litsinger (12)
Hydrellia philippina	Ferino (33)
Nymphula depunctalis	Sison (98)
Spodoptera exempta	Rimando (87)
B. mutica	
Spodoptera exempta	Rimando (87)
B. reptans	
Dryopeia hirsuta	Tan (100)
Leptocorisa acuta	Uichanco (103)
Chrysopogon aciculatus	
Chaetocnema basalis	Barrion and Litsinger (12)
Coix sp.	
Chilo suppressalis	Gabriel (35)
Commelina benghalensis	
Scotinophara latiuscula	Barrion and Litsinger (14)
Cynodon dactylon	
Hydrellia philippina	Ferino (33) Karim (49)
Nymphula depunctalis	Bandong and Litsinger (8)
Cyperus difformis	
Hydrellia philippina	Karim (49)
Marasmia patnalis	Joshi *et al.* (47)
Scotinophara latiuscula	Barrion and Litsinger (14)
C. iria	
Scotinophara latiuscula	Barrion and Litsinger (14)
C. rotundus	
Euscyrtus concinnus	Barrion and Litsinger (10)
Spodoptera exempta	Rimando (87)
Cyperus sp.	
Leptocorisa acuta	Gabriel (35)
Leptocorisa oratorius	Gabriel (35)
Cyrtococcum patens	
Nymphula depunctalis	Sison (98)

TABLE 8. Continued.

Alternate host and insect pest	Reference

Dactyloctenium aegyptium
 Chaetocnema basalis — Barrion and Litsinger (12)
 Euscyrtus concinnus — Barrion and Litsinger (10)
 Nymphula depunctalis — Bandong and Litsinger (8)

Dichanthium annulatum
 Spodoptera exempta — Rimando (87)

Digitaria ciliaris
 Chaetocnema basalis — Barrion and Litsinger (12)

Digitaria compacta
 Dryopeia hirsuta — Tan (100)

D. sanguinalis
 Euscyrtus concinnus — Barrion and Litsinger (10)

D. setigera
 Leptocorisa acuta — Uichanco (103)

Digitaria sp.
 Cnaphalocrocis medinalis — Lim (54)
 Leptocorisa acuta — Gabriel (35)
 Leptocorisa oratorius — Gabriel (35)

Echinochloa colona
 Dryopeia hirsuta — Tan (100)
 Hydrellia philippina — Ferino (33)
  Karim (49)
 Leptocorisa acuta — Quisumbing (80)
  Uichanco (103)
 Leptocorisa oratorius — Taylo *et al.* (101)
 Nephotettix nigropictus — IRRI (39)
  Khan (50)
 Nymphula depunctalis — Bandong and Litsinger (8)
  Sison (98)
 Scotinophara latiuscula — Barrion and Litsinger (14)

E. crus-galli
 Hydrellia philippina — Karim (49)
 Leptocorisa acuta — Uichanco (103)
 Leptocorisa oratorius — Taylo *et al.* (101)
 Nephotettix nigropictus — Gabriel (35)
  IRRI (39)
  Razzaque *et al.* (81)
 Scotinophara latiuscula — Barrion and Litsinger (14)

TABLE 8. Continued.

Alternate host and insect pest	Reference
E. crus-galli ssp. *hispidula*	
Nephotettix nigropictus	Khan (50)
Echinochloa glabrescens	
Leptocorisa oratorius	Taylo *et al.* (101)
Marasmia patnalis	Joshi *et al.* (47)
Nephotettix malayanus	Rezaul Karim and Saxena (86)
Nephotettix nigropictus	IRRI (39)
	Khan (50)
	Rezaul Karim and Saxena (86)
Nephotettix virescens	Rezaul Karim and Saxena (86)
Nymphula depunctalis	Bandong and Litsinger (8)
Scotinophara latiuscula	Barrion and Litsinger (14)
Echinochloa spp.	
Euscyrtus concinnus	Barrion and Litsinger (10)
E. stagnina	
Nymphula depunctalis	Sison (98)
Eleusine indica	
Cnaphalocrocis medinalis	Lim (54)
Dryopeia hirsuta	Tan (100)
Euscyrtus concinnus	Barrion and Litsinger (10)
Hydrellia philippina	Karim (49)
Nephotettix nigropictus	Khan (50)
Nymphula depunctalis	Bandong and Litsinger (8)
Eleusine sp.	
Leptocorisa acuta	Gabriel (35)
Leptocorisa oratorius	Gabriel (35)
Eragrostis japonica	
Dryopeia hirsuta	Tan (100)
Nymphula depunctalis	Sison (98)
Fimbristylis miliacea	
Hydrellia philippina	Karim (49)
Scotinophara latiuscula	Barrion and Litsinger (14)
Imperata cylindrica	
Cnaphalocrocis medinalis	Lim (54)
Ischaemum rugosum	
Chaetocnema basalis	Barrion and Litsinger (10)

TABLE 8.   Continued.

Alternate host and insect pest	Reference
Leersia hexandra	
Hydrellia philippina	Ferino (33)
Marasmia patnalis	Joshi *et al.* (47)
Nephotettix malayanus	Khan (50)
	Rezaul Karim and Saxena (86)
Nephotettix nigropictus	Khan (50)
	Razzaque *et al.* (81)
	Rezaul Karim and Saxena (86)
Nephotettix virescens	Rezaul Karim and Saxena (86)
Nilaparvata lugens	Sogawa (99)
Leptochloa chinensis	
Hydrellia philippina	Ferino (33)
Marasmia patnalis	Joshi *et al.* (47)
Nephotettix nigropictus	Khan (50)
Nymphula depunctalis	Bandong and Litsinger (8)
Scotinophara latiuscula	Barrion and Litsinger (14)
Melinis minutiflora	
Spodoptera exempta	Rimando (87)
Miscanthus sp.	
Chilo suppressalis	Gabriel (35)
Panicum auritum	
Leptocorisa acuta	Cevallos (21)
Panicum maximum	
Leocopholis irrorata	Otanes (69)
Spodoptera exempta	Rimando (87)
P. repens	
Hydrellia philippina	Ferino (33)
Nymphula depunctalis	Sison (98)
Spodoptera exempta	Rimando (87)
Stenchaetothrips oryzae	Andres (2)
Panicum sp.	
Cnaphalocrocis medinalis	Lim (54)
Leptocorisa acuta	Gabriel (35)
Leptocorisa oratorius	Gabriel (35)
Paspalidium flavidum	
Euscyrtus concinnus	Barrion and Litsinger (10)
Leptocorisa acuta	Uichanco (103)
Leptocorisa oratorius	Taylo *et al.* (101)

TABLE 8. Continued.

Alternate host and insect pest	Reference
Paspalum conjugatum	
Chaetocnema basalis	Barrion and Litsinger (10)
Nymphula depunctalis	Bandong and Litsinger (8)
Spodoptera exempta	Rimando (87)
P. distichum	
Nymphula depunctalis	Bandong and Litsinger (8)
P. scrobiculatum	
Hydrellia philippina	Karim (49)
Leptocorisa oratorius	Taylo *et al.* (101)
Nymphula depunctalis	Sison (98)
Paspalum sp.	
Cnaphalocrocis medinalis	Lim (54)
Pennisetum clandestinum	
Hydrellia philippina	Ferino (33)
Pennisetum polystachion	
Chaetocnema basalis	Barrion and Litsinger (12)
P. purpureum	
Spodoptera exempta	Rimando (87)
Pennisetum sp.	
Leptocorisa acuta	Gabriel (35)
Leptocorisa oratorius	Gabriel (35)
Phragmites sp.	
Chilo suppressalis	Gabriel (35)
Rottboellia cochinchinensis	
Cnaphalocrocis medinalis	Lim (54)
Euscyrtus concinnus	Barrion and Litsinger (10)
Leptocorisa acuta	Cevallos (21)
Saccharum spontaneum	
Cnaphalocrocis medinalis	Lim (54)
Setaria sp.	
Cnaphalocrocis medinalis	Lim (54)
Tripsacum laxum	
Spodoptera exempta	Rimando (87)

REFERENCES

1. Alvenda-Bernardo, M., Histopathology of rice and weed host responses to the blast fungus, *Pyricularia oryzae* Cavara. M.S. thesis, University of the Philippines at Los Baños, College, Laguna, Philippines, 1989, 48 pp.

2. Andres, F.L., The life history of the rice thrips *Chloethrips oryzae* (Williams) Thripidae; Thysanoptera. Paper prepared for Entomology 290, University of the Philippines at Los Baños, College, Laguna, Philippines, n.d., 22 pp.

3. Anjaneyulu, A., Aguiero, V.M., Mesina, M.E., Hibino, H., Lubigan, R.T. and Moody K., Host plants of rice grassy stunt virus (GSV). Int. Rice Res. Newsl., 1988a, 13, 37.

4. Anjaneyulu, A., Daquioag, R.D., Mesina, M.E., Hibino, H., Lubigan, R.T. and Moody, K., Host plants of rice tungro (RTV)-associated viruses. Int. Rice Res. Newsl., 1988b, 13, 30-31.

5. Anjaneyulu, A., Salamat Jr., G.Z., Mesina, M.E., Hibino, H., Lubigan, R.T. and Moody, K., Host plants of ragged stunt virus (RSV). Int. Rice Res. Newsl., 1988c, 13, 32-33.

6. Anon., Agronomical experiments performed by the Plant Industry Division, Bureau of Agriculture, during 1923. Philipp. Agric. Rev., 1924, 17, 167-193.

7. Babatola, J.O., Studies on the weed hosts of the rice root nematode, *Hirschmanniella spinicaudata* Sch. Stek. 1944. Weed Res., 1980, 20, 59-61.

8. Bandong, J.P. and Litsinger, J.A., Plant hosts of rice caseworm. Int. Rice Res. Newsl., 1984, 9, 20-21.

9. Bandong, J.P. and Litsinger, J.A., Egg predators of rice leaffolder (LF) and their susceptibility to insecticides. Int. Rice Res. Newsl., 1986, 11, 21.

10. Barrion, A.T. and Litsinger, J.A., *Euscyrtus concinnus* (Orthoptera:Gryllidae) -- a new rice pest in the Philippines. Int. Rice Res. Newsl., 1980, 5, 19.

11. Barrion, A.T. and Litsinger, J.A., Effect of azolla on insect predators. Int. Rice Res. Newsl., 1983, 8, 16.

12. Barrion, A.T. and Litsinger, J.A., Flea beetle *Chaetocnema basalis* (Baly) (Coleoptera: Chrysomelidae), a pest of slash-and-burn upland rice in the Philippines. Int. Rice Res. Newsl., 1986, 11, 31-33.

13. Barrion, A.T. and Litsinger, J.A., A larval parasite of swarming caterpillar and common cutworm in the Philippines. Int. Rice Res. Newsl., 1987a, 12, 34-35.

14. Barrion, A.T. and Litsinger, J.A., The bionomics, karyology and chemical control of the node-feeding black bug, *Scotinopharalatiuscula* Breddin (Hemiptera: Pentatomidae) in the Philippines. J. Plant Prot. Trop., 1987b, 4, 37-54.

15. Basilio R.P. and Litsinger, J.A., Host range and feeding preference of golden apple snail. Int. Rice Res. Newsl., 1988, 13, 44-45.

16. Buckle, A., An integrated scheme for rice field rat management in South-east Asia. Paper (AGP:IPC/11/WP/8) presented at the FAO/UNEP Panel of Experts Meeting, 5-10 March 1982, Kuala Lumpur, Malaysia, 1982.

17. Castillo, M.B., Alejar, M.S. and Litsinger, J.A., Effect of crop rotation on nematode populations in dryland rice-based cropping systems of Batangas, Philippines. Int. Rice Res. Newsl., 1978a, 3, 18-19.

18. Castillo, M.B., Arceo, M.B. and Litsinger, J.A., Effect of geomorphic field position, flooding and cropping pattern on plant parasitic nematodes of crops following rainfed wetland rice in Iloilo, Philippines. Int. Rice Res. Newsl., 1978b, 3, 27.

19. Catindig, J.L.A., Barrion, A.T. and Litsinger, J.A., Life history and hosts of *Sogatodes pusanus* (Distant) (Hemivtera: Delphacidae). Int. Rice Res. Newsl., 1989, 14, 41-42.

20. Cendaña, S.M. and Calora, F.B., Insect pests of rice in the Philippines. In The Major Insect Pests of the Rice Plant, International Rice Research Institute and John Hopkins Press, Baltimore, Maryland, USA, 1964, pp. 591-616.

21. Cevallos, F.O., Control of diseases and pests by cultural methods. Philipp. Agric. Forester, 1911, 1, 86-88.

22. CIAT (Centro Internacional de Agricultura Tropical), Rice program for 1978. Cali, Colombia, 1979, 30 pp.

23. Claridge, M.F., den Hollander, J. and Morgan, J.C., The status of weed-associated populations of the brown planthopper, *Nilaparvata lugens* (Stål) - host race or biological species? Zool. J. Linnean Soc., 1985, 84, 77-90.

24. Custodio, H.A., Pablo, S.J., Olivares Jr., F., Cortado, R.V., Bergonia, H.T., Bueno, A. and Cornello, D., Critical review of pests, diseases and weed complexes in high yielding varieties under intensified agricultural practices in the Philippines. In A Review of Pest, Disease and Weed Complexes in High Yielding Varieties in Asia and Pacific. Food and Agriculture Organisation, No. RAPA 45, FAO Regional Office for Asia and the Pacific, Bangkok, Thailand, 1981, pp. 52-72.

25. de Braganca Pereira, E., *Sclerotium* disease of rice. Philipp. Agric., 1922, 10, 331-345.

26. Dalmacio, S.C. and Exconde, O.R., Host range of *Xanthomonas oryzae* in the Philippines. Philipp. Agric., 1967, **51**, 283-289.

27. Drost, D.C. and Moody, K., Rat damage in weed control experiments in rainfed transplanted rice. Trop. Pest Manage., 1982, **28**, 295-299.

28. Dyck, V.A., The role of ecology in insect pest control. IRRI Saturday seminar, 3 June 1973, International Rice Research Institute, Los Baños, Laguna, Philippines, 1973.

29. Elliot, P.C. and Moody, K., Weed control studies in upland rice-based cropping systems. Paper presented at the Claveria Cropping Systems Annual Review, 6-7 March 1986, Cagayan de Oro City, Philippines, 1986, 14 pp.

30. Elliot, P.C., Navarez, D.C., Fajardo, F.F. and Moody, K., Farmers concepts of weeds and their weed control practices in transplanted rice in Guimba, Nueva Ecija. Paper presented at the Cropping Systems Research Design Workshop, 7-8 May 1984, Muñoz, Nueva Ecija, Philippines, 1984, 7 pp.

31. Escuro, P.B. and Aquino, R.C., Comparison of Margate, Masagana and traditional methods of lowland rice culture. Philipp. Agric., 1963, 47, 215-237.

32. Fall, M.W., Rodents in tropical rice. Technical Bulletin No. 36, Rodent Research Centre, Los Baños, Laguna, Philippines, 1977, 37 pp.

33. Ferino, M.P., The biology and control of the rice leaf-whorl maggot, *Hydrellia philippina* Ferino (Ephydridae, Diptera). Philipp. Agric., 1968, **52**, 332-383.

34. Forno, I.W., How quickly can insects control salvinia in the tropics? In Proceedings of the 10th Asian-Pacific Weed Science Society Conference, Chiangmai, Thailand, 1985, pp. 271-276.

35. Gabriel, B.P., A review of the major insect pests of some upland crops in the Philippines. In Reviews on Pest, Disease and Weed Problems in Rainfed Crops in Asia and the Far East, ed. D.B. Reddy, FAO Regular Programme No. RAFE 23, Bangkok, Thailand, 1975, pp. 86-101.

36. Guh, J.O. and Lee, H.S., Successive growth of weeds as affected by soil fertility and light intensity in paddy field fertilized differently for many years. Seoul National University Faculty Paper. Biological Agric. Ser. E, 1974, 3, 84-115.

37. Heinrichs, E.A. and Medrano, F.G., *Leersia hexandra*, a weed host of the rice brown planthopper, *Nilaparvata lugens* (Stål). Crop Prot., 1984, 3, 77-85.

38. Hoque, M.M. and Olvida, J.L., Rodent activity and damage in clean and weedy cropfields. Philipp. Agric., 1986, **69**, 329-340.

39. IRRI (International Rice Research Institute), Annual report for 1969. Los Baños, Laguna, Philippines, 1970, 266 pp.

40. IRRI (International Rice Research Institute), Annual report for 1972. Los Baños, Laguna, Philippines, 1973, 246 pp.

41. IRRI (International Rice Research Institute), Annual report for 1981. Los Baños, Laguna, Philippines, 1982, 548 pp.

42. IRRI (International Rice Research Institute), Insecticide evaluation for 1982. Entomology Department, International Rice Research Institute, Los Baños, Laguna, Philippines, 1983.

43. IRRI (International Rice Research Institute), Insecticide evaluation for 1983. Entomology Department, International Rice Research Institute, Los Baños, Laguna, Philippines, 1984.

44. IRRI (International Rice Research Institute), Annual report for 1985. Los Baños, Laguna, Philippines, 1986, 555 pp.

45. IRRI (International Rice Research Institute), Annual report for 1986. Los Baños, Laguna, Philippines, 1987, 639 pp.

46. Ishii, S. and Hirano, C., Growth response of the larvae of rice stem borer to rice plants treated with 2,4-D. Entomol. Exp. Appl., 1963, 6, 257-262.

47. Joshi, R.C., Medina, E.B. and Heinrichs, E.A., Host plants of rice leaffolder (LF) *Marasmia patnalis* Bradley. Int. Rice Res. Newsl., 1985, 10, 29-30.

48. Joy, P.J., Sathesan, N.V., Lyla, K.R. and Joseph, D., Successful biological control of the floating weed *Salvinia molesta* Mitchell using the weevil Cyrtobagous salviniae Sands in Kerala (India). In Proceedings of the 10th Asian-Pacific Weed Science Society Conference, Chiangmai, Thailand, 1985, pp. 622-626.

49. Karim, R., Bionomics of rice leaf whorl maggot (*Hydrellia philippina* Ferino) (Diptera, Ephydridae) at the International Rice Research Institute. Los Baños, Laguna, Philippines, 1969, 38 pp. (unpublished)

50. Khan, M.A., Weed hosts of rice tungro-associated viruses and leafhopper vectors. Ph.D. thesis, University of the Philippines at Los Baños, College, Laguna, Philippines, 1988, 115 pp.

51. Kim, S.C. and Moody, K., Effect of seedling number per hill and seedling age on the competitive ability of rice (*Oryza sativa* L.) grown at different plant spacings. Philipp. Agric., 1982, 65, 177-194.

52. Kroeck, T. and Watanabe, I., Growth of azolla under rice canopies. IRRI Saturday seminar, 13 July 1985, International Rice Research Institute, Los Baños, Laguna, Philippines, 1985.

53. Lacson, J.L., The response of early maturing rice varieties to different concentrations of 2,4-dichlorophenoxy acetic acid. B.S. thesis, Central Philippine University, Iloilo City, Philippines, 1968, 41 pp.

54. Lim, L.L., The biology of *Cnaphalocrocis medinalis* Guen. (Lepidoptera, Pyralidae) with a consideration of its natural enemies. B.S. thesis, University of the Philippines College of Agriculture, College, Laguna, Philippines, 1962, 29 pp.

55. Litsinger, J.A., Barrion, A.T. and Soekarna, D., Upland rice insect pests: their ecology, importance and control. IRRI Res. Pap. Ser. No. 123, International Rice Research Institute, Los Baños, Laguna, Philippines, 1987a.

56. Litsinger, J.A., Kenmore, P.E. and Saxena, R.C., Cultural control of rice insect pests. Lecture presented at the Regional Training Seminar on Integrated Pest Control for Irrigated Rice in South and Southeast Asia, 16 October-18 November 1978, Manila, Philippines, 1978.

57. Litsinger, J.A., Mochida, O., Guevarra, H.T. and Basilio, R., Golden apple snail *Pomacea canaliculata*: an introduced pest of rice. Paper presented at the 11th International Congress of Plant Protection, 5-9 October 1987, Manila, Philippines, 1987b.

58. Mackill, A.O. and Bonman, J.M., New hosts of *Pyricularia oryzae*. Plant Disease, 1986, 70, 125-127.

59. Manas, Y., Cruz, M. and Galang, F.G., The rice, sugar cane, tobacco, and corn projects of the Division of Plant Investigations of the Bureau of Agriculture. Philipp. Agric. Rev., 1929, 22, 43-70.

60. Manila, C.E. and Lapis, D.B., Severity of rice blast, bacterial blight and sheath blight in rice after application with herbicides and insecticides. Philipp. Agric., 1977, 61, 1-11.

61. Manwan, I., Resistance of rice varieties to yellow borer, *Tryporyza incertulas* (Walker). Ph.D. thesis, University of the Philippines at Los Baños, College, Laguna, Philippines, 1975, 185 pp.

62. Mew, T.W., Fabellar, N.G. and Elazegui, F.A., Ecology of the rice sheath blight pathogen: parasitic survival. Int. Rice Res. Newsl., 1980, 5, 16.

63. Mew, T.W., Vera Cruz, C.M. and Reyes, R.C., Effect of nitrogen on lesion development in rice resistant to bacterial blight. Int. Rice Res. Newsl., 1979, 4, 12-13.

64. Moody, K., Weed control in rice and sugarcane cropping systems. In Weed Control in Tropical Crops, ed. K. Moody, Weed Science Society of the Philippines, College, Laguna, Philippines, 1979, pp. 56-74.

65. Moody, K., Weed-fertiliser interactions in rice. IRRI Res. Pap. Ser. No. 68, International Rice Research Institute, Los Baños, Laguna, Philippines, 1981, 35 pp.

66. Moody, K., Weeds: definitions, costs, characteristics, classification and effect. In Weed Management in the Philippines. PLITS 1(1), ed. H. Walter, Institut fur Pflanzenproduktion in den Tropen und Subtropen. Universitat Hohenheim, Stuttgart, Federal Republic of Germany, 1983, pp. 11-32.

67. Nono, A.M. and Mane, A.M., Biology of cohol (*Ampullaria luzonica* Reeve) a common Philippine Fresh-water snail. Philipp. Agric., 1931, 19, 675-695.

68. Omura, T., Saito, Y., Usugi, T. and Hibino, H., Purification and serology of rice tungro spherical and rice tungro bacilliform viruses. Ann. Phytopath. Soc. Japan, 1983, 49, 73-76.

69. Otanes, F.Q., Some observations on root grubs (*Leucopholis irrorata* Chevr.) in the Philippines and suggestions for their control. Philipp. Agric. Rev., 1924, 17, 109-119.

70. Otanes, F.O., Circular No. 159, The rice stem borer (Schoenobius incertellus Walker). Philipp. Agric. Rev., 1925a, 18, 81-82.

71. Otanes, F.O., Circular No. 160, The rice bug (*Deptocarisa acuta* Thunberg). Philipp. Agric. Rev., 1925b, 18, 83-85.

72. Otanes, F.Q., Circular No. 83 (Revised), Rice cutworms (*Spodoptera mauritia* Boisd, Prodenia Litura Fbr.). Philipp. Agric. Rev., 1925c, 28, 551-554.

73. Pablico, P.P. and Moody, K., Effect of fentin acetate on wet-seeded rice (*Oryza sativa*), *Pistia stratiotes* and *Azolla pinnata*. Paper presented at the 19th annual conference of the Pest Control Council of the Philippines, 3-7 May 1988, Cebu City, Philippines, 1988, 8 pp.

74. Paje, E.P., Exconde, O.R. and Raymundo, S.A., Host range of *Piricularia oryzae* in the Philippines. Philipp. Agric., 1964, 48, 35-48.

75. Palo, M.A., *Rhizoctonia* disease of rice. I. A study of the disease and of the influence of certain conditions upon the viability of the sclerotial bodies of the causal fungus. Philipp. Agric., 1926, 15, 361-375.

76. Pantua, P.C. and Litsinger, J.A., Notes on sampling methods, monitoring time and ecological areas for the yellow stem borer moth, *Scirpophaga incertulas*. Philippine Association of Entomologists, Abstracts of Papers, 19th Anniversary and Annual Convention, Pest Control Council of the Philippines, 3-7 May 1988, Cebu City, Philippines, 1988.

77. Parducho, M.A., Arida, G.S. and Shepard, B.M., Effect of flooding on Malayan black bug (MBB) egg hatching and parasitoid emergence. Int. Rice Res. Newsl.. 1988, 13, 39.

78. Pathak, M.D., Insect pests of rice. International Rice Research Institute, Los Baños, Laguna, Philippines, 1969, 77 pp.

79. Pathak, M.D. and Dyck, V.A., Developing an integrated method of rice insect pest control. PANS. 1973, 19, 534-547.

80. Quisumbing, E., General characters of some Philippine weed seeds. Philipp. Agric. Rev.. 1923, 16, 298-351.

81. Razzaque, Q.M., Heinrichs, E.A. and Rapusas, H.R., Comparative levels of resistance of rice cultivars and weeds against *Nephotettix nigropictus* (Stål) and *Nephotettix virescens* (Distant). J. Plant Prot. Trop.. 1987, 4, 75-84.

82. Rentz, D.C., The ecology, behaviour and description of a new species of cricket from the Osa Peninsula of Costa Rica. Entomol. News. 1973, 84, 237-246.

83. Reyes, G.M., A preliminary report on the stem-rot of rice. Philipp. Agric. Rev.. 1929, 22, 313-331.

84. Reyes, G.M. and Palo, A.V., Nematode disease of rice. Araneta J. Agric.. 1956, 3, 72-77.

85. Reynaud, P.A. and Franche, C., *Azolla pinnata* var. *africana*. ORSTOM, n.d., 15 pp.

86. Rezaul Karim, A.N.M. and Saxena,R.C., Feeding behavior of three *Nephotettix* species on selected and graminaceous weeds. Int. Rice Res. Newsl.. 1989, 4, 28.

87. Rimando, L.C., A biological study of the black armyworm, *Laphygma exempta* Walker, Noctuidae, Lepidoptera. B.S. thesis, University of the Philippines College of Agriculture, College, Laguna, Philippines, 1954, 40 pp.

88. Rivera, C.T., Ling, K.C. and Ou, S.H., Suscept range of rice tungro virus. Philipp. Phytopathol.. 1969, 6, 16-17.

89. Rodrigo, P.A., The effect of spacing on tillering and production of three varieties of rice. Philipp. Agric.. 1924, 13, 5-28.

90. Room, P.M. and Thomas, P.A., Nitrogen and establishment of a beetle for biological control of the floating weed Salvinia in Papua New Guinea. J. Applied Ecol., 1985, 22, 139-156.

91. Rowar, A.A., The rice borer (Schoenobius incertellus Walker). Philipp. Agric., 1923, 12, 225-237.

92. Rubia, E.G., Biological and toxicological studies of the cricket, Metioche vittaticollis (Stål) (Orthoptera: Gryllidae), a predator of rice insect pests. M.S. thesis, University of the Philippines at Los Baños, College, Laguna, Philippines, 1986, 79 pp.

93. Salamat, G.Z. Jr., Parejarearn, A. and Hibino, H., Weed hosts of ragged stunt virus (RSV). Int. Rice Res. Newsl., 1987, 12, 30.

94. Sanchez, F.F., Pest and disease complexes in high yielding varieties in Asia and Pacific. In A Review of Pest, Disease and Weed Complexes in High Yielding Varieties in Asia and Pacific. Food and Agriculture Organisation, No. RAPA 45. FAO Regional Office for Asia and the Pacific, Bangkok, Thailand, 1981, pp. 73-79.

95. Sankaran, S. and De Datta, S.K., Weeds and weed management in upland rice. Adv. Agron., 1985, 38, 283-337.

96. Saxena, R.C. and Puma, B.C., Effect of trans-aconitic acid, a barnyard grass allelochemic, on hatching of eggs of brown planthopper and green leafhopper. Paper presented at the 10th annual Conference of the Pest Control Council of the Philippines, 2-5 May 1979, Manila, Philippines, 1979.

97. Scharpenseel, H.W. and Knuth, K., Use and importance of Azolla-Anabaena in industrial countries. In Azolla Utilization. International Rice Research Institute, Los Baños, Laguna, Philippines, 1987, pp. 153-167.

98. Sison, P.L., Some observations on the life history habits and control of the rice caseworm, Nymphula depunctalis Guen. Philipp. J. Agric., 1938, 9, 273-301.

99. Sogawa, K., The brown planthopper, Nilaparvata lugens (Stål) (Homoptera, Delphacidae) at the IRRI, the Philippines. Technical Bulletin No. 15, Tropical Agricultural Research Centre, Tsukuba, Ibaraki, Japan, 1982, 51 pp.

100. Tan, J.P., The rice root aphis (Dryopeia hirsuta A.C. Baker). Philipp. Agric., 1924, 13, 277-288.

101. Taylo, L.D., Litsinger, J.A. and Cadapan, E.P., Plant host range of the rice bug (RB). Int. Rice Res. Newsl., 1987, 12, 36.

102. Teodoro, N.G. and Bogayong, J.R., Rice diseases and their control. Philipp. Agric. Rev., 1926, 26, 237-241.

103. Uichanco, L., The rice bug, Leptocorisa acuta Thunberg, in the Philippines. Phillip. Agric. Rev., 1921, 14, 87-112.

104. Uichanco, L.B., Insects in Philippine folklore. *Philipp. Agric.*, 1937, **26**, 485-499.

105. Valdez, R.B., Sheath spot of rice. *Philipp. Agric.*, 1955. 39, 317-336.

106. Viajante, V.D. and Heinrichs, E.A., Influence of water regime on rice whorl maggot oviposition. *Int. Rice Res. Newsl.*, 1985a, 10, 24.

107. Viajante, V.D. and Heinrichs, E.A., Oviposition of rice whorl maggot as influenced by azolla. *Int. Rice Res. Newsl.*, 1985b, 10, 23.

108. Waibel, H. and Meenakanit, P, Economics of integrated pest control in rice in Southeast Asia. In *Pesticide Management and Integrated Pest Management in Southeast Asia*, eds. P.S. Teng and K.L. Heong, Consortium for International Crop Protection, United States Agency for International Development, College Park, Maryland, USA, 1988, pp. 103-111.

109. Wathanakul, L., A study on the host range of tungro and orange leaf viruses of rice. M.S. thesis, University of the Philippines at Los Baños, College, Laguna, Philippines, 1964, 35 pp.

110. ZDSDP (Zamboanga del Sur Development Project), Second Agricultural Report, 1980. Mindanao, Philippines, 1981, 248 pp.

111. Zhang, Z.T. and Saxena, R.C.. Variations in courtship signals between the rice and *Leersia*-infesting biotypes of brown planthopper, *Nilaparvata lugens* (Stål). Philippine Association of Entomologists, Abstracts of Papers, 19th Anniversary and Annual Convention, Pest Control Council of the Philippines, 3-7 May 1988, Cebu City, Philippines, 1988.

YIELD LOSSES IN TROPICAL RICE
AS INFLUENCED BY THE COMPOSITION OF WEED FLORA
AND THE TIMING OF ITS ELIMINATION

ANDREAS ZOSCHKE

CIBA-GEIGY Limited, Agricultural Division, Research & Development,
CH-4002 Basle, Switzerland

ABSTRACT

Yields from 140 rice trials conducted at CIBA-GEIGY's Agricultural Research and Development Station in Indonesia were compiled with the aim of determining yield losses in both irrigated transplanted and wet-sown rice in relation to the timing of the elimination of the annual weed flora and its composition.

Relative yield losses due to weeds averaged 50% and were independent of rice production system or season. Actual yield losses were greater in wet season and transplanted rice. While all weed species present contributed to yield losses, the contribution from both grasses and sedges was more significant than that of broadleaved weeds.

Yields increased irrespective of the timing of weed elimination. In particular, elimination of weeds during the period 0-9 DAT/DAS lead to significant yield gains, also when compared with the handweeded check. Such early elimination of weeds was especially beneficial in dry seasons and wet-sown rice.

It is suggested that future studies on rice/weed-competition should pay particular attention to the weed group of sedges as well as to seasonal influences.

INTRODUCTION

Weeds compete with the rice crop for a variety of reasons. Weeds serve as hosts for pests and diseases and thus intensify these problems, they interfere with operational techniques (harvest, irrigation), and can also negatively influence rice grain yield and quality. This is mainly attributed to competition for nutrients, water, and light [1, 3, 7, 15]. Weed competition is complicated because various factors affect the extent to which it occurs [1, 15]. Such factors include environment, weed density, composition of the weed flora and agronomic conditions such as fertilization, water management and variety.

The first 30-40 days after rice planting are considered to be most critical with regard to rice/weed-competition and during this period the rice crop should be kept weed free [5]. Weed control is more difficult in wet-sown rice than in transplanted rice. Handweeders moving through broadcast rice may damage or destroy some rice plants and often have difficulties in distinguishing between young grassy weeds and young rice [3]. However, early weed control is important to obtain high yields in both broadcast wet-sown and transplanted rice [4]. Yield losses are greater in wet-sown rice compared to transplanted rice [4, 11, 13], and can vary in relation to season [11].

Grasses reduce yields more than broadleaved weeds and sedges [6, 9, 15], and this is even more so when competition is from more than one weed species [4, 16]. The longer the duration of competition, the more serious are yield losses [1]. Once the rice plants are well established, weeds no longer affect rice yield [1, 8, 9].

In South-East Asian rice fields the potential grain yield is cut down by 11% due to pests, diseases and weeds; about 20% of such losses can be attributed exclusively to weeds [2]. One estimate at IRRI suggests that uncontrolled weed growth reduces yields in lowland rice by 34-45% [3].

However, the degree of yield losses is not consistent as can be seen, for instance, from the overview table published by Smith [15]. Over several years CIBA-GEIGY at its Agricultural Research and Development Station in Indonesia has carried out trials in irrigated rice designed to test agrochemicals for their properties as weed control agents. These trials were conducted under fairly consistent conditions with regard to climate and agronomic practice. The impact of such factors on rice yield is therefore considered to be relatively low. By compiling a large number of observations from such trials, this paper aims to determine grain yield losses in the two major production systems of transplanted and wet-sown rice relative to the composition of the weed flora and the timing of elimination of the weed competition.

MATERIALS AND METHODS

Trial location
The trials were carried out at CIBA-GEIGY's Agricultural Research and Development Station in Cikampek, West Java, Indonesia (6º 20' S, 107º 30' E).

Temperatures at Cikampek Station, located in the humid tropics, vary little during the course of the year (daily mean of about 25-27º C). Mean annual precipitation is about 1400 mm; there is a distinct wet (rainy) season lasting from October/November to April/May, while the dry season lasts from May/June to September/October. Mean relative humidity is about 85-90%. There are about 2500 sunshine hours per year.

The soil at Cikampek Station is alluvial with 25% sand, 34% silt and 41% clay (sandy clay loam), an organic matter content of 3.9%, and a pH (KCl) of 4.4.

Rice crop details
In general, irrigated rice at Cikampek Station is cropped according to local practice in the region. This also refers to field and water management, fertilization, and crop protection measures. Soil preparation consists of 1-3 puddlings and/or hoeings followed by 1-2 levellings. In transplanted rice, three seedlings per hill (at development stage of 4-6 leaves) were planted by hand 2-3 cm deep and spaced 20-25 x 20-25 cm. In wet-sown rice, seeding rate was 60-80 kg/ha; rice seeds were water-soaked and pre-germinated for 24 hours each.

Rice (*Oryza sativa L.*) **varieties** were of the Indica-type. In most of the trials, cv. *IR-36* was used; in few trials, cvs. *Cisadane, Citandui, IR-64,* or *Pelita* were planted. Mean duration of growth period in wet and dry season was 99 and 85 days, respectively, in transplanted rice (# ORYSP) and in wet-sown rice (# ORYSW) it was 109 and 103 days, respectively.

Weed flora details

At Cikampek Station, weed flora consists of annuals of the families of Cyperaceae (C), Gramineae (G), Marsileaceae (M), Onagraceae (O), Pontederiaceae (P), Scrophulariaceae (Sc), and Sphenocleaceae (Sp).

The **dominant weed species** in the trials were (C) *Cyperus difformis* L. (# CYPDI), (C) *Cyperus iria* L. (# CYPIR), (G) *Echinochloa crus-galli* (L.) Beauv. (# ECHCG), (C) *Fimbristylis miliacea* (l.) Vahl (# FIMMI), (P) *Monochoria vaginalis* (Burm. f.) Presl (# MOOVA), and (C) *Scirpus juncoides* Roxb. (# SCPJO). However, the degree of infestation of each weed species varied from trial to trial.

Weed species with a low degree of infestation or occurring less frequently were (G) *Leptochloa chinensis* (l.) Nees, (Sc) *Lindernia procumbens* (Krock.) Philcox, (O) *Ludwigia adscendens* (L.) Hara, (O) *Ludwigia octovalvis* (Jacq.) Raven (# LUDOC), (M) *Marsilea crenata* Presl, and (Sp) *Sphenoclea zeylanica* Gaertn.

In order to facilitate both data presentation and statistical analysis, **weed species** were divided into **three subgroups**: Gramineae [mainly ECHCG: from now on called GRA], Cyperaceae [CYPDI, CYPIR, FIMMI, SCPJO: CYP], and broadleaved weeds [mainly MOOVA, some LUDOC: BLW].

The degree of weed infestation was homogeneous between rice production systems and seasons. On average, 76-83% of the soil surface that was not occupied by rice was covered with weeds (range: 32-97%, evaluated at canopy closure at 45-65 DAT/DAS [days after transplanting/sowing]). The mean figures of weed cover at 10-15 and 20-30 DAT/DAS were 35% and 54%, respectively, in wet seasons, and in dry seasons 42% and 61%, respectively.

Calculation of grain yield losses

A total of 174 observations were entered into the calculations. They were collected from 140 trials carried out in 8 seasons during 1984 and 1989. Of the total, 110 observations (84 and 26 in transplanted and wet-sown rice, respectively) were derived from wet seasons, while 64 observations (43 and 21 for transplanted and wet-sown rice, respectively) were derived from dry seasons.

Rice grain yield losses in each trial were computed by deducting the yield in the untreated (weedy) check from the yield recorded for the corresponding herbicide treatment or the yield recorded for the handweeded check (weeded twice at 21 & 42 DAT and included only in transplanted rice). Such differences are expressed in relative [%] or actual terms [decitons/hectare, dt/ha].

The herbicide treatment used for the calculations was the one giving the highest yield within each trial. The herbicides were always applied at their recommended application timings and dosages and included only compounds specifically developed for use in transplanted and/or wet-sown rice. They were predominantly those from CIBA-GEIGY Limited (such as piperophos, piperophos/dimethametryn [AVIROSAN®], piperophos/2,4-D [RILOF-H®], pretilachlor [RIFIT®], pretilachlor/fenclorim [SOFIT®] or CGA 142464 [SETOFF®]). In a few cases, development compounds of CIBA-GEIGY Limited or compounds of 3rd parties (such as bensulfuron, benthiocarb, butachlor, oxadiazon, propanil, quinclorac, or 2,4-D) were also used alone or in a mixture. Since, in this context, it is not relevant which treatment was the best yielding and how it compared to the others, the term **herbicide** is used to indicate that best treatment. It has also to be stated that this yield level in some cases may not be equivalent to the maximal yield.

In order to facilitate both the presentation of data as well as statistical analysis, **herbicide treatments** were divided into **four subgroups**, those applied 0-4 DAT/DAS [from now on called <u>pre</u>-application], 5-9 DAT/DAS [<u>early post</u>-application], 10-15 DAT/DAS [<u>post</u>-application], and 16-25 DAT/DAS [<u>late post</u>-application].

Trial designs and Evaluations

All trials were laid out as randomised complete blocks; plot size was 15-20 sqm. Data on both **weed cover** and **infestation of weed species** are expressed as % actual soil cover (neglecting the rice crop). 5-20 sqm were harvested for **rice grain yields** and are compared following adjustment to a 14% moisture content. Every single value that entered the calculations represented the mean from **3-4 replications**.

Biometrics and statistics

The loading of canonical **correlation coefficients** to serve as indicators of the competitive ability of the different weed species to reduce rice grain yield (tab. 1) was calculated according to Mardia *et al.* [10]. The data for wet-sown rice were not divided into seasons due to the rather small number of observations. The coefficients were also used in fig. 5 and fig. 6 to adapt the actual weed covers accordingly ('adjusted weed cover').

Both GRA and CYP were found to be the major species determining yield losses. Therefore, data in fig. 5 and fig. 6 are arranged in the order of increasing weed cover of GRA plus CYP.

For data in fig. 4, analysis of variance was carried out using a model including the factors 'rice production system', 'season', 'application timing' and 'production system x season-interaction'. Tukey's method was then used to compare possible treatment differences (i.e. differences in system, season, and application timing). Significant differences (at 5%-level) were found for the production system factor: LSD - 3.4736 [overall means: transplanted rice - 24.01 dt/ha, wet-sown rice - 17.38 dt/ha] and for the season factor: LSD - 3.1883 [overall means: wet season - 25.63 dt/ha, dry season - 16.51 dt/ha]. Mean differences between application timings (comparing pre- with late-post-application: 6.56 dt/ha; comparing early-post- with late-post-application: 9.63 dt/ha) are also significant at the 5%-level; due to the different number of observations in treatment combinations calculation of a LSD is meaningless.

In fig. 7 the ratios between herbicide treatment and handweeded check as well as the corresponding 95%-confidence interval of the average ratios are indicated separately for application timings and seasons.

RESULTS

Development of weed species and composition of weed flora

Fig. 1 shows that grasses (GRA, mainly ECHCG) are the first weeds to germinate immediately following seeding of the rice grains. Plant tips of GRA are already visible at 5 DAT/DAS. While an obvious growth of sedges (CYP) begins around 5-10 DAT/DAS [FIMMI slightly later], broadleaved weeds (BLW, mainly MOOVA) show visible growth only at 10-15 DAT/DAS [in the case of LUDOC even later]. After 20-25 DAT/DAS, GRA attain the same height as rice plants in both transplanted and wet-sown rice. Similarly to

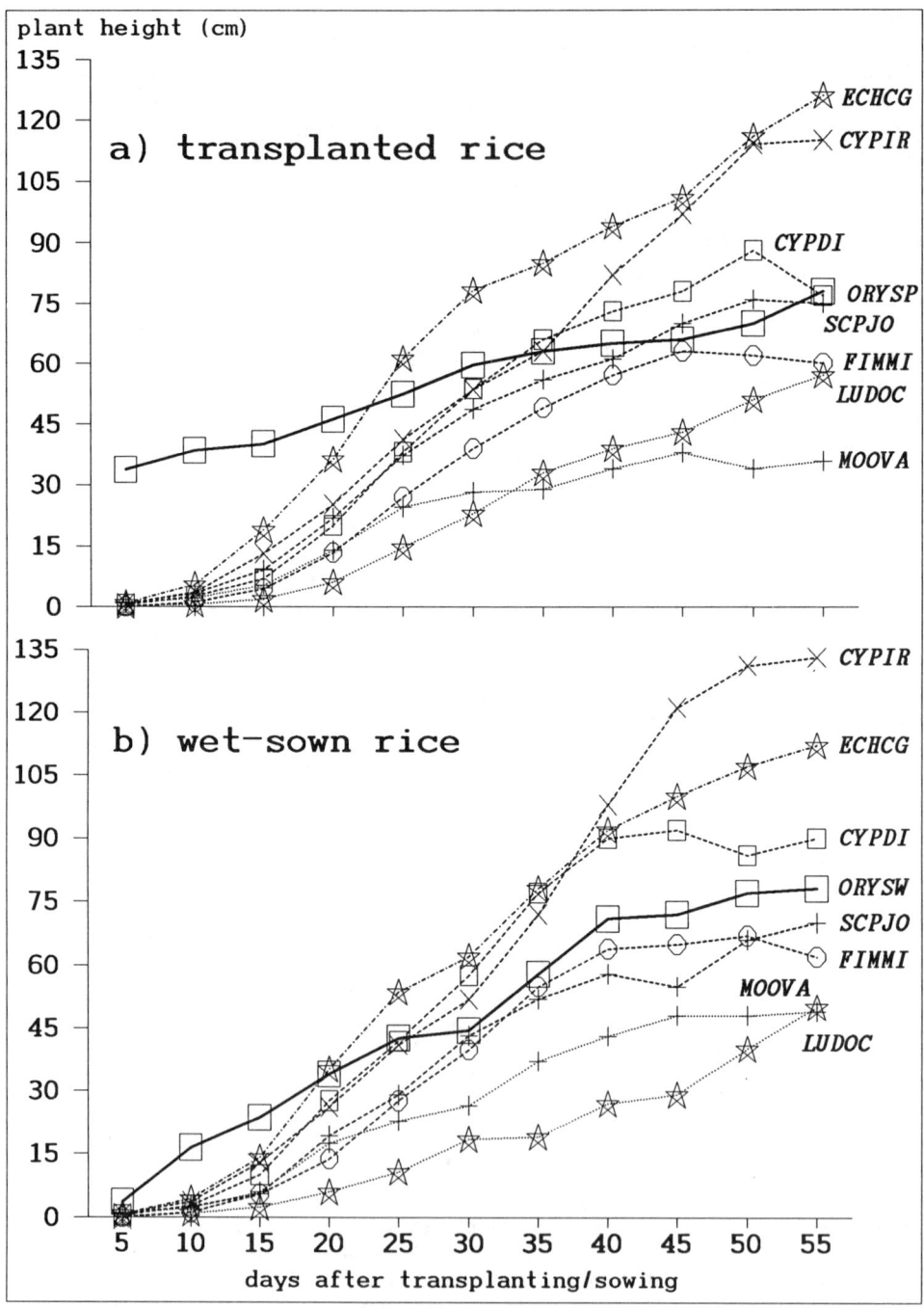

Figure 1. Development of crop and weeds in transplanted and wet-sown rice at Cikampek (wet season).

GRA, weeds out of the species *Cyperus* (CYPDI, CYPIR) will also eventually overgrow the rice plants. Even though the weed species show similar growth patterns in both rice production systems, their speed of height development is different. During the first weeks of development, weeds tend to grow faster in transplanted rice. This is possibly due to the fact that transplants immediately shade the soil surface, and, therefore, force the weeds to grow taller to become competitive for light.

At rice canopy closure (45-65 DAT/DAS), mean overall weed cover in our trials was about equal, irrespective of the rice production system and season (~80%; fig. 2). Also the composition of weed flora was homogeneous with the exception of dry season trials in wet-sown rice, in which BLW and CYP were less dominant (in favour of GRA).

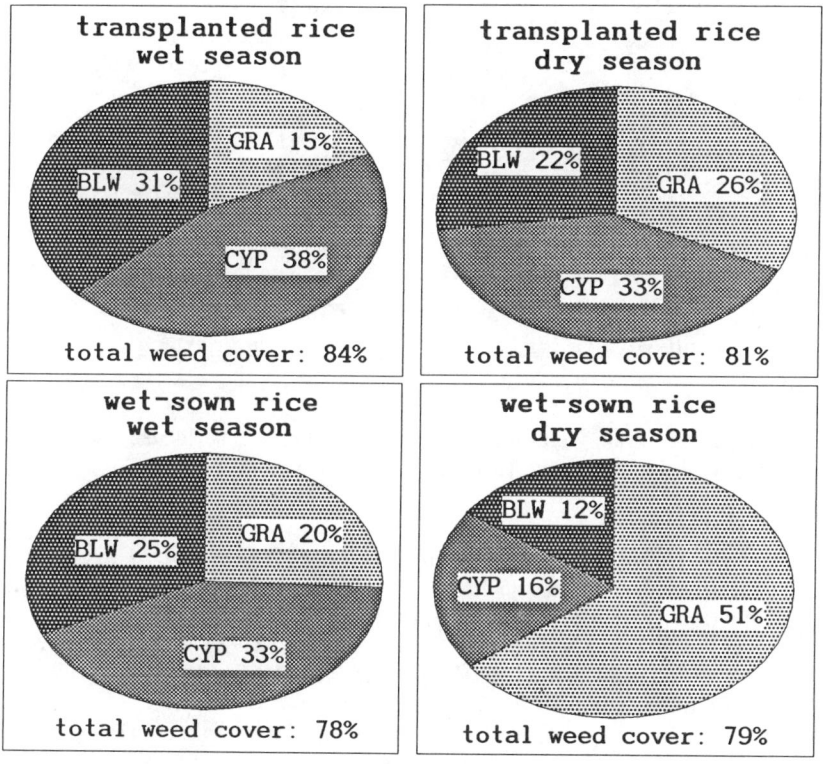

Figure 2. Mean actual weed cover and mean composition of weed flora in relation to rice production system and season (at 45-65 DAT/DAS).

Level of rice grain yield

In general, rice yielded more in wet than in dry seasons and yields were higher in transplanted compared to wet-sown rice (fig. 3). The performance of untreated (weedy) check, herbicide treatments, and handweeded check was the same in this respect. Actual yield losses (difference between herbicide and weedy check) were also higher if the rice crop was grown in a wet season, or if it was transplanted. On average, the handweeded check yields were less than with the herbicide.

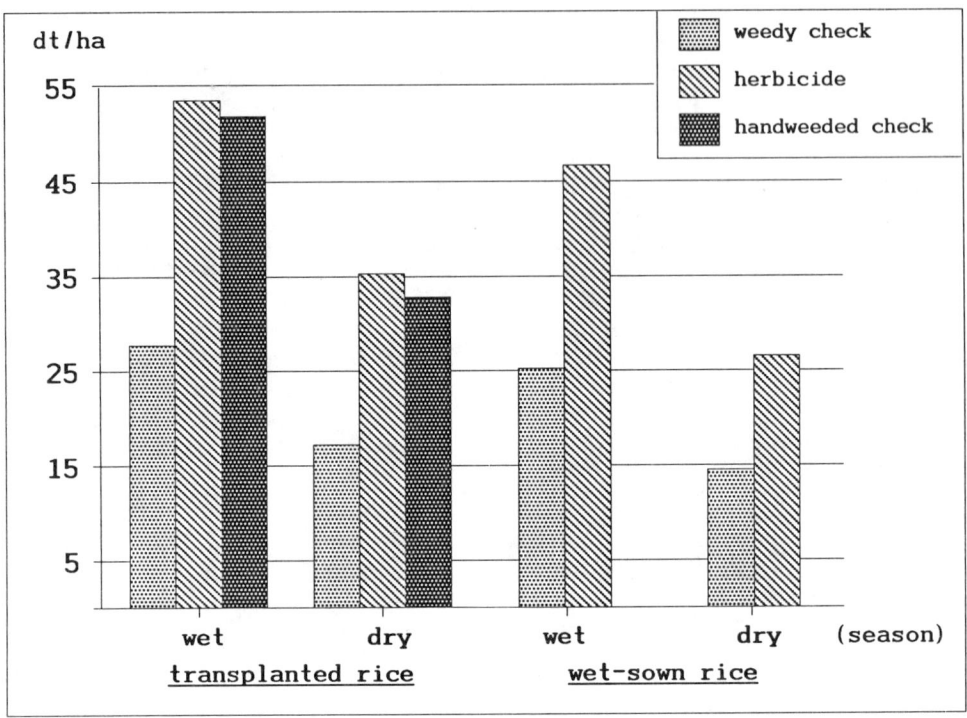

Figure 3. Mean rice grain yield of untreated (weedy) check, herbicide, and handweeded check in relation to rice production system and season.

Influence of the timing of weed elimination on rice grain yield losses

Expressed in relative terms, *mean yield losses ($\bar{x}$)* showed little variation. Irrespective of the rice production system and season, such losses reached 46-51% (fig. 4). In actual terms, however, yield losses were higher in wet seasons (21-26 dt/ha) compared with dry seasons (12-18 dt/ha), and higher in transplanted rice (18-26 dt/ha) compared with wet-sown rice (12-21 dt/ha; compare fig. 3 and fig. 4).

In general, the influence of the timing of weed elimination on rice grain yield was visible in both seasons, but more pronounced in dry seasons (fig. 4). In all cases, elimination of weeds at pre- to early-post-applications will fill yield gaps better than post- or late-post-applications (the rather unexpected result due to late-post-application observed in a dry season cropped wet-sown rice is possibly due to both the high infestation of GRA in this segment as shown in fig. 2 and the small number of observations). Even though this **positive influence of early elimination of weed competition on** yield holds true for both rice production systems, gains were higher in transplanted rice (45-67% or 22-31 dt/ha) as compared with wet-sown rice (42-54% or 10-27 dt /ha). With regard to season, possible yield gains derived from early elimination of weeds were higher in wet season (43-57% or 20-31 dt/ha) than in dry season (42-67% or 10-25 dt/ha). The delay in the timing of the elimination of weed competition towards later application timings of the herbicide reduced the maximal achievable yield gains both in relative and actual terms. However, in late-post-application, they still attained 10-22 dt/ha.

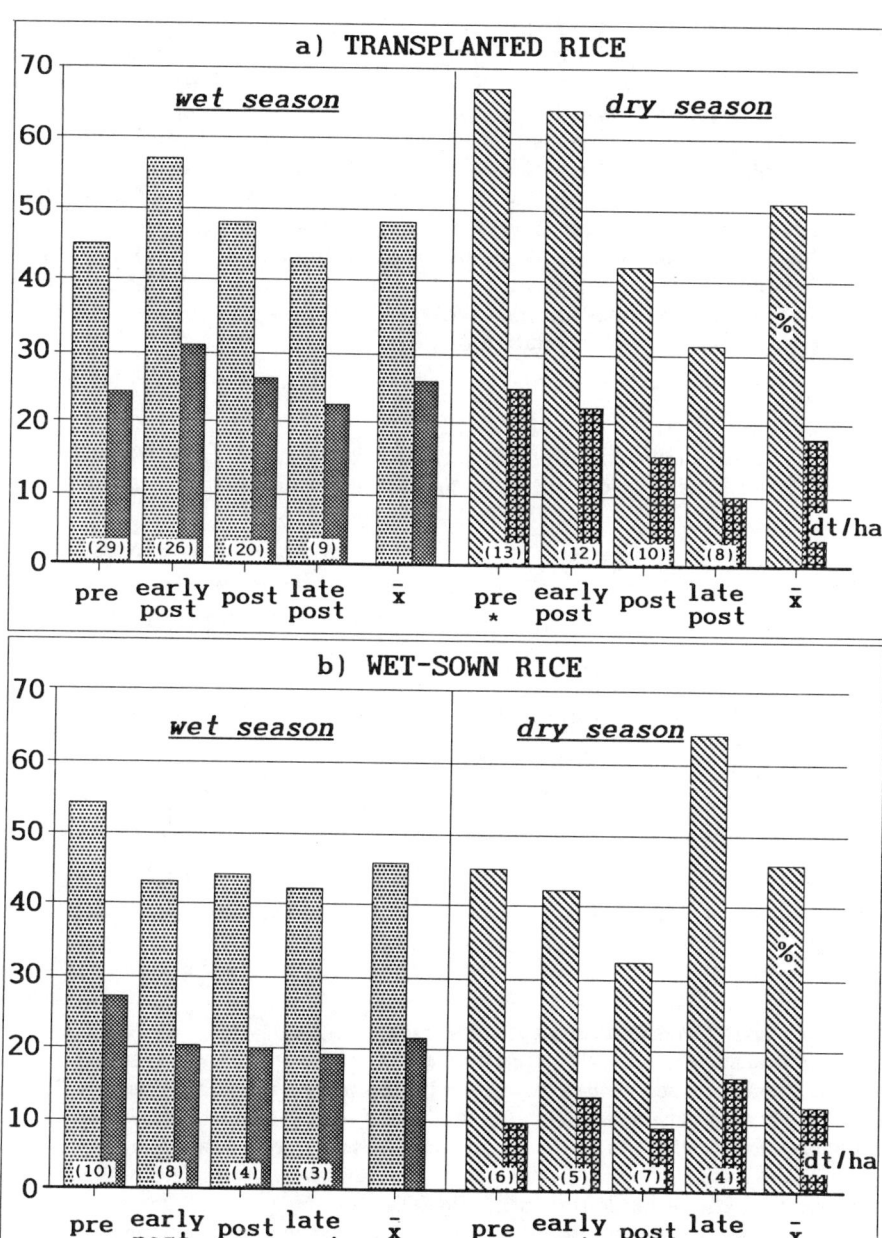

Figure 4. Relative and actual grain yield losses in relation to rice production system, season, and timing of elimination of weed competition.

Influence of composition of weed flora on rice grain yield losses

In general, there was a clear trend that yield losses in both transplanted and wet-sown rice became larger with increasing weed cover (fig. 5, fig. 6). The adjusted weed cover (for loading of correlation coefficients, see tab. 1) also showed that the group of weed species is significant. Depending on the composition of the weed flora differences in yield losses were observed. While BLW had little influence, and especially so in dry seasons, both GRA and CYP contributed about the same high degree in causing enhanced yield losses. Even though the yield losses were highest if GRA were the dominant components of a given mixed weed flora, a pure stand of GRA did not necessarily lead to maximal yield losses. This was since weed infestations dominated by CYP also lead to a comparable yield loss (see fig. 5a, fig. 5b). A mixed infestation of GRA plus CYP tended to cause yield losses at least as high as those from a pure infestation of GRA.

TABLE 1

Loading of the canonical correlation coefficients indicating the competitive ability of the different weed species to reduce grain yield in rice. (calculated after [10])

group of weed species

		GRA	CYP	BLW
transplanted rice	wet season	1.00	0.80	0.53
	dry season	0.96	1.00	0.36
wet-sown rice	wet & dry season	0.98	1.00	0.64

Yield of handweeded check relative to herbicide treatment

The yield in the handweeded check [weeded twice at 21 & 42 DAT], like that of the untreated check and herbicide treatment, revealed a dependency on season (see fig. 3). The yield in the handweeded check was significantly lower in dry seasons. Averaged over all trials and application timings, in both dry and wet seasons the handweeded check yielded significantly more compared with the untreated check. However, it yielded slightly less when compared with the herbicide treatment (fig. 3).

Compared at late-post-application (≈ recommendation for handweeding), the more detailed analysis of yield data (fig. 7) shows that irrespective of the season, rice yields did not differ between the herbicide and the handweeded check. With earlier application timings, however, the yields in the herbicide treatments increased to a greater extent over the hand- weeded check, so that, at pre-application, the herbicide gave a maximum yield over the handweeded check (up to 20%). With regard to seasons, the yield advantage of the herbicide over the handweeded check was more pronounced in dry seasons (mean 8-18%) than in wet seasons (mean 2-8%).

(A) WET SEASON (84 OBS.)

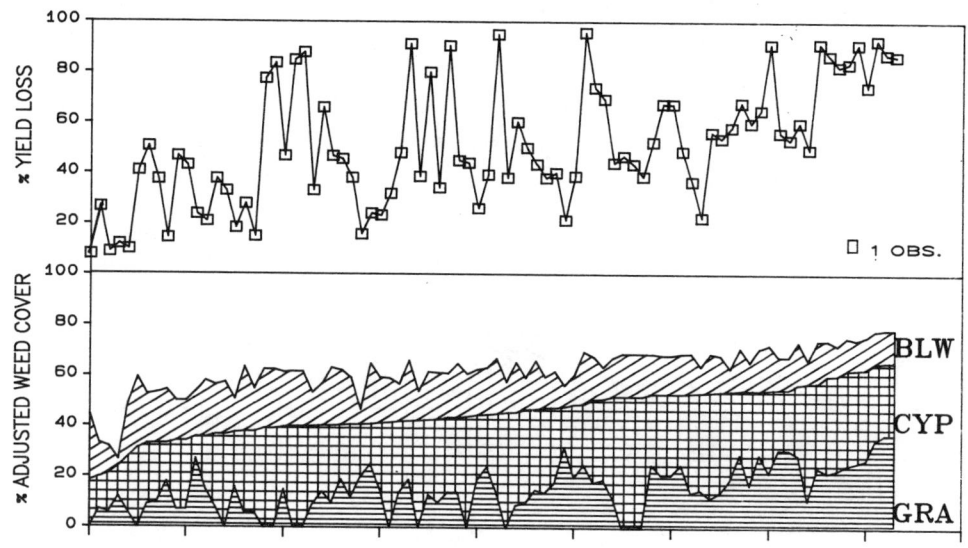

(B) DRY SEASON (43 OBS.)

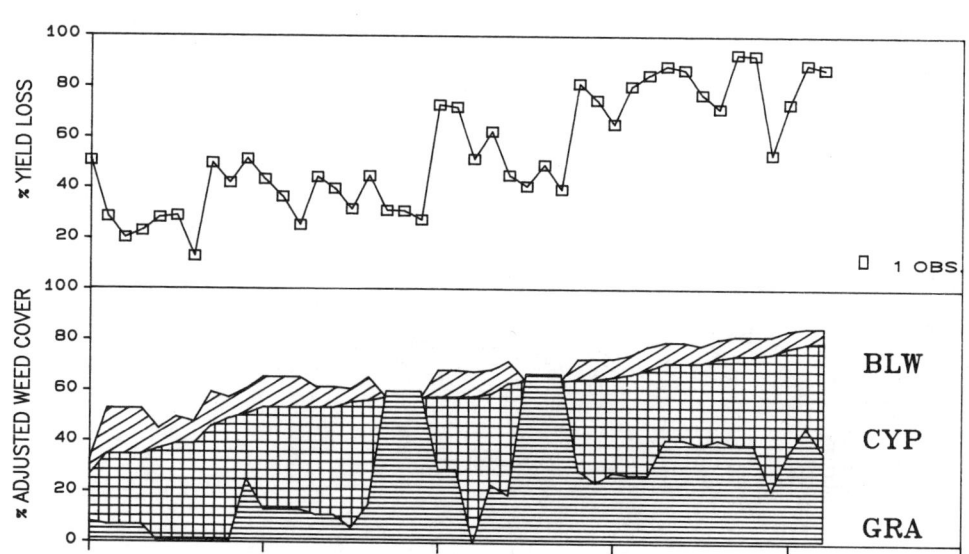

Figure 5. Grain yield losses in transplanted rice in relation to weed cover
and composition of weed flora

(A) WET SEASON (25 OBS.)

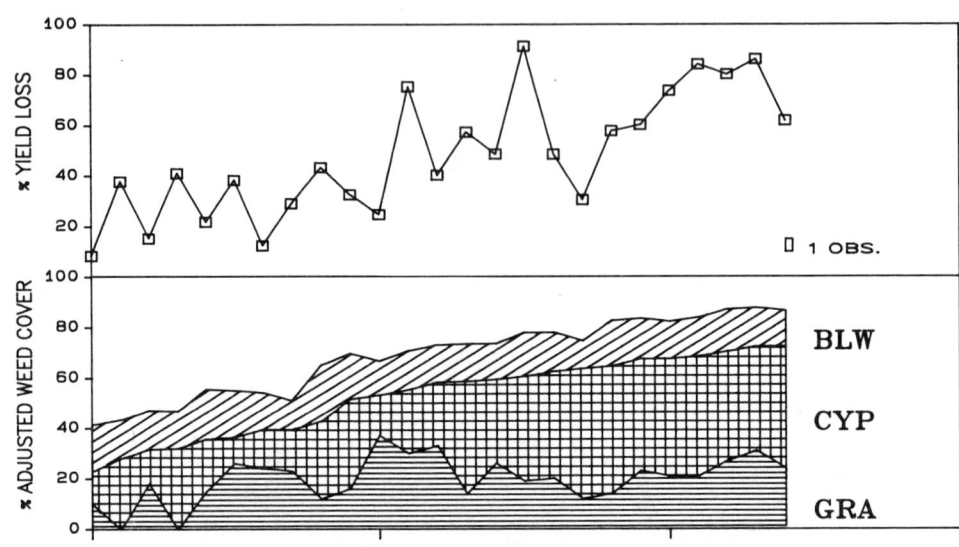

(B) DRY SEASON (22 OBS.)

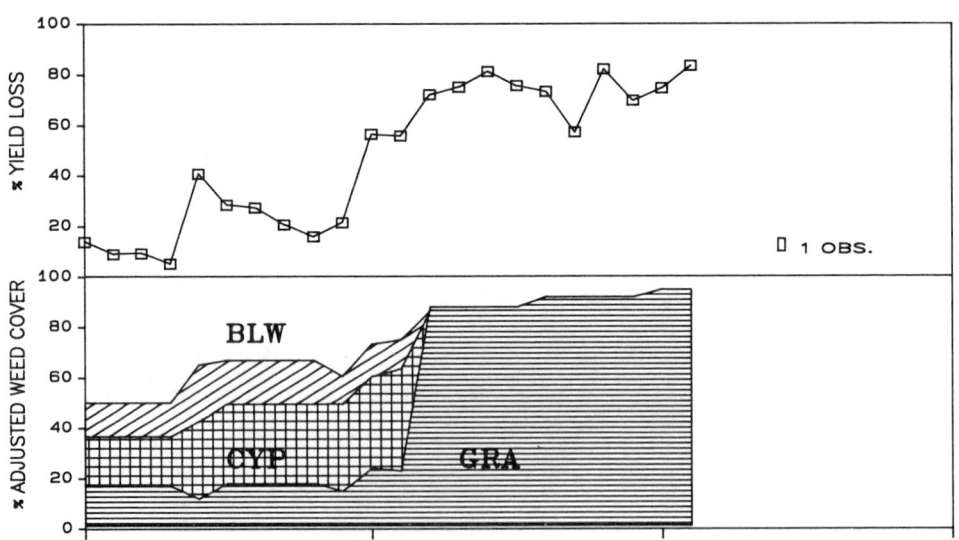

Figure 6. Grain yield losses in wet-sown rice in relation to weed cover
and composition of weed flora

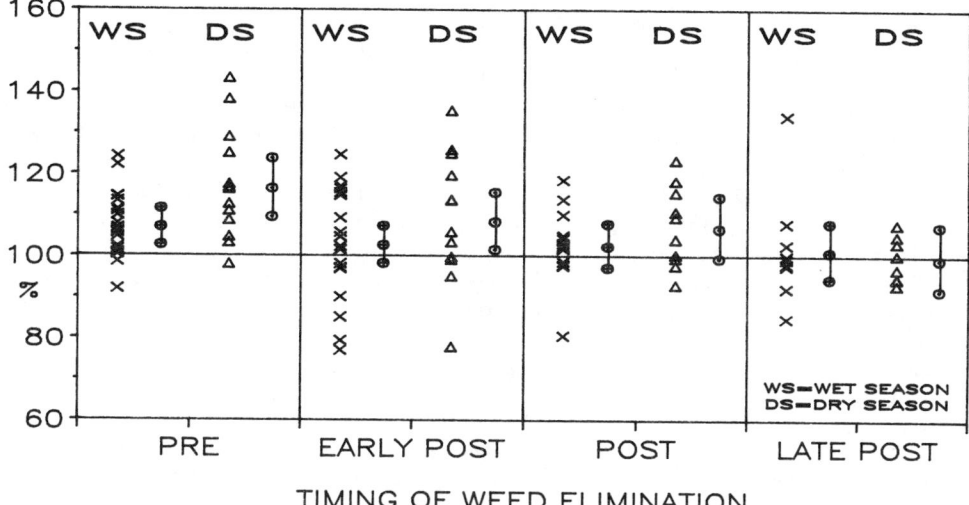

Figure 7. Ratio between grain yield of herbicide treatment and handweeded check
in relation to the timing of weed elimination and season in transplanted rice.

DISCUSSION

In our trials, **yield losses** in rice due to weeds averaged about 50%, which is about in line
with the published data [3, 11, 16]. Influences due to rice production system or season
were not found to be significant. In actual terms, however, the picture changed. From our
data, the season factor was found to contribute more to yield losses than the production
system factor. Yield losses were greater in wet season as compared with dry season (24
vs. 15 dt/ha) and greater in transplanted rice than in wet-sown rice (22 vs. 17 dt/ha).

Compared with the extensive data of Moody [11], our figures of actual yield losses
differ in so far as they are generally lower, especially in dry seasons, and they are also
higher in wet than in dry seasons. The degree of weed infestation as well as the composition of weed flora were quite homogeneous throughout our trials. Therefore, the shorter
rice growth period in dry seasons at Cikampek (up to 2 weeks; compare in Materials and
Methods) may explain why the yield potential of wet seasons is not reached. With regard
to the contrasting magnitude of losses relative to season no obvious explanation exists.
However, the reasons may be due to factors not mentioned in detail (such as agronomic
and climatic conditions). Nevertheless, with the exception of Moody [11], very little
information is available on the significance of the factor season on weed competition and
yield losses in rice.

Since the actual yield losses in transplanted rice were higher than those in wet-sown
rice, our data are not in line with those given previously [4, 11, 13]. The lower yield level
of wet-sown rice is probably the reason for such a result. In spite of homogeneous weed
infestations, it is possible that some herbicide treatments used in this production system

were not effective in completely eliminating the weed competition, so that the maximal yield level was not achieved. The raw data, however, do not support this idea.

The **degree of weed infestation** as well as the **composition of the weed flora** have a clear influence on yield losses in rice. It was to be expected that yield losses would become greater with increasing weed cover. While all groups of weed species contribute to yield losses, both grass weeds and sedges do so more significantly than broadleaved weeds. Our results confirm the contrasting competitiveness of *Echinochloa* and *Monochoria* [1, 6, 9, 14]. Sedges appeared to be about as competitive as grasses. In other studies, *Fimbristylis* [8, 14, 16] and *Cyperus* [4, 12, 17] have been recognized as harmful weed species. We cannot conclude whether any species out of those present in our trials (*Cyperus difformis, Cyperus iria, Fimbristylis miliacea, Scirpus juncoides*) is particularly competitive. However, our results indicate that from the point of view of rice/weed-competition more attention should in future research be paid to the group of sedges.

The magnitude of yield losses clearly depends on the **timing at which the competition of weeds was eliminated**. Both grasses and sedges show visible growth before 5-10 DAT/DAS. Like grasses, sedges commence to compete at an early growth stage of rice [17], and, therefore, they should be weeded before 3 weeks after planting [8]. If the weeds were controlled at pre- to early-post-emergence (equivalent to 0-9 DAT/DAS), yield gains were highest. Applications eliminating the weeds at post- or late-post-emergence (10-25 DAT/DAS) were still able to reduce yield losses, but not to the same extent. While at late-post-application herbicide and handweeded check yielded about the same, earlier elimination of weeds lead based on the yield results to the herbicide being clearly better. In general, the effect of early elimination of weeds was more pronounced in dry seasons, but did not show too much variation in relation to the rice production system. If the rice crop develops faster, such as in dry seasons, the weeds would be expected to behave similarly. This means earlier competition and consequently the need to eliminate them early in order to minimize yield losses, or, in other words, to obtain the optimal yield.

In summary, it can be stated that weeds compete with rice starting from the early stages of rice development. In our trials, such competition lead to mean yield losses of 50%. Highest rice grain yields were obtained when a mixed weed flora was eliminated up until about 10 DAT/DAS. An early elimination of weeds was particularly beneficial in wet-sown rice and in dry seasons. This indicates that the factor season is an important element and should be considered more closely. Besides grass weeds, sedges were found to be highly competitive and warrant further attention. Broadleaved weeds were the least competitive.

ACKNOWLEDGEMENTS

The author would like to thank Mr. Harris Burhan and his colleagues at CIBA-GEIGY's Agricultural Research and Development Station in Cikampek, Indonesia, for their accurate trial evaluations over the past years and the investigations on rice/weed-phenology. The author would also like to express his gratitude to Dr. Amy Racine, who kindly provided the expertise necessary to analyse the large set of data biometrically as well as to plot some of the graphs.

REFERENCES

1. Chisaka, H., Weed damage to crops: yield loss due to weed competition. In <u>Integrated control of weeds</u>, eds. J.D. Fryer and S. Matsunaka, Univ. Tokyo Press, Tokyo, 1977, 1-16.

2. Cramer, H.H., Pflanzenschutz und Welternte. <u>Pflanzenschutz-Nachrichten Bayer</u>, 1967, **20**, 99-133.

3. Datta, S.K. de, <u>Principles and practices of rice production</u>. Wiley & Sons, New York, 1981, 618 pp.

4. Datta, S.K. de, Moomaw, J.C. and Bantilan, R.T., Effects of varietal type, method of planting and nitrogen level on competition between rice and weeds. <u>Proc. 2nd Asian-Pac. Weed Control Interchange</u>, Los Baños, 1969, 152-63.

5. Dominicata, J. and Alcala, E., Critical stages of rice (IR-54) as influenced by weed control duration. <u>Newsl. Weed Sci. Soc. Phil.</u>, 1983, **11**, 4-5.

6. Guyer, R. and Koch, W., Competitive effects of *Echinochloa crus-galli* and *Monochoria vaginalis* in tropical irrigated rice. <u>Proc. 12th Conf. Asian-Pac. Weed Sci. Soc.</u>, Seoul, 1989, 195-202.

7. Koch, W., Beshir, M.E. and Unterladstatter, R., Crop losses due to weeds. <u>FAO Plant Prod. & Prot. Paper</u>, 1982, **44**, 153-65.

8. Kruijf, H.N. de and Pons, T.L., Effect of period of coexistance and population density on competition of weeds with transplanted rice. <u>Biotrop Bull. Trop. Biol.</u>, 1985, **23**, 37-46.

9. Lubigan, R.T. and Vega, M.R., The effect on yield of the competition of rice with *Echinochloa crus-galli* (L.) Beauv. and *Monochoria vaginalis* (Burm. F.) Presl. <u>Phil. Agriculturist</u>, 1971, **55**, 210-5.

10. Mardia, K.V., Kent, J.T. and Bibby, J.M., <u>Multivariate analysis</u>. Academic Press Ltd., London, 1988.

11. Moody, K., Weeds: definitions, costs, characteristics, classification and effects. In <u>Weed management in the Philippines: report of seminars</u>, ed. H. Walter, <u>PLITS</u>, 1983, **1**, 11-32.

12. Pons, T.L., Growth rates and competitiveness to rice of some annual weed species. <u>Biotrop Bull. Trop. Biol.</u>, 1985, **23**, 13-21.

13. Pons, T.L. and Kruijf, H.N. de, Competition between *Echinochloa* species and rice in replacement series experiments. <u>Biotrop Bull. Trop. Biol.</u>, 1985, **23**, 3-11.

14. Pons, T.L. and Utomo, I.H., Competition of four selected weed species with rice. <u>Biotrop Bull. Trop. Biol.</u>, 1985, **23**, 23-35.

15. Smith, J.R. jr., Weeds of major economic importance in rice and yield losses due to weed competition. <u>Proc. Conf. Weed control In Rice</u>, Los Baños, 1983, 19-36.

16. Suriapermana, S., Weed competition in transplanted rice. <u>Intern. Rice Res. Newsl.</u>, 1977, **2**, 9-10.

17. Swain, D.J., Nott, M.J. and Trounce, R.B., Competition between *Cyperus difformis* and rice: the effect of time of weed removal. <u>Weed Res.</u>, 1975, **15**, 149-52.

WEED CONTROL TECHNOLOGY IN U.S. RICE

ROY J. SMITH, JR.
U.S. Department of Agriculture - Agricultural Research Service and
University of Arkansas Rice Research and Extension Center
Stuttgart, Arkansas
and
JAMES E. HILL
Department of Agronomy and Range Science
University of California, Davis, California

ABSTRACT

The objectives of weed control in a rice production system are: (a) to minimize losses in grain yield due to weed competition and interference; (b) to prevent or minimize quality losses and subsequent lower value of rough and milled rice; (c) to permit highly efficient use of costly production inputs such as high yielding cultivars, fertilizers, insect and disease control and irrigation; (d) to prevent weed buildup in crops rotated with rice; (e) to lower water and energy requirements for production of rice; and (f) to minimize potential damage to the environment and beneficial nontarget organisms. Effective weed control programs for rice integrate preventive, cultural, mechanical, chemical and biological practices. Although nonchemical methods are important in an effective weed control program, chemical methods are essential for weed control in rice. Chemical methods involve the use of herbicide treatments as single, mixture or sequential applications that, correctly applied, selectively control weeds. Biological methods include use of endemic fungi and the use of wild ducks. Various types of weed management methods are combined in the weed control programs, and the weed control inputs are integrated with other pest management and production practices for rice.

WEED ECONOMICS AND INTEGRATED CONTROL

Weeds are the major pests of rice because they reduce yield and quality by an estimated 17% in the U.S. (1), compared with about 8 and 7% for insects and diseases, respectively (2, 3). Losses due to weeds were estimated at 34% in Texas, 12% in California and Missouri, and 17% in Arkansas, Louisiana, and Mississippi; total losses for the U.S. were 1.4 million metric tons of rough rice valued at $269 million (4).

Weeds interfere with rice production and processing in the following ways (5): (a) reduce rice yields and quality, (b) intensify problems with insects, diseases and other pests by serving as hosts, (c) reduce harvesting and processing efficiency, (d) lower efficiency of irrigation systems by restricting the availability and flow of water to reservoirs, canals and ditches, (e) cause consumption of energy for their control, (f) reduce the value and productivity of land, and (g) interfere with normal marketing strategies of the crop from weed seeds being present in the grain.

Weed control technology for rice integrates preventive, cultural, mechanical, chemical and biological practices (6, 7). Although all of these are important in developing successful weed management strategies, the use of herbicides is the backbone of the weed management system for rice. Almost every hectare of rice grown in the U.S. is treated at least once with a herbicide and 80% receives multiple herbicide treatments.

Most farmers integrate cultural and mechanical weed control practices with at least one herbicide treatment that is targeted for control of grass weeds, principally barnyardgrass[1] (watergrass) or sprangletop. However, where grass reinfestations occur or where broadleaf, aquatic, sedge or other hard-to-control weeds occur, farmers may apply as many as four herbicide treatments--two during the early season and two later in midseason. Additionally, each treatment may combine two herbicides in a tank mixture. Herbicide programs for rice may cost as little as $45/ha to as much as $168/ha, depending on weed species or rate and number of herbicide applications required (8). However, the 6-year average cost of weed control was $86/ha in verification trials conducted from 1983 to 1988.

Integrated control systems for hard-to-kill weeds, including red rice and perennial grass and broadleaf weeds, are complex and, in those hectares where rice is rotated with other crops, require attacking the weeds in rotated crops as well as in rice (9). Cropping-herbicide-cultivation systems are essential for the control of red rice where present control technology is very limited (10).

WEED COMPETITION IN RICE

Weed species vary in their competitive ability. Weed competition experiments in Arkansas indicated that season-long competition of red rice or barnyardgrass reduced rice grain yields more than other grass weeds such as bearded sprangletop and broadleaf signalgrass or broadleaf and aquatic weeds such as eclipta, ducksalad, hemp sesbania, spreading dayflower or northern jointvetch (11).

Rice yields decreased with increasing durations of weed-rice interference. Barnyardgrass, red rice, bearded sprangletop, broadleaf signalgrass, hemp sesbania, northern jointvetch, eclipta and spreading dayflower reduced drill-seeded rice yields of standard cultivars with increasing durations of interference (11).

The density of weeds in drill-seeded rice affects rice yields. Research indicates that rice yields decreased with increasing density of

---

[1]For common and scientific names of weeds see Composite List of Weeds, Weed Sci., 1984, 32 (Suppl. 2), 137 pp.

barnyardgrass, red rice, bearded sprangletop, broadleaf signalgrass, hemp sesbania or northern jointvetch (11).

Weed interference varies with rice cultivar (11). Lemont, a semi-dwarf cultivar (plants 82 cm tall) that requires high rates (200 kg/ha) of nitrogen competed less with barnyardgrass or red rice than Newbonnet, a short-statured cultivar (plants 96 cm tall) that requires 150 kg/ha of nitrogen.

Differences in the biology of weeds influence their interference with rice. Blackhull red rice biotypes emerged earlier after seeding, tillered 27% more, produced 18% more straw, were 29% taller, and matured 2 weeks later than strawhull biotypes (12). Because red rice germinates and emerges sooner and seedlings grow more rapidly than rice cultivars, this weed is a severe competitor with rice cultivars. Broadleaf signal-grass, which germinates and emerges later than and accumulates biomass slower than barnyardgrass, interferes less with rice (11).

WEED CONTROL TECHNOLOGY

Chemical Control

Herbicides used in combination with good management practices are impor-tant components of an integrated weed program (5, 6, 7, 13). Their effi-cacy is improved by including proper management of water before and after herbicide application, good fertilizer practices, and the use of preventive and cultural weed control methods. Herbicides rarely give complete weed control by themselves, just as nonchemical methods alone fail to give satisfactory weed control.

Guidelines for using herbicides in rice in the U.S. are published frequently by the U.S. Department of Agriculture (14) in which herbi-cides, rates, times of application, weeds controlled, and special instructions are presented. Individual state guidelines for control of weeds in rice that provide details on herbicide practices for specific weed problems in each state are available and are usually revised annu-ally (13, 15, 16, 17, 18, 19). Specific citations generally are not pre-sented in this section, but all statements are based on the above cita-tions unless other publications are cited.

Propanil[2]: The weed-control program for rice in the southern U.S. is built around propanil. This herbicide is also effective in water-seeded rice but has limited use in California because of injury to deciduous orchard crops. Drift of propanil spray may also injure nontar-get crops including cotton and soybean. Propanil at 3.4 to 5.6 kg/ha is applied postemergence when grass weeds are at the 1- to 4-leaf stages. Sequential applications are made to about 70% of the rice in the south-ern U.S. The first treatment is applied when the rice has been estab-lished and grass weeds are at the 1- to 3-leaf stages. The second treat-ment is applied after grasses have reinfested and are in the 1- to 3-leaf stages, immediately before applying the permanent floodwater. Propanil is usually applied aerially in spray volumes of 50 to 95 L/ha or may be applied by ground equipment in volumes of 140 to 190 L/ha.

---

[2]Commercial and chemical names of herbicides are given on cover pages of Weed Sci. 37(5), 1989 or in Weed Science Terminology, Weed Sci., 1985, 33 (Suppl. 1), 23 pp.

Proper water management is essential for satisfactory weed control with propanil. If the soil is dry, irrigation 2 to 5 days before propanil application increases its efficacy. Because propanil is absorbed through the foliage and not translocated rapidly, good spray coverage is essential to contact the weed. In water-seeded or flooded rice, the floodwater is drained to expose weeds to propanil sprays. Flooding 8 to 10 cm deep within 1 to 5 days after application prevents germination of more grass weeds.

The activity of propanil is influenced by rain and temperature. At least 8 hours without rain after propanil application are required for effective weed control. Best control of weeds occurs when daily maximum air temperatures range from 21 to 32C and daily minimums are above 10C. Rice may be severely injured by propanil if temperatures are above 38C.

Propanil interacts with carbamate and organophosphate insecticides to injure rice severely. Rice plants contain an enzyme, aryl acylamidase, that rapidly detoxifies propanil. These insecticides inhibit the action of this enzyme in rice which prevents rapid detoxification of propanil. Rice is injured severely when carbofuran (2,3-dihydro-2,2-dimethyl-7-benzofuranyl methylcarbamate) is applied before propanil, but this insecticide does not injure the crop when applied after propanil treatments. Carbaryl (1-naphthyl N-methylcarbamate) or ethyl parathion [O,O-diethyl O-(p-nitrophenyl)phosphorothioate] applied from 13 days before to 13 days after propanil treatments injure rice; applications before or after the 26-day period do not injure rice. Methyl parathion [O,O-dimethyl O-(p-nitrophenyl)phosphorothioate] (emulsifiable or encapsulated formulations) applied within 6 days of propanil treatments injures rice, but rice is not injured when this insecticide is applied before or after the 12-day period. Fungicides used in rice do not interact adversely with propanil to injure rice.

Propanil is frequently applied in tank mixtures with other herbicides to increase the weed species controlled or to give residual weed control. Herbicides frequently mixed with propanil include acifluorfen, bentazon, molinate, pendimethalin, thiobencarb, MCPA or 2,4-D. When propanil is applied in tank mixtures with these herbicides, timing and rate adjustments frequently are required.

Molinate: Molinate controls species of the Echinochloa genus, including barnyardgrass or watergrass. In water-seeded rice, molinate at 2.2 to 5.6 kg/ha is applied preplant incorporated before flooding or can be metered into the floodwater (water-run) preplant before weeds germinate. Water-run molinate can also be applied postemergence to dry-seeded rice after it is large enough to tolerate the flood; or it can be applied postemergence to dry- or water-seeded rice when weeds are in the 1- to 5-leaf to early-tillering growth stages. Molinate is absorbed by weed seedlings during early stages of germination through roots, coleoptiles, and basal shoot tissue. Once molinate is absorbed, it rapidly translocates throughout the plant to accumulate in the meristems of the root and shoot. Some rice cultivars and breeding lines are injured more than others by molinate (20). Dry- and water-seeded rice is tolerant to molinate. Postemergence applications into the floodwater control weeds better and injure rice less than preplant applications. Granular formulations applied into the floodwater control weeds better than emulsifiable formulations. The half-life of molinate in flooded soil is about 4 days and effective residual activity is about 2 weeks.

Water management is critical for effective weed control when using molinate. Submergence of weeds for at least 12 days after molinate treatment controls barnyardgrass, broadleaf signalgrass and other annual grasses. Drainage before 12 days may result in poor control. As grass weeds get larger, longer submergence is required to kill or suppress weeds. Water flow into treated fields is regulated to maintain at least 5 cm of water on the soil. Exposed soil provide sites for reinfestation of germinating barnyardgrass, sprangletop and other grass weeds. In California, floodwater is held on the field for prescribed times following molinate application to prevent pollution of public waterways. Cool temperatures (less than 10C) reduce weed growth and uptake of molinate and reduce control. High temperatures (greater than 38C) warms the floodwater which may increase crop injury; adding cool water to fields during periods of high temperatures helps reduce floodwater temperatures and reduces injury.

Residual herbicides: Thiobencarb is similar to molinate in activity and mechanism of action, but controls more weed species than molinate. This herbicide provides excellent residual control, lasting from 2 to 4 weeks after application. In dry-seeded rice, thiobencarb at 2.2 to 4.5 kg/ha may be applied preemergence as a spray after the soil is wet by rain or irrigation, and it may be applied postemergence in single or sequential treatments. Frequently, it is applied postemergence tank mixed with propanil.

In water-seeded rice, thiobencarb granules are applied postemergence at the 2-leaf stage of rice primarily to control the Echinochloa complex and smallflower umbrellaplant. Applications before the 2-leaf stage of rice may thin crop stands. In California, the floodwater is held after application to prevent pollution of downstream waters by runoff from rice fields. Holding periods vary depending on the water management system, but a minimum of 6 to 7 days are required for effective weed control.

Pendimethalin at 0.8 to 1.1 kg/ha is applied postemergence in a tank mixture with propanil for control of barnyardgrass, bearded and Amazon sprangletop, and broadleaf signalgrass in dry-seeded rice. Water-seeded rice may be injured severely by pendimethalin. However, dry-seeded rice in the spiking to 3-leaf stages of growth is tolerant to this herbicide. Pendimethalin controls weeds residually for up to 2 weeks after application by absorption through roots of weeds, but flushing, flooding or heavy rainfall may reduce residual life to less than 1 week.

Bensulfuron controls many species of sedge and broadleaf weeds in rice. In California, bensulfuron at 70 g/ha is applied into the floodwater at the 2- to 4-leaf stages of water-seeded rice for control of smallflower umbrellaplant, roughseed bulrush, and several broadleaf aquatic weeds. Bensulfuron may be applied in combination with molinate or thiobencarb for control of grass weeds as well as sedges and aquatic weeds. In the southern U.S., bensulfuron at 40 to 70 g/ha is applied into the floodwater to control ducksalad, redstem, waterhyssop, and rice flatsedge in water-seeded rice (21). It may also be applied to dry-seeded rice, but aquatic weed infestations do not usually occur at threshold levels in this culture. Bensulfuron is applied aerially to flooded rice fields as a dry-flowable formulation in a minimum of 50 L/ha of water-mixed spray. The herbicide is absorbed through the emerging weed shoots and its residual activity is 35 to 45 days after application. It is most effective on actively growing weeds that are submerged with 3 or fewer leaves.

Phenoxy Herbicides: MCPA and 2,4-D, the only two phenoxy herbicides registered for use in rice, are applied at 0.6 to 1.7 kg/ha as postemergence sprays for control of broadleaf, aquatic and sedge weeds but not grass weeds. MCPA and 2,4-D are translocated, systemic herbicides.

Phenoxy herbicides, at rates necessary for weed control, frequently injure rice plants at almost any stage of growth but injury is particularly severe before tillering and after panicle initiation. The tolerant stage can be positively identified when the basal internode begins to elongate and up to 1.3 cm long. Rice may be injured when the internode is longer than 1.3 cm. Commercial U.S. rice cultivars are tolerant at the early jointing stage (internodes 1.3 cm long) regardless of maturity-group classification. A computer program based on degree days has been developed for accurate timing of MCPA and 2,4-D applications to prevent crop injury in the southern U.S.

MCPA at low (0.6 kg/ha) rates, applied as early as the 4-leaf stage of rice growth, are used to control early weed growth in water-seeded rice in California. Rice plants that have well-developed root systems and are growing vigorously are most tolerant to early applications of MCPA. A shallow (5 cm deep) flood reduces phytotoxicity and exposes weeds for optimum efficacy. Early weed control with such treatments of MCPA reduces early season weed competition and improves rice growth and grain yield.

Rapidly growing weeds, stimulated by high levels of nitrogen, are more susceptible than stressed plants to MCPA or 2,4-D. Rapidly growing rice may be injured temporarily by MCPA or 2,4-D even when applied at the tolerant stage of growth.

Water management may affect the response of rice and weeds to phenoxy herbicides. If water covers low-growing aquatic weeds such as arrowhead, burhead, ducksalad, falsepimpernel, flatsedges, redstem or waterhyssop at spraying time, the weeds may not be controlled because the herbicide does not contact them. Lowering or draining the water from rice fields to expose weed foliage to phenoxy herbicides improves control. If the field is drained and the soil dries to stress the weeds, control may be reduced. Therefore, applying herbicides before the weeds become stressed improves control.

Phenoxy herbicides control weeds effectively if rain occurs no sooner than 6 hours after treatment. Even when rice is treated during the tolerant stage, high temperature (above 35C) may increase rice injury by phenoxy herbicides. Dry, hot winds also enhance rice injury. Temperatures below 15C one week before treatment may slow weed growth and reduce control.

Inorganic or amine salts of MCPA or 2,4-D are used in rice. These formulations injure rice less and are less volatile than ester formulations.

MCPA or 2,4-D applied in mixtures with propanil at midseason control weeds better on rice field levees than any one herbicide alone. These treatments suppress and control grasses, broadleaves, aquatics and sedges that grow on and along levees. Treatments are made with special power-driven ground applicators that broadcast spray the earth levee and the flooded levee ditch. The mixtures are applied in 140 to 190 L/ha of total spray.

Bentazon: Bentazon at 0.8 to 1.1 kg/ha controls certain broadleaf, aquatic, sedge and other weeds in rice by absorption and translocation through foliage. It is especially effective on spreading dayflower, smartweed, redstem and other weeds during the early season before rice will tolerate phenoxy herbicide applications. Bentazon is especially effective for control of river bulrush when applied after about 50% of the weeds are flowering; earlier applications do not control this weed as well. Spray drift of bentazon does not injure nontarget crops such as soybean or cotton as do phenoxy herbicides. Although bentazon is translocated in plants it primarily acts as a contact herbicide that requires good spray coverage of plant foliage for successful weed control. Translocation of bentazon in plants is less than that of MCPA or 2,4-D. The spectrum of weed species controlled is increased by tank mixing bentazon with propanil. Propanil controls many of the grass weeds and adds to the activity of bentazon on many of the weeds moderately susceptible to bentazon. Bentazon alone or in tank mixtures with propanil is applied aerially in 95 L/ha.

Water management is critical when using bentazon. The floodwater is lowered to expose young weed foliage to sprays of bentazon, often requiring complete drainage at early stages of growth. A shallow flood is maintained on fields treated with bentazon to prevent reinfestation of weeds. Bentazon provides no residual weed control. Floodwater is restored to normal depth (about 10 cm) after bentazon is absorbed by plant foliage beginning 1 day after application.

Environment affects activity of bentazon. Rain sooner than 12 hours after application may wash-off the herbicide spray and reduce weed control. Cool temperature (minimum and maximum below 10 and 21C, respectively) 1 to 2 days before spraying reduces weed control. High temperature (greater than 35C) 1 to 2 days before spraying enhances activity of bentazon which may injure rice or rapidly burn weed foliage to reduce weed control.

Acifluorfen: Acifluorfen at 0.14 to 0.28 kg/ha is a special use herbicide for controlling hemp sesbania in the southern U.S. It is not used for weed control in California rice. Addition of propanil to acifluorfen in a tank mixture also controls annual morningglory species. Although acifluorfen is translocated in plants, it principally acts as a contact herbicide absorbed by foliage and requires good coverage of plant foliage for successful weed control. Acifluorfen translocates in plants less than phenoxy herbicides. Acifluorfen or mixtures with propanil are applied aerially in 95 L/ha.

Acifluorfen and mixtures with propanil may be applied over a wider range of rice growth stages than phenoxy herbicides. Rice treated from the 4-leaf to the late jointing stages with acifluorfen alone or with propanil is not reduced in yield. Although it may cause temporary chlorosis and bronzing on rice foliage, rice recovers within 2 weeks after application.

Tank mixtures of acifluorfen and propanil applied at midseason suppress or control many grass, broadleaf, and aquatic weeds growing on levees. Although a tank mixture of acifluorfen plus propanil does not control as many weed species as mixtures of phenoxy herbicides plus propanil, it reduces weeds on levees. Rates of propanil used in mixtures with acifluorfen are slightly higher for levees than for paddy rice. Special ground machines are used to spray the earth levee and the adjacent levee ditches. Sprays are broadcast in a total volume of 95 to 190 L/ha.

Fenoxaprop: Early postemergence applications of fenoxaprop at 0.17 kg/ha control grass weeds such as barnyardgrass, broadleaf signalgrass and sprangletop species by absorption through foliage. Rice is tolerant to fenoxaprop when applied to plants in the 4-leaf to early tillering growth stages and before flooding while weeds are in the 1- to 4-leaf growth stages. Rice in the 4-leaf to early tillering stages is injured slightly as manifested by temporary chlorosis and inhibition of growth. However, the rice recovers as soon as the early-season nitrogen and irrigation water are applied. Floodwater applied before 7 days of applying fenoxaprop increases injury of rice seedlings. Delaying the flood until 7 days after applying fenoxaprop prevents or reduces the phytotoxicity. Furthermore, fenoxaprop applied after flooding or nitrogen topdressing or during periods of cloudy, rainy weather increases phytotoxicity. Applications of fenoxaprop during periods of stressed or excessively succulent plant growth may also injure rice.

Fenoxaprop is generally used in a program with other herbicides such as propanil alone or in combinations with molinate, thiobencarb or pendimethalin. These herbicides are applied first when rice and weed plants are in the 1- to 3-leaf stages and fenoxaprop follows after weed reinfestations are in the 1- to 3-leaf stages and rice plants are in the 4-leaf to early-tillering stages.

Sprangletop in the 1- to 4-leaf or early-tillering stages, less than 10 cm tall, are especially susceptible to fenoxaprop. They are controlled with fenoxaprop at rates of 0.08 to 0.11 kg/ha, which are less phytotoxic to rice than higher rates.

Triclopyr: Triclopyr, active on broadleaf weeds and a partial substitute for MCPA and 2,4-D in the southern U.S., is applied postemergence alone or in tank mixtures with propanil from 4-leaf to internode elongation (1.3 cm long) stages of growth of rice (22). Triclopyr alone at 0.28 to 0.42 kg/ha combined with a surfactant controls or suppresses many broadleaf weeds including northern jointvetch, hemp sesbania, morningglory species, eclipta and common cocklebur. However, tank mixtures of triclopyr and propanil at 0.28 plus 3.4 to 4.5 kg/ha control weeds better than either herbicide alone. This mixture, in addition to the above weeds, controls smartweed, ducksalad, redstem, waterhyssop, gooseweed, falsepimpernel, rice flatsedge and spikerush. Triclopyr is absorbed and translocated by weed foliage. The activity of triclopyr in rice is similar to phenoxy herbicides, but the stage of application is less critical than with MCPA or 2,4-D. Triclopyr does not injure rice when applied from the early-tillering to the 1.3-cm internode elongation stages. Applications to rice after panicle initiation severely injure rice. Triclopyr alone and in mixtures with propanil is applied by airplane in 50 to 95 L/ha or by ground sprayers in 95 to 190 L/ha. Uniform application is essential to provide effective weed control and prevent excessive rice injury. Applications streaked through the field may severely injure rice in the high-rate strips and weeds may not be controlled in the low-rate strips. Although spray drift of triclopyr is less injurious than MCPA or 2,4-D to nontarget crops such as cotton or soybeans, preventing even small amounts of spray drift is essential to avoid injury to these crops.

Bromoxynil: Bromoxynil, active on broadleaf weeds and a partial substitute for MCPA or 2,4-D in the southern U.S., is applied postemergence alone or in tank mixtures with propanil from the 4-leaf stage of rice until 3 days before field flooding (23). Bromoxynil at 0.28 to 0.42

kg/ha alone or in tank mixtures with propanil at 2.2 to 3.4 kg/ha controls seedling plants (3- to 4-leaf growth stages) of hemp sesbania,
morningglory species, smartweed species, common cocklebur and texasweed.
The tank mixtures are more effective than either herbicide alone on some
weeds including texasweed and pale smartweed. Bromoxynil is absorbed
and translocated by foliage. Bromoxynil is applied to unflooded fields
and the subsequent floodwater is maintained in the rice field for 7 days
to prevent contamination of surface water in natural drainage systems.
Bromoxynil, alone and in mixtures with propanil, is applied aerially in
50 to 95 L/ha. Although spray drift of bromoxynil is less injurious
than phenoxy herbicides to cotton and soybeans, it may injure these
crops even if small amounts of spray drift occur.

Copper herbicides: Copper sulfate (pentahydrate) at 2.8 to 4.2
kg/ha controls filamentous green and blue-green algae in rice fields and
irrigation systems. Algae are especially severe when temperatures are
high during early rice stand establishment or when phosphorous fertilizer has been applied to the soil surface without incorporation. Fine
granules or crystal treatments, are applied into the floodwater when
algae colonies form on the soil and begin to emit gas bubbles. Once
algae float on the surface forming large mats, they are difficult to control with copper herbicides and can decimate a young rice stand. Surface mats submerge small rice plants and reduce tillering. Algae mats
on the surface are best controlled by draining or lowering the floodwater.
Copper herbicides are most effective when applied to acid soils low
in organic matter or to water low in carbonate or bicarbonate salts.
Salts in the soil-water medium inactivate copper by forming insoluble
basic copper compounds.
Control of algae with copper herbicides prevents the necessity of
draining for control. During the drained period, grass and other weeds
may reinfest the field which require application of other costly herbicides for control. Draining and reflooding rice fields increase irrigation costs and loss of nitrogen through nitrification and denitrifica-
tion.

Endothall: Endothall, which is registered only for use in
California rice, controls several submersed aquatic weeds but is used
primarily for American pondweed control. Endothall prevents formation
of winter buds that grow in future rice crops. A granular formulation
is applied postemergence at 2.2 to 3.4 kg/ha into the floodwater during
the 6-week period after rice emergence and while submersed weeds are
small, 2- to 8-cm tall. Irrigation water flow through treated rice
fields is stopped for a 5-day period after application to prevent herbicide dilution and movement in the field. Cool temperatures (minimum and
maximums below 10 and 21C, respectively) before and after treatment reduce efficacy.

Glyphosate: Glyphosate, a nonselective herbicide, at 0.42 to 1.3
kg/ha is applied before seeding rice for control of weeds that were not
controlled during seedbed preparation. It is applied aerially in 50 to
95 L/ha or with tractor sprayer in 140 to 190 L/ha of water carrier.
Good spray coverage of plant foliage is essential for successful weed
control. Glyphosate spray that drifts onto nontarget beneficial plants
will severely injure or kill them.

Biological Control

An endemic anthracnose disease of northern jointvetch, incited by the fungus Colletotrichum gloeosporioides (Penz.) Sacc. f. sp. aeschynomene (C.g.a.) controls this weed in rice fields (24). This microbial herbicide, registered for use in rice in 1982, is the first of this type cleared for weed control in an agronomic crop. C.g.a., applied at 187 billion spores/ha postemergence at mid-season as weeds emerge through the crop canopy, controls northern jointvetch; the microbial herbicide requires 5 to 10 days to show symptoms on weeds and 4 to 5 weeks to kill them. Tank mixtures of C.g.a. and acifluorfen control northern jointvetch and hemp sesbania. Many herbicides and fungicides applied in tank-mixture or sequential treatments injure spores of C.g.a. Herbicides, including propanil, MCPA and 2,4-D, and fungicides injure C.g.a. when applied together in tank mixtures. However, timely sequential applications of C.g.a. followed by these toxic pesticide treatments do not inhibit activity of the mycoherbicide. Advantages of C.g.a. over chemical herbicides are: (a) C.g.a. does not injure rice in any stage of growth, (b) nontarget crops such as cotton, soybean or other economical plants are not injured by the mycoherbicide, and (c) the mycoherbicide is not residual in the soil or environment. Disadvantages include: (a) C.g.a. controls only one, compared with several weed species, for a chemical herbicide, (b) the mycoherbicide action is slower on weeds than chemicals, (c) weeds are susceptible to C.g.a. during narrower stages of growth than they are to chemicals, and (d) C.g.a. activity is inhibited by chemical herbicides and fungicides.

Integrated Weed Control Systems

Integrated weed control systems combine: (a) weed control technology in rice with other crop and pest management practices, and (b) weed control technology in all crops in the rotation with pest management practices used in all crops (6, 7).

Integrated weed control in rice: The complex of weed species that infests rice is controlled by integrating weed control practices (5, 7). Early-season control of annual grass, broadleaf, sedge and aquatic weeds is accomplished by combining: (a) preplant tillage or herbicide treatments to kill all weed growth at the time of seeding rice, (b) seeding rice in such a way as to obtain stands of fast-growing plants that compete with and shade the weeds, (c) the use of timely tank-mixture or sequential treatments of propanil, molinate, fenoxaprop, pendimethalin, thiobencarb, bensulfuron, triclopyr, bromoxynil or other appropriate herbicide treatments to kill the weed plants during the early season before they compete with the crop, and (d) timely flooding or proper water management before and after herbicide treatments.

Annual grass weeds that escape these early treatments are controlled by timely postemergence treatments of molinate after flooding the rice. Aquatic, broadleaf and sedge weeds that germinate after flooding the crop are controlled by early season postemergence treatments of bensulfuron, bentazon or acifluorfen either alone or combined with other appropriate herbicides. Weeds present at midseason, when rice internodes are 1.3 cm long or less, are controlled with single herbicide treatments of 2,4-D, MCPA, bentazon, acifluorfen, triclopyr, or C.g.a. Also, 2,4-D, MCPA or triclopyr alone or in tank mixtures with propanil or a tank mixture of acifluorfen and propanil control weeds when applied at midseason. Control of many aquatic weeds in the paddy is enhanced when the soil is dried after the herbicide treatment.

Weed control programs are integrated with insect and disease control
practices. For example, the control of weeds on levees reduces overwintering habitat for the rice water weevil. Some fungicides and insecticides interact antagonistically with specific weed control technologies.
For example, carbamate and organophosphate insecticides (carbaryl, carbofuran and methyl parathion) applied untimely before or after propanil
may injure rice severely. These insecticides are applied by specific
timing guidelines to prevent adverse interactions with propanil and
subsequent phytotoxicity to rice. If the insecticide has already been
applied to rice, alternate herbicide programs that do not interact adversely with the insecticides are used. Molinate, thiobencarb, fenoxaprop, bentazon or acifluorfen do not interact adversely with carbamate
and organophosphate insecticides.

Integrated weed control in the cropping system: Integrated weed control systems for hard-to-kill weeds, such as red rice, combine complex
cropping-herbicide-cultivation systems. A well planned program for red
rice includes: (a) use of weed-free crop seed, irrigation water free of
red rice seed, and clean equipment; (b) crop rotation with control of
red rice in all crops; (c) mechanical cultivation; (d) careful crop and
water management; and (e) herbicides (5, 9, 10, 25).
   The most effective way to control red rice is by crop rotation combined with effective cultivation and herbicide treatments. Rotations
include one or two years of either soybeans or grain sorghum followed by
rice the second or third year; or one year of grain sorghum followed by
soybeans the second year, then rice the third year (6, 10).
   During the years in which soybeans are grown several herbicide treatments control red rice. Preplant soil-incorporated herbicides, which
are usually more effective than preemergence treatments, are combined
with followup postemergence directed or over-the-top herbicide treatments. Effective preplant incorporated herbicide treatments include
alachlor or metolachlor either alone or tank mixed with trifluralin,
pendimethalin, metribuzin, or imazaquin (10). Several postemergence
treatments including over-the-top sprays of fluazifop, quizalofop or
sethoxydim or directed sprays of paraquat control red rice missed by
preplant treatments (15). Cultivation is essential to complete the
program in soybeans.
   A rotation with grain sorghum is another effective approach to red
rice control. Grain sorghum is treated preplant incorporated with
alachlor or metolachlor or postemergence directed with paraquat (15).
The grain sorghum seed is treated with a chemical to protect the crop
from injury by alachlor or metolachlor. Grain sorghum is grown for 2
years or followed with soybeans the second year. Cultivation is re-.
quired to control red rice missed by the herbicide treatments in both
crops. Rice grown during the third year after a successful red rice control program in grain sorghum or soybean, produces good yields of
high-quality grain.
   In the rice crop, registered herbicide treatments are not active on
red rice. However, infestations are reduced by applying molinate preplant incorporated, water seeding the rice and maintaining the floodwater or keeping the soil moist by frequent irrigations for several
weeks after seeding (25).
   Field research has identified a successful red rice control program
in dry-seeded rice culture (26). Rice seed treated with calcium perox-
ide to supply oxygen to the crop seed will emerge through soil and water
whereas the untreated red rice seed will not survive. A chemical is

also applied to the rice seed to protect the crop from injury by molinate. Additional components include molinate, incorporated preplant,
drill-seeding rice about 1.3 cm deep in the soil, flooding immediately
after constructing levees, and maintaining the floodwater for 4 to 5
weeks. After this period the rice may be drained for controlling
straighthead, a physiological disorder of rice, and applying herbicides
for control of aquatic weeds that germinated and infested the rice during the flooded period. This practice controlled 96% of the red rice
and produced 6,020 kg/ha of U.S. grade 4 grain (4% red rice grain) with
a 63% head rice yield (26). Untreated rice produced 3,160 kg/ha of U.S.
sample grade (74% red rice grain) with a 55% head rice yield.

Other field research indicated that red rice is controlled in
water-seeded rice by combination treatments of a herbicide and a plant
growth regulator (27). In Arkansas research, molinate applied preplant
incorporated and mefluidide applied postemergence controlled 95% of the
red rice, compared with 67% for molinate alone (28). The molinate treatment killed germinating red rice seed and seedling plants; mefluidide
suppressed height, panicle emergence and seed production of red rice
plants. Mefluidide injured the rice crop similar to red rice when applied before the late-booting growth stage. A very-short season rice
cultivar such as Bond or Labelle which matures well ahead of the red
rice was used to obtain differences in stages of the crop and red rice.
Mefluidide, applied to rice in the late booting stage, injured rice
slightly. At this time, red rice growth in the panicle initiation stage
was suppressed. Rice treated with molinate plus mefluidide produced
4,860 kg/ha of U.S. grade 2 rough grain, compared with 5,640 kg/ha of
U.S. grade 4 grain for molinate alone and less than 1,350 kg/ha for
mefluidide alone (28). Although mefluidide alone reduced red rice
panicles by 96%, vegetative competition reduced the grain yield.

REFERENCES

1. Chandler, J.M., Estimated losses of crops to weeds. In Handbook of
   Pest Management in Agriculture, Vol. 1, ed. D. Pimentel, CRC Press,
   Inc., Boca Raton, FL, 1981. pp. 95-109.

2. James, W.C., Estimated losses of crops from plant pathogens. In
   Handbook of Pest Management in Agriculture, Vol. 1, ed. D. Pimentel,
   CRC Press, Inc., Boca Raton, FL, 1981. pp. 79-84.

3. Schwartz, P.H. and Klassen, W., Estimate of losses caused by insects
   and mites. In Handbook of Pest Management in Agriculture, Vol. 1,
   ed. D. Pimentel, CRC Press, Inc., Boca Raton, FL, 1981. pp. 15-77.

4. Chandler, J.M., Hamill, A.S. and Thomas, A.G., Crop Losses Due to
   Weeds in Canada and the United States. Special Rpt. Losses Due to
   Weeds Committee, Weed Sci. Soc. Amer., 309 W. Clark St., Champaign,
   IL, 1984, 22 pp.

5. Smith, R.J. Jr., Integrated weed management in rice in the USA.
   Korean J. Weed Sci., 1983, 3, 1-13.

6. Eastin, E.F., Weed management systems for U.S. rice. In Handbook of
    Pest Management in Agriculture, Vol. 3, ed. D. Pimentel, CRC Press,
    Inc., Boca Raton, FL, 1981, pp. 539-47.

7. Smith, R.J. Jr., Flinchum, W.T. and Seaman, D.E., Weed Control in
    U.S. Rice Production. U.S. Dep. Agric. Handb. 497, U.S. Gov.
    Printing Office, Washington, DC, 1977, 78 pp.

8. Chaney, H.M., Helms, R.S., Dodgen, W.H. and Huey, B.A., Six-Year Sum-
    mary of Rice Research Verification Trials. Arkansas Agr. Exp. Stn.
    Special Rpt. 139, 1989, 24 pp.

9. Baker, J.B. and Sonnier, E.A., Red rice and its control. In Proc.
    Conf. Weed Control in Rice (Aug. 31 - Sept. 4, 1981), Int. Rice Res.
    Inst., Los Banos, Laguna, Philippines, 1983, pp. 327-33.

10. Smith, R.J. Jr., Cropping and herbicide systems for red rice (Oryza
    sativa) control. Weed Tech., 1989, 3, 414-19.

11. Smith, R.J. Jr., Weed thresholds in southern U.S. rice, Oryza
    sativa. Weed Tech., 1988, 2, 232-41.

12. Diarra, A., Smith, R.J. Jr. and Talbert, R.E., Growth and morpholog-
    ical characteristics of red rice (Oryza sativa) biotypes. Weed
    Sci., 1985, 33, 310-14.

13. University of California, Integrated Pest Management for Rice. Pub.
    3280, California Div. Agric. Sciences, Berkeley, CA, 1983, 94 pp.

14. U.S. Department of Agriculture, Suggested Guidelines for Weed Con-
    trol. U.S. Dep. Agr., Agr. Handb. 565, 1988, 222 pp.

15. Arkansas Cooperative Extension Service, Recommended Chemicals for
    Weed and Brush Control. Misc. Pub. No. 44, Little Rock, AR, 1989,
    103 pp.

16. Hill, J.E., Bayer, D.E., Scardaci, S.C., Williams, J.R., Wick, C.M.
    and Fischer, B.B., Weed Control in Rice. California Cooperative
    Ext. Serv., Davis, CA, 1990, 19 pp.

17. Louisiana Cooperative Extension Service, Louisiana's Suggested Chemi-
    cal Weed Control Guide for 1989. Louisiana State Unviersity, Baton
    Rouge, LA, 1989, 193 pp.

18. Mississippi Cooperative Extension Service, Weed Control Guidelines
    for Mississippi. Mississippi State Univ., Miss. State, MS, 1989,
    227 pp.

19. Texas Agricultural Extension Service, Rice Production Guidelines.
    Texas A&M University, College Station, TX, 1989, 72 pp.

20. Richard, E.P. Jr. and Baker, J.B., Response of selected rice (Oryza
    sativa) lines to molinate. Weed Sci., 1979, 27, 219-23.

21. Arkansas Cooperative Extension Service, Londax, a New Herbicide for Aquatic Weed Control in Rice. Rice Information Pub. 100, Little Rock, AR, 1989, 3 pp.

22. Arkansas Cooperative Extension Service, Rely, a New Herbicide for Broadleaf Weed Control in Rice. Rice Information Pub. 103, Little Rock, AR, 1989 3 pp.

23. Arkansas Cooperative Extension Service, Broadleaf Weed Control with Blazer, Tackle, Buctril, and Brominal. Rice Information Pub. 89, Little Rock, AR, 1987, 3 pp.

24. Smith, R.J. Jr., Biological control of northern jointvetch (Aeschynomene virginica) in rice (Oryza sativa) and soybeans (Glycine max) -- a researcher's view. Weed Sci., 1986, 34 (Suppl. 1), 17-23.

25. Baker, J.B., Sonnier, E.A. and Shrefler, J.A., Integration of molinate use with water management for red rice (Oryza sativa) control in water-seeded rice (Oryza sativa). Weed Sci., 1986, 34, 916-22.

26. Diarra, A., Smith, R.J. Jr. and Talbert, R.E., Red rice (Oryza sativa) control in drill-seeded rice (O. sativa). Weed Sci., 1985, 33, 703-07.

27. Dunand, R.T., Red rice control in rice production by seedhead suppression. Proc. So. Weed Sci. Soc., 1985, 38, 35.

28. Smith, R.J. Jr. and K. Khodayari. Control and bioregulation of red rice in rice. Proc. So. Weed Sci. Soc., 1985, 38, 444.

CURRENT WEED CONTROL STRATEGIES IN LOUISIANA RICE

S. H. CRAWFORD, J. B. BAKER, and D. E. SANDERS
Louisiana State University Agricultural Center
Baton Rouge, Louisiana 70803, USA

ABSTRACT

Rice was introduced into Louisiana in the late 1800's and is currently grown on 162,000-243,000 hectares annually. Production is centered in two diverse areas: the southwestern coastal prairie and the Mississippi River floodplain of northeast Louisiana. Economic losses due to weeds, including weed control costs and direct losses are estimated to be $172 (US) per hectare annually. Red rice, barnyardgrass, and junglerice are common and troublesome weeds from the family Gramineae. Alligatorweed, dayflower, ducksalad, purple ammania, and eclipta are major aquatic or wetland broadleaf weeds. Problem terrestrial broadleaf weeds include hemp sesbania, Indian jointvetch, texasweed, and palmleaf morningglory. Crop rotation, cultural management, and chemical control each play significant roles in current weed control strategies. A rotation of one year in rice and two years in soybean is most common. Early planting and establishing a permanent flood as soon as possible in both dry- and water-seeded systems are cultural practices that are used to suppress terrestrial weeds. Herbicides are used on essentially all rice produced in Louisiana. Where red rice is the major weed of concern, molinate (4.5 kg/ha) is most often used preplant incorporated prior to early water-seeding with brief drainage. Where barnyardgrass and junglerice are most troublesome, foliar applications of propanil at rates of 3.4-4.5 kg/ha are used when weeds reach the 2-3-leaf stage. Thiobencarb or molinate (3.4 kg/ha) are increasingly used to enhance the performance of propanil and provide residual weed control, which tends to reduce the need for successive propanil treatments. Broadleaf weeds that are not controlled by molinate or propanil treatments are generally controlled by mid-season application of 2,4-D (0.6-1.7 kg/ha) in southwest Louisiana. In northeast Louisiana where 2,4-D use is restricted, aquatic and wetland broadleaf weeds are most often controlled with bentazon (1.1 kg/ha). Terrestrial broadleaf escapes can be controlled with triclopyr or bromoxynil (0.28-0.42 kg/ha). Rice flatsedge and yellow nutsedge escapes are controlled with bentazon (1.1 kg/ha).

INTRODUCTION

Louisiana is a small boot-shaped state located in the south central United States of America. A major segment of its eastern boundary is the Mississippi River, and it is bordered on the south by the Gulf of Mexico. Flood-irrigated rice (<u>Oryza sativa</u> L.) is produced on 162,000-243,000 hectares annually, with fluctuations determined in large part by control programs initiated by the United States Department of Agriculture. Rice was produced on approximately 210,000 hectares in 1989 (1). Both long-grain indica and medium-grain japonica rice types are produced.

Rice was first grown in the southwestern coastal prairie region of the state beginning in the late 1800's, and approximately 80% of the total annual production remains in this region. Beginning in the late 1970's rice production in the Mississippi River floodplain of northeastern Louisiana began to expand significantly and now comprises approximately 20% of the state's acreage (2). Because of differences between the two regions in soils, climates, cultural practices, adjacent crops, indigenous weeds, and herbicide restrictions, weed control strategies often vary greatly.

LOSSES DUE TO WEEDS

Economic losses due to weeds are quite high. According to recent estimates, weed costs average $172 (US) per hectare per year in Louisiana rice. Approximately $62 of this cost is for herbicides and application, and the remaining $110 is due to direct losses as a result of reduced yield and quality, and increased costs for land preparation and harvesting (3).

SPECIFIC PESTS

Weed pests in Louisiana rice are numerous and formidable (4). Three weeds in the Gramineae family are particularly troublesome: red ·rice (<u>Oryza</u> <u>sativa</u> L.), barnyardgrass (<u>Echinochloa</u> <u>crus-galli</u> [L.]Beauv.), and junglerice (<u>Echinochloa</u>

colonum [L.]Link). Aquatic, wetland, and terrestrial broadleaf
weeds are also significant pests. Alligatorweed (Alternanthera
philoxeroides [Mart.]Griseb.), dayflower (Commelina spp.),
ducksalad (Heteranthera limosa [Sw.]Wild.), purple ammania
(Ammania coccinea Rottb.), and eclipta (Eclipta prostrata L.)
are aquatic or wetland broadleaf weeds that are of major
importance. Numerous terrestrial broadleaf weeds are common;
however, hemp sesbania (Sesbania exaltata [Raf.]Rydb.), Indian
jointvetch (Aeschynomene indica [L.]B.S.P.), texasweed
(Caperonia palustris [L.]St. Hil.), and palmleaf morningglory
(Ipomea wrightii Gray) are of greatest significance since they
effectively compete with rice prior to and following permanent
flood. Several weeds in the sedge family, including rice
flatsedge (Cyperus iria L.) and yellow nutsedge (Cyperus
esculentus L.), appear to be increasing as problems.

CONTROL METHODS

Crop rotation, cultural management, and chemical control
measures are all extremely important aspects of weed management.
Each of these tools is employed to some extent on virtually all
of the rice produced in Louisiana.

Crop Rotations

Rotation of rice acreage to other crops where weeds can be more
easily controlled, thereby reducing weed seed populations in the
soil, is a weed management strategy of long standing.

Before the rapid increase in the acreage of soybean
(Glycine max [L.] Merr.) which occurred in the southern United
States in the 1960's, the most common rotation of rice acreage
was into semi-permanent grass pastures where heavy grazing
suppressed reseeding of red rice, Echinochloa spp., and
terrestrial broadleaf weeds. The non-irrigated pasture culture
also reduced the proliferation of various aquatic and wetland
species. A rotation of one year in rice and two or three years
in pasture was common (2).

With the dramatic rise in soybean acreage in Louisiana came
a shift in rotation practices from rice and cattle or sheep to

the now most common rotation of one year in rice and two years in soybean. With the current arsenal of extremely effective soybean herbicides for control of virtually all indigenous grass, broadleaf, and sedge weeds--most of which may be used in the cropping season preceding seeding of rice--this rotation provides a superb opportunity for greatly reducing weed seed reserves in the soil before the area is returned to rice production. However, the low profit potential for soybeans in recent years and the tendency for producers to, at best, reduce weeds only to near economic loss thresholds in the soybean crop, have greatly limited the practical accomplishments of this rotation.

Cultural Management

All of the rice produced in Louisiana is grown in a flooded culture, and manipulation of water is a major component of weed control strategies. Once a permanent flood is established, germination of weeds that are not aquatic is prevented. In addition, many weeds that are competitive with rice prior to flooding either die or lose their ability to compete soon after a flood is established.

Suppression of weed germination and competition from water-susceptible weeds is utilized to the maximum extent in water-seeded rice where a continuous flood is maintained until near harvest. Although approximately 60% of the rice produced in Louisiana is water-seeded, since rice varieties grown in the state are not easily established in a continuous flood, other variations in water-seeding are more common (2). The most widely utilized water-seeding technique for suppression of red rice and other susceptible weeds is that of pinpoint flooding, which employs a very brief drainage period that allows rice seedlings to become anchored before a permanent flood is established. Minimizing the duration of drainage has been demonstrated to reduce red rice emergence and seed production (5).

While water-seeding is highly beneficial for suppression of terrestrial weeds, the practice is highly conducive to the development of aquatic and wetland weeds. This problem is intensified in southwest Louisiana where winters are often very mild and water is frequently collected in the fields throughout the winter in preparation for early spring seeding.

In dry-seeding, which is used on approximately 40% of the rice produced in the state, a permanent flood is typically timed to closely follow the final contact herbicide application. This prevents germination of further flushes of terrestrial weeds and complements the effects of herbicide treatment. Rice is typically in the 4-leaf to early tillering stage at this point and aquatic weeds are suppressed if rice stand density is adequate.

Optimum seeding dates for rice are the last week of March through April in southwest Louisiana and mid-April through mid-May in northeast Louisiana (2). However, water-seeding commercial rice approximately three weeks earlier than optimum, when soil and water are relatively cool, has been demonstrated to reduce red rice emergence (5). This practice is widely used in southwestern Louisiana where red rice is the major weed pest.

Another cultural modification for suppression of red rice in water-seeded rice is that of increased seeding rates. Seeding rates of 224 kg/ha are often used, compared to a normal rate of 151 kg/ha in an effort to provide greater early competition from commercial rice. This is most commonly used in conjunction with early seeding. Increased seeding rate has not been shown to reduce red rice stand or panicle density; however, reductions in seed production per unit area due to pronounced reduction in grains per panicle have been demonstrated (5).

Chemical Weed Control

Herbicides are used in virtually all rice produced in Louisiana, and most are aerially applied. Herbicide selection and use patterns vary with major weed pests, seeding methods, soil types, and herbicide restrictions.

Red rice is the predominant weed in the southwestern rice growing area of Louisiana. Most weed control programs in this area concentrate on red rice, with other weeds considered to be somewhat secondary. Both strawhull and blackhull red rice biotypes have been shown to be more sensitive to molinate (Ordram) than domestic rice cultivars (6). Molinate has been demonstrated to provide effective suppression of red rice when used in water-seeded production systems that utilize a continuous flood or brief drainage following seeding (5). Similar effects have been demonstrated with thiobencarb (7). Application of molinate 4.5 kg/ha has become the standard treatment for suppression of red rice populations in domestic rice. The most common practice is that of shallow preplant incorporation of molinate (Ordram 8E or 15G) in a weed-free seedbed immediately prior to establishment of the seeding flood, followed by pinpoint flooding to further inhibit germination of red rice. An alternate method of application of molinate which has been found effective is that of application of the granular formulation directly into the flood after the soil has been puddled, immediately prior to water-seeding (5). Thiobencarb (Bolero 8E) applied at 4.5 kg/ha to a weed-free soil surface immediately prior to establishing the seeding flood has been found to be similarly effective, but to date has not achieved wide usage. Both molinate and thiobencarb also offer excellent control of Echinochloa spp. as well. Molinate also controls annual sedges and suppresses many annual broadleaf weeds. Thiobencarb offers the added benefit of excellent control of emerging aquatic annuals.

The Echinochloa spp., barnyardgrass and junglerice, are second in importance as weeds only to red rice. Barnyardgrass populations of as few as 5 plants per square foot have been shown to reduce rice yield by nearly half when allowed to compete season-long (8). Barnyardgrass is distributed throughout the state, while junglerice is more common in southwestern Louisiana. Where the Echinochloa spp. are the predominant weed targets, postemergence application of propanil (several formulations) 3.4-4.5 kg/ha when weeds are in the 2-3-leaf stage is the most common choice of herbicide treatments. In addition to barnyardgrass, timely postemergence applications of propanil provide control of many other seedling grasses, broadleaves, and sedges. Efficacy of propanil is dependent upon environmental conditions being conducive to active growth of the barnyardgrass and timeliness of application with respect to weed size. Timeliness of application is often compromised due to unfavorable conditions for application or conditions that might cause drift onto adjacent crops of soybean and cotton (Gossypium hirsutum L.). When propanil cannot be applied at the optimum timing, rates can be increased to a maximum of 6.7 kg/ha to improve control of the larger weeds.

Yield losses from barnyardgrass are probably greatest in dry-seeded rice produced on the clay soils in northeast Louisiana. Rice is commonly seeded at a shallow depth in soil that has dried to the depth to which it was tilled. Moisture for rice germination must come from irrigation or rainfall; however, barnyardgrass and other weeds commonly begin emerging from the moist soil below the depth of the rice seed immediately after tillage. Thus, in order to achieve proper timing with respect to weed size, propanil applications are often required long before a permanent flood is possible. In these situations, multiple applications of propanil are frequently used. Combining a residual herbicide such as thiobencarb or molinate 3.4 kg/ha with propanil has been demonstrated to enhance propanil activity and reduce emergence of barnyardgrass in succeeding flushes (9,10). These herbicide combinations are being used more widely; however, in practice the benefits of the residual herbicide component of such treatments are difficult to gain since moisture requirements for activation of these herbicides are rather exacting, and weeds frequently germinate below the herbicide depth and emerge through cracks that develop if the soil is allowed to dry.

Application of granular molinate 3.4-4.5 kg/ha shortly after a permanent flood has been established is a common practice for control or suppression of barnyardgrass that has escaped previous treatment or emerged between treatment and flooding. When the grass can be two-thirds covered without covering the rice, and the flood held essentially static until the grass dies, this practice proves highly effective. However, it is often difficult to achieve these requirements, and even if accomplished, the grass has already lowered rice yield potential due to early season competition.

Another weapon for dealing with barnyardgrass escapes is postemergence foliar application of fenoxaprop (Whip 1EC) at 0.17-0.22 kg/ha (11). Fenoxaprop may be applied either prior to or following permanent flood and is very effective on barnyardgrass and other graminaceous weeds, except red rice. Fenoxaprop has been positioned in the marketplace to compete with later timings of propanil pre-flood, or post-flood as a substitute for granular molinate. While it is generally very effective in controlling grasses, problems with rice injury and exacting water and fertility management prior to and following application have greatly limited acceptance of this herbicide.

Aquatic and wetland broadleaf weeds are most common in southwest Louisiana where rice is water-seeded. Alligatorweed, a perennial, is no doubt the most notorious of these; however, damaging infestations of dayflower, ducksalad, purple ammania, and eclipta are also frequently found. In areas where the use of phenoxy herbicides is not severely restricted, aquatic broadleaf weeds are most commonly controlled by foliar postemergence application of 2,4-D amine (several formulations) at 0.6-1.7 kg/ha after rice has tillered but before panicle initiation. In northeast Louisiana where 2,4-D application is restricted by law because of potential damage to cotton, foliar postemergence application of bentazon (Basagran 4EC) 1.1 kg/ha plus crop oil concentrate at 1-2 L/ha acre effectively controls many annual aquatic and wetland broadleaf weeds when the flood is removed and the weeds are small at treatment. Bensulfuron (Londax 60DF) was recently introduced for control of aquatic and wetland broadleaf weeds and sedges. It is applied as a dilute spray at 42-70 g/ha into the rice flood. Weeds must be in the very early seedling stages and covered by the flood. Control of many aquatic and wetland weeds, including purple ammania, ducksalad, pickerelweed (Pontederia cordata L.), gooseweed (Sphenoclea zeylandica Gaertn.), and waterhyssops (Bacopa spp.) has been excellent. However, alligatorweed control has been only moderate (12). Application of bensulfuron when weeds are very small and holding the flood over the weeds until their demise are critical requirements for effective performance. The relatively high cost of bensulfuron compared with that of 2,4-D has limited acceptance. However, the opportunity to remove weeds with bensulfuron, as with bentazon, before they have reduced the yield potential of the rice, would suggest greater acceptance for both of these herbicides in the future.

Control of terrestrial broadleaf weeds that escape early season herbicide applications is routinely accomplished with mid-season applications of 2,4-D in areas where use of this material is not restricted. However, where 2,4-D use is restricted, several of these weeds present particular problems. Texasweed, hemp sesbania, Indian jointvetch, and palmleaf morningglory often escape early season applications of molinate, propanil, or propanil combinations and flourish in the aquatic environment after rice is flooded. Acifluorfen (Blazer 2L) 0.14 kg/ha effectively controls hemp sesbania when it is applied prior to flowering, and rice tolerance is acceptable at this rate. An emergency exemption was granted in 1989 for triclopyr (Rely 3EC) at 0.28-0.42 kg/ha for postemergence control of Indian jointvetch, texasweed, and palmleaf morningglory. Performance at the grower level was excellent when weeds were treated in the seedling stages. However, some concerns remain with regard to rice tolerance in several varieties, especially at early treatment timings.

Weeds in the sedge family, including rice flatsedge and yellow nutsedge, are usually effectively controlled or suppressed by molinate 4.5 kg/ha preplant incorporated or timely foliar postemergence applications of propanil 3.4-4.5 kg/ha. When these treatments are not successful, application of bentazon 1.1 kg/ha with the addition of 1-2 L/ha crop oil concentrate often provides control of rice flatsedge and top-kill of yellow nutsedge.

Several other herbicides are used to a limited extent in Louisiana rice. These include MCPA (several formulations) as a 2,4-D substitute, pendimethalin (Prowl 4EC) as a residual herbicide for combination with propanil, glyphosate (Roundup 4EC) for preplant burndown, and bromoxynil (Buctril 2EC) which has been granted several emergency exemptions for control of broadleaf weeds.

SUMMARY

Louisiana rice producers face many and varied weed pests that are potentially very costly in terms of expenditures for control and losses in yield and quality of the crop. Management practices which are used to combat these pests include crop rotation, cultural management, and chemical control. Losses due to weeds, however, remain substantial. Weeds can be effectively controlled only by combination of these practices into weed management programs that fit individual farm situations.

REFERENCES

1. Louisiana Agricultural Statistics Service, LA Farm Reporter, 1989, **21**, 2.
2. Linscombe, Steve (editor), Louisiana Rice Production Handbook, 1987, pp. 63.
3. Sanders, D. E., Estimated southern state losses due to weeds in rice. Proc. Sou. Weed Sci. Soc., 1988, **41**, 390.
4. Sanders, D. E., The ten most common and ten most troublesome weeds in Louisiana rice in 1987. Proc. Sou. Weed Sci. Soc., 1988, **41**, 403.
5. Griffin, J. L., J. B. Baker, R. T. Sonnier, Red rice control in rice and soybeans in southwest Louisiana. LA Agri. Exp. Sta. Bull. 776, 1986, pp. 20-21.
6. Henry, C. S., Alteration of rice tolerance to several thiolcarbamate herbicides by seed treatment with 1,8-naphthalic anhydride. M. S. Thesis LA State Univ. Agri. and Mech. Coll., Baton Rouge, LA, 1971, pp.94.
7. Baker, J. B. and J.W. Shrefler, Thiobencarb and water management for red rice control in water-seeded rice, Proc. Sou. Weed Sci. Soc., 1988, **41**, 282.
8. Smith, R. J. and D. E. Seaman, Weeds and their control in rice production. U. S. Dept. Agr. Handb. 292, 1971, pp. 135-136.
9. Carroll, K. R. and S. H. Crawford, Performance of propanil and propanil-residual herbicide combinations in dry-seeded rice. Proc. Rice Tech. Work. Group, 1986, **21**, 39.
10. Crawford, S. H. and K. R. Carroll, Herbicide performance in drill-seeded rice. Proc. Sou. Weed Sci. Soc., 1984, **37** 39.
11. Crawford, S. H. and K. R. Carroll, Rice weed control studies with HOE-33171 (fenoxaprop). Proc. Sou. Weed Sci. Soc., 1985, **38**, 37.
12. Sanders, D. E. and S. D. Linscombe, Evaluation of bensulfuron for broadleaf weed control in rice. Proc. Sou. Weed Sci. Soc., 1989, **42**, 67.

PYRAZOSULFURON-ETHYL, A NEW SULFONYLUREA HERBICIDE FOR PADDY RICE

K. SUZUKI, Y. SHIRAI, H. HIRATA
Shiraoka Research Station of Biological Science
NISSAN CHEMICAL INDUSTRIES, LTD.
1470 Shiraoka, Minamisaitama, Saitama pref., Japan 349-02

ABSTRACT

Pyrazosulfuron-ethyl(code name NC-311, trade name Sirius^{TM}) is a new sulfonylurea herbicide for rice, being developed and launched in major rice producing countries.

In greenhouse studies, NC-311 demonstrated excellent herbicidal activity with both pre and post emergence applications. I_{75} values of NC-311 pre-emergence were below 10 g a.i./ha except for Echinochloa crus-galli. However, E. colona was much more sensitive. Selectivity of NC-311 between rice seedlings and Cyperus serotinus was sufficient, while that between rice seedling and E. crus-galli was marginal in water culture study.

In field study, however, NC-311 alone demonstrated a good control of E. crus-galli in China and other countries. It is necessary for NC-311 to be combined with grass herbicides in dry seeded rice or in countries where they require higher performance and a wider application window.

NC-311 proved to have little possibility of carry over problems, and no influence was observed on neighboring crops by volatility.

INTRODUCTION

Pyrazosulfuron-ethyl(NC-311)is a new rice herbicide , being developed or launched in rice producing countries by NISSAN CHEMICAL INDUSTRIES, LTD.. A large number of trials have been conducted over the last five years.

In this paper, we will discuss the activity of NC-311 in the glasshouse and its field performance in major rice producing countries.

MATERIALS AND METHOD

1) The herbicidal spectrum of NC-311 was determined by pot tests under glasshouse conditions. Thirteen weed species and alga were used for these tests. Weed seeds, tubers, run-

ners and soil containing algal spores were planted in 1.3 liter plastic pots filled with clay loam soil(clay 18.0 %, total carbon 0.15%, pH 6.18) paddy conditions. Wettable powder suspensions of NC-311 and the reference herbicide DPX-84 (bensulfuron-methyl) were applied to the surface water using a pipette. Water depth was kept at 4 cm and no leaching was allowed. Twenty five to thirty five days after treatment, fresh weight of weeds or alga were measured and dosages for 75 % control were calculated.

2) A water culture study was conducted to determine the sensitivity difference between rice seedlings and weeds. Rice seedlings(Oryza sativa cv."Nihonbare"), germinated tubers of Cyperus serotinus and seeds of Echinochloa crus-galli were planted in plastic pots which contained quartz sand. Pots were filled with Kasugai's solution containing various concentrations of NC-311. Shoot and root weights were recorded after fifteen days incubation in a greenhouse.

3) A number of field trials have been conducted in major rice producing countries in commercial fields and at experimental stations. In Japan, 2856 applications of NC-311 single or in combinations have been made in official trials through JAPR(the Japan Association for Advancement of Phytoregulators) besides private trials. Generally the plot sizes were about 10 m^2 with two replications, rice seedlings were transplanted by machine three to seven days after puddling and leveling. Granular formulations of NC-311 single, or in combination with grass herbicides were applied from five to twenty five days after transplanting. Weed control levels were examined from thirty to forty five days after transplanting by measuring weed dry weights. In other countries, the trial method varied in each case.

4) The relationship between NC-311 concentration in soil and growth reduction of upland crops was examined to predict the possibility of carry over . NC-311 was incorporated into air dried clay loam soil shown above and 500 g of this soil was filled into 500 ml plastic pot. Radish, carrot, Chinese cabbage, spinach, onion, soybeans or red beans were seeded in pots individually. These crops were incubated for 21 days under glasshouse conditions. Finally, crops were weighed and I_{50} values were calculated.

6) The influence of NC-311BCG(a combination granule of NC-311, quinclorac and pretilachlor) application on the following crops was tested in farmers fields in 1989. The standard rate of the granules were applied in three locations which were located in north, central and south Japan. Soil

samples were taken at harvest time. Some portion of the soil was used for chemical analysis and the rest was air dried and sieved for bioassay. Fifty grams soil was extracted with methanol/water(2:1). The extract was purified using a SEP-PAK C_{18} silica column, a silica gel column(5 % water contained) and a SEP-PAK silica column. NC-311 was determined by HPLC. A bioassay was conducted as follows. Two hundred grams of soil samples were placed into plastic pots and twenty carrot seeds were planted. Plants were incubated for 14 days in a glasshouse and weighed.

7) The possibility of NC-311 vaporizing from paddy water was examined under a simulated condition. A plastic tunnel, 8 m in length, 1.2 m in width and 0.9 m in height, was placed in a field. Both ends were open and an electric fan was placed at one end. A plastic container(52 X 74 cm) with water 4 cm in depth was put in front of the fan and ten times the standard rate of granular herbicides were applied. Vegetables grown in plastic pots, including cucumber, soybeans, tomatoes and okra, were put at 1.5 m and 4.5 m in front of the container. The tunnel was fanned(1.1 m/s) for 5 days. The pots were then moved to a greenhouse and were grown on for 9 days. Visual assessments of phytotoxicity were made, rating from 0 to 9. Maximum temperature in the plastic tunnel was 24-30°C and the minimum temperature was 19-21°C.

RESULTS

1) Both pre and post emergence applications of NC-311 showed high herbicidal activity against major paddy weeds and alga (table-1). In pre emergence application, I_{75} values of NC-311 were below 10 g a.i./ha, except for Echinochloa crus-galli. Broad leaf weeds and sedges were especially sensitive to NC-311. In post-emergence application, equal or higher rates were required to control these weeds than with pre-emergence. It was noticeable that Echinochloa colona was much more sensitive to NC-311 or DPX-84 than E. crus-galli.

In both applications, NC-311 was 1.4-14 times more active than DPX-84 against these weed species, especially against cyperaceous weeds.

2) NC-311 completely inhibited Cyperus serotinus growth at 1 ppb under water culture conditions (Fig-1), while inhibition of rice seedlings was minimal at 10 ppb. Therefore, NC-311 has a very high selectivity between C. serotinus and

rice seedlings. On the other hand, its selectivity against <u>E. crus-galli</u> was marginal.

    3)   In Malaysia, the 10 % WP of NC-311 has been mainly tested (Table-2). NC-311 showed excellent weed control

TABLE-1

Herbicidal activities of NC-311 against paddy weeds
with pre-emergence or post-emergence application
under glasshouse condition

weed names	application timing	I_{75} values(g a.i./ha) NC-311	DPX-84
<u>Echinochloa</u> <u>crus-galli</u>	pre	19.2	69
	1.5L a)	14.2	66.2
<u>Echinochloa</u> <u>colona</u>	pre	3.7	13.1
	1.5L	14.2	19.2
<u>Scirpus</u> <u>juncoides</u>	pre	8.3	25.6
	2 L	12.0	35.0
<u>Scirpus</u> <u>nipponicus</u>	pre	9.1	77.5
(tuber)	2 L	16.0	33.0
<u>Scirpus</u> <u>planiculmis</u>	pre	9.6	35.4
(tuber)	20-25 cm	22.2	50.1
<u>Monochoria</u> <u>vaginalis</u>	pre	2.7	5.8
	2 L	2.1	6.7
<u>Alisma</u> <u>canaliculatum</u>	pre	2.1	3.0
	3-4 L	1.4	3.6
<u>Sagittaria</u> <u>pygmaea</u>	pre	1.9	9.0
(tuber)	2 L	2.7	9.4
<u>Sagittaria</u> <u>trifolia</u>	pre	2.3	4.8
(tuber)	4-5 L	2.7	2.4
<u>Cyperus</u> <u>serotinus</u>	pre	0.9	9.1
(tuber)	3-4 L	1.9	26.6
<u>Eleocharis</u> <u>kuroguwai</u>	pre	1.9	27.1
(tuber)	10-20cm	7.4	56.0
<u>Oenanthe</u> <u>javanica</u>	post	1.9	6.8
(runner with two leaves)			
<u>Spirogyra</u> <u>arcla</u>	pre	4.3	17.4

a) L=leaf stage

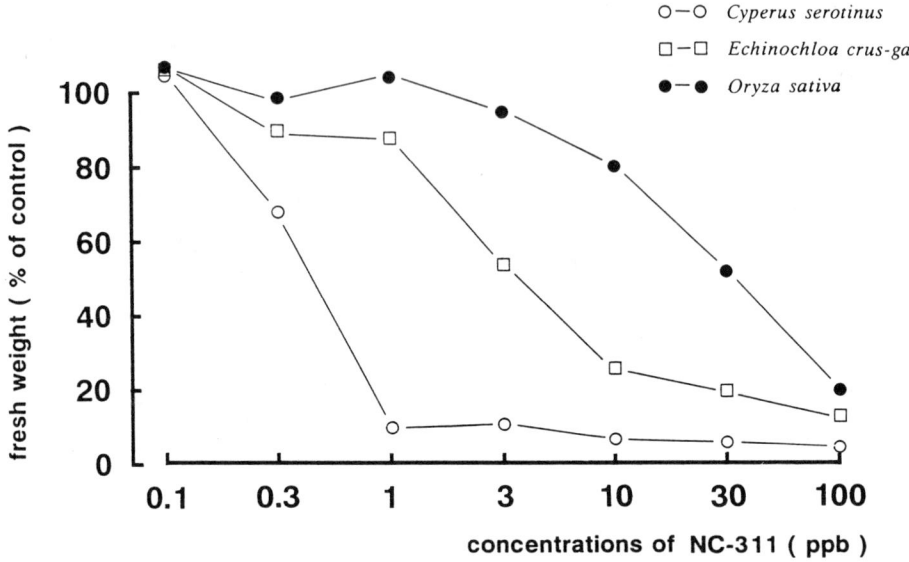

Fig-1   **Growth response of weeds and rice seedling to NC-311
under water culture conditions (15 days)**

against <u>Sagittaria guyanensis</u> even at 14 g a.i./ha. <u>E. crus-</u>
<u>galli</u> was also well controlled at these rates until 59 DAT
with the number of panicles also highly reduced at 84 DAT.

   <u>E. crus-galli</u> is widely spread and noxious in China. In
1988, six trials were conducted where it was the predominant weed. In Guangdong (Table-3), NC-311 gave good control of
<u>E. crus-galli</u> and gave perfect control of other weeds at all
rates and timings. In these trials, the control of <u>E. crus-</u>
<u>galli</u> varied to some extent, particularly with 9 DAT applica-
tion. Thus the average control was 81-87 % at 5 DAT and 64-73
% at 9 DAT application.

   In South America, trials were carried out against both
water seeded and dry seeded rice in major rice producing
areas. In the trial in dry seeded rice , <u>Cyperus serotinus</u>
was the predominant weed (Table-4). Foliage application of
NC-311 at 20-30 g a.i./ha resulted in excellent <u>C. serotinus</u>
control without causing any phytotoxicity to rice equal to
bentazon at 960 g a.i./ha.

   NC-311 single applications were also conducted in official trials in Japan(Table-5). Application with 21 g a.i./ha

at 3-5 DAT resulted in an average of 84 % E. crus-galli control. To improve this efficacy, several combinations with
grass herbicides have been tested, NC-311BCG(
+quinclorac+pretilachlor) and NC-311T (+mefenacet-). These
demonstrated high levels of overall weed control and good
application timing flexibility against E. crus-galli and other
weeds (Table-5,6). Yield was not influenced by these herbicide
applications.

4) In glasshouse tests, the influence of NC-311 soil
incorporation on upland crops was examined(Table-7). Spinach
was one of the most sensitive crops to NC-311 followed by
onion, Chinese cabbage, carrot, and radish. Soybeans, red
beans, and wheat were much less sensitive than the vegetables.

5)   Carry over trials have been conducted for five
years. In 1989 three fields were treated with
NC-311BCG(Table-8). No significant difference was observed
with fresh weight of carrot grown in the treated and untreated
soil. Furthermore NC-311 was not detected by chemical analysis

TABLE-2
Trial results of NC-311 in direct seeded rice
in Malaysia (1988)

	a.i. g/ha	SAGGU 27 DAT	ECHCG 59 DAT	84 DAT
		- weed control rate (%)-		
NC-311	14	96	99	85
(10%WP)	21	90	100	88
	42	96	98	72
molinate 2000 (10 % Gr.)		0	100	98

SAGGU = Sagittaria guyanensis   ECHCG = Echinochloa crus-galli
location   :   Pinang Tunggal, Malaysia
seeding   :   November 6,1988 ; cv. MR-84
application :   November 18,dripping with a pipette (10 1/ha)
plot size   :   50 m^2

in these soils. This result is supported by many other trial observations.

6) Under simulated conditions, NC-311 caused no damage against neighboring crops while molinate, a carbamate herbicide caused young leaf malformation on cucumber, tomato, and soybeans(Table-9). Okra was not influenced by molinate.

TABLE-3

Trial result of NC-311 in transplanted rice in China(1988)

	Timing	a.i. g/ha	Efficacy (%)				
			ECHCG	SCPJU	MOOVA	LIDPR	SAGTR
NC-311 (10% WP)	5 DAT	10	99 (81)	100	100	100	100
		20	90 (85)	100	100	100	100
		30	100 (87)	100	100	100	100
	9 DAT	10	87 (64)	100	100	100	100
		20	94 (63)	100	100	100	100
		30	98 (73)	100	100	100	100
butachlor	5 DAT	900	84 (95)	94	53	100	93
	9 DAT	900	70 (86)	53	65	100	100

Average of ECHCG control in 6 trials in this year are shown in(
location   : Plant Protect Institute,
                 Guangdong Academy of Agr. Sci., China
transplanting : April 24, 1988
application   : 5, 9 DAT
plot size   : 20 m^2 , three replications

TABLE-4

Field trial results of NC-311 in Argentina (1986/1987)
in dry seeded rice

treatment	rate a.i. g/ha	phytotoxicity		CYPES control (%)	
		10 DAT	20 DAT	20 DAT	30 DAT
NC-311	20	0	0	95	80
	30	0	0	95	90
	40	0	0	95	90
	60	0	0	95	90
bentazon	960	0	0	95	90

location   : Chaco, Argentina
sowing   : January 6, 1987
application: January 31, sprayed at 140 1/ha
weeds   : Cyperus esculentus 3-4 leaves, 164/m^2

DISCUSSION

E. crus-galli needed the highest rate of NC-311 for adequate control while E. colona, which is common in South Asian countries, was much more sensitive to NC-311 in pre-emergence application. It is possible that NC-311 single application at lower rates could give overall control of weeds in these area.

Although the water culture study indicates that selectivity of NC-311 between rice and E. crus-galli is not sufficient, the selectivity improves in the field. This is because the roots of transplanted rice are located deep in the soil and those of E. crus-galli are shallow, being easily exposed to herbicides.

In Malaysia and China it was proved that NC-311 single application can give overall control of weeds including E. crus-galli. The rate of 20 g a.i./ha at 5 DAT is considered to be economically acceptable. Similar results were obtained in Japan. However, the market in Japan requires a very high

Thus combinations with grass herbicides are essential. In dry seeded rice, E. crus-galli is generally one of the most competitive weeds. Foliage application of NC-311 in upland conditions is not as effective against E. crus-galli as in paddy conditions and therefore it is necessary for NC-311 to be combined with grass herbicides like propanil.

In upland conditions some sulfonylurea herbicides have

TABLE-5

Mean weed control and yield in field trials of NC-311 alone and in combinations in transplanted rice in Japan (1986-1989)

	– weed control (%) –						
	ECHCG	CYPDI	MOOVA	B.L.	ELOAC	SCPJU	SAGPY
NC-311 (0.07% Gr.)	84	97	97	98	99	97	88
NC-311BCG (0.07+0.9+1.5% Gr.)	99	99	99	98	99	98	92
NC-311T (0.07+3.5 % Gr.)	99	100	100	97	100	95	90

	– weed control (%) –			yield (%)	No. of treat-ments
	CYPSE	ALACA	overall		
NC-311	99	100	90	99	38
NC-311BCG	95	100	98	100	250
NC-311T	97	99	97	100	251

NC-311	21 g a.i./ha, 3-5 DAT application
NC-311BCG	21g + 270g quinclorac + 450g pretilachlor /ha 5-20 DAT application
NC-311T	21g + 1050g mefenacet a.i./ha 5-18 DAT application

ECHCG=Echinochloa crus-galli
B.L. =broad leaf weeds
ELOAC=Eleocharis acicularis
SAGPY=Sagittaria pygmaea
yield= % of weeding check

CYPDI=Cyperus difformis
MOOVA=Monochoria vaginalis
SCPJU=Scirpus juncoides
CYPSE=Cyperus serotinus

carry over problems. However, there are few reports of the same problems with sulfonylurea herbicides in paddy rice. In the case of NC-311, no residual activities against vegetables has been observed at harvest time. Lack of moisture is of course never a limiting factor under paddy conditions.

Vaporized herbicides from paddy water sometimes cause crop injury against neighboring crops in Japan. In this study, NC-311 had no such action while molinate under similar conditions caused effect on some vegetables. This result is supported by the vapor pressure difference between two herbicides, that of NC-311 and molinate being 2.5×10^{-7} mmHg and 5.6×10^{-3} mmHg (25°C), respectively.

TABLE-6

The flexibility of application timing in controlling weeds with NC-311BCG and NC-311T in Japan(1988-1989)

	E. crus-galli				overall			
application timing (DAT)	0 -5	6 -10	11 -15	16 -20	0 -5	6 -10	11 -15	16 -20
	- mean control (%) -							
NC-311BCG	100	99	99	99	99	98	98	98
NC-311T	100	99	99	95	98	98	98	94

TABLE-7

Influence of NC-311 soil incorporation on upland crops after 21 days incubation

crops	I_{50} values (ppb in soil)
radish	5.65
carrot	5.50
Chinese cabbage	4.22
spinach	2.62
onion	3.97
soybeans	> 10
red beans	> 10
wheat	> 10

TABLE-8

Influence of NC-311 application on a following
crop and the residue levels in soil from
field trials at three locations in Japan

locations	days from treatment to sampling	fresh weight of carrot(%)	residue in soil (ppb)
Mizusawa, Iwate	142	98	2 >
Shiraoka, Saitama	118	105	2 >
Fukuoka, Fukuoka	113	100	2 >

TABLE-9

Influence of herbicide vapor on neighboring crops by visual
assessment of phytotoxicity at 9 DAT

a.i. g/ha	cucumber − distance −		tomato		soybeans		okra	
	1.5m	4.5m	1.5m	4.5m	1.5m	4.5m	1.5m	4.5m
NC-311 210	0	0	0	0	0	0	0	0
molinate 24000	1.3 c	0.5 c	2.0 c	2.0 c	0.3 c	0.8 c	0	0

* visual assessment rating from 0 (no damage) to 9 (killed)
** c=curling of new leaves

REFERENCES

1. Suzuki, K., Nawamaki, T., Watanabe, S., Ikai, T., NC-311, a new sulfonylurea herbicide in rice. 11th Asian-Pacific Weed Science Society, 1987, 461-468
2. Suzuki, K., Watanabe, S., Shirai, Y., Endo, T., Hirata, H., Crop safety of NC-311, pyrazosulfuron-ethyl, in paddy rice. 12th Asian-Pacific Weed Science Society, 1989, 141-148

CINMETHYLIN - A NEW HERBICIDE DEVELOPED FOR USE IN RICE.

R.G.JONES
Shell International Chemical Company Ltd.,
Shell Centre, London, U.K.

ABSTRACT

Cinmethylin (trade mark ARGOLD) is a novel herbicide representing chemistry in the cineole family. It has been developed as a herbicide for use in transplanted rice at low application rates of 25 to 100 g ai/ha.

The development of cinmethylin took account of the difficulties of using a herbicide in rice - of ensuring crop safety while controlling weeds which may be similar to the crop and also the environmental constraints associated with herbicides in aquatic systems. Consideration was taken of agronomic and environmental conditions, and their interaction with activity of the herbicide.

Development in the field was matched by an extensive programme of research in controlled environment conditions to help predict and interpret field performance.

Soil type, and particularly organic matter content, emerged as a key factor in managing the safe and effective use of cinmethylin, particularly in machine transplanted rice. Temperature was of secondary importance.

INTRODUCTION

Cinmethylin is a novel molecule, representing new chemistry in the cineole family.

The physical, chemical and toxicological features of cinmethylin make it attractive for use in paddy rice conditions (1). It is composed of only carbon, hydrogen and oxygen, has a low order of mammalian toxicity and is readily degraded by animals, plants and soil. The toxicological effects of the compound have been investigated in a wide range of species and show that it should have minimal effect on the environment.

Cinmethylin has been shown to control some of the more difficult weeds in rice; including Echinochloa spp., Cyperus spp. and Monochoria vaginalis.

It is active at dose rates between 25 and 100 g ai/ha, choice of dose rate being dependent on factors such as temperature, soil type and weed size, and it is compatible with a wide range of other herbicides.

Herbicidal effect

Cinmethylin herbicidal effects result primarily from disruption of meristematic development in shoots and roots of susceptible species by inhibiting the entry of cells into mitosis. Major sites of action are the stem and root apices. Physiological effects on mature plant parts have not been observed.(2)

Uptake of cinmethylin occurs through the shoots and roots of germinating or emerged weeds. The physical and chemical properties of cinmethylin ensure that movement within the plant is limited to the apoplast (ie. xylem and cell wall). Movement is therefore upwards in the plant and it must enter the weed below the growing point to be effective.

These properties of cimethylin were understood when development in rice started but a considerable research programme ensued to understand mechanisms of activity as a rice herbicide and to develop recommendations for safe use. This required an extensive field programme coupled with laboratory, glasshouse and controlled environment studies to help predict and interpret field performance

MATERIALS AND METHODS

Since 1985 cinmethylin has been evaluated in field trials in paddy rice in most Asian countries, and in research trials in both Asia and Europe.

The methods of evaluation have been varied, but included:

1. Small plot glasshouse tests to evaluate primary efficacy and weed spectrum.

2. Outdoor pot tests primarily designed to evaluate crop selectivity and spectrum under carefully controlled conditions of water depth, rate of water loss etc.

3.   Small scale trials in growth rooms to gain understanding of
      metabolism and activity under conditions of varying temperature
      and soil type.

4.   Replicated field trials with small plots ($1m^2$ - $50m^2$)
      independently irrigated. Herbicide applications were as diluted
      liquid formulations or as granules. Herbicide applications were
      made over the crop, and weeds if emerged, in plots containing a
      nominal water depth of 3-5 cm. Water depth was managed after
      application.

5.   Non replicated demonstration plots where application was made
      according to local practice.

Herbicide treatments were applied to rice crops at a range of timings
from zero up to eighteen days after transplanting. At these timings
the growth stage of Echinochloa spp. ranged from pre-emergence to early
tillering.

Assessments of crop injury and activity on weed species were made at
intervals following herbicide application. The efficacy data presented
in the tables of results were derived from assessments based on counts,
estimates of % cover or fresh weights of weeds.

RESULTS

Performance - Application rate

Echinochloa spp. were well controlled by cinmethylin applied pre- or
post-emergence at growth stages up to 2.5 to 3 leaves when used in the
range 25-100 g ai/ha. This range in dose rates reflected the
geographical distribution and agronomic practices of the trial
locations.

In the Philippines good control of Echinochloa spp. was given by
cinmethylin at 100 g ai/ha in the period zero to ten days after
transplanting which covered pre-emergence to the two to three leaf
stages of the grass weed. Control of later growth stages required a
higher dose of cinmethylin (Table 1).

     In trials in the Peoples Republic of China, similarly effective
control of Echinochloa spp. was given by cinmethylin at lower doses of
25 - 35 g ai/ha. This followed applications made to the grass weed at
the 1 to 1.5 leaf stage, four to seven days after transplanting. The
crops were transplanted by hand at the four to five leaf stage and
20 cm in height. They rapidly formed a crop canopy that was highly
competitive to weeds so allowing the use of lower rates of cinmethylin
for Echinochloa spp. control since long persistence of effect was not
necessary.

TABLE 1

Percentage control of Echinochloa spp. following treatment with
cinmethylin at different application timings in The Philippines.

Application timing - days after transplanting	0-4	5-7	8-10	11-13
Echinochloa spp. leaf number at treatment	pre-1	1-2	2-3	3-early tillering
Cinmethylin g ai/ha 50	76	78	55	19
75	91	84	-	28
100	95	97	94	42
200	95	98	97	94

In similar work in Japan a rate of 45 g ai/ha gave effective
control up to the 2 leaf stage of Echinochloa crus-galli, but 60 g was
necessary for effective control at the three leaf stage (Table 2).

TABLE 2

Performance in pot trials in Japan - % Control of Echinochloa spp.

Leaf stage at Application	1	2	3
Cinmethylin 90 g ai/ha	100	100	95
60	100	100	95
45	100	100	80
30	100	80	40

Results from field trials have thus shown that the optimum dose
rate does vary between countries. This reflects differences in both
cultural practice and environmental factors.

Use of controlled environment growth chambers has demonstrated
that the activity of cinmethylin on Echinochloa spp. in simulated paddy
conditions was influenced by temperature. Under cool conditions of
25°C day/10°C night, E. crus-galli, treated at the one leaf stage was
very sensitive to cinmethylin at a low dose of 30 g ai/ha. To achieve
similarly effective control under higher temperatures increases in dose
of up to 60-90 g ai/ha were needed (Table 3).

TABLE 3
The influence of temperature on the activity of cinmethylin on
Echinochloa spp. in simulated paddy conditions

Temperature regime (day / night)	PERCENTAGE WEED CONTROL Dose g ai/ha of cinmethylin		
	30	60	90
Cool (25°C/10°C)	87	100	100
Warm (35°C/20°C)	59	82	98

Water depth also influences performance with reduced activity at depths below 3 cm. Deeper water allows greater absorption of herbicide into the weed stem close to the apical meristem where it can readily inhibit growth.

Performance - Weed Spectrum

Some other important rice weeds including Monochoria vaginalis and annual Cyperus spp. proved susceptible to cinmethylin. Data from The Philippines presented in Figure 1 show that susceptibility of these weeds varied with application timing. While Monochoria vaginalis was susceptible to 100 g ai/ha of cinmethylin applied up to 7 days after transplanting; Echinochloa was susceptible up to 10 days and Cyperus difformis up to 13 days.

Figure 1

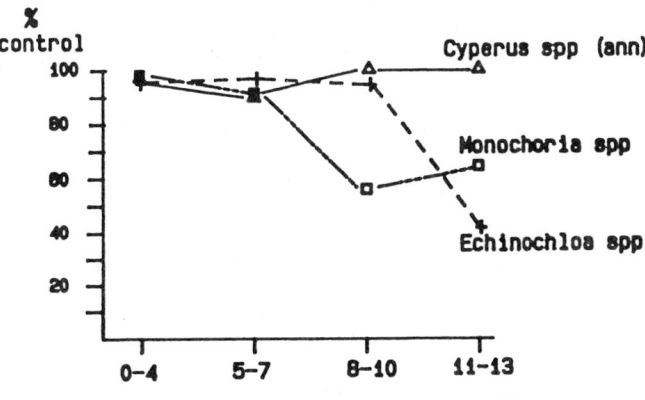

Effect of application timing on activity.
Philippines - 100 g ai/ha.

Application - Days after transplanting.

Based on trials to date the following list of weed
susceptibilities are proposed at a dose rate of 30-100 g ai/ha.

Efficacy	Annual	Perennial
Excellent	- Echinochloa crus-galli - Echinochloa oryzicola - Echinochloa glabrescens - Monochoria vaginalis - Cyperus difformis	
Good	- Dopatrium junceum - Elatine triandra - Elatine triandra - Rotala indica - Cyperus iria	- Eleocharis acicularis - Scirpus junciodes - Scirpus hoturai

Performance - Mixtures

The excellent performance of cinmethylin against Echinochloa species
and a limited range of other important weeds indicated its suitability
as a mixture partner in herbicide combinations. Partners being
evaluated include well established products such as 2,4-D or butachlor
as well as more recently developed candidates including sulphonyl
ureas, pyrazols, phenoxys such as clomeprop, and acetamides such as
pretilachlor.

For example, by adding 2,4-D ester to cinmethylin in trials in The
Philippines the consistency of performance on M. vaginalis was
enhanced, (Table 4).

TABLE 4
Summary of the performance of cinmethylin + 2,4-D applied four to
ten days after transplanting of rice in The Philippines.

	g ai/ha	Percentage weed control (7)			Grain yield(5) % of untreated
		Echinochloa spp.	Monochoria vaginalis	Cyperus spp.	
Cinmethylin	100	96	88	94	146
+ 2,4-D	100+500	98	96	100	161

( ) number of trials.

Similarly, mixtures with pyrazols have extended the spectrum of annual broadleaf weeds whilst with sulphonyl ureas the spectrum has been extended to control perennial sedges.

Crop tolerance

Crop tolerance has been influenced by both climate and cultural practices in different rice growing areas.

In The Philippines, rice seedlings hand transplanted at the three to four leaf stage were initially sensitive to treatment with cinmethylin. However, as seedlings became established they rapidly became tolerant such that applications more than four days following transplanting were safe at 200 g ai/ha (Table 5). Seedlings were sensitive while roots were close to the soil surface, but as the root-system penetrated more deeply into the soil the sensitivity declined.

TABLE 5
Crop effects on EWRC 1-9 scale in transplanted rice in The Philippines following applications made at intervals after transplanting.

Application - D.A.T.		0	1	2	4	6-8	9-11	12-18
WL95481g ai/ha	50	2.7	1.9	1.0	1.0	1.0	1.0	1.0
	100	3.2	1.5	1.0	1.0	1.0	1.0	1.0
	200	5.0	2.2	1.3	1.5	1.0	1.1	1.0
Number of trials			1	2	1	4	12	4	2

(1 = no effect, 9 = total crop loss)

Under Japanese conditions younger rice seedlings, normally at the 2 to 3 leaf stage, are transplanted by machine and are particularly sensitive to herbicides. Here cinmethylin exhibited good crop selectivity at 60 g ai/ha in applications made from five days after transplanting. This delay in application allowed an adequate period for root establishment of the younger rice transplant. Seedlings transplanted less than 3 cm deep required a longer period to become tolerant. The effect of leaching has been investigated in pot and concrete tank tests which showed acceptable crop tolerance with water loss rates of up to 1 cm per day.

Improved understanding of crop selectivity followed work in controlled environment facilities studying interdependence of dose rate, water depth, temperature and soil type. Soil organic matter (OM) appeared paramount; phytotoxicity followed application at high temperature with shallow water depths in low organic matter soils (<2%). At higher OM's there was little problem indicating a buffering effect of the OM in soil preventing movement of toxic levels of herbicide into the rice crop.

No evidence emerged that rice is physiologically tolerant to cinmethylin, but that tolerence depends on placement selectivity based on the position of the apical meristem - the site of action. In rice the apical meristem is below the soil surface and recommendations for use must minimise the amount of herbicide which moves to this depth in the soil from solution in the paddy water. In susceptible grass species the apical meristem is above the soil surface and easily accessible to herbicide diffusing into the stem from the paddy water.

Crop Yield

Grain yields from trials treated with cinmethylin have shown large increases over untreated controls, reflecting effective control of heavy infestations of weeds and safety to the crop (Table 6). In trials in the Peoples Republic of China, against light weed infestations, cinmethylin gave small but consistent yield increases at doses up to 50 g ai/ha.

TABLE 6
Grain yields expressed as a percentage of the untreated control in The Philippines (heavy weed infestation).

Dose, g ai/ha of cinmethylin	0	50	75	100	200
Yield - % of untreated	100	183 (7)	143 (8)	160 (18)	206 (10)

( ) number of trials.

DISCUSSION

The physical, chemical and toxicological features of cinmethylin make the product attractive for use in paddy conditions. The compound has a low order of mammalian toxicity. In addition it has moderate persistence and shows no tendency to accumulate in the environment.

Cinmethylin at low doses of 25-100 g ai/ha provided outstanding control of Echinochloa spp. up to the 3 leaf stage. The compound was also effective against several other important rice weeds including Monochoria vaginalis, and annual Cyperus spp. Cinmethylin was safe to large hand transplanted rice plants at doses up to 200 g ai/ha when applied 4 days after transplanting. Smaller, machine transplanted rice required a longer period of establishment to acquire tolerance i.e. 7 days.

Development of an understanding of the properties of cinmethylin in rice paddy has allowed recommendations for safe use to be made in the many and varied conditions under which transplanted rice is grown in Asia. Basic recommendations are:
- do not use on low organic matter soils.
- do not use with a leaching rate greater than 1 cm per day.
- ensure a minimum depth of 3 cm of water in the paddy.

- paddy should remain flooded for two weeks after application.
- plant rice 3 cm deep.
- allow plants to become established before treatment, 4 days for hand planted rice and 7 days for machine planted.
- apply when majority of <u>Echinochloa spp</u> plants are 1.5 to 2.5 leaf.
- dose rate to be based on local development trial experience.

In some areas the spectrum of weeds controlled by cinmethylin covers the principal weeds of paddy rice and the compound could therefore be used alone. Where other weeds are encountered cinmethylin provides a sound base of performance on which mixture combinations can be built to suit local needs.

REFERENCES

1.   Jones, R.G., Iinuma, T. and Murphy, M.W., WL95481 - A novel herbicide for use in flooded, transplanted rice. Proceedings of 11th Asian-Pacific Weed Science Society Conference, 1987, pp. 407-413.

2.   May, J.W., Goss, J.R., Moncorge, J.M. and Murphy, M.W., SD95481 - A versatile new herbicide with wide spectrum crop use. Proceedings of 1985 British Crop Protection Conference, 1985, pp. 265-270.

ACKNOWLEDGEMENTS

The Author thanks colleagues in Shell companies and Shell Research Limited, and co-operators in Asia for the production of technical information used in this report.

BENSULFURON-METHYL, METSULFURON-METHYL AND THEIR COMBINATION FOR WEED CONTROL IN RICE

LARRY W. PETERSON
E. I. Du Pont De Nemours & Company
Stine-Haskell Laboratory, P.O. Box 30,Newark, Delaware 19711

WALTER T. REED, TOSHITAKA YAMAGUCHI
E. I. Du Pont De Nemours & Company
Walker's Mill, Barley Mill Plaza, Wilmington, Delaware 19898

ABSTRACT

Two new herbicides are being developed for world rice markets. The first, metsulfuron-methyl, is used at 2-10 gai/ha for the control of a broad range of broadleaf and sedge weeds. Metsulfuron methyl can be applied when weeds are small and submerged or directly to aerial plant parts after they break the paddy surface. For many weeds metsulfuron-methyl can be used in a manner similar to 2,4-D without the need to drain the paddy or concern for damage to off-site species due to volatilization. The second, DPX-70248, is a 5:1 combination product of bensulfuron-methyl and metsulfuron-methyl. This combination gives improved performance on several difficult to control species. DPX-70248 is used at 10-20 gai/ha and gives the advantages of a wider application window offered by metsulfuron-methyl and an expanded weed spectrum with bensulfuron-methyl.

INTRODUCTION

Bensulfuron-methyl (methyl 2-[[[[[(4,6-dimethoxypyrimidin-2-yl)amino]carbonyl] amino]sulfonyl]methyl]benzoate or Londax® Herbicide) has been introduced as a paddy rice herbicide into various areas of the world over the last three years [1,2,3,4,5,6]. In this short period, bensulfuron-methyl has gained wide recognition for its herbicidal activity on a wide range of annual and perennial broadleaf weeds and sedges found in continuously flooded paddy rice. Bensulfuron-methyl has also proven to be highly flexible with a broad window of application timing, good residual activity [7,8], and is compatible with a wide range of other commercial rice herbicides. The unique physical [9,10] and toxicological properties of bensulfuron-methyl have also helped it gain wide acceptance as a premier, environmentally sound rice herbicide. Bensulfuron- methyl has gained popularity in Japan, Korea, USA, Spain, Portugal, Italy, Australia and Chile where difficult to control perennial and annual weeds are a problem.

Since these areas do not represent all of the major rice growing regions of the world, Du Pont is attempting to extend its technology with sulfonylurea chemistry into major production areas in other parts of the world. The objective is to provide low use rate herbicides that are environmentally safe and provide flexible weed control programs for a wide range of rice culture techniques.

Two products have been field tested over the past 3 years in many locations with good results. The first is metsulfuron-methyl (methyl 2-[[[[4-methoxy-6-methyl-1,3,5-triazin-2-yl) amino]carbonyl]-amino]sulfonyl] benzoate and the second is DPX-70248. DPX-70248 is a combination product containing 5 parts bensulfuron-methyl to 1 part metsulfuron-methyl and will be marketed as Sindax® Herbicide 10% WP. The intent of this paper is to discuss the advantages of these two new products as rice herbicides.

MATERIALS AND METHODS

Bensulfuron-methyl, metsulfuron-methyl, and DPX-70248 have been field tested in dozens of trials throughout the rice growing areas of the world. Trials have included a wide range of rice cultural conditions and herbicide treatment conditions. No attempt will be made to give the details of these trials but a selection of representative trials that show the trends in the activity that are important will be discussed. Necessary details of these tests are given in the appropriate tables. Environmental concerns that arise because of the special nature of rice culture will also be discussed.

RESULTS AND DISCUSSION

Biology

Bensulfuron-methyl has become a popular treatment for the control of a wide range of broadleaves weeds and sedges in paddy rice. Because of its low application rate, favorable toxicology, spectrum of activity, residual activity, environmental compatibility, and ability to be premixed with a wide range of other herbicides bensulfuron-methyl has captured significant shares of the rice markets in Japan, Korea, Spain, Portugal, Italy, Australia, Chile and the U.S.A. Some of the most important weeds controlled in these various areas are summarized in Table 1.

One application of 20-75 gai/ha of bensulfuron-methyl is recommended for the control of most annual broadleaves and sedges. Application timing can range from preemergence through the 3 leaf-stage of the weeds. Best results are obtained with early postemergence applications with the seedlings completely submerged. Control of perennial weeds requires rates of 40-75 gai/ha. Timing can range from preemergence to 2-3 leaf stage of the weeds. As with the annuals species, early post applications, when most of the leaf tissue is still submerged, usually give the best results.

Metsulfuron-methyl was initially developed for the post emergence control of broadleaf weeds in wheat in Europe and the United States. Recently it has been found that metsulfuron-methyl can be used effectively for the control of weeds in rice (see Table 1). Metsulfuron-methyl is more active than bensulfuron- methyl and is used at 2-10 gai/ha with 4 gai/ha being the most commonly recommended rate. Metsulfuron-methyl has a wider application window than bensulfuron-methyl and can be used either when the weeds are still submerged or later after leaves are out of the water. Unlike bensulfuron-methyl, metsulfuron-methyl does not require the presence of paddy water in the field at the time of application. Applications can be made either early, similar to propanil timing, or later, similar to 2,4D and bentazon timing. Metsulfuron-methyl can also be used in true upland growing conditions.

DPX-70248 has been developed to take advantage of the herbicidal characteristics of both bensulfuron-methyl and metsulfuron-methyl. DPX-70248 will be marketed as a 10% wettable powder containing 5 parts of bensulfuron- methyl to 1 part of metsulfuron-methyl. For most applications, it is recommended that DPX-70248 be applied at 10-20 gai/ha to paddy water either before or shortly after weeds appear above the paddy surface. The metsulfuron-methyl in the combination gives DPX-70248 a broad application window similar to metsulfuron-methyl. Bensulfuron-methyl gives DPX-70248 excellent control of a variety of submerged broadleaves and sedges (Table 1). Depending on the cultural conditions, treatment timing, and the weed spectrum that must be controlled, ratios other than 5:1 may be considered.

TABLE 1

Important rice weeds controlled by bensulfuron-methyl, (50 gai/ha) metsulfuron-methyl, (4 gai/ha) and DPX-70248 (20 gai/ha). The level of control may vary depending on leaf stage, water level, and growing conditions at the time of treatment.

Scientific Name	Habit	Ben-sulfuron	Met-sulfuron	DPX-70248
Aeschynomene spp.	A-B*	M**	E	E
Alisma canaliculatum	P-B	E	M	M-
Alisma plantago	P-B	E	M	M-E
Alternanthera philoxeroides	P-B	M-E	E	E
Amaranthus dubius	A-B	M	E	E
Butomous umbellatus	P-B	E		
Cassia obtusifolia	A-B	M	E	E
Corchous spp.	A-B	P	E	E
Cynodon dactylon	P-G	P	P	P
Cyperus difformis	A-S	E	P-M	E-M
Cyperus iria	A-S	E	P-M	E
Cyperus serotinus	P-S	E		
Echinochloa spp.	A-G	P	P	P
Eclipta alba	A-B	E	E	E
Eleocharis acicularis	P-S	E		
Euphorbia spp.	A-B	M	E	E
Fimbristylis littoralis	A/P-S	E	E	E
Fimbristylis miliacea	A/P-S	E	M	E
Gratiola juncea	A-B	E		
Heteranthera limosa	A-B	M	E	
Heteranthera reniformis	A-B	M	E	E
Ipomea sp.	A/P-P	M	E	E
Ludwigia linifolia	A/P-B	M	E	E
Ludwigia repens	P-B	E	M	-
Ludwigia adscendens	P-B	E	E	E
Ludwigia octovaluvis	A/P-B	E	E	E
Limnocharis flava	P-S	E	E	E
Lindernia procumbes	A-B	E	E	E
Marsilea crenata	P-B	M-E	E	E
Monochoria vaginalis	A-B	E	E	E
Potamogeton distinctus	P-B	E	M	E
Rotala indica	A-B	E	E	E
Sagitaria pygmaea	P-S	E	E	E
Sagitaria guyanensis	P-S	E	E	E
Sagitaria trifolia	P-S	E	E	E
Scirpus juncoides	A/P-S	E	P	M
Scirpus maritimus	P-S	M	P	P
Scirpus mucronatus	A/P-S	E	P	M

TABLE 1

Scientific Name	Habit	Ben-sulfuron	Met-sulfuron	DPX-70248
Scirpus nipponicus	P-S	M	P	P
Scirpus supinus	A-S	E	P	M
Sesbania exaltata	A-B	M	E	E
Sphenoclea zeylanica	A-B	E	E	E

* A - Annual   ** E - Excellent control
  P - Perennial   M - moderate control
  B - Broadleaf   P - poor control
  S - Sedge   Blank - insufficient data
  G - Grass

     The benefits realized from the use of DPX-70248 can be demonstrated from two representative field trials in Thailand (Tables 2). In these tests bensulfuron-methyl at 20 gai/ha did not give satisfactory control of *S. zeylanica* and was weak on *M. crenata*. Metsulfuron-methyl at 4 gai/ha gave excellent control of M. crenata but was also weak on *S. zeylanica*. (The rates of 20 and 4 gai/ha were selected for these treatments because they are similar to their component part rate at the 20 gai/ha application rate of DPX-70248.) The 20 gai/ha application of DPX-70248 resulted in acceptable control of all weed species in the test. Weed control dropped some, especially for *S. zeylanica*, with lower DPX-70248 rates. However, in both tests the yield of all DPX-70248 treatments was statistically higher than for bensulfuron-methyl alone. Excellent weed control would have been expected with bensulfuron-methyl alone in these tests if the application rate had been 30 to 40 gai/ha. However, using DPX-70248 makes it unnecessary to use higher rates of bensulfuron-methyl.

     Because of the spectrum of activity and the ability to control weeds by direct post contact, metsulfuron-methyl can be used in a manner similar to 2,4-D. Many tests have been performed in Southeast Asia and Central and South American countries to evaluate this type of use. Since bensulfuron-methyl is the 'base' ingredient of DPX-70248 it is important to consider whether 2,4-D or metsulfuron-methyl would be the best partner in the combination. Data from many tests suggest that metsulfuron-methyl is generally the better choice. Again as an example, data from a test in Indonesia is presented in Table 3. In this test the rate of bensulfuron-methyl is held constant and either 2.4 gai/ha of metsulfuron-methyl (DPX-70248) or 500 gai/ha of 2,4-D is added to bensulfuron-methyl. As would be expected, the weed control with bensulfuron-methyl alone at 12.6 gai/ha was very poor. The combination with 2,4-D at 500 gai/ha gave some improvement in activity, but the treatment was still unacceptable. The combination with 2.4 gai/ha of metsulfuron-methyl in the DPX-70248 treatment resulted in perfect weed control and the highest yielding treatment. Using higher rates of bensulfuron methyl and/or 2,4-D would have resulted in better weed control for these two treatments.

TABLE 2

DPX-70248 compared to metsulfuron-methyl and bensulfuron-methyl in a field trial near Muang, Thailand on a clay soil type pH 5.6 and 4.4% organic matter. The application was made 10 days after direct seeding and evaluated 6 weeks after treatment. No crop injury was observed with any treatments.

Treatment	Rate (gai/ha)	Weed Species Code* Sz	Mc	Yield (kg/15M^2)
DPX-70248	10	80	92	5.8 abcde**
	15	82	97	6.3 abc
	20	95	98	6.1 abcd
metsulfuron-methyl	4	88	100	5.5 bcde
bensulfuron-methyl	20	78	92	5.4 cde
UNTREATED		0	0	4.1 f

* Sz - *Sphenoclea zeylanica*
  Mc - *Marsilea crenata*
** Yields with the same letter are not significantly different
   at the 95% confidence level.

TABLE 3

DPX-70248 compared to bensulfuron-methyl and bensulfuron-methyl+2,4-D amine for the control of major weeds in transplanted rice in Indonesia. The applications were made 10 days after transplanting and evaluated 6 weeks later.

Treatment	Rate (gai/ha)	Mc	Weed Species Code* Mv	Lf	Ci	Fl	Yield (kg/ha)
DPX-70248	15.0	100	100	100	100	100	6300**
bensulfuron-methyl	12.6	45	75	30	65	40	5692
bensulfuron-methyl plus 2,4-D	12.6+ 500	50	95	40	70	50	5468
UNTREATED		0	0	0	0	0	4542

* Mc - *Marsilea crenata*
  Mv - *Monochoria vaginalis*
  Lf - *Limnocharis flava*
  Ci - *Cyperus iria*
  Fl - *Fimbristylis littoralis*
** LSD 0.05 = 289, LSD 0.01 = 480

The question of using metsulfuron-methyl alone as a direct replacement for 2,4-D has also been considered in several field trials. The results of three trials in Thailand are summarized in Table 4. In these test metsulfuron-methyl alone at 4 gai/ha was compared to 1500 gai/ha of 2,4-D and with DPX-70248 at 24 gai/ha. Metsulfuron-methyl did not control *C. iria* or *F. miliacea* but did control *J. linifolia* and *S. zeylanica*. Just the opposite was found with 2,4-D with acceptable control of the first two species but marginal and unacceptable control of the last two.

These results show that the weed spectrum controlled by each compound should be taken into consideration before a choice is made. There are species that metsulfuron-methyl and 2,4-D will control equally well. Depending on the weed spectrum to be controlled, there will be circumstances where one product will perform better than the other. Each compound has its special strengths and weaknesses. When there is a mixture of broadleaf and sedge weeds, as in the tests reported in Table 4, metsulfuron-methyl can be tank mixed with 2,4-D with good results. Where expected weed control is similar, metsulfuron-methyl offers several advantages over 2,4-D. The much lower application rate , flexibility in water management programs, and the lack of volatility problems make metsulfuron-methyl a better choice. In these tests, as with those discussed above, DPX-70248 gave superior performance with excellent control of all weed species in the test.

TABLE 4

Comparison of DPX-70248, metsulfuron-methyl, and 2,4-D amine in field trials at three locations in Thailand. The values are the means of the three locations. All three locations had a clay soil texture with 1.3-1.5 % organic matter and pH 4.5-5.8. Evaluations were taken 40-43 days after treatment.

Treatment	Rate (gai/ha)	Weed Species Code*				Yield (T/ha)
		Ci	Fm	Ll	Sz	
DPX-70248	24	96	97	100	100	4.9
Metsulfuron-methyl	4	3	0	100	99	4.5
2,4-D	1500	93	93	88	58	4.1
Untreated Check	0	0	0	0	0	3.5

* Ci - *Cyperus iria*   Ll - *Jussiaea linifolia*
  Fm - *Fimbristylis miliacea*   Sz - *Sphenoclea zeylanica*

Grasses, especially Echinochloa species, are generally poorly controlled by bensulfuron-methyl. High rates of bensulfuron-methyl (50-75 gai/ha) are required to give a reasonable level of suppression, complete control is seldom realized. In areas such as Japan, Korea, Spain, Portugal, Italy and the U.S.A., the level of control even at high rates is generally considered insufficient. Similarly, metsulfuron-methyl and DPX-70248 will not give control of barnyardgrass.

In Japan and Korea, bensulfuron-methyl has been premixed with a range of grass herbicides to give commercial products with broad spectrum, one-shot control of grasses, broadleaves, and sedges. Premix products with bensulfuron-methyl and molinate have become available in Spain in 1988 and Portugal in 1989. Recent trials with DPX-70248 indicate similar strategies could be successful with it as well. Field trials in Egypt last summer showed that DPX-70248 plus molinate was as good as bensulfuron-methyl plus molinate for *Echinochloa crus-galli* control. Further trials are required to identify appropriate timing and water management techniques for tank mix combinations with popular grass control materials before specific recommendations can be made.

Toxicology

Bensulfuron-methyl and metsulfuron-methyl belong to the sulfonylurea class of chemistry. The mode of action of these compounds is the inhibition of acetolactate synthase (ALS) enzyme [13]. ALS is a key enzyme in the branched-chain amino acid biosynthetic pathway of bacteria, fungi, and higher plants [14]. Sulfonylureas are very active in sensitive plant species leading to the need for very low application rates. Since animals do not have the ALS enzyme and the activity of sulfonylureas on other metabolic systems is generally low the mammalian toxicity of sulfonylureas is usually low. As can be seen from Table 5 both bensulfuron-methyl and metsulfuron-methyl have excellent toxicological profiles.

TABLE 5

Comparison of the toxicological profile of bensulfuron-methyl and metsulfuron-methyl.

Administration	Test Organism	Bensulfuron-methyl	Metsulfuron-methyl
Acute Oral	LD50 Rat	> 5000 mg/kg	>5000 mg/kg
Acute Dermal	LD50 Rabbit	>2000 mg/kg	>2000 mg/kg
Chronic Feeding	Rat 2-year	NOEL 750 ppm not oncogenic	NOEL 500 ppm not oncogenic
	Mouse 18-Mo/2-year	NOEL 2500 ppm not oncogenic	NOEL 5000 ppm not oncogenic
Reproduction	Rat 2-generations	NOEL 250 ppm	NOEL 500 ppm
Mutagenicity	Ames test	Negative	Negative
	Unscheduled DNA synthesis	Negative	Negative
	Chinese hamster ovary	Negative	Negative
	Bone marrow chromosome studies	Negative	Negative
Teratogenicity	Rat and Rabbit	Negative	Negative
Aquatics	LC50 (48 hrs.) Carp	>1000 ppm	>1000 ppm
	LC50 (48 hrs.) Daphnia	>100 ppm	>150 ppm
	LC50 (96 hrs.) Bluegill Sunfish	>150 ppm	>150 ppm
	LC50 (96 hrs.) Rainbow Trout	>150 ppm	>150 ppm

TABLE 5

Administration	Test Organism	Bensulfuron-methyl	Metsulfuron-methyl
Wildlife	Oral LD50 Mallard duck	>2510 mg/kg	>2510 mg/kg
	Dietary LC50 Mallard duck	>5620 ppm	>5620 ppm
	Bobwhite quail	>5620 ppm	>5620 ppm
Eye Effects	Rabbit	Slight irritant	Moderate irritant

Since these compounds are applied to a water environment their effects on fish and ducks are of particular interest. Again from Table 5, the toxicity of bensulfuron-methyl and metsulfuron-methyl to carp, daphnia, bluegill sunfish, trout, and ducks is low. The highest labelled application rate for bensulfuron-methyl would result in a paddy water concentration of less than 0.3 ppm with 3 cm of water in the field. The concentration of metsulfuron-methyl applied at 4 gai/ha would be less than 0.02 ppm. These paddy water levels are far below the highest no-effect levels reported in Table 5, making the margin of safety with these compounds extremely favorable.

Acute oral toxicity studies in mice with DPX-70248 compared to bensulfuron-methyl and metsulfuron-methyl have shown neither toxic symptoms nor mortality at the 5000 mg/kg dosage level. This suggests there will be no difference in the toxicology between DPX-70248 and that of its component parts bensulfuron-methyl and metsulfuron-methyl.

CONCLUSION

Two new herbicides have been identified for use in a variety of rice cultural practices. Metsulfuron-methyl is a non volatile, very low use rate herbicide (2-10 gai/ha) which can be used to control several important broadleaf weeds. Applications can be made to the paddy water or by direct foliar contact after weeds emerge from the paddy water. Metsulfuron-methyl can be used in many different rice cultural conditions, including upland rice, for the control of broadleaf weeds.

DPX-70248 has been developed for use in paddy rice when conditions require control of a broader range of weeds than can be achieved by metsulfuron-methyl alone. DPX-70248 at 10-20 gai/ha can be used for either early or late post applications for the control of a wide range of broadleaf and sedge weeds.

Toxicological studies have shown that both metsulfuron-methyl and DPX-70248 are environmentally safe with low toxicity to mammals, birds, and fish.

REFERENCES

1. Takeda, S., Yuyama, T., Ackerson, R.C., Weigel, R.C., Sauers, R.F., Neal, W., Gibian, D.G., and Tseng, P.K., Herbicidal activities and selectivity of a new rice herbicide DPX-F5384. Weed Research, Japan, 1985, 30, 284-2889.

REFERENCES

2. Bernasor, P.C. and De Datta, S.K., Chemical and cultural control of bulrush *Scirpus-maritimus* and annual weeds in lowland rice *Oryza-sativa*. Weed Research, 1986, 26, 233-244.

3. Pacheco, J.L. and Pope, C.L., LONDAX - a new, highly flexible, broad spectrum herbicide for California rice. Western Soc. of Weed Sci. Proc., 1986, 39, 147-157.

4. Rapparini, G., Weed control in rice fields: new departures. Informators Agrario, 1988, 44, 93-100.

5. Clampet, W.S., LONDAX - a new herbicide for aerial sown rice. Farmers Newsletter (Australia), 1988, 132, 21.

6. Bozarth, G.A., Ackerson, R.C., and Rabalais, M.J., Bensulfuron methyl: a new herbicide for southern rice production. Proc. Southern Weed Sci. Soc., 1988, 41, 74.

7. Chiang, Y.J., and Leu, L.S., Effect of application timing and residual period of LONDAX on main paddy weeds in Taiwan. Proc. Eleventh Asian-Pacific Weed Sci. Soc. conf., 1987, CO-63, 223-231.

8. Hill, J.E., Pacheco, J.L., Brandon, B.W., Holzer, M.J. and Bayer, D.E., The response of weeds in water-seeded rice to the rate and timing of bensulfuron-methyl. Research progress report - Western Soc. of Weed Sci., 1987, pp. 230-231.

9. Yuyama, T., Ackerson, R.C., Takeda, S., and Watanabe, Y., Soil and water relationships on the behavior of bensulfuron methyl (DPX-F5384) under the paddy field condition. Weed Research. Japan, 1987, 32, 282-291.

10. Jang, I.S., Moon, Y.H., and Ryang, H.S., Adsorption, movement and decomposition of new herbicide bensulfuron-methyl in soils. Korean J. of Weed Sci., 1987, 7, 165-170.

11. Takeda, S., Herbicidal properties, mode of action and selectivity mechanism of DPX-F5384. Chemical Regulation of Plants, 1987, 22, 47-48.

12. Takeda,, S., Erbes, D.L., Sweetser, P.S., Hay, J.V., and Yuyama, T., Mode of herbicidal and selective action of DPX-F5384 between rice and weeds. Weed Research, 1986, 31, 157-163.

13. Takeda, S., The primary site of action and selectivity mechanism of sulfonylurea herbicides. J. Pestic. Sci., 1987, 12, 759-768.

14. Beyer, E.M. Jr., Duffy, M.J., Hay, J.V. and Schlueter, D.D., Sulfonylurea herbicides. In Herbicides: chemistry, degradation and mode of action, ed. P.C. Kearney and D.D. Kaufman, Marcel Dekker, Inc., 1988, pp. 117-189.

QUINCLORAC, A NEW HERBICIDE FOR THE USE IN VARIOUS PRODUCTION SYSTEMS IN SEEDED RICE

Dr. Ulrich Kießling, Matthias Pfenning
BASF Aktiengesellschaft, Agricultural Research Station
D-6703 Limburgerhof, FR Germany

ABSTRACT

Quinclorac (ISO-proposed, trade name FACET), a new chinoline-carboxylic-acid compound of BASF, has been developed as a rice herbicide with strong action on Echinochloa species. Due to its good crop selectivity and its wide application window, quinclorac can be integrated into many rice cultivation systems.

In drilled dry seeded rice quinclorac can be used in pre- as well as in post-emergence applications, whereas in water seeded rice or in dry broadcast seeded rice applications should not be made while radicles are exposed or until the rice plant has not reached the 2-leaf-stage.

With rates of 0.25 - 0.75 kg a.i./ha quinclorac controls Echinochloa species and other weeds like Brachiaria platyphylla, Sesbania exaltata, Aeschynomene spp. or Ipomoea spp.

The most consistent performance is achieved when applications are made on moist soil or in shallow-flooded fields. Under these conditions Echinochloa spp. can be consistently controlled until the start of tillering. There was no difference whether quinclorac was applied by ground or air equipment.

Tank mixes of quinclorac with carbamates or organo-phosphates had no negative influence on crop tolerance.

INTRODUCTION

Rice is one of the most important food crops in the world agriculture and is practically planted on all continents. There exist different production systems, mainly devided due to the planting-system in

transplanted and seeded rice

and, due to water availability, in

natural rainfed rice and irrigated or flooded rice.

Cultivation systems differ from each other mainly in field preparation, water management and sowing technique.

In almost all of these rice cultivation systems Echinochloa is the most important graminaceous weed. The ecological requirements of Echinochloa spp. and rice are very similar; this leads to high competition and as a result to potentially great yield losses. Studies from California have shown that 30 Echinochloa plants/sqm can reduce the rice yields by more than 50 %, depending on ecotype of the weed, rice variety and growing conditions.

Deep flooding has traditionally been used as one method for grass control. Low water levels lead to higher competition between the crop and Echinochloa compared to deeper water levels. However, in areas with either natural water shortage, special regulations, administration-restrictions or because of certain varietal requirements like the new semi-dwarf varieties, deep flooding cannot be used to control specific weeds. Therefore new herbicides, which give consistent weed control also under shallow-flooding-conditions, are required.

Many herbicides for use in seeded rice are on the market. However,
in spite of the impressive list of products for Echinochloa control there
still remain some niches, which must be filled -

- good selectivity under all growing conditions

- broad application window

- less dependence on water management

Recently BASF Aktiengesellschaft has developed quinclorac, a new
rice herbicide with strong action on Echinochloa species and some other
weeds.

This paper describes the suitability of Quinclorac for being
integrated into different rice cultivation systems under the aspect of
water management and sowing techniques practiced in Europe, the United
States and Latin America.

MATERIALS AND METHODS

All results presented are from field trials with plot sizes ranging from
10 - 25 sqm. All trials have been carried out in a randomized block
design with 3 replications. When applications were made into the flooded
field, plots were separated by plastic sheets or levees.

Quinclorac was applied as a 50 % WP or an 150 g/l SC formulation.
All products used in this study were sprayed with a pneumatic knapsack-
sprayer with a pressure of 2 - 4 bar and a spray volume between 100 and
400 l/ha.

Selectivity and herbicidal efficacy are expressed as % crop injury and % weed control compared to the untreated check.

RESULTS

Crop Tolerance

The selectivity of quinclorac was tested in numerous trials all over the world. In general, quinclorac was well tolerated until the end of tillering of the rice plant in dry drill seeded rice. Some slight temporary crop injury may occur in pre-emergence application when the seed placement was uneven and the seeds or the radicles were exposed on the soil surface.

In broadcast seeded rice as well as in water seeded rice applications before the 2 leaf stage of the crop should be avoided (Table 1). No significant difference in varietal response to Quinclorac was observed.

TABLE 1
Selectivity of quinclorac in seeded rice

Application timing	Dry seeded rice		Water seeded rice (pre-germinated)
	drilled	broadcast	
Pre-emergence			
- before seeding	+(+)	-	-
- after seeding	+(+)	-	-
Post-emergence			
- 1 leaf stage	+ +	+(+)	+
- 2-3 leaf stage	+ +	+ +	+ +
- tillering	+ +	+ +	+ +
- after tillering	+ +	+ +	+ +

+ + no injury
+ slight injury
- moderate - heavy injury

Quinclorac can also be applied in tankmix with insecticides. Post-emergence applications with organo-phosphates and carbamates did not harm the crop at recommended application rates (Table 2).

TABLE 2
Selectivity of quinclorac in tankmix applications with insecticides in seeded rice (Brazil)

	% injury	
	7	21 DAT
QUINCLORAC		
+ Monocrotophos	0	0
+ Dimethoate	0	0
+ Chlorpyriphos	0	0
+ Endosulfan	0	0
+ Carbaryl	0	0
Standard treatment		
+ Carbaryl	43	5

Rate of quinclorac: 0.375 kg a.i./ha
Application: 3 leaf stage up to 3 tillers of rice
Temperature at application: 32 C
Insecticides have been used at recommended rates

Efficacy

Weed spectrum (Table 3). In field tests and demonstration trials quinclorac has shown strong consistent activity on tested Echinochloa species. Besides Echinochloa quinclorac has also good activity on other economically important rice weeds, such as Aeschynomene spp., Alternanthera philoxeroides, Ammania multiflora, Brachiaria platyphylla, Elatine triandra, Ipomoea spp. and Sesbania spp. To complete the weed spectrum on other weeds and Cyperaceae combinations with other herbicides like bentazon, propanil, sulfonylureas, benthiocarb are possible.

TABLE 3
Weed spectrum of quinclorac in seeded rice

Susceptible weeds	Moderately susceptible
Echinochloa crus-galli v. caudata	Aeschynomene rudis
Echinochloa crus-galli v. formosensis	Aeschynomene virginica
Echinochloa crus pavonis	Alternanthera philoxeroides
Echinochloa crus-galli oryzicola	Ammania multiflora
Echinochloa crus-galli v. zelayensis	Brachiaria platyphylla
Echinochloa colonum	Elatine triandra
Echinochloa oryzoides	Ipomoea spp.
Echinochloa crus-galli	Sesbania spp.

Control of Echinochloa spp. Echinochloa crus-galli can be controlled effectively in pre plant incorporated, pre-emergence and post-emergence applications. The rate to be used in pre-emergence application depends on the soil type and the soil moisture content.

In light soils lower rates of quinclorac can be used compared to heavy soils (Table 4). In postemergence applications quinclorac is consistently effective from the first leaf stage up to mid-tillering (Table 5). Application rates to be used in the various countries for Echinochloa control are presented in Table 6.

TABLE 4
Influence of soil type on the activity of quinclorac in a pre-emergence application (USA)

Product	kg a.i./ha	soil type				
		loamy sand	sandy loam	silty loam	loamy clay	clay
Quinclorac	0.2	90	98	98	48	65
	0.25	89	97	100	78	65
	0.3	92	99	99	90	85
	0.35	92	97	95	93	88
	0.4	95	99	97	97	93

Numbers in % control Echinochloa

TABLE 5
Efficacy of quinclorac on Echinochloa spp. dependent on the growth stage
(Spain, Italy)

Growth stage of Echinochloa	% control of Echinochloa
2 - 3 leaves	96
3 - 5 leaves	98
Early to mid-tillering	96

Rate of quinclorac: 0.6 - 0.75 kg a.i./ha
Number of trials: 13

TABLE 6
Activity of quinclorac on Echinochloa in different rice growing areas
(pre vs. post-emergence)

Country/Region		Rate kg a.i./ha	% Control (Post/Pre)
Southern Europe	(31)	0.6	96/95
Latin America	(78)	0.37	97/90
USA	(98)	0.37	97/94

Post-emergence application: First leaf up to the start of tillering of
Echinochloa
() = number of trials

Water management (Table 7). Quinclorac is absorbed via leaves and roots.
These facts make the product relatively independent of the water
management. Quinclorac can be applied on dry, moist, drained and shallow-
flooded fields. Rainfall or a flush following a pre-emergence application
will improve the activity . Application into shallow-flooded fields gives
consistently better weed control compared to deep-flooding conditions.
For best control the water level should not exceed 5 cm.

TABLE 7
Influence of water management on the activity of quinclorac on
Echinochloa (Brazil)

	% Control
Dry soil	95
Moist soil	97
3 - 5 cm water level	96
~ 10 cm water level	85

Application rate: 0.375 kg a.i./ha
Timing: 3 leaves up to start of tillering

Influence on yield. All the quinclorac treatments resulted in an
significant increase in rice grain yield compared to the untreated
check. There was no significant difference in rice yield when compared to
the handweeded check (Tab. 8).

TABLE 8
Influence of quinclorac on yields in drilled seeded rice (Brazil)

	weedy check	hand-weeded check	standard treatment	quinclorac 0.5 kg a.i./ha
		yield (kg/ha)		
Pre-emergence	3360	6620	5586	6549
Postemergence	3052	6851	6691	6978

Weed infestation: Echinochloa crus-galli
                  Aeschynomene rudis
Number of trials: 5

Crop Rotation / Drift. Plant back studies have shown that crops like
barley, wheat, cotton, corn, soybeans and sugarbeet are not influenced
negatively if they are planted in a crop rotation after rice which has
been treated with quinclorac. Other crops are under investigation.

Quinclorac is not selective in tomatoes, safflower, alfalfa, cotton,
melon, artichoke, tabacco, carrots, spinach, eggplant, cucumbers,
sunflower, peas, lettuce and peanuts. Therefore, the physical drift of

the spray solution to these crops or irrigation of these crops with treated water has to be avoided.

CONCLUSIONS

Quinclorac, the new rice herbicide of BASF Aktiengesellschaft can be safely used in broadcast dry seeded and water seeded rice, when application is made post-emergent from the 1 - 2 leaf stage onwards. In drill seeded rice also pre-emergence applications are possible. However, good seeding technique with even sowing depth are necessary. In general, applications with quinclorac should be avoided as long as radicles are exposed.

Quinclorac has a very strong activity on Echinochloa species in pre-emergence as well as in postemergence applications until mid-tillering. The activity is relatively independent of the water management and means that quinclorac can be applied under dry and flooded conditions. In applications on dry fields care has to be taken that flooding or irrigation should occur within 7 days after treatment.

For optimum control the water level should not exceed 5 cm and field outlets should be closed at least for 4 - 5 days after application.

Owing to its good selectivity and the wide application window quinclorac can be incorporated into many of the seeded rice growing systems around the world. Quinclorac needs low water levels for optimum control. Therefore it also fits well to rice growing areas where, water use or supply is for some reason restricted, shallow-flooding is required for economical reasons or, for example, when new semi-dwarf-varieties are planted.

To achieve a broad spectrum weed control combinations with other herbicides are recommended. Because of its flexibility regarding selectivity in rice, the growth stage of Echinochloa at application and the water management quinclorac can be mixed with a wide range of current commercially available herbicides and may become an important herbicide in the control of the most troublesome grass weed in rice.

A SOLUTION TO THE MAJOR WEED PROBLEMS IN WET-SOWN RICE: EXPERIENCES WITH PRETILACHLOR/FENCLORIM IN SOUTH-EAST ASIA

JEAN-LOUIS ALLARD and ANDREAS ZOSCHKE
CIBA-GEIGY Limited, Agricultural Division, Research & Development,
CH-4002 Basle, Switzerland

ABSTRACT

Wet-sown rice has become increasingly important in South-East Asia. Among other problems, weed control is a critical element in tropical wet-sown rice. Good crop tolerance and control of grasses are considered important characteristics to be fulfilled for a suitable herbicide in wet-sown rice.

Pretilachlor/fenclorim has been discovered and developed by CIBA-GEIGY Ltd. for broad spectrum weed control in wet-sown rice; it contains pretilachlor as the active ingredient and fenclorim as a safening agent. Field trials carried out during 1983-88 in wet-sown rice in Malaysia, the Philippines, Sri Lanka and Thailand are summarized with the aim of establishing an overall picture about the performance of the product in South-East Asia.

The results show that the product is safe to the wet-sown rice crop and provides excellent control of the most important annual weed species, including *Echinochloa*, *Leptochloa*, *Cyperus*, *Fimbristylis*, *Scirpus*, *Monochoria* and *Ludwigia*. Positive effects on rice grain yield support its favourable herbicidal efficacy. It is concluded that pretilachlor/fenclorim is very suitable for solving the major weed problems in wet-sown rice in South-East Asia.

INTRODUCTION

Development of wet-sown rice in South-East Asia

In most tropical Asian countries, cultivation of direct seeded rice was not widely practised before the late seventies due to difficulties in weed control, rat and bird damage, snail infestations, and poor irrigation water control [1, 2]. Since then, however, wet-sown rice has also gained importance in several countries of South-East Asia; e.g. Malaysia, the Philippines, Sri Lanka, and Thailand (tab. 1). Major reasons for this expansion are increasing labour costs and labour shortage, improvement and extension of irrigation systems [4, 7], and also the development of agrochemicals, especially for weed control.

TABLE 1
Area of wet-sown rice in South-East Asia.

country	total rice area 1985-87 (1'000 ha) based on [3]	wet-sown rice		
		area (1'000 ha)	year	source
Indonesia	9'889	negligible	1986	[4]
Malaysia	647	101	1988	[5]
Philippines	3'426	≃ 270	1986	calculated from data of [1]
Sri Lanka	851	440	1983	calculated from data of [6]
		650	1983	[4]
Thailand	9'378	> 800	1985	[1]
		1'400	1986	[4]

Problems in tropical wet-sown rice

The yield potential of a well managed wet-sown rice crop is equal to that of transplanted rice [8]; but for the farmer a successful switch to wet-sown rice means the adoption of new crop management practices: use of varieties resistant to lodging, better seed quality, good soil preparation and levelling, exact water management as well as more intense fertilization are required [1].

Weed infestations are generally more severe in wet-sown rice as compared to transplanted rice [9, 10]. The weed species predominant in wet-sown rice in South-East Asia are listed in fig. 1. Heavy infestations of annual grasses (mainly *Echinochloa spp.*) and sedges are common in wet-sown rice [1, 10], because the shallow water level at and after sowing (necessary for good germination and rooting of the rice seedlings) also favours weed germination and development. In addition to *Echinochloa spp., Leptochloa* is another grass species becoming increasingly frequent in wet-sown rice in Thailand and Malaysia [5, 9, 17].

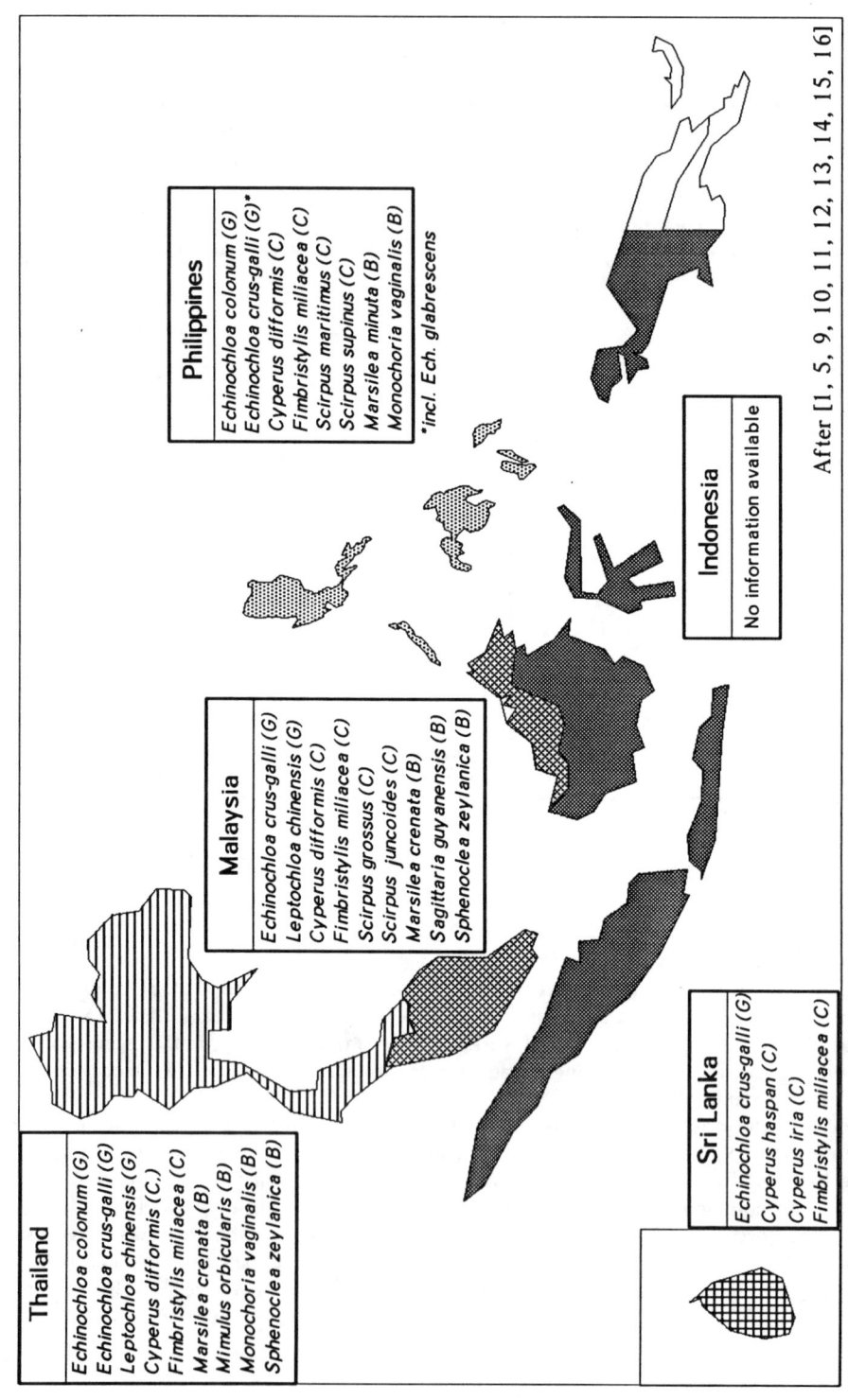

Thailand

Echinochloa colonum (G)
Echinochloa crus-galli (G)
Leptochloa chinensis (G)
Cyperus difformis (C.)
Fimbristylis milliacea (C)
Marsilea crenata (B)
Mimulus orbicularis (B)
Monochoria vaginalis (B)
Sphenoclea zeylanica (B)

Malaysia

Echinochloa crus-galli (G)
Leptochloa chinensis (G)
Cyperus difformis (C)
Fimbristylis milliacea (C)
Scirpus grossus (C)
Scirpus juncoides (C)
Marsilea crenata (B)
Sagittaria guyanensis (B)
Sphenoclea zeylanica (B)

Philippines

Echinochloa colonum (G)
Echinochloa crus-galli (G)*
Cyperus difformis (C)
Fimbristylis milliacea (C)
Scirpus maritimus (C)
Scirpus supinus (C)
Marsilea minuta (B)
Monochoria vaginalis (B)

*incl. Ech. glabrescens

Indonesia

No information available

Sri Lanka

Echinochloa crus-galli (G)
Cyperus haspan (C)
Cyperus iria (C)
Fimbristylis milliacea (C)

After [1, 5, 9, 10, 11, 12, 13, 14, 15, 16]

G= Gramineae (grasses), C= Cyperaceae (sedges), B= broadleaved weeds [all annuals except *Scirpus maritimus*]

Figure 1. Main weed species in wet-sown rice in South-East Asia

Weed control in wet-sown rice

Weeding is a problem if broadcast seeding is used [2]. Weed control in wet-sown rice is highly critical because handweeding is not as effective as in transplanted rice. Handweeders moving through wet-sown rice are likely to cause damage to some rice plants and also distinction between young grassy weeds and young rice is difficult [18]. All such problems have limited a broader spread of this production system in the region until recently [7].

In wet-sown rice, annual sedges and broadleaved weeds can be controlled to some extent with late-postemergent applications of hormonal herbicides, such as 2,4-D. In contrast to this, it is more difficult to control grasses sufficiently. Certain rice herbicides have a limited weed control spectrum (e.g. molinate has no activity against *Leptochloa*), or require a strict water management (e.g. propanil). Other herbicides, primarily developed for transplanted rice, have been introduced in wet-sown rice in South-East Asia. These compounds, such as acetanilides (butachlor, pretilachlor) have a broader weed control spectrum. However, their crop tolerance may be critical in wet-sown rice, especially when applied just after sowing [1, 12]. To overcome this problem, use recommendations (especially use rate and application timing) have to be adapted, which, in turn, affects their herbicidal performance. Therefore, there is a need to develop safer herbicides providing an optimal level of weed control for use in wet-sown rice [19].

The herbicide SOFIT® (pretilachlor/fenclorim, ratio 3:1) has been specifically developed for weed control in rice nurseries [20] and wet-sown rice and it has been registered in several countries in South-East Asia. In the following, biological data obtained in wet-sown rice from a large number of field trials with this product in Malaysia, the Philippines, Sri Lanka, and Thailand are summarized. It is the aim of this paper to analyse the extent to which this herbicide can solve current major weed problems in this rice production system in South-East Asia.

MATERIALS AND METHODS

Pretilachlor/fenclorim contains the active ingredient pretilachlor plus a safening agent. This safener, fenclorim, protects the young rice plants from pretilachlor-damage caused if pretilachlor is applied without safener. The role of the safener has been described in earlier publications [21, 22]. The structures of the compounds, their physico-chemical properties and information about the mode of action and toxicity of both pretilachlor and fenclorim have been published elsewhere [23]. In the current paper, *pretilachlor/fenclorim is abbreviated to pretilachlor*.

Pretilachlor was field tested from 1983 to 1988 in a total of 51 trials in irrigated wet-sown rice (*Oryza sativa* L.) in South-East Asia. Trial sites were in regions in which cultivation of wet-sown rice is already common: the Philippines (Luzon, Visayas), Thailand (Central plains, Suphanburi province), Malaysia (Muda scheme), and Sri Lanka. In Indonesia, trials were carried out at CIBA-GEIGY's Agricultural Research and Development Station in Cikampek (West Java). Each trial included 3-4 replications arranged in completely randomized blocks. Plot size varied from 14 to 20 sqm and plots were separated by earth dikes or metal sheets.

Pre-germinated (soaked and incubated) rice was sown on puddled soil under saturated water conditions. Fertilization and plant protection measures were carried out according to local recommendations. Further details on agronomic parameters such as seeding rate and rice varieties are summarized in tab. 2.

Pretilachlor was always applied at a *use rate of 400-450 g a.i./ha* with a spray volume of 450-500 l/ha (equipped with a spray boom or a single flat-fan or conical nozzle). *Application timing was always 3-4 DAS [days after sowing].* According to local practice in wet-sown rice, the water was progressively re-introduced a few days after the herbicide treatment.

The effects of the herbicide treatments were evaluated several times during the rice growth period [DAA = days after application] relative to the untreated check as well as small check strips arranged between the treated plots. The criteria investigated were 'crop tolerance' (measured as % phytotoxicity), 'weed control' (measured as % per species) and 'weed cover' (measured as % soil surface covered by weeds, excluding the area covered by the rice crop). The evaluations were based on the reduction of biomass of both the rice crop and the weeds present. As a reference compound, butachlor (applied 6-7 DAS, use rate: 600-1050 g a.i./ha) was included in several trials (fig. 6).

Rice grain yields were taken from several trials and comparisons made following adjustment to a 14% moisture level.

TABLE 2

Details on rice trials carried out with pretilachlor/fenclorim in South-East Asia.

country	year	nb of trials	rice variety	sowing rate (kg/ha)
Indonesia	1983-86	19	IR-36	60-80
Malaysia	1983, 1988	4	IR-42, MR-84	50
Philippines	1984-86	13	IR-36, IR-54, IR-64	150
Sri Lanka	1983	2	BG-34.8	60-90
Thailand	1983-88	13	RD-7, RD-23	75

RESULTS

Crop tolerance

In general, an excellent crop tolerance to pretilachlor was demonstrated (fig. 2). Not a single case of unacceptable crop tolerance was observed with the product. A slight growth retardation was observed in a few trials in Indonesia. However, in most cases, such inhibition was only transient. In all the trials carried out in Malaysia, the Philippines, Sri Lanka, and Thailand, pretilachlor was perfectly tolerated by the rice crop.

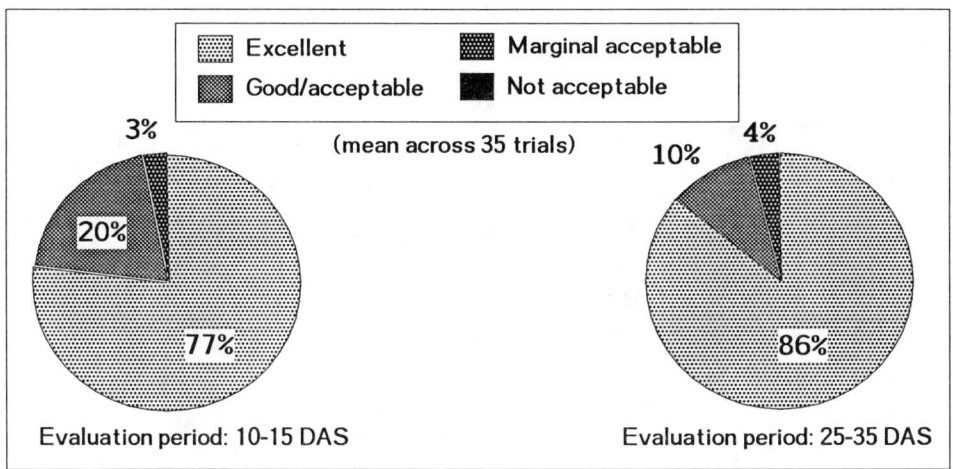

Figure 2. Crop tolerance of pretilachlor/fenclorim in wet-sown rice in South-East Asia.

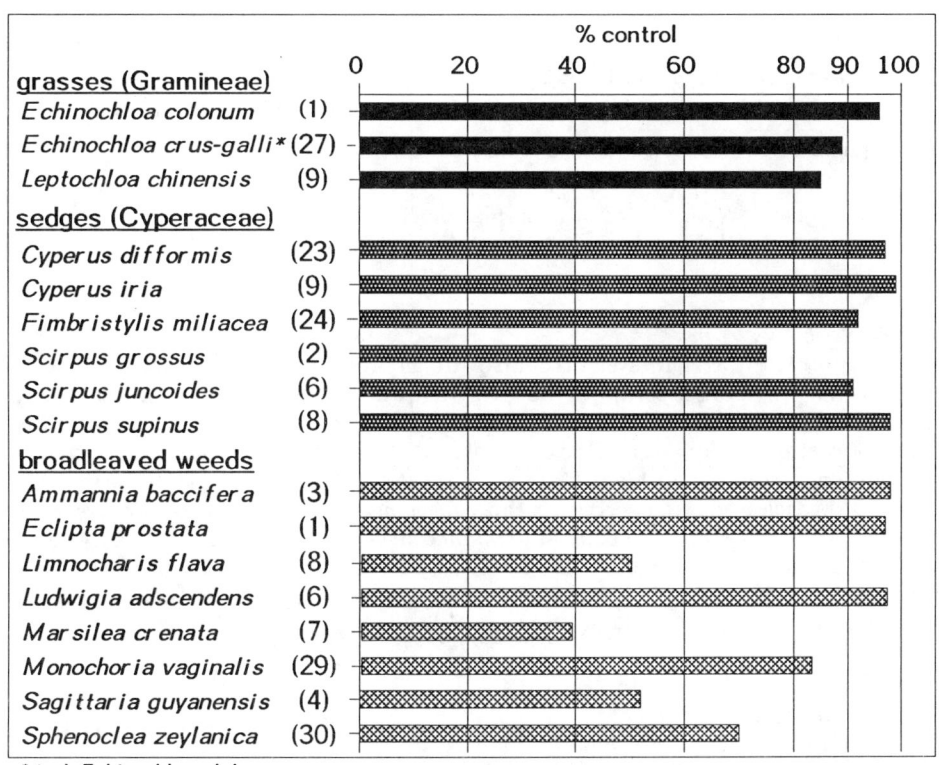

Figure 3. Level of weed control with pretilachlor/fenclorim in wet-sown
rice in South-East Asia.

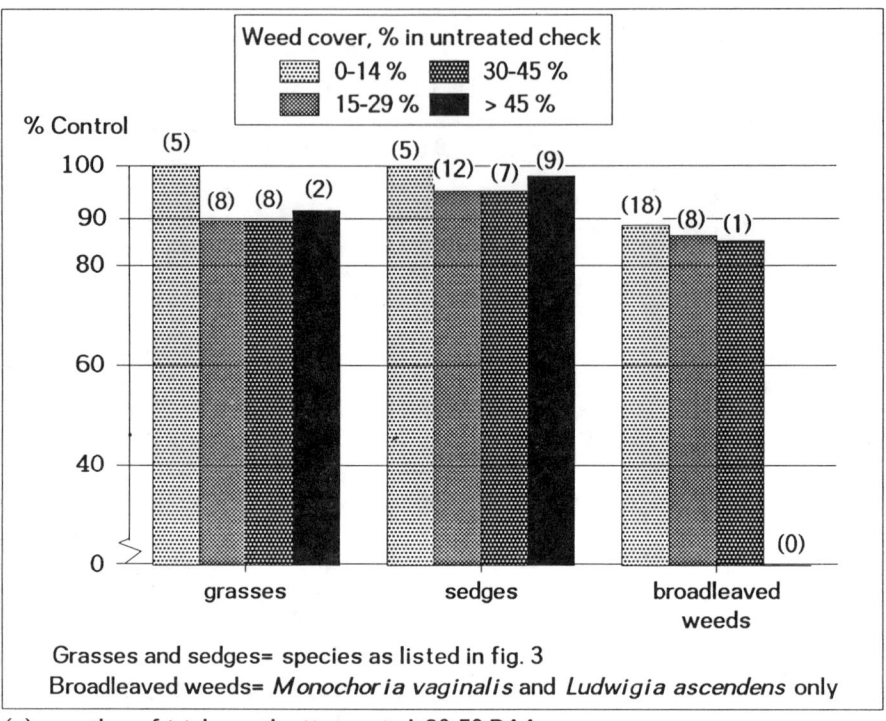

Figure 4. Weed control with pretilachlor/fenclorim in wet-sown rice
in South-East Asia at different degrees of weed infestation.

Weed control spectrum

Pretilachlor provided a high level of control of all weed species of grasses and sedges present in the trials for a period of 50 days following application (fig. 3). The level of control was consistent, so that under both high and low weed infestations a sufficient degree of weed control was achieved   (fig. 4). Besides *Echinochloa spp.*, another important grass weed, *Leptochloa chinensis*, was excellently controlled by the product.

Several broadleaved weed species, such as *Ammania* and *Eclipta prostata*, were very well controlled. On average, the control of *Monochoria* was also good. However, in a few situations, season long control of this weed species was not achieved. Other broadleaved weeds, such as *Limnocharis flava, Marsilea crenata, Sagittaria guyanensis*, and *Sphenoclea zeylanica*, were only partially controlled by the product.

Effect on rice grain yield

Depending on the degree of weed infestation, the treatment with pretilachlor increased the mean rice grain yields by 890 to 2570 kg/ha over those obtained from untreated check plots (tab. 3). The highest yield increases of up to 4.2 t/ha were observed in trials heavily infested with grasses and sedges (90-100% weed cover in untreated plots, compare fig. 5).

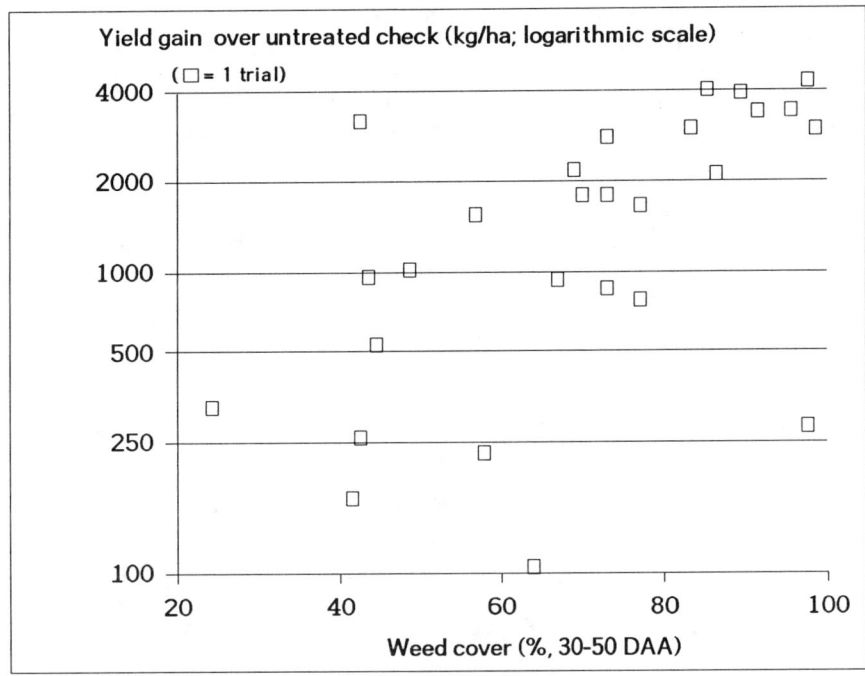

Figure 5. Increase of rice grain yield by pretilachlor/fenclorim over untreated check in relation to the degree of weed infestation in wet-sown rice in South-East Asia.

TABLE 3
Average yield increase with pretilachlor/fenclorim at different degrees of weed infestation in wet-sown rice in South-East Asia.

weed infestation (% cover at 30-50 DAA)	nb of trials	yield gain over untreated check (kg/ha)
0 - 24	-	-
25 - 49	7	890
50 - 74	9	1260
75 - 100	11	2570

Average yield in untreated check: 3600 kg/ha

Performance relative to the reference compound
When compared with the reference compound (butachlor) applied at 6-7 DAS (fig. 6), pretilachlor provided a better rice crop tolerance, a superior level of control against *Echinochloa* and sedges and also resulted in higher yield increases. In regard to the control of sedges and *Ludwigia*, the difference between both herbicides was less obvious.

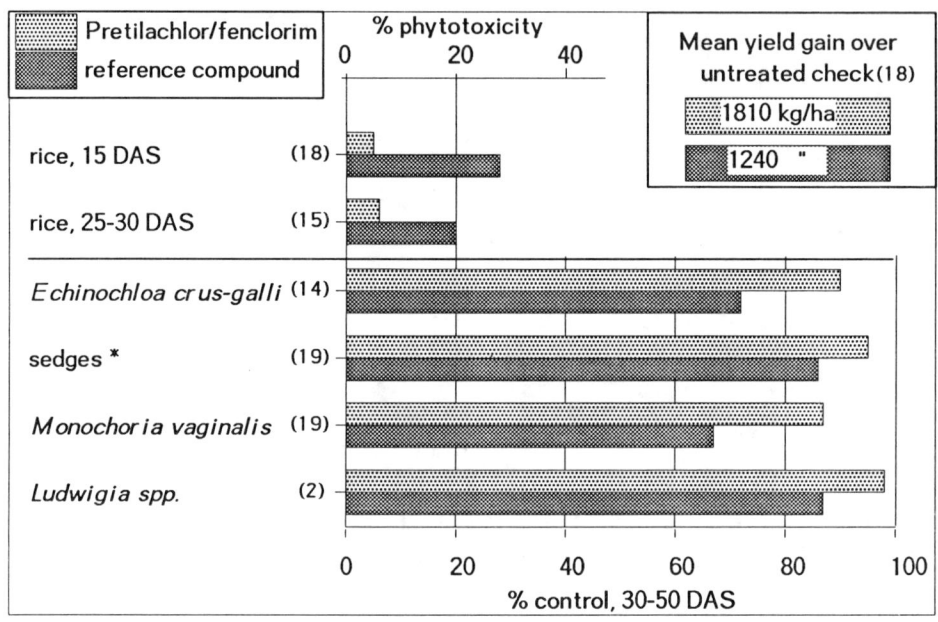

(n)= number of trials; * species as listed in fig. 3

Figure 6. Performance of pretilachlor/fenclorim as compared with with the reference compound in wet-sown rice in South-East Asia.

DISCUSSION AND CONCLUSIONS

The results presented in this paper show that crop tolerance of pretilachlor applied at 3-4 DAS is excellent. The application timing can be safely advanced to 0-1 DAS by using appropriate water management (shallow water level at and during a few days following application) and proper pre-germination of rice seeds [23, 24]. Throughout this range of application timings (from 0-4 DAS) pretilachlor provides an even and good control of susceptible weeds [19].

The product's weed control spectrum, including both *Echinochloa* and *Leptochloa spp.*, overlaps very well with the weed species encountered as problems in wet-sown rice in South-East Asia (as indicated in fig. 1). Applications in farmers' wet-sown rice fields in Malaysia [5], in Thailand [24] and in the Philippines [19] confirm this finding.

The early elimination of weeds, and the high weed control level attained, results in a substantial rice grain yield increase. This finding confirmed earlier results [24]. In addition, appreciable yield gains with pretilachlor were even observed at low or medium degrees of weed infestation (25-50% weed cover, fig. 5).

While it has no activity against perennial *Scirpus maritimus*, pretilachlor provides partial control of some broadleaved weeds such as *Limnocharis flava, Marsilea spp., Sagittaria guyanensis*, and *Sphenoclea zeylanica*. In order to also provide a solution for those situations in which heavy infestations of such broadleaved weeds occur in conjunction with grasses, a combination of pretilachlor with CGA 142'464 is currently under development [17].

When compared with the reference compound, which was applied at 6-7 DAS (fig. 6), pretilachlor clearly demonstrates a better crop tolerance. This result is in line with observations reporting variable or marginal tolerance of butachlor in wet-sown rice [1, 12]. The higher level of *Echinochloa* control obtained with pretilachlor (as compared to the reference compound) together with its broader weed control spectrum (as compared to other herbicides) including *Leptochloa* are further positive characteristics of the product providing an interesting alternative to the farmer planting wet-sown rice in South-East Asia.

Due to its excellent crop tolerance and its high level of control of the most important weed species, pretilachlor/fenclorim fulfils the most important features required for a herbicide in wet-sown rice. Therefore, it can be concluded that pretilachlor/fenclorim is a suitable product to support the further successful development of this rice production system in South-East Asia.

REFERENCES

1. Datta, S.K. de and Flinn, J.C., Technology and economics of weed control in broadcast-seeded flooded tropical rice. In Weeds and the environment in the tropics, eds. K. Noda and B.L. Mercado, Chiang Mai, 1986, 51-74.

2. Mabbayad, B.B. and Obordo, R.A., Transplanting rice vs. direct seeding. World Farming, 1971, 13, 6-7.

3. anonymus, IRRI rice facts 1988. The International Rice Research Institute, Manila, 1988.

4. Matthews, L.J., Weed control component of integrated pest control in rice in South and South-East Asia. Int. Rice Com. Newsl., 1986, 35, 53-60.

5. Ho, N.K. and Md. Zuki, I., Status of weed control under direct seeded condition in the Muda area. MADA/MARDI Quart. Meeting, Alor Setar, Malaysia, 1989, 31 pp.

6. Choudary, D.B.B., Ramakrishnan, L. and Assiri-Yagi, A., Evaluation of oxyfluorfen in pregerminated broadcast lowland rice in Sri Lanka. Proc. 9th Conf. Asian-Pac. Weed Sci. Soc., Manila, 1983, 198-202.

7. Datta, S.K. de, Technology development and the spread of direct-seeded flooded rice in Southeast Asia. Expl. Agric., 1986, 22, 417-26.

8. Datta, S.K. de, Principles and practices of rice production. John Wiley & Sons Inc., New York, 1981, 618 pp.

9. Ho, N.K., Comparison of weed flora and farmers' weed control practices in the transplanted and direct-seeded rice in the Muda area of Malaysia. Proc. 2nd Int. Conf. Pl. Prot. Trop., Penang, 1986, 233-4.

10. Kittipong, P., Weed control in farmers' fields in Thailand. Proc. Conf. Weed Control In Rice, Los Banos, 1983, 193-200.

11. Janiya, J.D. and Moody K., Weed populations in transplanted and wet-seeded rice as affected by weed control method. Trop. Pest Manag., 1989, 35, 8-11.

12. Migo, T.R. and Datta, S.K. de, Improvement in herbicide application technique and application timing in transplanted and broadcast-seeded flooded rice. Proc. 9th Conf. Asian-Pac. Weed Sci. Soc., Manila, 1983, 162-75.

13. Noda, K., Teerawatsakul, M., Prakongvongs, C. and Chaiwiratnukul, L., Major weeds in Thailand. Project Manual No. 1, Dep. Agric., Bangkok, 1985, 142 pp.

14. Pablico, P.P. and Moody, K., Lowland rice field weeds in Nueva Ecija, Philippines. Int. Rice Res. Newsl., 1986, 11, 29.

15. Pablico, P.P. and Moody, K., A survey of weeds in transplanted and wet-seeded rice under rainfed and irrigated conditions. Int. Rice Res. Newsl., 1987, 12, 23.

16. Weerakoon, W.L. and Gunewardena, S.D.I.E., Rice weed flora of Sri Lanka. Trop. Agric., 1983, 139, 1-14.

17. Hare, C.J., Chong, W.C., Ooi, G.T., Bhandhufalck, A., Nawsaran, S. and Chanprasit, P., SOFIT® Super: Broad sprectrum weed management for wet sown rice in S.E. Asia. Proc. 12th Conf. Asian-Pac. Weed Sci. Soc., Seoul, 1989, 165-70.

18. Datta, S.K. de and Bernasor, P.C., Chemical weed control in broadcast-seeded flooded tropical rice. Weed Res., 1973, 13, 351-4.

19. Mabbayad, M.O. and Moody, K., Effect of time of herbicide application on crop damage and weed control in wet-seeded rice. Int. Rice Res. Newsl., 1984, 9, 22.

20. Quadranti, M. and Guyer, R., New Possibility in weed control in the nursery-beds. Proc. 10th Conf. Asian-Pac. Weed Sci. Soc., Chiang Mai, 1985, 277-81.

21. Rufener, J. and Quadranti, M., Early weed control in wet-sown rice: the role of the safener CGA 123'407. Proc. 10th Int. Congr. Plant Prot., Manila, 1983, 332-8.

22. Christ, R.A., Effect of CGA 123'407 as a safener for pretilachlor in rice (Oryza sativa L.). Recordings of elongation rates of single rice leaves. Weed Res., 1985, 25, 193-200.

23. Quadranti, M. and Ebner, L., SOFIT, a new herbicide for use in direct-seeded rice (wet-sown rice). Proc. 9th Conf. Asian-Pac. Weed Sci. Soc., Manila, 1983, 405-12.

24. Bhandhufalck, A. and Hare, C.J., SOFIT® 300 EC, Practical considerations and benefits from its use in wet sown rice in Thailand. Proc. 10th Conf. Asian-Pac. Weed Sci. Soc., Chiang Mai, 1985, 168-76.

ESPROCARB HERBICIDE MIXTURES: USE IN JAPANESE PADDY RICE

G KADOTA, S MATSUMOTO, S NAKAMURA, R F S GORDON
ICI Japan Agricultural Research Station
780 Kuno-cho, Ushiku, Ibaraki, Japan
and J HAYAKAWA
ICI Agrochemicals, Fernhurst, Haslemere, Surrey. GU27 3JE.

ABSTRACT

Esprocarb[S-benzyl-N-ethyl-N-(1,2-dimethylpropyl)thiocarbamate], code
name ICIA-2957 (formerly SC-2957) is a pre- and post-emergence herbicide
giving effective control of Echinochloa crus-galli and annual weeds in
paddy rice. Official field tests throughout Japan demonstrated that
combinations with bensulfuron-methyl provided broad spectrum control of
annual and perennial weeds with considerable flexibility of timing and
crop tolerance in transplanted rice. New combinations of esprocarb with
bensulfuron-methyl and third compounds are under development aimed at
offering longer residual control of annual weeds, particularly
Echinochloa crus-galli, in the cooler regions of Japan.

INTRODUCTION

The main weed species, regional variations and traditional rice herbicide
practices in Japan were described by Mizutani in 1986 [1]. But since
1982, when the first so called 'one-shot' herbicides were introduced,
farmer practice has changed rapidly. Single applications of a one-shot,
or a one-shot application followed by a specific annual weed killer have
become the norm, replacing the traditional 2,3 or 4 applications made
previously. The time and cost savings achieved with one-shot treatments
have led to a rapid expansion of their use. In 1988 58% of the planted
area was treated with one-shots and by 1989 this had risen to 74% of the
total 1.55 million hectares of paddy rice.

Current one-shot herbicides can be classified into those which are applicable at the early stage of Echinochloa crus-galli (pre-emergence to 1.5 leaf stage) and those which are applicable at the early to middle stage (pre-emergence to 2.5 leaf stage).

Since 1986, new types of one-shot herbicides have been tested in official Japanese trials. These are aimed at providing the farmer with longer, residual control of later emerging annual weeds, hence eliminating the need for any follow up application.

Esprocarb is known to be highly effective on Echinochloa crus-galli and annual broadleaved weeds [2]. Bensulfuron-methyl (DPX-84) was chosen as a mixture partner because it provides excellent control of perennial weed species [3]. This combination of esprocarb and bensulfuron-methyl was found to give good broad spectrum weed control and good crop tolerance [4]. In Japan, the esprocarb-based combinations have been tested since 1985 by national and prefectural agricultural stations in co-operation with JAPR (Japan Association for the Advancement of Phyto-Regulators).

This paper describes the use of esprocarb in one-shot herbicide mixtures developed and under development for transplanted rice in Japan. The mixtures that include esprocarb, bensulfuron-methyl and a third active ingredient have been included specifically to evaluate residual control of later emerging annual weeds. These latter studies have been carried out in special co-operation with Nihon Nohyaku Co.

MATERIALS AND METHODS

a. .Herbicides
Code name: ICIA-2957 (formerly SC-2957)
Common name: Esprocarb
Chemical name: [S-benzyl-N-ethyl-N-(1,2-dimethylpropyl)thiocarbamate]
Chemical Structure

$$\text{C}_6\text{H}_5-\text{CH}_2\text{SCN} \underset{\text{O}}{\overset{\text{O}}{\|}} \begin{array}{l} \text{CH}_2\text{CH}_3 \\ \text{CH}-\text{CH}-\text{CH}_3 \\ \text{CH}_3\text{CH}_3 \end{array}$$

Chemical and physical properties as well as toxicological data were previously presented by Seaman et al [2].

Combination partners

CODE	COMMON NAME
DPX-84	bensulfuron-methyl
TSH-888	piributycarb
MON-72	dithiopyr
BAS-514	quinchlorac

CODE	COMBINATIONS	A.I. (%)	TRADE NAME*
DPX-84 SC	ICIA-2957 + DPX-84	7 + 0.25	Fujigrass 25
DPX-84 SC L	ICIA-2957 + DPX-84	7 + 0.17	Fujigrass 17
NH-727	ICIA-2957 + DPX-84 + TSH-888	5 + 0.25 + 3	
NH-728	ICIA-2957 + DPX-84 + MON-72	5 + 0.25 + 0.3	
NH-8911	ICIA-2957 + DPX-84 + BAS-514	5 + 0.25 + 1	

b) Field trials

Field trials to evaluate esprocarb combinations in transplanted rice were conducted at the ICI Japan Agricultural Research Station in 1989 and in official experiments from 1985 to 1989. The official trials were designed to evaluate weed control and crop safety under different climatic conditions (cool northern region, warm central region and hot southern regions of Japan). Rice seedlings at the 2-3 leaf stage were transplanted by machine at a planting depth of about 3 cm.

Pre-mixed combinations were applied to paddy water by hand as granules, at a product rate of 30 kg/ha. Applications were made at 3 different timings (0.5-1.0, 1.5-2.0 and 2.5 leaf stage of _Echinochloa crus-galli_) using 2-3 replications in a randomised block design.

Soil types at the different locations included clay, clay loam, loam and sandy loam. Water leakages of the paddies ranged from 0.1 to 2.0 cm/day. Field and crop management as well as water management were carried out according to local practices.

* registered trade mark of Imperial Chemical Industries PLC

Evaluations of weed control and crop safety were made by visual estimation or direct measurement (dry weight) up to 44-55 days after application. To determine residual activity or persistence, the number of newly emerged Echinochloa crus-galli with grain panicles was counted 91 days after application.

Target weeds tested included:

Latin name	Bayer code used in Tables/Figures
Echinochloa crus-galli	ECHCG
Cyperus difformis	CYPDI
Monochloria vaginalis var. plantaginea	MOOVP
Annual broadleaved weeds	BLWS
Eleocharis acicularis var. longiseta	ELOAL
Scirpus juncoides spp. juncoides	SCPJO
Sagittaria pygmaea	SAGPY
Cyperus serotinus	CYPSE
Elecocharis kuroguwai	ELOKU
Sagittaria trifolia	SAGTR
Alisma canaliculatum	ALSCA
Potamogeton distinctus	PTMDI
Oenanthe javanica	OENJA

Abbreviations:
DAT = days after transplanting   DAA = days after application

RESULTS

Esprocarb/DPX-84 combinations
Results were obtained from a total of 37 official trials carried out between 1985 and 1988.

(1) Weed control: The high DPX-84 content formulation (0.25%) was tested in the cool region and the low DPX-84 content formulation (0.17%) in warm and hot regions. These combinations provided good control of major Japanese paddy weeds tested under different climatic conditions, when applied at 0.5-2.5 leaf stage of Echinochloa crus-galli (Table 1). In the cooler regions, the higher DPX-84 content is required in order to provide adequate control of the slower-growing weeds. In the hot regions, the appropriate application timings were 0.5-2.0, 2-2.5 and 2.5 leaf stagefor Cyperus serotinus, Sagittaria trifolia and Eleocharis kuroguwai, respectively.

TABLE 1

Weed control activity of esprocarb/bensulfuron-methyl combination under different climatic conditions. Granular appliction of 30 kg poduct/ha at different timing in transplanted rice (JAPR official trials, 1985-1988)

Region	Appin. (DAT)	Leaf stage (ECHCG)	% Weed control*													
			ECHCG	CYPDI	MOOVP	BLWS	ELOAL	SCPJO	SAGPY	CYPSE	ELOKU	SAGTR	ALSCA	PTMDI	OENJA	Average
Cool	7	0.5-1.5	98	100	100	100	100	99	97	100	100	83	100	100	-	98
Cool	15	1.5-2.0	98	100	100	100	100	98	99	100	100	93	100	97	-	98
Cool	19	2.5	95	100	100	100	100	98	98	99	100	100	100	99	-	96
Warm	5	0.5-1.0	99	100	100	99	100	99	98	98	92	100	90	100	99	99
Warm	12	2.0-2.5	100	100	100	97	100	98	99	93	97	100	92	100	100	99
Warm	14	2.5-3.0	99	100	100	98	100	96	97	89	94	86	85	100	99	98
Hot	5	0.5-1.0	100	100	100	100	100	99	89	100	82	66	-	99	-	94
Hot	9	2.0	100	100	100	100	100	98	98	99	83	90	-	100	-	97
Hot	12	2.5	99	100	100	100	100	96	96	86	90	94	-	99	-	96

Means of 37 official trials

*Final evaluation: 45-55 days after application

DAT = days after transplanting

BLWS = annual broadleaved weeds

Species codes: see text for explanation

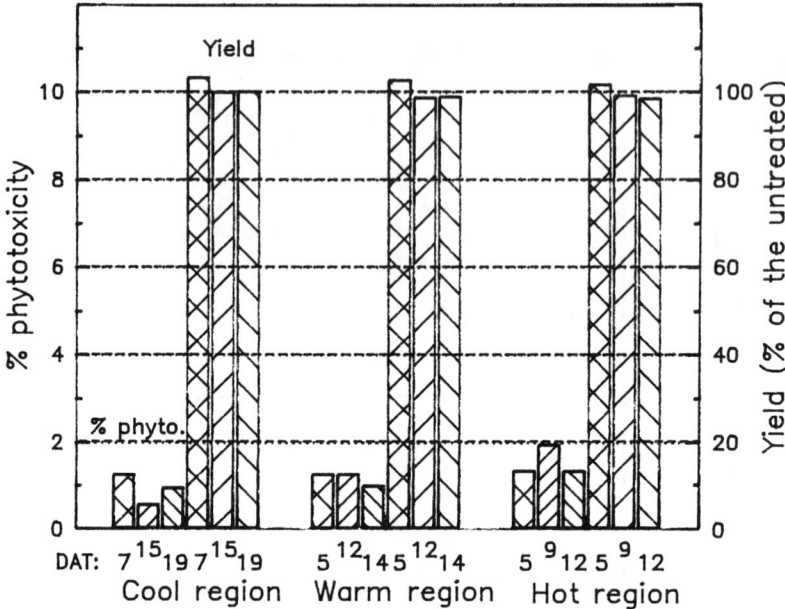

Figure 1. Rice crop tolerance of esprocarb/bensulfuron-methyl combination under different climatic conditions at different application timing (30 kg product/ha)

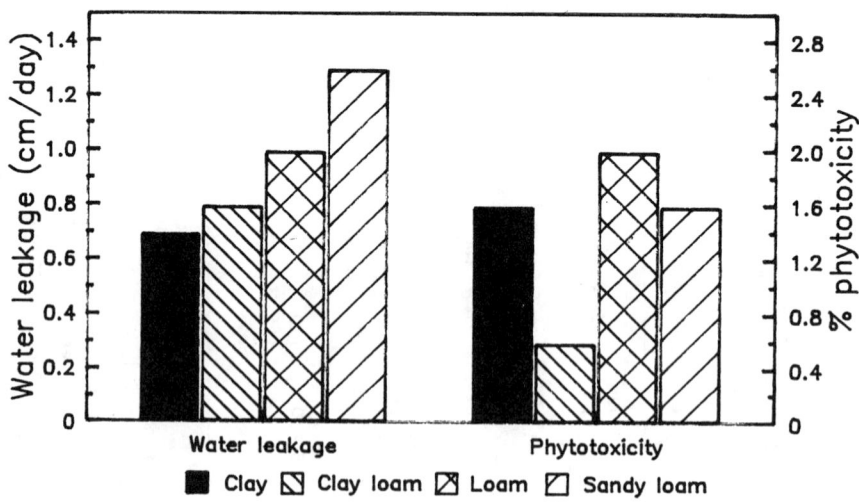

Figure 2. Rice crop tolerance of esprocarb/bensulfuron-methyl combination on different soil types in warm region.

(2) Crop tolerance: Both of these mixture ratios give excellent crop selectivity, ie. no adverse effect on the growth or yield of rice under any climatic condition at all application timings. (Fig. 1).

The effects of soil type and water leakage upon crop tolerance are shown in Figure 2. These data clearly demonstrate that even in paddy soils where the water leakage rate is high, crop selectivity is excellent.

Esprocarb/other partner combinations

(1) Weed control activity and crop tolerance: Table 2 shows the weed control activity and crop tolerance of transplanted rice to the combination of esprocarb with other partners. The results were obtained from 8-13 official trials (depending on the combination tested) in the cool region of Japan between 1987 and 1989. By 44-55 DAA, depending on the timing of application, (0.5-2.5 leaf stage of Echinochloa) all combinations (NH-727, NH-728, and NH-8911) gave good control of all weed species tested when applied at 30 kg product/ha. No adverse effects on the growth or yield of rice was observed.

(2) Flexibility of application timing: Figure 3 shows the performance of various esprocarb containing mixtures against Echinochloa when applied at different growth stages. NH-727, NH-728 and NH-8911 showed very good Echinochloa control at all timings tested. (NB. NH-8911 only tested at the 0.5-1.0 and 2.5 leaf stage, while the other combinations were tested at all 3 weed growth stages).

(3) Speed of activity: When applied at the 0.5 leaf stage, the peak of activity for all mixtures was observed at the second evaluation timing (15 DAA) (Fig. 4).

TABLE 2

Weed control activity and selectivity to transplanted rice of esprocarb-based combinations. Granular appliction of 30 kg product/ha at different timings
(JAPR official trials. 1987-1989)

Products	Application (DAT)	Leaf stage (ECHCG)	% Weed control*							Phyto-toxicity	Yield(% of untreated)
			ECHCG	CYPDI	MOOVP	BLWS	p-sedge	p-blws	Average		
NH-727	6	0.5	100	100	100	100	98	98	100	3	103
	14	2	99	100	100	100	99	99	99	1	100
	16	2.5	99	100	100	100	98	100	99	1	102
NH-728	7	0.5	100	100	100	100	97	97	100	2	103
	14	2	98	100	100	100	99	97	98	0	101
	18	2.5	95	100	100	100	97	97	96	1	100
NH-8911	6	0.5	99	100	100	100	97	98	99	1	100
	16	2.5	100	100	100	100	97	98	100	1	100

p-sedge: perennial sedges - mean of 4 species (ELOAL, SCPJO, CYPSE and ELOKU)
p-blws : perennial broadleaved weeds - mean of 5 species (SAGPY, SAGTR, ALSCA, PTMDI and ORNJA)
Means of 8-13 trials
*Final evaluation: 45-55 days after appliction
Species and product codes: see text for explanation
Phytotoxicity scale: 0-100 (0 = no effect, 100 = completely killed)
DAT = days after transplanting

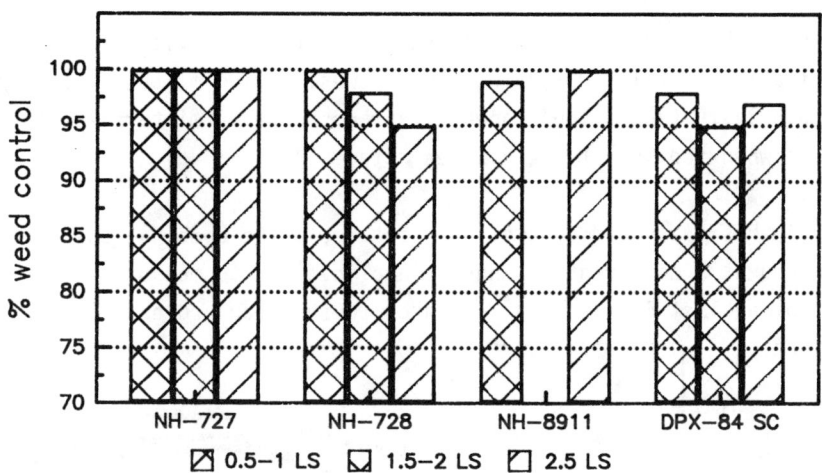

Figure 3. Weed control of esprocarb-based combinations on _Echinochloa_ at different timing (mean of 2-3 trials, 45-55 DAA)

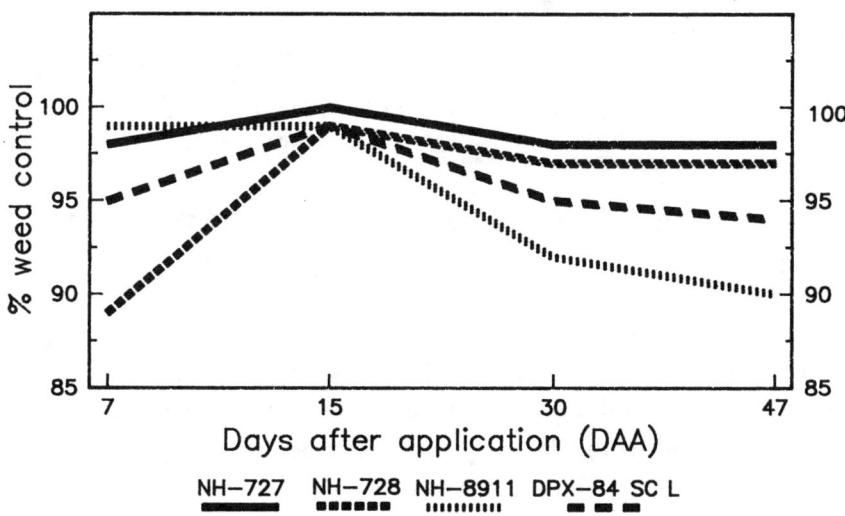

Figure 4. Longevity of _Echinochloa_ control by esprocarb-based combinations applied at 5 DAT (0.5 leaf stage _Echinochloa_)

However, when application was made at the 2.5 leaf stage, peak activity was observed at the third evaluation timing (30 DAA) (Fig. 5).

At both timings, NH-8911 showed the fastest activity of all the combinations, whilst NH-728 was the slowest acting mixture when applied at the 0.5 leaf stage. The standard esprocarb/DPX-84 mixture was the slowest to act when applied at the 2.5 leaf stage.

(4) Residual weed control (persistence): The combinations were examined at 2 different timings of application, to measure persistence of control against Echinochloa, compared to the standard esprocarb/DPX-84 mixture.

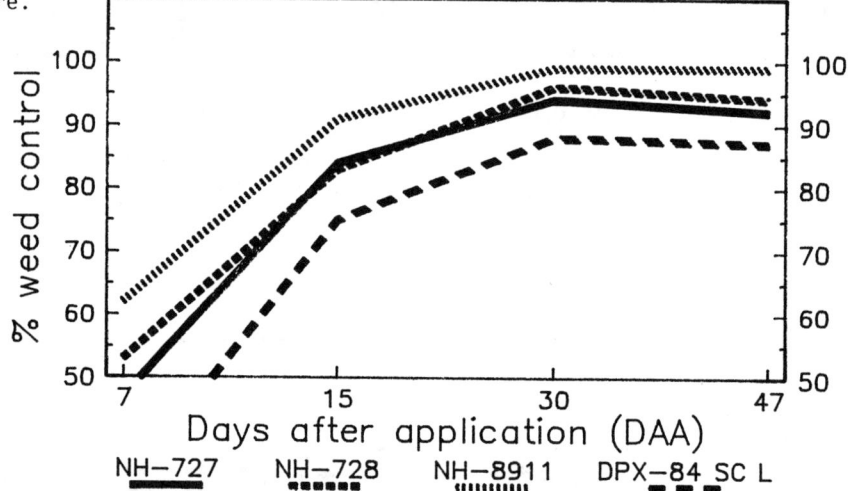

Figure 5. Longevity of Echinochloa control by esprocarb-based combinations applied at 2.5 leaf stage Echinochloa)

When applied at the 0.5 leaf stage, NH-727 and NH-728 provided good weed control, ie. 95% control, 47 DAA (Fig. 4). While at the 2.5 leaf stage, NH-8911 showed the highest level of control, with nearly 100% control 30-47 DAA (Fig. 5). Although some newly emerged Echinochloa were observed by 47 DAA following application with NH-8911 at the 0.5 leaf stage, none of these survived to produce panicles by 91 DAA (Figure 6). At the 2.5 leaf stage NH-8911 completely suppressed panicle-forming Echinocloa.

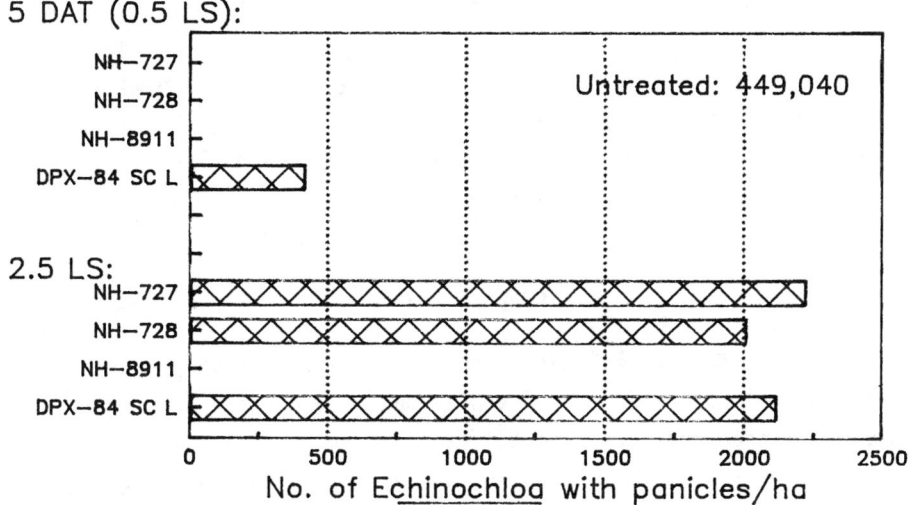

5 DAT (0.5 LS):

2.5 LS:

No. of Echinochloa with panicles/ha

Figure 6. Effect of esprocarb-based combinations at two different applicaton timings on later emerging Echinochloa that produce panicles (91 DAA)

DISCUSSION

The addition of esprocarb to bensulfuron-methyl can provide not only enhanced activity against paddy weeds but also a reduction of phytotoxicity to transplanted rice compared to bensulfuron-methyl alone [4]. It was also demonstrated in official tests that this combination was applicable throughout Japan under variable conditions, eg. climate, soil type and water leakage in the paddy field. However, a higher bensulfuron-methyl content formulation is required in the cool region of Japan.

Both esprocarb plus bensulfuron-methyl mixtures provide good broad-spectrum weed control with a high degree of application flexibility, which can largely be attributed to esprocarb's wide potency window against Echinochloa (pre-emergence to 2.5 leaf stage). Hence these mixtures are classed as early to middle stage one-shots. However, in the hotter regions of Japan, perennial weed species can grow away fast if not controlled early, so a slightly narrower application range is recommended under these conditions.

Combinations of esprocarb and DPX-84 with a third chemical (TSH-888, MON-72 or BAS-514), also offer good overall control of annual and perennial paddy weeds. Against Echinochloa crus-galli, the dominant weed, NH-727 (with TSH-888), NH-728 (with MON-72) and NH-8911 (with BAS-514) provide better application flexibility and longer residual control than the two-way esprocarb, DPX-84 mixture.

CONCLUSIONS

The combination of esprocarb with bensulfuron-methyl provides good broad spectrum control of annual and perennial paddy weeds with a wide application window and good crop tolerance in transplanted rice in Japan. To provide longer residual control of annual weeds, including Echinochloa crus-galli and Scirpus juncoides, in the cool region of Japan, this combination benefits from mixture with a third compound such as, TSH-888, MON-72 or BAS-514.

ACKNOWLEDGEMENTS

The authors wish to thank the national and prefectural agricultural stations of Japan and JAPR for their co-operation. We also wish to thank our esprocarb combination partner Nihon Nohyaku Co. Ltd. for their kind co-operation.

REFERENCES

1. Mizutani, M., Weed control and growth regulation (in Japan). Chemistry and Industry, 1986, 2, pp 51-54.

2. Seaman, D., Foss, S.C., Hsu, J.K., Hyzak, D.L., Randolph, B.J., Walker, F.H., and Nakamura, S. Esprocarb: S-benzylthiocarbamate co-herbicide for rice. Weed Research, Japan, 1988, 33, pp 35-26.

3. Takeda, S., Yuyama, T., Ackerson, R.C., Weigel, R.C., Sauers, R.F., Neal, W., Gibian, D.G., and Tseng, P.K.: Herbicidal activities and selectivity of a new rice herbicide DPX-F5384, Weed Research, Japan, 1985, 30, pp 284-289.

4. Hikawa, M., Hachitani, Y., Matsuba, K: Herbicidal properties of thiocarbamate herbicide, esprocarb (in publication in Japanese).

5. Beck, J., Ito, M., Kashibuchi, S.: Quinchlorac (BAS-514) and its herbicide combinations in transplanted rice in Japan. In Proc. 12th Conf. of Asia-Pacific Weed Science Society, 1989, pp 235-244.

6. Sugasawa, Y., Kibuoka, T., Mouri, A., Nishihara, T., and Wakasa, F.: MON-7200: A rice herbicide mixture candidate for wide-application window treatment. In Proc. 12th Conf. of Asia-Pacific Weed Science Society, 1989, pp 157.

A Novel Use for Benfuresate as a Paddy Rice Herbicide

R Rees, T Matsui and B Hanisch
Schering AG, Berlin

ABSTRACT

Practical uses of benfuresate are presently confined to pre-plant incorporated treatments in cotton. Recent developments are described in paddy rice from the Far East where the compound shows outstanding control of several important perennial weeds notably, Eleocharis kuroguwai and Cyperus serotinus, together with good crop safety.

Envisaged recommendations are rates of 300 - 750 g ai/ha applied in early to mid post-emergence applications after rice transplanting.

Benfuresate has relatively low soil mobility which contributes greatly to ensuring good levels of residual activity.

INTRODUCTION

Weed control in paddy rice today is approached with a wide
variety of herbicide programmes depending on local weed problems and economics. Traditionally, annual weeds, especially
Echinochloa oryzicola (ECHOR) have been well controlled by
acetamides and diphenyl ethers, whilst triazines and phenoxy
compounds gave additional activity on perennial weeds.

During the last decade, pyrazolate and sulphonyl urea chemistry was introduced giving wide spectrum broad-leaved weed
control with good timing flexibility. Despite these rapid in-
novations, further improvements are sought not only for the
control of Sagittaria trifolia but also for the control of
Eleocharis kuroguwai and Cyperus serotinus. Initial control
of perennial sedges is relatively easy to achieve at early
growth stages, but a strong regrowth and a resultant propagation of daughter bulbs or tubers often occurs. In Japan, no
official information is available on the extent of infesta-
tation but it is reported in Hokuriku, Tohoku and Kyushu that
approx. 300,000 ha and 160,000 ha are infested annually with
Cyperus and Eleocharis (1). This geographical area con-
stitutes only about 50 % of the Japanese rice hectarage.

Benfuresate, 2,3-dihydro-3,3-dimethyl-benzofuran-5-yl-
ethanesulphonate, was originally discovered and developed by
Fisons Limited and later, FBC Ltd., in the early 1980's as a
pre-plant incorporated or pre-emergence herbicide for the
control of perennial Cyperacae (sedges) in cotton, especially
Cyperus rotundus and C. esculentus (2, 3). Currently,
benfuresate under the trade name of Cyperal, is being
developed for use on cotton in several countries of Central
and South America. Typical dose rates are ca. 3 kg ai/ha,
applied pre-plant incorporated. Further to continued
development in cotton, major efforts are now being given to
the evaluation of benfuresate in paddy rice by Schering AG,
Berlin.

Previous investigations in direct seeded or paddy rice from
pre-emergence or pre-plant applications had shown adequate
crop safety, when the correct application technique was used.
This paper describes results of benfuresate applied to paddy
rice post transplanting at low dosages and gives suggestions
on the future application of the compound to present day weed
problems.

MATERIALS AND METHODS

Benfuresate was tested in the Far East on paddy rice in early 1987, primarily in Japan, South Korea, Philippines, Thailand and Taiwan. The great majority of work was conducted in Japan and South Korea. Initial work began under glass in Japan where basic pot studies were conducted by the Japan Association for Advancement of Phyto Regulators (JAPR). This led to replicated small field plot trials (0.8 - 1.0 m²) being established to confirm the weed control spectrum and level of crop selectivity relative to application timing and weed size. Subsequent pot, concrete frame and field trials have been conducted by Schering personnel, third party companies and JAPR to further optimise application recommendations.

Applications were made of less than 1 kg ai/ha at pre- and post-emergence of the weed from 3 - 25 days after transplanting (DAT) with either a 40 EC formulation or more typically hand-applied granules of 1.5 - 2.0 % w/w. Assessments for efficacy and selectivity were generally made 14 - 28 days after application. Later assessments after more than 100 days were made to assess residual activity on perennial sedges. In most of the work reported, a weed population (seeds, tubers and bulbs) was artificially established in each plot or pot, at or around rice transplanting time, to ensure a homogeneous weed spectrum and to encourage even germination. Further methods are tabulated with the results.

RESULTS

Efficacy

The effective weed control spectrum of benfuresate was confined to perennial sedges, showing primarily activity on Eleocharis kuroguwai and the annual grass Echinochloa oryzicola Vasing (ECHOR). Control of broad-leaved weeds (both annual and perennial) and additionally the annual, Cyperus difformis (CYPDI) was moderate. Best level of control was achieved when weed targets were still in the pre-emergence to early post-emergence stage. Applicable dose rates range from 300 - 1000 g ai/ha depending on weed target and application timing.

Similar relative efficacy shown by the 40 EC and a number of extruded granule formulations under glass, have not shown any significant differences. Granule formulations were mostly employed in field trials.

Initial dose and weed spectrum determination studies, conducted in glasshouse pot trials and field work, showed a wide variation in weed susceptibility within the main species found in Japan, Korea and the Philippines (Table 1).

Table 1

Weed control spectrum of benfuresate in paddy rice

		Maximum weed/leaf stage (LS) controlled (90 %)	
		500 g ai/ha	1000 g ai/ha
Perennial sedges			
Scirpus juncoides	(SCPJU)	up to 1.0 LS	up to 2.5 LS
Eleocharis acicularis	(ELOAC)	post-emergence before flowering	post-emergence before flowering
Eleocharis kuroguwai	(ELOKU)	5-10 cm height	10-15 cm height
Cyperus serotinus	(CYPSE)	10-15 cm height	15-20 cm height
Annual grass			
Echinochloa oryzicola Vasing	(ECHOR)	1.5 LS	2.5 LS
Perennial grasses			
Leersia oryzoides	(LEROR)	) pre to early	early to mid
Paspalum distichum	(PASDS)	) post-emergence	post-emergence
Annual broad-leaved weeds/ Cyperaceae			
Fimbrystylis miliacea	(FIMMI)	4 LS	–
Lindernia dubia	(LIDDU)	–	–
L. pyxidaria	(LIDPY)	–	–
Elatine triandra	(ELTTR)	–	–
Rotala indica	(ROTIN)	–	–
Monochoria vaginalis	(MOOVA)	–	–
Cyperus difformis	(CYPDI)	–	–
Cyperus iria	(CYPIR)	2 LS	–
Scirpus supinus	(SCPSU)	3 LS	–
Perennial broad-leaved weeds			
Sagittaria trifolia	(SAGTR)	) from pre-em	from pre-em
S. pygmaea	(SAGPY)	) right through; stunting only	right through; stunting only
Alisma canaliculatum	(ALSCA)	–	–
Potamogeton distinctus	(PTMDI)	–	–
Oenanthe javanica	(OENJA)	early regrowth stage (inhibition) 3-4 cm	early regrowth stage (moderate control) 40-50%, 3-4 cm

A field trial result from Japan confirmed the weed control spectrum from applications made up to the 2.1 leaf stage of ECHOR. The strong performance of benfuresate on perennial sedges and ECHOR, and low levels of other annual weed control is illustrated in Table 2.

Table 2

Herbicidal activity under Japanese field conditions
Visual score (0 - 100 % weed control)

	Weed and leaf stage (LS) at application		Benfuresate 40 EC		Bensul-furon-methyl	Pre-tila-chlor
			500 g ai/ha	1000 g ai/ha	75 g ai/ha	600 g ai/ha
	ECHOR	<0.5	90	100	–	100
	CYPDI	<0.5	25	30	–	100
Pre- and early	MOOVP	<0.5	15	30	–	90
post-emergence	SCPJU	<0.5	60	75	–	95
	SAGPY	<1.0	30	35	–	10
2 DAT	CYPSE	1.0cm	90	98	–	30
	ECHOR	1.1	90	95	55	95
	CYPDI	1.0	30	45	100	100
Early post-	MOOVP	1.0	35	45	95	80
emergence	SCPJU	1.1	70	70	83	70
	SAGPY	1.0	30	35	98	10
4 DAT	CYPSE	5cm	93	98	87	30
	ECHOR	2.1	80	100	65	93
	CYPDI	1.5	35	50	100	100
Post-	MOOVP	2.0	35	45	98	75
emergence	SCPJU	2.1	65	83	85	70
	SAGPY	2.0	30	35	95	10
6 DAT	CYPSE	10cm	90	100	90	30

Location: Ibaragi Pref., Japan, 1987
Transplanting: 02 June, '87
Assessments: Annual weeds, 21-25 days after treatment
                 Perennial weeds, 46-50 days after treatment
Soil: Clay loam

Field work was carried out on perennial sedges in 1988, especially in relation to application timing and dose rate.
Results showed outstanding control of ELOKU far superior to
bensulfuron-methyl (Table 3), together with very high levels
of ELOAC, CYPSE and SCPJU control. Of particular note in
these results is the wide application window offering weed
control up to relatively large growth stages. Additionally it
became clear that ECHOR susceptibility rapidly decreased
after the 2.0 leaf stage even at 800 g ai/ha.

Table 3

Control of perennial sedges and ECHOR in comparison
to bensulfuron-methyl
Visual score (0 - 100 % weed control)

	Weeds and leaf stage (LS) at application		Benfuresate		Bensulfuron-methyl
			600g ai/ha (39 DAT)	800g ai/ha (39 DAT)	51g ai/ha (39 DAT)
5 DAT	ECHOR	1.0	100	100	30
	ELOAL	pre-em.	78	73	90
	SCPJU	1-2	80	78	80
	CYPSE	1-3 cm	53	80	78
	ELOKU	pre-em.	92	90	68
9 DAT	ECHOR	2.0	93	90	20
	ELOAL	early post-em	100	100	88
	SCPJU	2-3	93	100	88
	CYPSE	3-8 cm	73	80	73
	ELOKU	early post-em	100	100	58
13 DAT	ECHOR	3.1	30	40	10
	ELOAL	2-3 cm	100	90	90
	SCPJU	3-4	93	90	80
	CYPSE	7-15 cm	47	68	40
	ELOKU	10-15cm	100	93	30

Location: Shizuoka Pref., Japan
Transplanting: 11 June, '88
Assessments: 26-34 days after treatment
Soil: Sandy loam

As mentioned in the introduction, current control problems
with perennial sedges are associated with residual activity.
Weed size at application time is also of critical importance.
Trial work in Japan has indicated that optimum control of
ELOKU with benfuresate is achieved with a weed size of less
than 5 cm. This proved more successful than a later treatment
and was particularly more beneficial than applications made
pre-emergence (Table 4). At all application stages, however,
benfuresate proved superior to established herbicide standards. In this work the compound showed good residual activity
after post-emergence applications.

Table 4

Dependence of Eleocharis control on application timing

	Dose g ai/ha	Weed No's/m²					
		Pre-emergence		Pre to early post-emergence of ELOKU (less than 5 cm)		Early to mid post-emergence of ELOKU (less than 10 cm)	
DAT		28	60	21	53	11	43
benfuresate	400	3	28	0	5	10	17
	600	4	21	2	3	12	8
	800	1	11	0	3	11	7
pretilachlor	600	10	50	11	55	14	47
chlornitrofen	2,700	2	34	8	44	7	39
bensulfuron-methyl	75	13	48	8	40	11	41
untreated	–	18	51	18	51	18	51

Location: Ibaragi Pref., Japan, 1988
Soil: Sandy loam
Weed: Populations of ELOKU were naturally occurring.

Results with benfuresate indicated that bulb production was strongly reduced in post-emergence applications (Table 5) and clearly correlated well to the visual efficacy result (Table 4). The great majority of daughter bulbs were produced at soil depths between 10 - 20 cm.

Table 5

The influence of application timing and dose rate
on daughter bulb production in Eleocharis kuroguwai

Timing	Benfuresate dose (g ai/ha)	Daughter bulb distribution and no's in a 30 cm soil profile						
Soil depth (cm)	-	0-5	5-10	10-15	15-20	20-25	25-30	Total
pre-emergence	400	0	3	2	16	5	2	28
	600	0	3	7	8	12	5	35
(01 June)	800	1	1	10	6	5	5	28
early post-	400	2	1	3	3	3	0	12
emergence	600	1	0	2	1	8	1	13
(<5cm height)	800	3	5	2	0	1	0	11
(08 June)								
mid post-	400	2	6	3	4	1	1	17
emergence	600	0	0	3	2	1	0	6
(<10cm height)	800	2	0	1	0	0	1	4
(18 June)								
untreated	-	1	3	10	11	4	5	34

Location: Ibaragi Pref., Japan, 1988
Soil: Clay loam, OM - 1.92 %, pH - 5.8 (H_2O)
Leaching: -0.5 cm/day
Weed: Populations of ELOKU were naturally occurring
Assessment: 6 months after application

Confirmation of good residual activity under leaching conditions (2 cm/day x 1 day) against ELOKU was obtained from pot
experiments (Figure 1). After initial chemical treatment,
successive plantings of germinating bulbs (1st linear leaf
emerged) were made for up to 63 days. Residual activity,
judged on weed emergence, differed substantially between
various standards applied and tended to fall off rapidly
21 - 35 days after treatment. Benfuresate showed excellent
levels of control, compared to the standards throughout the
duration of the test.

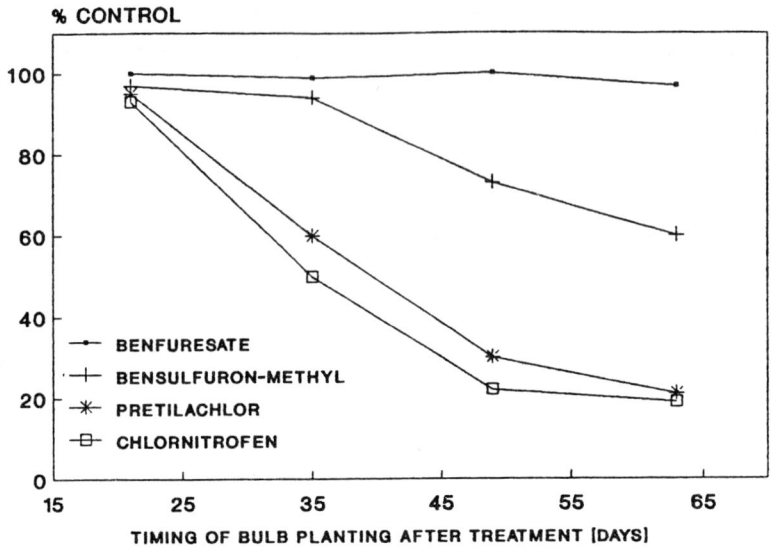

Figure 1

The effect of benfuresate, (750 g ai/ha), bensulfuron-methyl
(75 g ai/ha), pretilachlor (600 g ai/ha) and chlornitrofen
(2700 g ai/ha) upon the residual control of Eleocharis
kuroguwai.

Typically ELOKU mother bulbs will produce 3 - 5 shoots, each
of which can establish new plantlets with subsequent daughter
bulbs. Bulb size or weight is considered to be an important
factor in determining capacity for shoot production. It follows therefore that where larger bulbs are treated with herbicide, these bulbs have a greater capacity for replacing
killed shoots and regrowth occurs. Smaller or lighter bulbs
are killed more easily. Under pot test conditions, bulbs of
various weight were planted immediately before herbicide
treatment and the control period until ELOKU emergence was
observed. Benfuresate was able to control even the largest
bulbs, in contrast to standard materials (Figure 2).

411

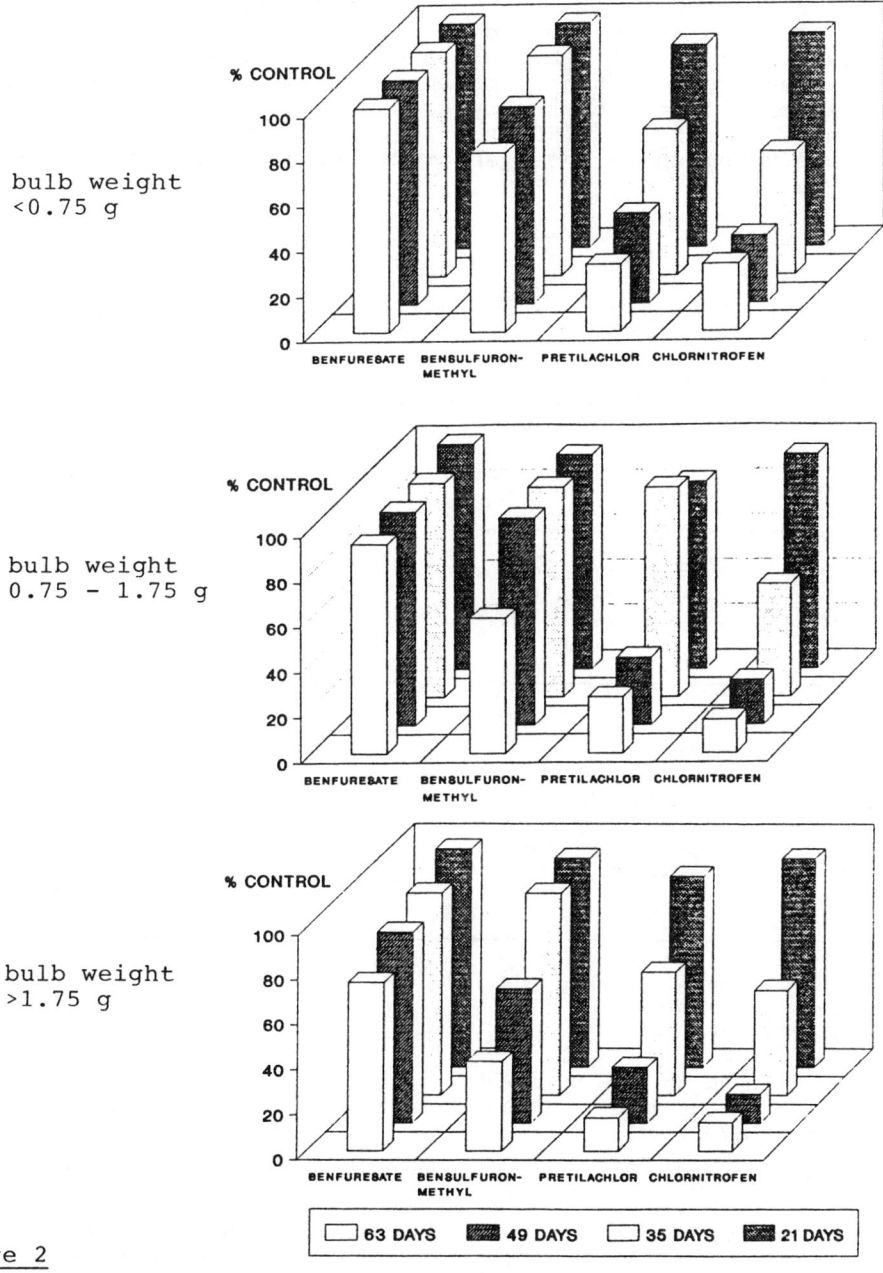

bulb weight
<0.75 g

bulb weight
0.75 - 1.75 g

bulb weight
>1.75 g

63 DAYS   49 DAYS   35 DAYS   21 DAYS

Figure 2

Influence of bulb weight on the control of Eleocharis
kuroguwai by ben- furesate (750 g ai/ha), bensulfuron-methyl
(75 g ai/ha), pretilachlor (600 g ai/ha) and chlornitrofen
(2700 g ai/ha).

In 1989, regional trials were arranged by JAPR throughout major rice growing areas in Japan to demonstrate the performance of benfuresate applied in both single or sequential applications (Table 6). Benfuresate demonstrated excellent and consistent weed control from both treatment approaches. As single treatments, there appeared no differences between benfuresate rates of 450 and 540 g ai/ha to current recommendations using up to three separate applications of standard materials.

Table 6

Control of ELOKU in sequential herbicide programmes

	First treatment	Second treatment	No.of trials	Efficacy (ELOKU survival as % percentage of control)			
				Assessed 64-72 DAT		Assessed 93-113 DAT	
	g ai/ha			stems/m²	weight/m²	hills/m²	stems/m²
ELOKU growth stage at application	pre- & early post-emergence (0-5cm height)	early post- & post-emergence (5-19cm height)					
Single application	benfuresate (450)		6	0.9	5	31	7.0
	benfuresate (540)		6	0.7	1	38	6.5
	5-24 DAT benfuresate (450)	25-50 DAT benfuresate (540)	5	5	1	6.5	6
Sequential application	bensulfuron +thiobencarb (51+2,100)	benfuresate (540)	3	1	1	12	5
	chlornitro-fen+dymuron (2,700+2,100)	benfuresate (540)	3	1		8.3	2

Control of CYPSE has additionally shown to be consistently high (Table 7). Initial glasshouse pot testing in Korea showed dose rates of 505 - 750 g ai/ha giving high activity irrespective of weed size (applied 5 - 15 DAT). In this respect, benfuresate showed distinct flexibility advantages over bensulfuron-methyl.

Table 7

Control of Cyperus serotinus by benfuresate under
glasshouse conditions

| Compound | g ai/ha | Timing | Visual Score (% weed control) | | |
			1)	g ai/ha	2)
benfuresate	505	5 DAT	98	450	100
		10 DAT	95		90
		15 DAT	80		–
	625	5 DAT	100	650	100
		10 DAT	100		100
	750	5 DAT	100	750	100
		10 DAT	95		100
		15 DAT	87		–
bensulfuron-methyl	39	5 DAT	98	.	89
		10 DAT	70		75
		15 DAT	37		–
untreated	–	–	0		0

Location	1) Seoul/Korea, 2) Berlin/FRG
Plot:	1/5000 a pot
Replication:	3
Assessment:	25-35 days after treatment
Weed size at treatment:	3 tubers or bulbs per pot were sown. The following growth stages were recorded at application time:

CYPSE	5 DAT:	1 LS
	10 DAT:	3 LS
	15 DAT:	3-4 LS

Field trial dose ranging studies have shown rates between
400 - 800 g ai/ha giving similarly high levels of control,
especially at early application times. Initial sensitivity at
the 5 - 6 leaf stage was significantly reduced compared to
earlier application timings, but higher levels of control
were recorded in assessments made after 42 days (Table 8).

Table 8

Influence of application timing on
Cyperus serotinus control

Benfuresate dose (g ai/ha)	Weed stage (LS) and timing (DAT)	Assessment (% of remaining weed in relation to untreated) Days after application		
		13-27	42-56	48-92
400		0	0	1
600	pre- and 1 LS	0	2	1
800	(5 DAT)	0	0	trace
400	1-2 LS	0	trace	1
600	(<10cm height)	0	trace	1
800	(10 DAT)	0	0	trace
400	5-6 LS	39	7	11
600	(10-20cm height)	38	9	10
800	(24 DAT)	18	6	6
untreated	no. of CYPSE/m²	71(100)	202(100)	282(100)

Location: Kagoshima Pref., Japan, 1988 (JAPR official trial)
Plot: 2 m²/plot
Replication: 2
Soil: Sandy loam
Leaching: -1cm - 1.5cm/day
Weed: Naturally infested

Selectivity

The susceptibility of transplanted paddy rice seedlings to
benfuresate depends on many factors. The main determinants
are dose rate, application timing, seedling planting depth,
soil type and temperature.

Selectivity in japonica rice is generally higher than in
japonica x indica hybrid varieties.

Where symptoms occur, these are mostly stunting and tiller
reduction. In extreme cases, individual hills can be killed.

The most critical studies on benfuresate selectivity have
been undertaken at JAPR in Japan. Pot studies showed that
adequate selectivity strongly relied on sowing depth
(Table 9). In commercial practice, rice seedlings are machine
planted to a target average depth of ca. 2.5 cm. At this
depth, seedlings showed good tolerance to benfuresate at
rates of 500 g ai/ha applied as early as 3 DAT. Slightly
higher but acceptable phytotoxicity occured at 750 g ai/ha.
At a depth of 1.0 cm, only 500 g ai/ha was safe at 7 DAT. Induced chemical leaching did not influence selectivity.
Pretilachlor was in contrast more susceptible to leaching and
generally showed higher levels of phytotoxicity than benfuresate, especially applied at 7 DAT.

JAPR pot trial results have been confirmed in field work
carried out in Japan and Korea (Table 10). Rates of
400 - 600 g ai/ha have shown acceptable selectivity applied
as early as 5 DAT although safety improves further when
application is delayed to 10 to 20 DAT. Dose rates of
750 - 800 g ai/ha appear practically feasible only after
10 DAT.

Table 9

Selectivity of benfuresate under glasshouse conditions

Sowing depth		Dose rate (g ai/ha)	3 DAT 1)	2)	Visual Score 7 DAT 1)	2)	10 DAT
	benfuresate	500	0.5	0	0.5	0	0.5
		750	0.8	1.0	0.8	0	0.8
		1,000	1.0	2.0	0.8	1.0	0.5
		1,500	1.3	–	1.3		0.8
2.5cm	pretilachlor	600	2.0	–	1.0	–	1.3
		1,200	2.8	1.5	1.5	1.0	2.0
	benfuresate	1,000	0.8*	–	0.8*	–	
	pretilachlor	600	1.8*	–	1.8*	–	
	benfuresate	500	1.0	1.0	0.8	0	0.5
		750	1.8	2.5	0.8	0	0.8
		1,000	3.5	2.5	0.8	1.5	0.8
		1,500	4.0		1.5		1.5
1.0cm	pretilachlor	600	3.0	–	1.3	–	1.0
		1,200	3.3	3.0	2.5	1.5	2.0
	benfuresate	1,000	0.8*	–	0.8*	–	
	pretilachlor	600	2.5*	–	1.8*	–	

Location: 1) JAPR, Ibaragi Pref., Japan, Variety: Koshihikari,
2) Berlin, FRG, Variety: Koshihikari
Plot size: 2×10^{-4} ha (pot)
Soil: Alluvial clay loam
*Leaching: 2-3cm water for one day after treatment

Assessment scale (0-5):
0 - no effect
1 - slight effect
2 - moderate effect
3 - severe phytotoxicity
4 - very severe phytotoxicity
5 - completely killed

Table 10

Selectivity of benfuresate in field trials

Dose of benfuresate	Visual score			
(g ai/ha)	5 DAT	10 DAT	12 DAT	14 DAT
400 - 450	0.9 (4)	0.7 (3)	0.5 (1)	
600	1.0 (4)	0.7 (3)	0.5 (1)	0 (1)
750 - 800	1.5 (3)	0.8 (3)	0.5 (1)	0 (1)

Location: Ibaragi Pref. (2), Kagoshina Pref. (1), Shizuoka Pref.
              (1), Japan and Seoul/Korea.
              ( ): No. of trials.

Assessment scale (0-5):
              0 - no effect
              1 - slight effect
              2 - moderate effect
              3 - severe phytotoxicity
              4 - very severe phytotoxicity
              5 - completely killed

Table 11

The soil mobility of benfuresate as evaluated in
a standard JAPR SAKI III test

	Dose	Depth (cm)	% Control (fresh weight) ECHCG	BRSCH
		0-1	60	35
		1-2	18	8
		2-3	8	0
	450 g ai/ha	3-4	0	0
		4-5	0	0
		5-6	0	0
Clay Loam (A)		6-7	0	0
3.28 % organic matter	900 g ai/ha	0-1	68	40
		1-2	20	20
		2-3	5	5
		3-4	0	0
		4-5	0	0
		5-6	0	0
		6-7	0	0

	Dose	Depth (cm)	% Control (fresh weight) ECHCG	BRSCH
		0-1	60	33
Loam (B)		1-2	13	10
		2-3	5	0
	450 g ai/ha	3-4	0	0
7.74 % organic matter		4-5	0	0
		5-6	0	0
		6-7	0	0
	900 g ai/ha	0-1	75	40
		1-2	28	10
		2-3	10	9
		3-4	0	0
		4-5	0	0
		5-6	0	0
		6-7	0	0

BRSCH = Brassica chinensis L. (chinese cabbage)

Soil mobility and safety to following crops

Applied as a granule formulation, benfuresate is known to be relatively immobile in paddy soil. In a standard JAPR soil mobility test, after 4 days of strong simulated leaching conditions, fractions of benfuresate were found largely restricted to the upper 1 cm of loam and clay loam soil profiles (Table 11).

Subsequent investigations into possible carry-over effects in following crops have shown no deleterious effects. In a range of crops including sweetcorn, wheat, legumes and various vegetables sown in clay loam paddy soil treated 198 days previously with 2 kg ai/ha benfuresate, these crops showed normal germination and subsequent establishment.

DISCUSSION

The recent introduction and widespread use of one-shot paddy rice herbicides in both Japan and Korea has encouraged the spread of perennial weeds. Generally these products lack the necessary residual control and outright killing ability required to control germination and regrowth occurring throughout the first 60 days after transplanting. The novel use of benfuresate to alleviate such weed problems has been confirmed. Benfuresate shows outstanding control of several perennial paddy weeds notably, ELOKU and CYPSE. The compound will prove particularly useful in herbicide programmes because of its flexibility in timing (weed size) and long residual control.

Another advantage of benfuresate use is to minimise the propagation of daughter bulbs produced at the end of the growing season.

Dose rate recommendations considering efficacy and selectivity suggest rates of 450 - 540 g ai/ha applied after 5 DAT, as optimal. A greater margin for selectivity is obtained 7 - 10 DAT.

There appear worthwhile opportunities for the further development of benfuresate as a single product for the specific clean-up of ELOKU within current herbicide programmes. Of greater practical usage, however, will be the development of benfuresate based mixtures. Such mixture products are currently being developed for broad spectrum weed control as early-mid post-emergence applications.

Acknowledgements

The authors thank their colleagues throughout the Schering Research and Development Department and wish to especially acknowledge the help of Nihon Schering, Japan, in preparing this paper.

Finally we thank JAPR, Japan for the generation of the trial data and for their generous help and sound advice in the early stages of benfuresate evaluation.

REFERENCES

1. Japan Association of Phytoregulators (Tohoku). Paddy weed appearance and herbicide usage within six pre-fectures of Tohoku (1986). Issue No. 22-1, 1986.

2. Chisholm, K. W., Welch, J. J. and Ekins, W.L. NC 20484 A new pre-emergence herbicide for cotton crops. Proc. Southern Weed Sci. Soc., 1980 33, 326.

3. Horne, S. D. and van Hoogstraten, S. D. (1980) NC 20484 A new selective herbicide for control of Cyperus spp. and other weeds in cotton. Proc. 1980 British Crop Protection Conference, 201-208.

APPLICATION TECHNIQUES FOR SMALL-SCALE FARMERS

G.A. Matthews and R.P. Bateman

International Pesticide Application Research Centre,
Imperial College at Silwood Park, Sunninghill, Ascot,
SL5 7PY, England.

ABSTRACT

Considerable efforts have been made to introduce insect resistant varieties and encourage biological control of insect pests, but farmers still use pesticides. In some countries large areas of rice may be treated aerially, and in Japan, even with relatively small farms, granules are applied with motorised equipment. In contrast, the small-scale grower in the tropics usually applies pesticides with a manually operated knapsack sprayer or applies granules by hand. Work rate with the small sprayers is low, many growers fail to apply the correct dosage and they get heavily contaminated with pesticide, particularly by walking into the spray and treated foliage. Some research has examined alternative techniques, such as root-zone application, but the work rate is still too low to be accepted by farmers. Clearly more research is needed to devise appropriate application techniques for rice pest management that take account of the differences in the behaviour of pests and time of infestation, so that if a pesticide application is needed, the farmer can treat his crop rapidly and efficiently when needed.

INTRODUCTION

Rice is grown under a wide range of farming and climatic conditions with agronomic practices changing as labour costs increase. Direct seeding instead of transplanting and flooding lowland rice to control weeds has increased the use of herbicides even on small-scale farms [8]. Pre-plant applications gave better weed control and higher yields [18]. Similarly more insecticides are now applied where there is intensive cultivation of high-yielding varieties, despite the crucial importance of growing resistant varieties, conserving important natural enemies, such as lycosid and micryphantid spiders and other predators including <u>Microvelia atrolineata</u>

and <u>Cyrtorhinus lividipennis</u> [6,11] and encouraging fish farming in an integrated management system. Farmers therefore require means of applying pesticides rapidly in response to estimates of pest populations rather than on a fixed calendar schedule.

The equipment needed by a small-scale farmer or by those managing large-scale production areas may be different, but the principles of delivering the correct dosage at the most appropriate time is similar whether the crop is rain-fed, irrigated or, in the case of deep-water rice, the crop is flooded.

The target for pesticides will change from the relatively small seedling with a far larger area of soil, irrigation water or weeds between the rice, when herbicides will be the main type of pesticide applied, to gradually a denser mass of rice plants, on which the insect pests may be confined to:-
a)   the upper canopy, e.g. <u>Nephotettix</u> and other leafhoppers, <u>Hispa</u>, rice-bugs.
b)   the lower stem and leaf sheath e.g. planthoppers, sheath blight.
c)   leaf surfaces (and subsequently the inside of stems), e.g. <u>Scirpophaga</u> and other stemborers, gall midge, protection against neck and panicle blast.

Early season applications are generally inefficient as only a small proportion of the active ingredient reaches the intended site of action, and much of the chemical is lost on the soil or in the irrigation water. The presence of paddy water, however has been exploited for herbicides such as oxadiazon by applying an emulsifiable concentrate formulation directly from its container through a special lid with holes, that is swung across the field as the operator is walking through it. Later the mass of densely packed leaf sheaths, especially when bundles of seedlings are planted together, provide a cluster of 'funnels', which can collect small granules and spray, which is applied with sufficiently large droplets to fall in a mainly vertical trajectory. The bending over of leaves presents a horizontal surface which is also favourable to droplets larger than 100 μm. diameter. Assessments of spray recovery by Pickin, <u>et al.</u> [21] indicated that droplets with a volume median diameter (VMD) of 127 μm. would be better than small droplets (<100μm.) that tend to be carried by air movement over the canopy and impacted only on the upper part of the leaves. Even with a VMD greater than 100 μm., proportionally more droplets are collected at the leaf tips, due to improved collection efficiency as the leaf lamina narrows [24]. This spray distribution is particularly effective against thrips, leafhoppers leaf rollers, and <u>Hispa</u>.

Very few assessments of spray distribution on rice by different application techniques have been reported, and so there has been very little guidance to farmers on how to treat their crops. The majority of small-scale farmers use knapsack sprayers, while in Japan motorised knapsack equipment is also very popular.

Knapsack spraying

A vast number of lever-operated knapsack sprayers are used by rice farmers, principally throughout S.E. Asia, although pressure-retaining hydraulic sprayers are used in some areas (e.g. Southern Viet Nam). Many of these machines are extremely cheap, but their construction is so poor that breakages occur and farmers often struggle to continue to use them even when much of the chemical is wasted by leakages and incorrect application. The majority of these machines are fitted with a simple hand lance with an on/off or trigger valve and one or more nozzles. Inevitably farmers walk through their rice crops waving this lance from side to side so that they go through the foliage treated with pesticide. Apart from walking into the spray cloud with the risk of inhalation of small droplets, the operator brushes his legs against the sprayed plants and gets heavily contaminated. Efforts were made by Fernando [6] to design a knapsack with a pair of nozzles mounted to the rear of the spray tank so that the operator walked away from the spray. Such equipment has never been promoted commercially despite its improved safety.

Many organisations have expressed particular concern when hazardous chemicals - most notably the more toxic cyclodiene and organophosphorus insecticides - are used without any protective clothing, simply because the chemical is cheap; but this problem continues even now in many rice growing areas. There was a significant correlation between increased mortality among economically active men and the widespread adoption of insecticides - notably endrin prior to 1982 - by smallholder farmers in the Northern Philippines [17].

Research trials with motorised equipment have often used extremely high volumes of spray in an attempt to completely wet the surface of rice plants, thus routine spraying at the International Rice Research Institute in the Philippines, used a portable line system operated from a tractor sprayer at the edge of the fields, to apply 1000 litres per hectare. Translation of such high volumes to knapsack spraying would mean having to refill a 15 litre sprayer tank approximately 66 times, each load treating only 150 square metres. Litsinger and Sanchez [15] estimated that 200-300 l/ha in rainfed areas, and 300-400 l/ha with irrigated rice, were more realistic volume application rates for smallholder farmers. Even if water is readily available on irrigated rice, it is time consuming and laborious to collect, so inevitably few farmers persevere with this, and if they used the recommended amount per tank-full i.e. the correct concentration, the amount applied per hectare was usually too little to achieve adequate control of the pest [14]. Even today there are no precise recommendations on how a pesticide should be applied on rice. Instructions on the label often leave it to the farmer to decide how to translate a recommended dosage per area into the correct amount to mix per sprayer load, but few farmers know how to calibrate their equipment and know what volume they are applying.

In areas of high rainfall, most of the spray deposited on plants is washed off, so persistence is also poor. It is not surprising, therefore, that with much of the chemical wasted farmers often affected the natural enemies of rice pests more than the pests and had serious pest resurgences when insecticides were used against planthoppers in Indonesia and other countries. In consequence in Indonesia the government banned the use of many insecticides and promoted the use of the more selective insect growth regulators such as buprofezin. Such a policy only partly addresses the problem, and there is still a need to study application technology to determine which nozzles should be used with knapsack sprayers, how they should be used and how improvements in preparation of sprays can be made to increase accuracy of application.

Motorised knapsack mistblowers

In Japan, many of those growing rice are working in a factory and need a quick method of treating their fields when they go to their farm. The knapsack mistblower is readily adapted to apply dry particles (microgranules) rather than sprays. Instead of projecting the granules across the rice plants, a long tube of polyethylene lay-flat is attached to the air delivery tube so that when it is inflated by the air from the blower, it extends across the whole field. This boom type blow head was invented by a Japanese farmer in 1965. An assistant holds the far, closed end of the tube and moves it vertically up and down so that the tube has a waving action as it is carried down the field. The granules carried in the air stream impinge on small obstructions at intervals inside the length of the tube so that they are deflected thorough small holes placed at regular intervals and are blown by the escaping air into the crop [9]. Tests indicated that 23% of the microgranules were deposited on the foliage. Similar rigid tubes can also be used for seeding and fertiliser application at the start of the season.

High work rates can be achieved with motorised mistblowers, and they probably constitute the most practical means of obtaining good pesticide droplet penetration into the crop after the tillering stage. Mistblowers have been used occasionally in developing countries, especially when supplied as part of aid programmes. They have provided an effective means of controlling widespread BPH outbreaks in Viet Nam [2]. The major constraints are maintenance and spare part supply, and mistblowers are often most appropriately used by co-operative ventures or district plant protection stations rather than smallholder farmers themselves.

Other Techniques

Nymphal populations of the brown planthopper are mostly confined to the vertical base of leaf-sheaths where spray is difficult to direct unless nozzles are placed under the crop canopy. For this reason, root zone application of systemic insecticides has been tried but placement of either a liquid, granules or small capsules is difficult due to the problem of moving an applicator through the soil especially with the close spacing of rice seedlings [1]. Root zone application has not been adopted by growers due to the very low work rate, despite the advantage of reducing the impact on natural enemies and enhancing the absorption by the plant. Experiments also assessed the use of a high spreading oil applied to the water surface so that when the water level was lowered the oil remained clinging to the plants [13].

The use of hand-held battery operated spinning disc sprayers has yet to be promoted commercially for rice, and research on controlled droplet application (CDA) techniques has been limited to pilot studies. Oil-based formulations are usually used at ultra-low volume (ULV) rates of application, but these have mostly been developed for cotton. In the most comprehensive work to date [20], water-based CDA techniques were assessed with insecticides suitable for rice, and the conclusion was that there was little scope for lowering mass application rates for control of BPH with acephate. Droplet penetration was relatively poor, especially after full tillering when the crop canopy closes. From a logistic point of view however, applications at volumes of 5 l/ha. could prove to be an attractive economic alternative to high volume knapsack spraying [21], but the use of CDA has not been promoted commercially on a wide scale. A boat-mounted boom sprayer was developed to overcome the difficulties of applying pesticides in flooded deepwater rice; an aqueous diazinon mixture, applied at 35-50 l/ha as 250μm droplets, reduced Scirpophaga incertulas infestations, and increased yields [3].

The cost and availability of batteries may be considered a constraint with spinning disc sprayers in many countries, although this problem has been alleviated in some cotton growing areas with an electrostatic sprayer, the 'Electrodyn'. Unfortunately, research on rice with this sprayer was discontinued although the pyrethroid insecticide, cypermethrin, was effective at volume rates as low as 0.5 - 1.0 litre per hectare against green leafhopper (Nephotettix virescens), whorl maggots: (Hydrellia philippina) and leaf folder (Cnaphalocrocis medinalis). Electrostatically charged droplets are very selectively deposited in the upper canopy after tillering, and no way was developed to significantly improve spray penetration. However control of brown planthopper with carbosulfan or carbophenothion equalled the performance of conventional high volume application [19].

Aerial application of ultra-low volume sprays was used in Indonesia briefly in 1968-69 [10], but requires a high standard of cultivation over a large area to be economic. Aerial application, principally with granules is therefore confined mainly to large commercial farms in the USA and Australia.

DISCUSSION

Recent research has stressed the need for synchronous planting [16], and wherever possible good pest monitoring with need based timing of sprays based on action thresholds and sequential sampling of fields [12,23]. The requirement is therefore for quick responses at the onset of action thresholds and a reasonably high work rate. If knapsack sprayers are to be more effective, there is a need to define the nozzles used to reduce volumes, while avoiding nozzle blockages. Furthermore greater attention is needed to reduce operator contamination. CDA techniques have at least one important implication for the implementation of IPM: namely the timeliness of application in response to positive scouting. Although BPH resurgences can be induced by granular insecticide formulations, this effect is usually much greater subsequent to foliar hydraulic spraying [7]. ULV spray droplet penetration through the crop canopy becomes relatively poor during the tillering stage; although this is disadvantageous for control of pests such as BPH, it is conceivable that a combination of such selectivity with the use of narrow spectrum insecticides could be considerably less deleterious to natural enemies than more conventional sprays, especially for less mobile predators, but this has yet to be demonstrated.

Some authors consider that equipment other than hydraulic knapsack sprayers would be unsuitable for rice-growing smallholders. The feasibility of combining CDA with granular treatments as a practical alternative could perhaps be profitably investigated for irrigated rice production. The use of conventional formulations mixed with water might be especially appropriate with this crop, since these could be more cost effective than using oil-based formulations; certain of the more specific insecticides, including acyl urea compounds and fungal pathogens [22], are insoluble, and are more readily applied as water-based suspensions. These more selective pesticides will become increasingly important if BPH resurgence, fish toxicity and other similar problems are to be avoided, and better integration of chemical with biological control is to be achieved.

Whether knapsack or alternative sprayers are recommended, it is vital that more attention is given to the quality of equipment, so that the risks of operator contamination are reduced. Where equipment is purchased by tender boards, a quality specification should be included, and their aim should not simply be to buy cheapest available. Their task would be made easier in this respect with regularly updated publication of sprayer evaluations, based on internationally recognised test procedures [e.g. 5]. Even with better quality equipment, leakages and other problems will arise during use, unless there is a good supply of spare parts, and farmers receive more education on the maintenance and use of equipment.

We have shown that very little independent research on application techniques has been carried out, in relation to the range in the quality and type of pesticides used by rice farmers. It is not surprising therefore that there have been many problems associated with the improper use of pesticides on rice, despite its global importance as a basic food crop. Whereas improved agronomic practices, genetic engineering and conventional plant breeding are most appropriate, the attainment of reliable high yields will continue to depend on the careful and timely application of pesticides when specifically needed. More accurate application techniques should therefore become increasingly important.

REFERENCES

1. Anon., 10-row liquid injector. In: IRRI Annual Report for 1979, p. 478, IRRI publns., Los Baños, Philippines.

2. Bateman, R.P., FAO Technical Report on Pesticides and Application Methods in The Socialist Republic of Viet Nam, 1985.

3. Catling, H.D., Thornhill, E.W. and Islam, Z., A boat-mounted spray boom for deepwater rice. Tropical Pest Management, 1980, 26 (1), 56-60.

4. Cook, A.G. and Perfect, T.J., The influence of immigration on population development of Nilaparvata lugens and Sogatella furcifera and its interaction with immigration of predators. Crop Protection, (1985), 4 423-433.

5. ESCAP, RNAM test codes and procedures for farm machinery. Technical Series No. 12, 1983, Economic and Social Commission for Asia and the Pacific, Regional Network for Agricultural Machinery, c/o UNDP, Pasay City, Philippines.

6. Fernando, H.E. A new design of sprayer for reducing insecticide hazards in treating rice crop. FAO Pl. Prot. Bull., 1956, 4 117-120

7. Heinrichs, E.A., in 'Brown planthopper: threat to rice production in Asia'. IRRI publns., 1979, Los Baños, Philippines: p.162.

8. Heinrichs, E.A., Palis, F.V., Moody, K and Aquino, G.B., The effects of timing of butachlor application on the economics of direct seeded rice. J. Pl. Prot. Tropics, 1987, 4, 95-100.

9. Inata, T., Rice insect control by five granular formulations of insecticides in Japan. Japan Pesticide Information, 1973, 14, 23-26.

10. Joyce, R.J.V., Recent developments in ULV spraying. Proc. 5th Br. Insectic. Fungic. Conf., 1969, 221-224.

11. Kenmore, P.E., Carino, F.O., Perez, C.A., Dyck, V.A. and Gutierrez, A.P., Population regulation of the rice brown planthopper (Nilaparvata lugens Staäl) within rice fields in the Philippines. J. Pl. Prot. Tropics, 1984, 1, 19-37.

12. Kenmore, P.E. and Mochida, O., Application techniques involved in efficient use of insecticides: Timing and frequency. FAO/IRRI Workshop on Judicious and Efficient Use of Insecticides on Rice, IRRI, Los Baños, Philippines; February 1983, pp. 63-64.

13. Lim, G.S., Control of rice insects ULV concentrate and high spreading oil insecticides in Malaysia. Malasian Agricultural Journal, 1973, 49, 122-130.

14. Litsinger, J.A., Price, E.D. and Herrera, R.T., Small farmer pest control practices for rainfed rice, corn and grain legumes in three Philippine provinces. Paper presented at the 9th National Conference of the Pest Control Council of the Philippines, Manila, Philippines, 1978.

15. Litsinger, J.A. and Sanchez, F.F., Formulation, dosage, and application techniques related to crop stages. Presented Paper: FAO/IRRI Workshop on Judicious and Efficient Use of Insecticides on Rice, IRRI, Los Baños, Philippines; February 1983, pp. 61-62.

16. Loevinsohn, M.E., 'The ecology and control of rice pests, in relation to the intensity and synchrony of cultivation'. PhD thesis, University of London, 1984.

17. Loevinsohn, M.E., Insecticide use and increased mortality in rural central Luzon, Philippines, The Lancet, 1987, 1359-1362.

18. Migo, T.R. and De Datta, S.K., Improvements in herbicide application technique and application timing in transplanted and broadcast-seeded flooded rice. Proc. 9th Asian-Pacific Weed Science Society Conference, 1983, Manila Philippines, 162-175.

19. Pascoe, R., Biological results obtained with the handheld 'Electrodyn' spraying system. BCPC Symposium Monograph, 1985, 28 75-85.

20. Pickin, S.R., Heinrichs, E.A. and Matthews, G.A., CDA as a technique for the optimising insecticide deposition on rice for control of brown planthopper. BCPC Symposium Monograph, 1980, 24 65-71

21. Pickin, S.R., Heinrichs, E.A. and Matthews, G.A. Assessment of water based controlled droplet application of insecticides on lowland rice. Trop. Pest Management, 1981, 27, 257-261.

22. Rombach, M.C., Aguda, R.M., Shepard, B.M. and Roberts, D.W. Entomopathogenic fungi (Deuteromycotina) in the control of the black bug of rice, Scotinophara coarctata (Hemiptera; Pentatomidae). J. Invertebrate Pathology, 1986, **48**, 174-179.

23. Shepard, B.M., Ferrer, E.R. and Kenmore, P.E., Sequential sampling of planthoppers and predators in rice. J. Pl. Prot. Tropics, 1988, 5, 39-44.

24. Tu, Y.Q., Lin, Z.M. and Zhang, J.Y., The effect of leaf shape on the deposition of spray droplets on rice. Crop. Protection, 1986, 5, 3-7.

AREAWIDE RICEFIELD RAT CONTROL BY SMALL SCALE LANDOWNERS IN S.E. ASIA THROUGH THE CO-ORDINATED APPLICATION OF FLOCOUMAFEN BLOCK BAIT

N.M. PEARMAN
Shell International Chemical Company Limited,
London, UK

ABSTRACT

Two large (approx. 200ha) trial sites of small holder ricefields were selected in Indramayu Regency, West Jara province, Indonesia. Flocoumafen 3.5g wax block baits were applied to one site as three area-wide pulses of 100 blocks/ha followed up by two selective spot treatments to marginal areas of particularly high infestation. Rodent control in the second area was by traditional farmer practice including hunting or chasing, fumigation and the use of acute poisons such as zinc phosphide and aldicarb. The co-ordinated applications of flocoumafen blockbait resulted in a reduction of damaged seedlings which failed to mature or yield from 28% (control) down to 0.4%, and an increase in yield from 2.5 MT (control) to 7.3 MT dry unhulled rice per ha. Similar trials in Laguna province, Philippines also gave excellent rodent control resulting in significant decreases in crop damage (% cut tillers) with no observable effects on domestic or wild non-target species.

INTRODUCTION

Several species of rat cause severe damage and yield loss to a variety of crops grown in South East Asia. Rice is particularly vulnerable as rats attack all stages of the crop from sowing to harvest, and also depredate and contaminate the stored grain. In the nursery, rats pull up the sprouting plants and eat the seeds. Throughout the vegetative growth stages rats cut the tillers to eat the developing heads, and in the final heading and ripening stages they continue to cause substantial yield loss through eating the young grain. The principle pest species are Rattus argentiventer, R.r. mindanensis, R. exulans, R.r. diardii, R. losea and in some parts of the region Bandicota indica and B. bengalensis are important.

Few accurate area-wide estimates of rice economic losses attributable to rat damage are available; this is especially true for smallholder farmers where individual losses may vary from 0 to almost 100%. Average estimates for typical area-wide losses are usually between 7 to 11% [1,2,3]. In the

Philippines, monetary losses in 1975 were estimated at P405 million from an average of only 4.5% cut tillers [4]. In Malaysia, assuming a conservative loss of only 5% national production the loss is estimated at M$55 million a year [5]. In Thailand, losses are put at more than 200 million Baht per year [6].

Many different methods have been devised to control ricefield rodents including non-chemical methods such as trapping, rodent proofing with exclusion fencing, habitat/harbourage elimination along field margins and drainage ditch banks, and nest destruction through digging/drowning. Although some of these methods have a role to play in reducing the risk of a major infestation, effective and economic control relies on the use of poison baits [2]. In view of the tremendous reproductive potential of rats (in ideal situations one breeding pair can lead to several thousand new individuals in one year), rodenticide baits and application techniques must be selected to give minimal survivors from a control campaign. The baits themselves need to be capable of delivering a lethal dose from a single feed, yet sufficiently selective to offer minimal risk to non target species including domestic animals and wild birds.

Rodent control campaigns need to be organised over wide areas (150 ha or more) involving large number of smallholder farmers, as individual control on small plots is of limited effectiveness due to re-invasion from neighbouring land. When these large areas are cultivated by large numbers of individual farmers (eg. typical average smallholding in some regions of Indonesia is less than 0.5ha), considerable organisational problems arise. Integrated schemes have been successfully developed [3,7] which incorporate surveillance, control and monitoring components. Synchronous planting needs to be agreed and implemented as far as possible across the campaign area to minimise the tendency for rats to move from harvested fields into the remaining unharvested fields in search of food and shelter.

This paper describes area-wide rodent control campaigns with a new rodenticide flocoumafen (STORM*) in Indonesia and the Philippines. This compound is a highly potent "second generation" 4-hydroxycoumarin anticoagulant that is active against a wide range of rodent pests, including strains known to be resistant to warfarin [8,9,10,11,12,13,14, 15,16]. The methods used, control levels achieved, yield benefits and environmental observations are given.

MATERIALS AND METHODS

Description of trial sites

Two trials sites of 200 ha of rice paddy including irrigation channels, dikes, paths and ditches were selected in Indramayu Regency, West Java Province, Indonesia (an area suffering a chronic infestation of <u>Rattus argentiventer</u>), each division being an administrative unit of the district/province. Each of these 200ha. sites comprised some 525-600 separate paddies, individually owned by a separate smallholder-farmer. Similarly, two sites of ca 160 ha were chosen in Laguna Province, Luzon, Philippines each comprising ca 154 ha of rice paddy and 6 ha of village

* STORM is a Shell registered trademark

housing interspersed with coconut groves, fruit trees and vegetable plots - this region having a history of chronic infestation by <u>Rattus r. mindanensis</u>. Each of these 160ha. sites involved 75-85 separate smallholder landowners.

Pretreatment census

To establish the extent of rat infestation, a pretreatment census was carried out shortly before flocoumafen bait application on all trial sites. In Indonesia the census involved laying unpoisoned bait (sweet potato) in ca. 40g units and recording consumption after three days. In the Philippines, tracking boards (vinyl tiles half smeared with printers ink) were placed along fixed transects and footprints/fresh droppings scored some 18 hours later.

Rodent control application

A commercially available flocoumafen block bait was used on one of each pair of trial sites, each block weighing 3.5g and containing 0.005% flocoumafen on a rice cereal base. Bait was applied on five occasions at 7-14 day intervals, comprising 2-3 applications of 80-100 blocks per ha to all paddy bunds/irrigation ditch banks/infested path verges and adjacent unplanted land, and 2-3 spot applications to remaining centres of heavy rat infestation. Treatment comprised placing one block in every observed rat hole, and one block every 10-15 metres along each bund. In addition, all householders were given 10 blocks and instructed to place half of them inside and half outside the house in such a way that they would be well concealed from children and domestic animals. The second of each pair of trial sites were treated by the farmer-owners using traditional, unco-ordinated rodent control techniques with acute poisons such as zinc phosphide and aldicarb, by fumigation of the rat burrows with sulphur and also by nest destruction/hunting/chasing.

Evaluation of efficacy

Bait acceptance was estimated along fixed transects in the rice fields which were counted during the first 3-5 days after application.

Crop damage was assessed by examining 25 rice hills along fixed transects across 20-50 paddies per trial site and recording the percentage of damaged tillers. In the Indonesian trials the percentage of rice plants which failed to mature due to root damage was similarly estimated.

Actual crop yield was measured in the Indonesian trials by individually harvesting 20 randomly selected sub plots each of 5 x 5 m^2 and recording the weight of unhulled rice.

Monitoring environmental safety (Philippines trials)

Prior to baiting, questionnaires were completed by each farmer in both flocoumafen-treated and control areas to gauge the number of domestic animals attached to the smallholder households. Bait attractiveness to the domestic animals was assessed by deliberately exposing baits and observing animal reactions/behaviour over a total of 86 hours.

Rice fields and village areas were carefully searched from 3-32 days after first bait application for any animal carcasses.

Wildlife population in the ricefields were monitored across three permanent 4km long transects, selected to give good coverage of both flocoumafen-treated and traditional former control sites. A total of 125 man hours were spent on seven separate monitoring occasions.

Rodent carcass residue analysis

To examine whether rats dying following STORM baiting contained residues sufficient to pose a threat to scavenging dogs, live rats were collected in snap-trays at various periods during the trial and their carcasses deep frozen and shipped to the UK for residue analysis, in accordance with the Shell Analytical Method Series (SAMS) 419-3.

RESULTS AND DISCUSSION

Pretreatment census

High infestation levels were apparent in both the Indonesian trial sites as the following data show (10% census bait take is the normal economic threshold for triggering a rodent control campaign):

TABLE 1
Census bait consumption in Indonesia

	Percentage of census bait eaten 1 day before planting out/9 days after	
Flocoumafen	98	93
Traditional farmer control	97	100

Tracking board scores in the Philippine trial sites indicated a low-medium infestation.

STORM bait acceptance

The cumulative bait take from the areawide applications are shown below:

TABLE 2
Flocoumafen bait acceptance

Percentage of flocoumafen blockbait eaten at applications I-V					
	I	II	III	IV	V
Indonesia	99	87	98	*	*
Philippines	78	85	*	*	*

*Not area-wide applications; bait only spot-applied to remaining centres of heavy rat infestation

These data indicate very high overall bait take, confirming the high palatability of the flocoumafen blockbait formulation to rice rats under field conditions with competing food sources available.

Crop damage assessment and yields

Damage levels to rice stalk clusters in the Indonesian trials, taken from four separate surveys, are given below:

TABLE 3
Damage to rice stalk clusters in Indonesia

	Percentage damage at days after planting (DAP)			
	14 DAP	28 DAP	42 DAP	54 DAP
Flocoumafen	20.1	10.4	6.6	3.4
Traditional farmer control	33.3	32.9	26.5	16.3

From these data the control levels resulting from flocoumafen baiting are clearly apparent. Similar trends were seen in the Philippines though here the results are less spectacular owing to a low general level of infestation;

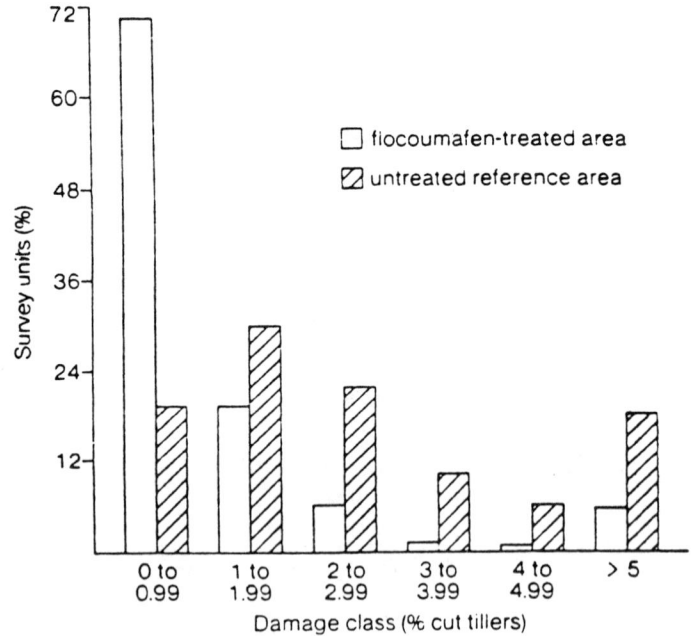

Figure 1. Frequency distribution of crop damage in the Philippines

Categorisation of the Philippine paddies into damage classes showed that 90% of the flocoumafen treated area had less than 2% cut tillers (and 6% had greater than 5% damage) compared to the untreated reference area with only 50% showing less than 2% damage and 20% with more than 5% damage.

The actual crop loss in Indonesia (recorded as the percentage of seedlings which failed to mature or failed to yield) and yield at harvest in terms of tons of dry unhulled rice per ha are shown below:

TABLE 4
Crop losses and yields in Indonesia

	Crop loss (%)	Crop yield (MT/ha)
Flocoumafen	0.4	7.29
Traditional farmer control	27.7	2.49

These data clearly demonstrate the crop yield losses that can be occasioned by a severe rat infestation and the inadequacy of traditional uncoordinated control measures taken by farmers. Three applications of flocoumafen blockbait with two follow-up spot applications in remaining areas of high infestation have resulted in a 3-fold increase in yield.

Environmental impact and effects on wildlife populations

The pre-treatment census in the Philippines showed that there were a large number of domestic animals associated with the smallholder households in both the flocoumafen-treated and the traditional farmer control sites:

TABLE 5
Domestic animals reported in the pre-baiting census in the Philippines

	Total Number of Animals	
	STORM-treated	Control
Dogs	190	95
Cats	185	114
Pigs	72	23
Chickens	485	633
Ducks	887	633
Goats	10	13
Doves	21	41
Water buffalo	40	5
Cattle	39	2
Horses	3	2
Turkey	2	2
Geese	2	0
Monkey	1	0

Results from the detailed observations on blocks deliberately laced in hazardous situations demonstrated that very few domestic animals were interested in the bait. Of the 120 animals recorded (6 species) only one dog, four chickens and one duck tried to eat the bait. None of the other animals appeared to even investigate the bait:

TABLE 6
Observations on behaviour of domestic animals around deliberately exposed baits

	Total no. observed	No. approaching the bait	No. investigating the bait	No. trying to eat the bait
Dogs	10*	2	1	1
Chickens	59	2	3	4
Ducks	36	3	3	1
Doves	11	0	0	0
Cats	2	0	0	0
Pigs	2	1	0	0

*One dog ate the bait

Although most baits in and around the village houses were well concealed and/or consumed quickly by rats, there were nevertheless ample opportunities for domestic animals to have access to and eat flocoumafen bait. No instances of confirmed poisoning were recorded however, even though the activity of domestic animals around most houses was moderate to high throughout the trial period. Furthermore, it is unlikely that any poisonings remained unreported, since a friendly relationship existed between researchers and farmers, the latter being assured of compensation for any losses to their livestock. The only incident involving domestic animals concerned one dog that was seen eating baits from the rice fields. This dog did not develop any symptoms of poisoning even though it received only a minimal treatment with vitamin K. It was therefore assumed to have eaten only a small number of blocks.

Additional information on the attractiveness of flocoumafen baits to domestic animals was obtained during the detailed studies carried out on blocks deliberately placed in hazardous situations. These observations showed that only one out of ten dogs attempted to eat the blocks, thereby suggesting that the majority of dogs did not find the baits attractive. A similar situation was found with other domestic animals; only 4 chickens and 1 duck were observed pecking at the bait. While a few individuals of these three species may be potentially at risk from exposed baits, adequate concealment of the blocks should reduce even this small risk to a negligible level. In this respect, it was interesting to note that where there was a high activity of domestic animals, householders tended to completely conceal the baits. It was thus concluded that the adequate concealment of blocks, coupled with the relative unattractiveness of flocoumafen blockbaits to most domestic animals, results in an acceptably low risk to non-target animals around houses. These results also support the conclusions from previous trials that blockbait is unattractive to birds [10].

Wildlife Populations

Of the 24 bird species seen during the 95-day observation period, only eight were recorded regularly. The fluctuations in the numbers of these eight species at both sites throughout the trial are shown below. With the exception of tree sparrows, which showed a marked increase in numbers in the flocoumafen-treated site, no statistically significant differences were observed between treated and reference areas in the abundance of any bird species.

In addition to birds, the only other wild vertebrates frequently observed during the trial were skinks (<u>Mabuya</u> spp.) frogs <u>(Rana</u> spp.) and toads <u>(Bufo</u> spp.). The numbers of these animals also varied with time but again no significant differences between sites were observed.

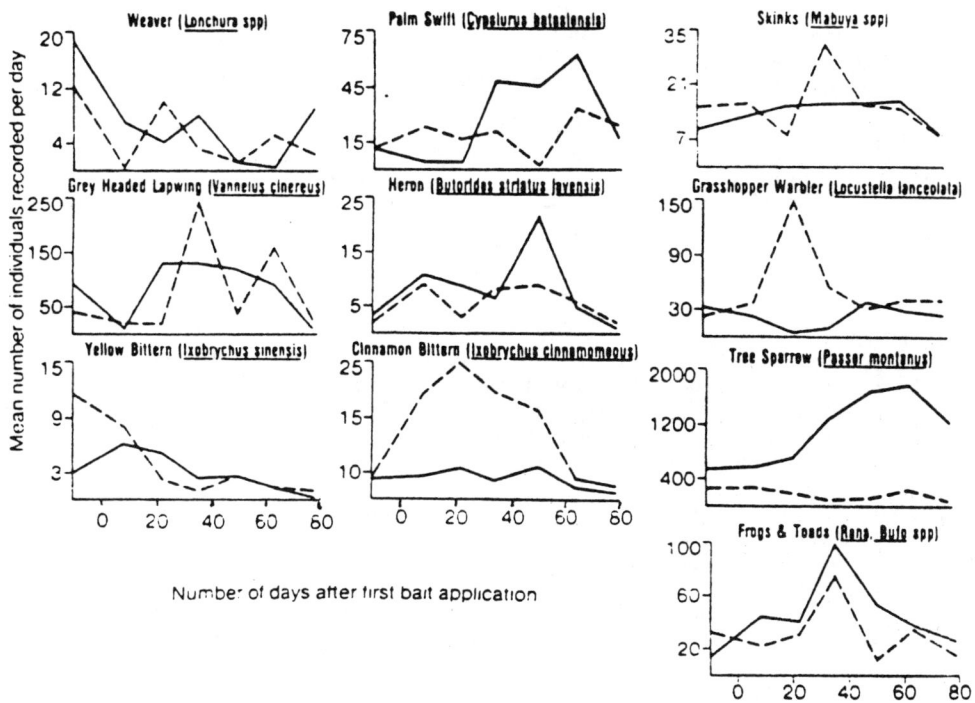

Figure 2. Temporal variations in wildlife numbers in the Philippines trials

None of the eight bird species, frogs, toads or skinks seen regularly during the trial showed any significant decreases after baiting with flocoumafen. Indeed, the only differences between sites was the significantly higher overall population of birds, frogs and toads in the treated area. Similarly, the general species richness of birds remained relatively unchanged throughout the trial in both sites. Additional evidence to support the lack of environmental impact of flocoumafen was the notable absence of bird carcasses in the trial area. The only non-target carcasses found during the study were two dead shrews <u>(Suncus</u> sp.).

None of the observed carcasses showed evidence of scavenging, although carcasses did tend to decompose rapidly in the hot humid climate, therefore affording little time for scavengers. The small number of rat carcasses found on the soil surface together with their quick

decomposition indicate that the secondary hazards associated with
flocoumafen baiting are also likely to be minimal.

Carcass Recovery and residue analysis

Only a few dead rodents were found either during the routine searches or
reported by farmers. All of these except for two shrews (Suncus spp),
were pest species, i.e., R. r. mindanensis, R. norvegicus and M. musculus.
No carcasses of non-target wild birds were found during the trials.

Live rodents trapped in the trial were Rattus r. mindanensis,
R. norvegicus and R. exulans. Total residue levels of flocoumafen in
these rodents were highest shortly after baiting (range from less than
0.02 - maximum of 1.2 mg/kg). Residues in rodents collected 10-25 days
after baiting were much lower (range from less than 0.02 to maximum of
0.03 mg/kg). Any secondary hazard of poisoning to scavenging species is
therefore likely to be limited to the period of a few days after baiting
and even then the risk is slight [11,12].

CONCLUSIONS

Rodent damage to paddy rice in S.E. Asia clearly can cause devastating
yield losses. Current traditional farmer procedures for rodent control
including fumigation with sulphur, baiting with acute poisons such as zinc
phosphide and aldicarb, and nest/habitat destruction are not sufficient on
their own to overcome the problem.

Providing smallholder farmers can be organised to co-operate together in
terms of synchronous planting and the coordinated applications of powerful
second generation rodenticides such as flocoumafen blockbait, yield
improvements by as much as 300% are achievable from only 5 applications of
bait. The use of these procedures has been demonstrated to have no
significant impact on non target species, domestic animals or wild birds.

LITERATURE CITED

1. Lam Y.M., Abdullah, H. (1975). A study of the rodent problem in
   district III of the Muda padi area Mardi report 35, 20pp.

2. Lam Y.M., (1982). Rats are rice field pests - their importance and
   control. Rodent Pests of Malaysia, K.C. Khoo et al (eds.) MAPPS,
   Kuala Lumpur pp. 9-17.

3. Buckle, A.P. F.P. Rowe, H. Abdul-Rahman, (1984). Field trials of
   warfarin and brodifacoum wax block baits for the control of the rice
   field rat Rattus argentiventer in Peninsular Malaysia. Tropical Pest
   Management 30, 51-58.

4. Schaffer, J. (1975). Crop gains - crop loss (views on the field rat
   situation in the Philippine rice crops). Plant Protection News, 4,
   12-16.

5. Shamsiah M., Idris A.G., Ngazizah I., Abd. Hamid B. (1990). The efficacy of flocoumafen (0.005%) in controlling rice field rat in Malaysia. Int. Cong. Pl. Prot. March 1990, Kuala Lumpur, Malaysia

6. Boonchanawiwat.S., K. Tongtavee, S. Somsook, S. Honkark, T. Artchawakom, K. Suasa-ard, (1987). Efficacy and environmental impact in Thailand of flocoumafen, a new rodenticide for rat control in rice fields. 11th Int. Cong. Pl. Prot. Oct 5-9, 1987, Manila, Philippines p146.

7. Ku, T.Y., (1982). Distribution and population fluctuation of field rodents and their control in Taiwan. Journal of the Agricultural Association of China, New Series No. 18.

8. Bowler, J.D. I.D. Entwistle, A.J. Porter, (1984). WL108366 - a potent new rodenticide. Proc. 1984 Brit. Crop Prot. Conf., 397-404.

9. Buckle, A.P. (1986). Field trials of flocoumafen against warfarin resistant infestations of the Norway rat (Rattus norvegicus Berk). Journal of Hygiene, Cambridge 96, 467-473.

10. Harrison, E.G., A.J. Porter, S. Forbes, (1988). Development of methods to assess the hazards of a rodenticide to non-target vertebrates. Environmental Effects of Pesticides, British Crop Protection Monograph 40, 89-96.

11. Hoque, M.M. and J.L. Olvida (1988). Rodent residue assessment in flocoumafen baited ricefields in the Philippines. Proc. 1988 British Crop Protection Conference 2, 721-726.

12. Hoque, M.M. and J.L. Olvida (1988) Efficacy and enviromental impact of flocoumafen waxblock baits used for ricefield rat control in the Philippines. Proc. 13th Vertebrate Pest Cont. Conf., Monterey, California, pp75-81.

13. Johnson, R.A. and Garforth, B. (1987). Performance and environmental studies with the new anticoagulant rodenticide flocoumafen. Stored Products Pest Control, British Crop Protection Council Monograph 37, 115-123.

14. Johnson, R.A. (1988). Performance studies with the new anticoagulant rodenticide flocoumafen against Mus domesticus and Rattus norvegicus OEPP/EPPO bulletin 18, 481-488.

15. Lazarus, A.B. (1986). Restricted application of flocoumafen bait blocks on six farms for the control of Rattus norvegicus. Ministry of Agriculture Fisheries and Food ADAS research and development services report SSD 289/86.

16. Rowe, F.P., A. Broadfield, and T. Swinney (1985). Pen and field trials of a new anticoagulant rodenticide flocoumafen against the home mouse (Mus musculus L.) Journal of Hygiene Cambridge University Press, 95, 623-627.

RECENT TRENDS IN FORMULATIONS FOR RICE

GORDON J MARRS

ICI Agrochemicals, Jealott's Hill Research Station
Bracknell, Berkshire, RG12 6EY

ABSTRACT

The development of granular formulations of pesticides for application to rice paddy is described, concentrating on the newer more specialised products aimed at controlling the persistence and/or placement of the pesticide. In the area of foliar application, formulations such as driftless dusts (DL) and microencapsulated suspensions (CS) designed to reduce drift and other potential environmental hazards are discussed. Some thoughts are offered on future trends in the research and development of formulated products for rice.

INTRODUCTION

Factors such as the increasing cost and complexity involved in the invention of new pesticide compounds and increasing demands for improved product safety, have contributed to a bourgeoning of R&D effort in the areas of novel and improved formulations. Rice, as a world crop of major importance, has been in the forefront of some of these developments so that at the present time a very diverse range of formulations are used (see table 1).

TABLE 1
Formulations used on rice

Formulation Type	Common Abbreviation	GIFAP Code
Dust	D	DP
Driftless Dust	DL	DP
Flo-dust	FD	DP
Fine Granule } micro	FG	MG
Fine Granule F } granules	FGF	MG
Granule	G	GR
Unusual Granule	UG	GR
Liquid	L	SL or UL
Solution	SoL	SL
Solution Concentrate	SL	SL
Emulsifiable Concentrate	EC	EC
Wettable Powder	WP	WP
Soluble Powder	SP	SP
Suspension Concentrate	FL	SC
Flowable Liquid	F or FL	SC

The major formulation types used for pesticides varies on a country by country basis according to local farming practice, preferred application technique, and bearable cost. For example, in Japan dusts and granules predominate, whereas in Indonesia spray formulations (such as ECs and WPs) are generally used. In most of the other rice growing countries usage of pesticide formulation types is quite diverse. Formulations such as "Driftless dusts" and "Fine granules" applied by means of pipe duster, and several types of granules for hand application, were developed specifically for use on rice. Recent research with spray formulations (using techniques such as microencapsulation and adsorbed suspensions) has concentrated on the reduction of toxicity to fish and other aquatic organisms, in order to render pesticide compounds "safe in practice".

The following sections profile the development of some of these novel formulations and indicate the direction of future trends, to enable the reader to have a fuller understanding of their uses and potential.

GRANULES

Granules are used to apply herbicides, insecticides and fungicides to paddy water in the Far East and South East Asia, particularly in Japan and South Korea. Insecticides and fungicides granules are also used as prophylactic treatments in "nursery boxes". The first granules to be introduced to the market place contained herbicides, broadcast mainly by hand either pre or post transplanting. A granular insecticide based on BHC (gamma - HCH) to control stemborers and plant hoppers was launched in 1963, followed in 1965 by the first fungicide, the blasticide iprobenfos.

The use of granular formulations has expanded considerably during the intervening years as the many advantages of this type of formulation (see table 2) became apparent.

TABLE 2
Advantages of Granular Formulations in Rice

1. Reduced toxicity

2. Reduced phytotoxicity

3. Prevention of drift

4. Suitable for hand application (no equipment needed)

5. Maximise xylem movement of compound

6. Controlled release/availability of compound

Thus by 1988 granular formulations had achieved a 40% share of the Agrochemicals market in Japan in volume terms, amounting to some 230,000 tons (see table 3).

TABLE 3
Japanese Agrochemical Market in 1988 (by Formulation type)

Formulation Type	Volume		Value	
	Tons	%	M¥	%
Dusts (DP/DL)	202,861	34.4	56,966	13.9
Granules	230,386	39.1	110,802	27.0
Liquids (EC/SL)	70,471	12.0	106,820	26.0
WP	41,776	7.1	113,537	27.6
WG	5,917	1.0	2,709	0.7
Others	37,951	6.4	19,923	4.8

Preparation Methods
Most granular formulations are prepared by one of the following techniques, or by a combination of more than one, with impregnation and extrusion being the most widely used:

1. Impregnation
2. Coating
3. Extrusion
4. Agglomeration

Direct Impregnation: A preformed inert absorbent granular carrier is impregnated with liquid toxicant (or a solution in solvent if the toxicant is a solid) by spraying the liquid onto the carrier. Suitable equipment would include a cement type mixer, double cone blender or "Nauta" Mixer. As a general guide a minimum of 5 to 10 litres of toxicant solution should be used for every 100kg of carrier, in order to obtain even distribution of the toxicant over the granules. A typical recipe and outline procedure is shown in table 4.

TABLE 4
Impregnated Granule: General Recipe and Procedure

(a) Recipe

Materials	Content % w/w
Active Ingredient	0.1 - 20.0
Solvent	0 - 10.0
Surfactant	0 - 5.0
Stabiliser	0 - 2.0
Polymeric Binder	0 - 5.0
Preformed Granular Carrier	To 100

(b)   Procedure

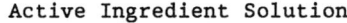

Active Ingredient Solution

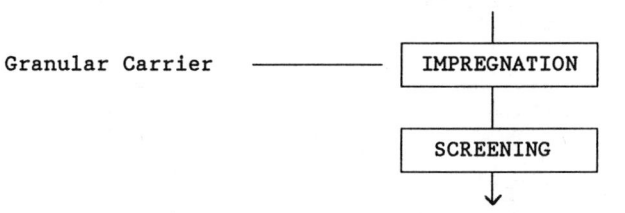

Granular Carrier ————————— | IMPREGNATION |

| SCREENING |

↓

Granular Product

<u>Coating Process</u>: A preformed non or low absorptive granular carrier is coated with powdered toxicant by means of pre and/or post spraying with a binder solution.   Suitable low absorptive carriers include sand, limestone or calcium carbonate and commonly used as binder are natural gums (eg. gum acacia) or water soluble polymers (eg. cellulose derivatives).

<u>Extrusion</u>: The pesticide, carrier powder and other ingredients are mixed together and kneaded with water to form a paste or "pug".   This is then extruded and dried, and the extrudate is crushed or spheronised and screened to obtain the desired granule size range.   Oversize and undersize material is recycled to the crusher and mixer respectively.   A general recipe and outline procedure is illustrated in Table 5.   Further information on extrusion and spheronisation techniques and equipment can be found in a recent review by Hicks and Freese [1].

TABLE 5
Extruded Granules: General Recipe and Outline Procedure

(a)   Recipe

Ingredients	Content % w/w
Active Ingredient	0.1 – 20.0
Dispersant/Surfactant	0.5 – 5.0
Binder	0 – 5.0
Lubricant	0.1 – 2.0
Stabiliser	0 – 2.0
Bentonite	0 – 50.0
Mineral Carrier (eg Clay)	To 100

(b)   Procedure

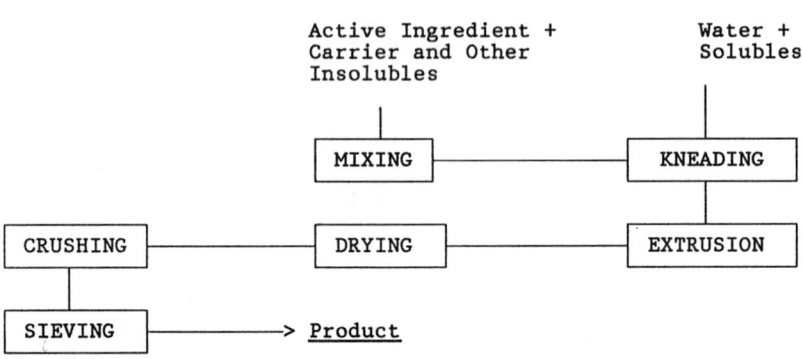

Agglomeration: Techniques available include pan granulation, "Matrix" mixing or the "Schugi" process, and are based on particle build-up induced by spraying a powder blend of toxicant and filler with water or aqueous binder solution, whilst agitating the mixture. The agglomerates produced are dried and screened to the desired granule size in the usual way.

Further information on particle size enlargement by granulation can be found in Sherrington and Oliver's book [2].

Product Specification
The MAFF and the National Federation of Agricultural Co-operative Associations (ZENNOH) in Japan have published specifications and approved test methods for granular products [3]. These cover quite comprehensively most of the parameters needed to define a satisfactory product for application to rice viz active ingredient content, granule size, apparent specific gravity, hardness (friability), flow properties (caking), disintegration in water, pH and moisture content.

A summary of the physical data obtained from a recently registered granular product, based on the plant growth regulator paclobutrazol, can be found in Table 6.

TABLE 6
Physical Data for "Smarect" Granules (6g/kg paclobutrazol)

Property	Specification	Test Result	Standard Deviation
Specific Gravity:	1.1 +/- 0.2	1.086	0.021
Hardness (friability):	< 10%	1.93%	0.53%
Water Content:	< 3%	1.3%	0.34%
Particle Size:	> 97% between 1700-500µm	99.8%	0.08%
pH:	9.0 +/- 0.5	9.04	0.14
Disintegration Rate: (in Water)	< 30 minutes	15'25"	1.47'

Developments in Granular Formulations

The use of granular formulations in combination with the flooded paddy conditions gives an opportunity to more fully exploit parameters such as release rate, placement, and systemic effects. As well as direct uptake of chemical via the plant root system, other lethal mechanisms such as capillary action between the leaf sheath and stem [4] and fumigant activity by vaporisation from the water surface [5] can be utilised.

An example of the increased fumigant effect by surface evaporation is illustrated in Table 7 which gives a comparison of the activity of different granular formulations of diazinon against plant hoppers.

TABLE 7
Biological Activity of Different Types of Diazinon Granules

	Formulation Reference		
	A	B	C
Formulation Type	Extruded Floating Granule	Impregnated Floating Granule	Coated Sinking Granule
Diazinon Content	50g/kg	50g/kg	50g/kg
Biological Activity	% Mortality of Plant Hoppers		
1 DAT*	100	100	100
3 DAT	100	100	50
5 DAT	95	90	0

* Days after Treatment

The superior effect of the two surface floating formulations over the conventional sinking granule is clearly demonstrated.

Controlled Release: By varying the granulation process and composition it is possible to prepare granules with different diffusion characteristics, thus controlling the rate of release of the active ingredient into water. Control of release can for example be used to reduce the phytotoxicity and give prolonged activity of herbicides [6] fungicides [7] and insecticides [8].

The major factors affecting rate of release are the properties of the active chemical (solubility, particle size of the technical material and affinity of the compound for the carrier) together with the characteristics of the granular product (eg. dispersibility, porosity, granule size and bulk density).

A comparison [9] between three extruded granular formulations of a herbicide (for control of weeds such as <u>Echinochloa crus-galli</u>) and an EC of the same compound, showed that the granules were more effective. The slower release formulations (granules B and C in Table 8) were the most effective, and granule C, which maintained the highest concentration of active ingredient in the water phase, showed the most biological persistence.

TABLE 8
Herbicide Granules: Physical Properties and Persistence in Paddy

(a) Physical Properties

	Granule 'A'	Granule 'B'	Granule 'C'
Size	0.7mm	0.7mm	0.7mm
No of Granules/g	1,700	1,800	1,900
Disintegration rate (mins)	1-2	6-8	Non-disintegrating
Dispersibility	Excellent	Moderate	Non-Dispersing

(b) Concentration profile of active ingredient in Paddy Water

Formulation	Approx. concentration of active ingredient in water (ppm)				
	1DAT	3DAT	5DAT	8DAT	15DAT
Granule 'A'	1.1	0.8	0.3	0.1	0.03
Granule 'B'	0.5	0.4	0.3	0.2	0.06
Granule 'C'	0.25	0.25	0.24	0.24	0.15
EC	2.5	0.3	0.15	0.06	0.025

<u>Controlled Placement</u>: There has been a lot of interest generated in producing special formulations that sink, float or form a surface film or oil, the objective being to place the toxicant more accurately where it is most needed. Sinking granules get the chemical onto the soil and into the root zone in order to maximise systemic effects. Floating formulations are targetted towards pests located near the surface of the water or for pests/pathogens on the aerial parts of the plant, working via fumigation (evaporation from the surface) or by local contact-systemic effects.

Granules are also placed in the "nursery box", to control seed/soil borne diseases (such as "damping off" and seedling blight) and early insect pests eg. rice leaf beetle, leaf miner and water weevil [10].

Surface/floating granules can be made by extrusion granulation of a mixture of active ingredient dispersed in a solution of a waxy substance (eg parafin wax) with carrier and binder [11]. Alternatively, impregnated granules on light carriers such as pumice or vermiculite can be surface coated with long chain fatty acids eg. stearic acid [12].

Recently a new type of granule, described as an "Unusual Granule" (UG) was introduced by Nippon Kayaku Co Ltd. This is an extruded granule of cycloprothrin based on a water soluble carrier/binder. According to the manufacturer when the formulation is applied to a paddy field it initially sinks to the bottom, then as it starts to dissolve it rises to the surface and releases the active ingredient which forms a film on the water surface [13]. Better control of rice water weevil <u>Lissorhoptrus oryzophilus</u> is claimed than with a conventional sinking granule, since adults and newly hatched larvae are killed by contact with the toxicant floating on the water or adhering to rice stems. Also, after a few days, the active ingredient sinks to the soil level and by this means is active against the larvae transferring to the roots of the rice plants.

The addition of low density, water insoluble oil to granular formulations assists the formation and maintainance of a surface film [14-15].

The potential advantages of the "floating oil" concept are illustrated by the results of a field trial in which a conventional sinking granule (Granule B in Table 9) and floating formulations (Granule A and floating oil) of a new insecticide are compared.

TABLE 9

Evaluation of granular formulations of an insecticide for control of rice water weevil (<u>Lissorhoptrus oryzophilus</u>)

Treatment	% of Damaged leaves (Mean of 10 hills/plot)		
	Pre-Appln	7DAT	21DAT*
Untreated	42.0	55.4	68.5A
Granule A (187.5g ai/ha)	47.1	29.4	23.3 DE
Granule B (187.5g ai/ha)	35.2	44.9	57.4 AC
Floating Oil (187.5g ai/ha)	47.5	38.2	28.0 B

* Treatment means with no letter in common are significantly different at the 5.0% probability level.

Safety Factors

Reduced operator hazard is one of the most often quoted advantages of granules over spray concentrate formulations. Granules are low strength, dust free, ready for use formulations and the absence of the need to mix or dilute the product combined with the low dermal toxicity, leads to a high degree of operator safety. The absence of drift is an environmental plus, together with the potential for improved safety to fish and other aquatic species through formulation design [16].

FORMULATIONS FOR FOLIAR APPLICATION

Research in the area of formulations for foliar application to rice has been concerned mainly with ameliorating the deficiencies of conventional dusts eg. drift, poor efficiency, and with improving the safety (especially to the environment) of aqueous sprays.

Dustable Products

The various categories of dustable product such as dust, driftless dust, fine granule etc. are most easily defined in terms of the particle size range. Table 11 lists products in descending order of size based on the specification of the Japanese MAFF.

TABLE 11
Relationship between particle size range and Registered Name

Formulation	Particle Size Range	
Description	JIS Standard (μm)	Sieve Size Tyler Mesh No.
Granule	300 - 1700	10 - 48
Fine Granule	106 - 300	48 - 150
Fine Granule F	63 - 212	65 - 250
Coarse Dust	45 - 106	150 - 300
Dust	< 45	Above 300
Driftless Dust	10 - 45	Above 300

Fine Granules

Fine granules, originally described as micro granules, were first introduced about twenty years ago as an attempt to combine the merits of dusts and granules. The original microgranule formulation (of the insecticide diazinon) was claimed [17] to have the following advantages over conventional products:

- reduced drift/contamination,
- adheres to the foliage and penetrates to the base of the plant,
- effective in paddy fields irrespective of irrigation conditions, and
- good fumigant activity.

Although quite effective with insecticides Fine Granules were considered to be too coarse for the application of fungicides, since good coverage of the crop is essential (due to the non-mobility of most pathogenic micro-organisms such as fungi) and most fungicides are less permeable to foliage and less volatile than most insecticides. Hence another new formulation termed Fine Granule F was developed for fungicides and fungicide/insecticide mixtures.

Both types of Fine Granule can be applied by hand, or more commonly by motorised knapsack applicator ("pipe duster") or by air (mainly helicopter). Field trials have demonstrated superior deposition on plants of Fine Granules compared with dusts or granules, applied both from the air and through a pipe duster [18]. Some supportive data taken from the paper by Iwata is given in Table 12.

TABLE 12
Deposition of Insecticide from different types of formulation
applied by motorised pipe duster [18]

Formulation Type	Application Rate (kg)	Average Insecticide Dep. (mg/Plant)	% of Applied Insecticide On Plant
Fine Granule	35	38.6	24.3
Granule	44	7.9	4.1
Dust	46	31.0	15.0

An added safety advantage with the use of Fine Granules for aerial applicaton is that the helicopter can fly at a higher altitude than when applying dusts or liquid sprays, whilst still maintaining reduced drift.

Fine Granules and Fine Granules F are prepared by the same processes used to make conventional granules, described in the previous section. The desired size range is obtained by crushing and sieving or some other means of classification. The need to classify the product incurs an add on cost. Several products based on organophosphate and carbamate insecticides were developed for control of rice stem borers and plant hoppers, and fungicides for control of rice blast and sheath blight.

To summarise the characteristics of Fine Granules, the advantages are perceived as:

(1)   less environmental pollution
(2)   reduced user hazard
(3)   labour saving - can be applied by hand, powered knapsack sprayer or helicopter (even in a slight wind)
(4)   superior deposition on the crop
(5)   better control of pests/diseases that attack the underneath parts of plants - particularly at heading stage.

The disadvantages are that it is difficult to see the deposit (and hence difficult to maintain uniform and constant volume application) and the manufacture is more complex and expensive.

Driftless Dusts
Driftless dusts (DL) are not strictly speaking "driftless" they drift less! They represent the other approach to resolving the drift problems of conventional dusts ie. removing most of the finest particles (those below 10 micrometers).

First marketted in Japan in 1974 they were slow to take off initially because of the ruling prevailing that only one formulation per pesticide was registerable. Pressure from farmers following the successful trials with DLs carried out by the Kyushu District Pest Control Promotion Association (KYUSHU BOKYO) and the National Federation of Agricultural Co-operative Associations (ZENNOH) led to an amendment of the regulations. This allowed the registration of standard dust formulations as DL dusts, by merely submitting a statement of change of inert ingredients. Since then although the total quantity of dustable products produced in Japan has declined, driftless dusts (DLs) account for an increasing proportion of the total (see Table 13).

TABLE 13
Total Quantities of Dustable Products sold in Japan
(Source : Society of Agricultural Chemical Industry of Japan)

	Product (Tons)			
	1980		1981	
Pesticides	DP	DL	DP	DL
Insecticides	88,458	8,482	73,176	3,903
Fungicides	73,037	1,332	60,038	4,551
Insecticide/ Fungicide Mixtures	62,864	4,985	55,459	4,532
Herbicides	265	–	186	–
Total	224,624	14,799	188,859	12,986

(Table 13 cont'd)

	Product (Tons)			
	1982		1983	
Pesticides	DP	DL	DP	DL
Insecticides	61,046	18,158	54,350	35,466
Fungicides	48,388	12,060	41,817	15,637
Insecticide/ Fungicide Mixtures	42,555	18,287	37,176	27,400
Herbicides	186	–	166	–
Total	152,175	48,505	133,509	78,503

Preparation and Properties
The active ingredient content of a DL is normally less than 5% by weight
so the most important component to consider when designing a formulation
is the diluent carrier and its physical and chemical properties. A DL
formulation has to meet a very tight specification eg. that set by ZENNOH
[19], especially with regard to the particle size distribution (see table
14).

TABLE 14
Driftless Dust (DL) - Physical Specification [19]

Property	Specification
Fineness	>95% passes 45μm
Apparent Gravity	0.7-1.10
Floatability Index	<15
Mean Particle Size	>20μm
% Particles less than 10μm	<20
Dispersibility Index	>20
Water Content	Not greater than 1%
Flowability	Less than 30 seconds

There are available special DL clay carriers that have been preground and classified by the supplier, to help the formulation chemist to meet these specifications. Most active ingredients can be incorporated by simple mixing or by dry grinding/mixing procedures.

Alternatively, a standard fine particle size carrier can be agglomerated to reduce the proportion of sub ten micrometer particles, by adding a dedusting agent such as dodecyl benzene or mineral oil. The effect of the addition of a dedusting agent is illustrated in Table 15, which compares some of the characteristics of a sample of talc before and after the addition of mineral oil, as measured using a "Malvern Instruments MASTER Particle Sizer".

TABLE 15
Particle Size data on Talc (S200 Grade) before and after
the addition of mineral oil

Property	Mineral Oil Added (g/kg)		
	0	5	10
% below 9.6μm	16.5	6.7	3.7
VMD (μm)	31.4	39.9	42.1
Specific Surface area (m²/g)	0.3667	0.2374	0.1981

The reduction in the proportion of fine particles and increase in the VMD (volume median diameter) with increase in the amount of oil addition is apparent.

Other ingredients often included in DL formulations are flow promotion aids such as fumed silica or magnesium stearate, surface active agents to improve foliar retention, and antistatic agents to prevent the build up of an electric charge during pipe dusting [20].

The Laboratory and field studies carried out during the development of DL formulations, together with suitable equipment for their application, were well summarised by Takehara [21].

The improvements offered by DL formulations compared with conventional dusts, such as:

(1)   safer operations (during loading and application)
(2)   better distribution (along the pipe)
(3)   less drift
(4)   better retention on the crop (up to 3 times that of a D) and hence
(5)   better biological efficacy, will ensure that the market for DLs will continue to grow at the expense of the D.

Aqueous Spray Formulations

Conventional Emulsifiable Concentrate (EC) Wettable Powder (WP) and Liquid (SL) formulations dominate the spray sector of the rice market and will continue to do so in the foreseeable future. However, a concern for safety, especially, to fish and other aquatic organisms, has generated a body of work aimed at producing safer spray formulations. Increasing acceptance of the concepts of "Safety in practice" of products based on intrinsically fairly toxic active ingredients has also helped to promote interest in the development of safer formulations.

Two different approaches have each met with some success, one based on the development of aqueous suspension concentrates and the other on controlled release techniques.

Suspension Concentrates (SC)

Suspension concentrates can be prepared from high melting point crystalline active ingredients with low solubility in water, by conventional wet milling processes. Further details concerning the preparation and properties of solid/liquid dispersions can be found elsewhere [22].

Active ingredients that are liquid or low melting solids can be formulated as suspensions by adsorption on to a suitable filler, followed by dispersion of the mixture in water [23]. The development and evaluation of an SC formulation for use in rice is covered in detail in the succeeding paper by Stephenson.

Controlled Release Formulations

Improved safety to fish is claimed for microencapsulated products prepared by techniques such as interfacial condensation polymerisation and phase separation (complex coacervation). Microencapsulation of permethrin in a crosslinked polyamide/polyurea copolymer made by interfacial condensation polymerisation reduced the acute toxicity to fish by more than 1,000-fold, compared to that of a standard EC [24]. The microcapsules maintained excellent biological activity against target insect species coupled with a low level of mammalian toxicity.

Microencapsulated insecticidal compositions prepared by complex coacervation using a positively chargeable colloid (gelatin) and a negatively chargeable colloid (gum arabic) are claimed to have decreased toxicity to fish and shellfish, without reduction in efficacy against target insects in paddy fields [25].

Microencapsulation is a potentially fruitful area for achieving reduction in both handling and environmental hazards associated with pesticide formulations. The pros and cons of the various techniques and their application to crop protection agents are discussed in a recent publication [26].

453

CONCLUSIONS

Much has already been accomplished in the development of granular formulations for application to rice paddy and improved products for application as dusts or sprays. There is however, scope for further research in many areas and as a personal selection, I opt for the following:

(1) further development in controlled release formulations to achieve eg. zero order kinetics (constant release rate), delayed or timed release.
(2) research to improve the application of pesticides (better distribution/retention and less wastage).
(3) more emphasis on formulations with improved operator and environmental safety, such as SC and CS types.
(4) increase in the number of one-shot mixture products.
(5) progress in the development of biological control agents as alternatives to chemical pesticides.

ACKNOWLEDGEMENTS

I am grateful to the following colleagues who allowed me to use some of their unpublished data: R.F.S Gordon, R.P Warrington and T. Nakahara (ICI Japan).

REFERENCES

1. Hicks, D.C. and Freese, H.L., Extrusion and Spheronising equipment. In Pharmaceutical Pelletisation Technology, ed. I. Ghebre-Sellassie, Marcel Dekker, New York, 1989, pp. 71-100.

2. Sherrington, P.J. and Oliver, R. Granulation, Hedyen and Son Ltd., London, 1981, pp. 7-59.

3. Physical Properties of granular formulations, 24/3/1978, National Federation of Agricultural Co-operative Associations (ZENNOH) Japan.

4. Pathak, M.D. International Pest Control, 1968, (6), pp. 12-17.

5. Koyama, T., Lethal mechanisms of granulated insecticides PANS, 1971, 17(2) 198-201.

6. Yanami, T., Enomoto, Y., Kubota, Y., Yoshimi, T. and Shimono, S. Formulation technique of granules for paddy field application. Abstracts of the Sixth International Congress of Pesticide Chemistry, IUPAC, Ottawa, Canada, August 10th-15th, 1986.

7. British Patent GB 2011788

8. Japanese Patent Application No. 61-249903

9. Wada, Y., Nakahara, T., Orii, T., Okano, Y., Aya, M., Yasui, K., Kamochi, A., Yamada, Y., Katsumata, O., Sakawa, S. and Kurahashi, Y. Physicochemical properties of formulations with respect to some specific biological effects, and methods for their determination. Proceedings of the 5th International Congress of Pesticide Chemistry, IUPAC, Kyoto, Japan, 29th August-4th September, 1982. Volume 4, pp. 257-269.

10. Hirao, J., Nursery-tray application of granular insecticides for the control of early season insect pests of rice in paddy fields. _Japan Pesticide Information_, 1984, No. 44, 10-16

11. Japanese Patent Application No. 55-154902

12. Japanese Patent JA 8600/69

13. Kirihara, S. and Sakurai, Y. Cycloprothrin, a new insecticide. _Japan Pesticide Information_, 1988, No. 53, 22-26

14. Japanese Patent Application No. 62-198602

15. Japanese Patent Application No. 63-17802

16. European Patent EP 189377

17. _Japan Pesticide Information_, 1971, No.6, p30.

18. Iwata, T. Rice insect control by Fine Granular formulations of insecticides in Japan, _Japan Pesticide Information_, 1973, No. 14, 23-6

19. Standards for physical properties of general dust and DL dust formulations to be supplied to ZENNOH, 30.8.85, Fertilisers and Pesticide Department, National Federation of Agricultural Co-operative Associations (ZENNOH) Japan.

20. Uejima, T. And Tanaka, F. On new Formulations of pesticides in Japan. _Japan Pesticide Information_, 1984, No. 45, 3-7.

21. Takehara, K. DL dusts application and deposit. No. 3 Symposium of Formulation and Application of Agrochemicals, 9th February 1983.

22. Tadros, Th.F., ed. _Solid/liquid Dispersions_, Academic Press, London, 1987.

23. European Patent EP 29626.

24. European Patent EP 183999.

25. Japanese Patent Application No. Jo-1066-104

26. Marrs, G.J. and Scher, H.B. Development and Uses of Microencapsulation.

In _Controlled Delivery of Crop Protection Agents_, ed. R.M. Wilkins, Taylor and Francis, London, in press.

A PYRETHROID INSECTICIDE FOR USE IN RICE - EFFECTS OF FORMULATION ON FISH TOXICITY AND ITS ASSESSMENT

RICHARD R. STEPHENSON
Environmental & Biochemical Toxicology Division,
Shell Research Ltd., Sittingbourne Research Centre,
Sittingbourne, Kent. UK ME9 8AG.

ABSTRACT

In the laboratory, 96 h LC_{50} values for rainbow trout (<u>Salmo gairdneri</u>) revealed that an SC formulation of the pyrethroid insecticide alphacypermethrin was much less toxic to fish than an EC formulation. Small scale enclosure tests carried out in the UK in an outdoor pond confirmed the lower hazard of the SC formulation.
Subsequently the acute toxicity of the SC formulation of alphacypermethrin to fish was assessed in the laboratory and in the field in West Java. In the field, small rice paddies stocked with common carp (<u>Cyprinus carpio</u>) and Java carp (<u>Puntius gonionotus</u>) were treated with a 'standard' insecticide or with alphacypermethrin SC; only the 'standard' insecticide caused significant fish mortality. Finally, experiments in which alpha-cypermethrin SC was used under a full season commercial spray regime revealed no adverse effects on fish growth or productivity.

INTRODUCTION

It is widely recognised that formulation is of great importance in determining the efficacy of pesticides. Less widely explored are the possibilities for the development of formulations which maintain efficacy against pests but improve environmental acceptability. This paper describes the methods used to demonstrate the benefits of a novel formulation of a pyrethroid insecticide (FASTAC*) which initial studies had indicated to be of relatively low toxicity to fish.

* FASTAC is a Shell registered Trade mark.

* RIPCORD is a Shell registered Trade mark.

The cultivation of rice in paddies involves the use of large volumes of water and in some parts of the world this has led to the simultaneous use of paddies for rice and fish cultivation. Because of this close association, it is important that pesticides used in rice should not pose a hazard to fish.

Laboratory studies can be used to determine the toxicity of technical and formulated pesticides to fish. If these tests indicate that there is a sufficient margin of safety between toxic concentrations and concentrations which might be achieved in shallow waters oversprayed at recommended application rates, then further testing may not be necessary. However, if the laboratory studies indicate that there is even a possibility of toxic effects in the field, the hazard to fish will need to be examined further. This more precise assessment of hazard to fish can best be achieved by a series of field experiments.

The sequential approach to hazard evaluation was recommended by FAO (1) and has been widely used in recent years. Such an approach was used to assess the hazard of the synthetic pyrethroid cypermethrin (RIPCORD*) to the aquatic environment (2, 3 and 4). More recently a series of studies with cypermethrin aimed at assessing its acute toxic hazard to fish when it was used for pest control in rice has been described (5). The sequence of studies with cypermethrin involved acute toxicity testing of the technical material in the laboratory, testing of formulated material in indoor tanks and finally cage tests in paddy rice.

INITIAL STUDIES IN THE UK

Laboratory tests
The acute toxicity of technical alphacypermethrin to the rainbow trout (Salmo gairdneri) was determined at Sittingbourne Research Centre in a semi-static water test with 12 hourly renewal of the test media made up in filtered (8 μm) mains tap water. Ten S. gairdneri (mean weight 3.3 g) were exposed to each of a series of concentrations of alphacypermethrin in 40 l glass aquaria at 15°C. Analysis of the fresh test media and the test media immediately prior to renewal (12 h later) by glc-ecd indicated that initial concentrations were approximately 80% of nominal values and that concentrations fell by some 25% during the 12 hours between renewals. The 96 h LC_{50} value for alphacypermethrin, based on nominal exposure concentrations, was calculated to be 2.8 μg l^{-1}. This high acute toxicity to fish is a characteristic shared by other synthetic pyrethroids (6).

Subsequent laboratory tests with two formulations of alphacypermethrin, an emulsifiable concentrate (EC) and a suspension concentrate (SC), revealed marked differences in their acute toxicity to S. gairdneri. In 96 h static (without renewal of the test media) water tests ten S. gairdneri (1-5 g) were exposed in 20 l of filtered (8 μm) mains tap water to each of a series of concentrations of the two formulations at 15°C. There were clear differences in the LC_{50} values for the two formulations (Table 1). The SC formulation with a 96 h LC_{50} of 240 μg a.i. l^{-1} was some 50 times less toxic than the EC formulation, which with a 96 h LC_{50} of 5 μg a.i. l^{-1} was of similar toxicity to the technical material.

TABLE 1

Acute toxicity of an EC and an SC formulation
of alphacypermethrin to S. gairdneri

| | LC_{50} μg ai 1^{-1} | | | |
	24h	48h	72h	96h
SC formulation	>500	380	270	240
EC formulation	5	5	5	5

Field experiments

The difference in the toxicity of the two formulations revealed in the
laboratory tests was further explored in a field study carried out in
enclosures in a small pond. The methods used were as described by Shires
(7 and 8) and involved introducing a series of open-ended stainless steel
enclosures with a capacity of ~1 m^3 into a mature experimental pond
located near Headcorn, Kent, UK. The enclosures were pushed into the pond
sediment, forming an effective seal, and the tops left open to the air.
Each enclosure therefore had a sediment/water interface and a water/air
interface. Twenty rainbow trout (~5 g) were introduced into each
enclosure. Four dosages of each formulation were tested by spraying
diluted formulation onto the water surface of different enclosures using
a hand-held aerosol sprayer. The fish were then monitored for mortality
for eight days. Temperatures during the experiment were low (~6°C).
Table 2 summarises the results of the experiment and shows that the SC
formulation, with only 5% mortality at an application rate of 300 g ai
ha^{-1}, had much less effect than the EC formulation, which caused mortality
at 30 g ai ha^{-1} but not at 10 g ai ha^{-1}.

TABLE 2

Mortality (%) of S. gairdneri after eight days
in pond enclosures treated with either the SC
or the EC formulation of alphacypermethrin

| | Dose rate (g ai ha^{-1}) | | | |
	10	30	100	300
SC formulation	0	0	0	5
EC formulation	0	30	100	100

These data from laboratory and field tests in the UK indicated that the
SC formulation of alphacypermethrin should provide a good margin of safety
for fish present in paddy rice. In view of this a further series of
experiments was carried out in West Java to fully assess the effects of the
SC formulation on fish under conditions more relevant to its use as a rice
insecticide. Simultaneous studies on the effects of the SC formulation on
important rice pests and beneficial organisms were carried out (9).

Studies In West Java

Laboratory tests

The acute toxicity of the two formulations to Cyprinus carpio (common carp) and Puntius gonionotus (Java carp) was determined in 96 h semi-static water tests with 24 hourly renewal of the test media. For each test substance seven glass vessels were filled with 20 1 of aerated tap water and quantities of a dispersion of one of the formulations added to six of them, the seventh received no test substance and served as a control. Ten C. carpio (3.5-4.0 g) or ten P. gonionotus (0.3-0.5 g) were introduced to each test vessel. The contents of the vessels were aerated and during the tests water temperatures ranged from 24 to 30°C. Under the conditions of these tests C. carpio and P. gonionotus were more susceptible to both the EC and SC formulations than had been S. gairdneri in the tests carried out in the UK and, P. gonionotus appeared to be more susceptible than C. carpio (Table 3). However the 24 h and 96 h LC$_{50}$ values for both species still indicated a marked difference in the toxic effects of the two formulations with the SC being 8 to 35 times less toxic than the EC, depending on the species tested and length of exposure (Table 3).

TABLE 3

Acute toxicity of an EC and a SC formulation
of alphacypermethrin to C. carpio and P. gonionotus

| | LC$_{50}$ (μg ai 1^{-1}) | | | |
| | C. carpio | | P. gonionotus | |
	24h	96h	24h	96h
SC formulation	460	11	20	3.2
EC formulation	4.5	0.8	0.7	0.4

In another experiment carried out in a series of 12 outdoor tanks the toxic effects of the SC formulation of alphacypermethrin were compared with those of two standard insecticides, one an EC and the other a granule (G), both widely used for pest control in rice. The bottoms of the tanks (215 x 75 cm) were partially covered with coarse gravel and then filled with tap water to a depth of 20 cm. For each insecticide three application rates were used, the recommended commercial rate (proposed rate for alphacypermethrin SC) and 1/2 and 1/4 of this rate. There was also a control tank for each insecticide which received no treatment.

Fifteen C. carpio (mean weight 6.3 g) were introduced into each tank. The fish were fed daily during the test and water temperature ranged from 23-26°C. In two of the three control tanks no mortality occurred; in the other, 3 fish died over the period day 5-7 (Table 4). Mortality in the tanks treated with alphacypermethrin did not exceed 20%, the same as the highest control mortality and was not dose-related. In all of the tanks treated with the standard EC and in the tank treated with the commercial rate of the standard granule there was a high mortality.

TABLE 4

Mortality of <u>C. carpio</u> 7 days after the application of
insecticides to outdoor tanks

Application Rate	% Mortality after 7 days		
	Alphacypermethrin SC	Standard EC	Standard granule
Control	20	0	0
1 x Commercial Rate	7	100	87
1/2 x Commercial Rate	20	100	7
1/4 x Commercial Rate	0	100	13

<u>Note</u>

Commercial rate for alphacypermethrin SC = 15 g ai ha^{-1}; standard EC= 250 g ai ha^{-1}; standard granule = 510 g ai ha^{-1}.

These results indicated that the SC formulation of alphacypermethrin was unlikely to be toxic at commercial rates in the field and a series of field experiments was therefore carried out to see if this was the case.

<u>Field Experiments</u>

<u>Acute toxicity study</u>

This study was designed to assess the acute toxic effects of the SC formulation of alphacypermethrin on fish under field conditions in rice paddies. The experiment was carried out in a series of purpose-built paddies using both caged and free fish, and compared the acute lethal effects of alphacypermethrin SC with that of the standard EC used in the outdoor tank tests. The experiment was carried out 12 days after trans-plantation when the rice was at an early stage of development.

Each plot of paddy rice was 5 m x 5 m and had a diagonal trench across it which was 20 cm deeper than the rest of the plot and approximately 50 cm wide. On the day prior to treatment with the insecticides the water depth in the plots was adjusted to 10 cm and the plots sealed for the 7 days of the study. There were three replicate plots of each of the following treatments:-

Control -	no insecticide
Alphacypermethrin SC	7.5 g ai ha^{-1}
Alphacypermethrin SC	15 g ai ha^{-1}
Alphacypermethrin SC	30 g ai ha^{-1}
Standard EC	200 g ai ha^{-1}

The rates for alphacypermethrin were chosen to bracket the likely commercial rate of 15 g ai ha^{-1}. All applications were made at 500 1 ha^{-1} by knapsack sprayer.

Prior to application of the insecticides 4 fish cages were placed in the trench in each plot. Two cages in each plot contained 20 C. carpio (2.5-5.0 g) and two 20 P. gonionotus (1.0-4.5 g). In addition 30 C. carpio and 30 P. gonionotus were released to swim freely in each plot. The plots were checked at least twice daily and fish found dead recorded. Water temperatures during the experiment were 25-34°C.

The percentage mortality of the caged and free fish at the end of the experiment is given in Table 5. Results for the two species were similar. For both, only the standard EC caused significant mortality, 70% or greater in all cases except for the free P. gonionotus where only 36% died. The mortality of caged or free fish in the plots treated with alphacypermethrin SC only exceeded 7% in one case (when it was 15%) and generally was less than 5%.

TABLE 5
Mortality of C. carpio and P. gonionotus during the seven days
after application of insecticides in an experiment in Bogor, West Java

| Treatment | Mean mortality (%) | | | |
| | Caged fish | | Free fish | |
(g ai ha^{-1})	C. carpio	P. gonionotus	C. carpio	P. gonionotus
Control	1	1	1	1
Standard EC (200)	71*	70*	82*	36*
Alphacypermethrin SC (7.5)	7	3	2	0
Alphacypermethrin SC (15)	15	5	1	1
Alphacypermethrin SC (30)	2	4	0	1

* Significantly higher mortality than in the control.

The lack of a significant effect of the SC formulation of alphacypermethrin was encouraging, particularly because the conditions under which the experiment was carried out were such as to have maximised the hazard. The application took place early in the growing season when crop-cover was minimal; the water was only some 10 cm deep; the plots were sealed during and after the insecticide applications and the C. carpio used were small. In the light of these promising results a further set of experiments was carried out the following year, again in West Java.

Effects on growth and productivity
    In these experiments, the effects of the alphacypermethrin SC and two widely used rice insecticides (the standard EC and the standard granule used in the outdoor tank experiment) on the survival and growth of C. carpio were examined under a season-long, commercially recommended spray regime (Fig. 1).

TABLE 6

Mean mortality (%) of C. carpio in the 48 hours after application
of insecticides at varying numbers of days after transplanting rice (DAT).
No application of granules of the standard insecticide were made 21 DAT or
50 DAT. This trial took place at Pusakanagara, West Java

| Treatment | Mean mortality % | | | |
| | Experiment 1 | | Experiment 2 | |
(g ai ha⁻¹)	(21 DAT)	(33 DAT)	(50 DAT)	(64 DAT)
Control	0	0	0	0
Alphacypermethrin SC (15)	1	0	0	0
Standard EC (250)	18	0	35	0
Standard G (517)	-	0	-	0

At the end of each experiment a significant proportion of the fish
introduced had not been recovered as corpses during the experiment or as
live fish at the end. This proportion ranged from 36% to 71%. These fish
are thought to have been taken by predators, probably snakes, considerable
numbers of which were found when the plots were drained. Despite the loss
of these fish interpretation of the data on fish growth and productivity
is clear. None of the treatments had a deleterious effect on fish growth
(Table 7), indeed in the first experiment the fish from plots treated with
the alphacypermethrin SC had grown significantly more than those in the
control plots. However, the total weight of fish harvested from the plots
treated with the standard EC was markedly less than that from the control
plots, plots treated with alphacypermethrin SC or those treated with the
standard granule. This was due to the fish mortality caused by the
applications of the standard EC.

TABLE 7

Growth of C. carpio in paddies treated with insecticides at varying
numbers of days after transplanting rice (DAT), following a commercial
spray regime. The mean weight of the fish at the start was 7.6 g in
Experiment 1 and 6.0 g in Experiment 2.

| Treatment | Experiment 1 (20-49 DAT) | | Experiment 2 (49-69 DAT) | |
| | Mean weight (g) | Total weight (g) | Mean weight (g) | Total weight (g) |
(g ai ha⁻¹)				
Control	23	800	14	330
Alphacypermethrin SC (15)	30*	1100	17	450
Standard EC (250)	29	410	-	-
Standard G (517)	21	1100	16	440

* Significantly different from control (p <0.05)

Figure 1
Spray regime used to study effects on fish survival
and growth in two experiments

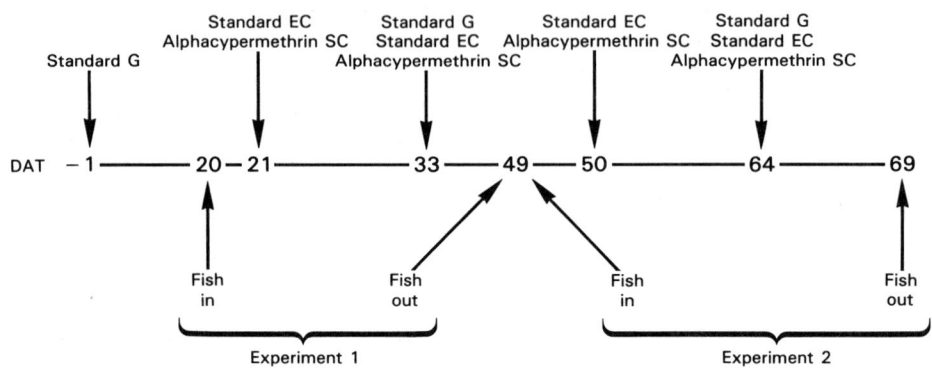

DAT = Days after transplantation

As in the previous field study a replicated experiment design was used, four plots of paddy rice (5 m x 10 m) were treated with each insecticide and four remained untreated as controls. Each insecticide was applied at the recommended rate (see Table 4) following a pre-established commercial spray regime.

Two experiments were carried out. During each, the plots received two applications of either alphacypermethrin SC or the standard EC and one application of the standard granule (Fig. 1). The alphacypermethrin SC and the standard EC were applied by calibrated knapsack sprayers and the standard granule by hand.

Twenty weighed <u>C. carpio</u> were released into each of the plots, 20 days after transplantation of the rice for the first experiment and 49 days after transplantation for the second. The mean weight of the fish in the first experiment was 7.6 g and in the second 6.0 g. The depth of the water in the plots at the time of spraying was less than 10 cm and the subsequent flow of water into the plots was limited to that required to replace losses resulting from evaporation and leakage. The plots were checked daily for dead fish and at the end of each experiment the plots were drained down and the surviving fish collected and weighed.

The mean values for mortality of <u>C. carpio</u> assessed on the basis of dead fish found during the 48 h following application of the insecticides are given in Table 6. Only the standard EC posed an acute lethal hazard to the fish, neither of the other treatments causing any significant mortality during the 48 h after treatment.

CONCLUSION

Taken together, the results of this series of experiments in the UK and West Java provide convincing evidence of the lack of hazard of alphacypermethrin SC to fish in rice paddies. What is more, the experiments indicate the value of a step-wise approach to evaluating the hazard posed by pesticides to fish in rice paddies. The initial experiments in the UK demonstrated that alphacypermethrin SC was less toxic to rainbow trout than alphacypermethrin EC. This was confirmed in simple enclosure experiments in the field.

Subsequent experiments in the laboratory and in the field in West Java confirmed that C. carpio and P. gonionotus were also less susceptible to alphacypermethrin SC than to alphacypermethrin EC. Field experiments were then used to show that there were no acute lethal effects on these fish when paddies were sprayed with alphacypermethrin SC at double the proposed commercial rate under conditions which posed maximum hazard. Finally, experiments under a season-long commercial spray regime showed no effects of alphacypermethrin SC on fish growth or productivity.

ACKNOWLEDGEMENTS

The help and support of the following is acknowledged;
Supomo Th. Wardoyo (BIOTROP Bogor W. Java); Dandi Soekarna (BORIF Bogor W. Java) and Santosa Koesoemadinata (Inland Fisheries Laboratory Bogor W. Java).

REFERENCES

1. FAO (1981) Second Expert Consultation on Environmental Criteria for Registration of Pesticides. FAO Plant Production and Protection, Paper 28 FAO, Rome.

2. Stephenson, R.R. (1982) Aquatic toxicology of cypermethrin. I. Acute toxicity to some freshwater fish and invertebrates in laboratory tests. Aquatic Toxicology 2, 175-185.

3. Crossland, N.O. (1982) Aquatic toxicology of cypermethrin. II. Fate and biological effects in pond experiments. Aquatic Toxicology 2, 205-222.

4. Crossland, N.O.; Shires, S.W; Bennett, D. (1982) Aquatic toxicology of cypermethrin. III. Fate and biological effects of spray drift deposits in freshwater adjacent to agricultural land. Aquatic Toxicology 2, 253-270.

5. Stephenson, R.R. (1984) Determining the toxicity and hazard to fish of a rice insecticide. Crop Protection 3, (2), 151-165.

6. Hill, I.R. (1985) Effects on non-target organisms in terrestrial and aquatic environments in The Pyrethroid Insecticides Ed. J.P. Leahey, Taylor and Francis, London and Philadelphia.

7. Shires, S.W. (1983) The use of small enclosures to assess the toxic effects of cypermethrin on fish under field conditions. Pesticide Science <u>14</u>, 475-480.

8. Shires, S.W. (1985) Toxicity of a new pyrethroid insecticide, WL85871, to rainbow trout. <u>Bulletin of Environmental Contamination and Technology</u> <u>34</u>, 134-137.

9. Shires, S.W. (1986) An integrated approach to developing a new rice insecticide. <u>Proceedings of the Second International Conference on Plant Protection in the Tropics</u>, Genting Highlands, Malaysia 198-201.

THE EFFECT OF ETHEPHON PLANT GROWTH REGULATOR UPON THE YIELD COMPONENTS OF RICE

T.G.SZOKE

RHONE-POULENC CROP PROTECTION DIVISION

LYON - FRANCE

ABSTRACT

The ability of ethephon to increase the yield of rice has been evaluated since 1981. Rates of (or close to) 240, 360, 480, 720 gai/ha have been used in both direct drilled and transplanted crops in a variety of territories including the USA, S. America and Asia.

During these years both critical replicated and large plot demonstration type trials were conducted to assess the suitability and the potential of ethephon for this use.
In the majority of these trials (more than 90 %) substantial yield increases (above 10 % and often 20-30 %) were obtained. The yield increases resulted from the increase of : panicle bearing stalks (tillers) ; the number of well filled grains and to a lesser extent the size and weight of panicles ; the weight of grains (TGW) and the decrease of sterile spikelets within the panicle.

The rates enabling consistently good results to be obtained were 240-360 gai/ha in the Far East whilst somewhat higher rates of around 480 gai/ha were more effective on the American continent.

Obtaining consistent positive results appears to be closely related to making applications during a limited critical period. The currently available data indicate that application should coincide with the start of panicle primordia differentiation in the main culms.

INTRODUCTION

Rice is a major food crop for over 1.5 milliard people and a secondary staple food for another 0.5 milliard. Phenomenal progress in rice production has already been made through breeding and improved production technologies. In spite of the apparent surpluses in some countries, it is estimated that a further substantial increase in rice production will be necessary to meet the needs of the growing world population (1).

Various plant growth regulators have been evaluated in rice during the last twenty odd years. Most of these, including the recently developed inabenfide (Seritard) were intended to prevent lodging, which could induce substantial yield losses in some areas and growing conditions.

Ethephon (2-chloroethyl phosphonic acid) is a plant growth regulator which is readily absorbed by plant foliage and, through a simple base-catalysed reaction releases ethylene plant hormone. It was also tested in the seventies to shorten the rice straw thus increase lodging resistance, and encouraging results were obtained. The interest remained in lodging control for a few years, but pinpointing the right timing happened to be difficult and necessitated conducting thorough timing studies. During this work yield, and in many instances the yield components, were also recorded and thus it was noticed that at specific application stages substantial yield increases of up to 30-40 % were obtained.

During the subsequent international development program, ethephon has been tested in all the major rice growing areas of the world for the purpose of increasing yield. The results confirmed that a positive yield response can be obtained with a good consistency i.e. in more than 90 % of the trials. No yield decrease occurred in any of the trials, thus the perfect selectivity of the compound has also been established. The following examples are intended to illustrate the effect of ethephon treatment on the yield and yield components of rice and the importance of right timing and rates of application. For convenience they are reported chronologically and by geographic areas.

RESULTS

The first indication we own (but others might exist) that etephon will increase rice yield in absence of lodging i.e.through direct physiological effects upon plant development was observed in 1981 (Louisiana USA). Whilst this trial was also laid out for the purpose of lodging control (180 kg excess nitrogen was applied to eventually provoke it) the treatment was 72 days after planting and besides lodging (which was actually minimal) and yield some of the yield components were also recorded :

TABLE 1

Treatments	TGW	Panicles/ 9 SQ.FT	Panicle Weight	Yield kg/ha
Untreated control	24.02 (b)	452.2	0.507 (b)	2622 (b)
Ethephon 0.56 kg/ha	25.15 (a)	462.4	0.656 (a)	3806 (a)

(* figures followed by different letters denote highly significant statistical differences).

In this trial the number of panicles has only been slightly and non significantly increased, therefore the substantial and highly significant yield increase was primarily due to the increase of panicle weight ant TGW.

During 1982 and 1983 the trials were conducted in Ecuador where paddy rice is grown the whole year round and therefore it was possible to study the effect of ethephon during both the dry and wet seasons.

The summarised results of this extensive trial work are presented below :

TABLE 2

	Ethephon rates kg ai/ha		
	0	0.24	0.48
Phytotoxicity	-	None	None
Tillers per m2	309	309	308
Panicles per m2	246	278	292
% Sterile florets	8.4	5.6	5.0
Yield kg/ha	5760	8080	8210
% Untreated control	100.0	140.3	142.5

Similar results i.e. substantial yield increases were obtained in 1983 with rates of 0.24-0.48 kgai/ha. These were seemingly coming from the increase of panicle numbers although a slight reduction of the sterile spikelets also occurred. The number of tillers as such was not increased.

The rather well educated hypothesis put forward at that time was that the temporary inhibition of the main stalks allowed an additional number of (existing) tillers to develop the capability of producing a fully mature panicle at harvest. In other words, directly or indirectly, ethephon promoted panicle development in those tillers which otherwise would not have had that capability. It thus made sense to adjust the timing of application so that it coincided with panicle primordium initiation in the main stalk. The recommendation was to apply ethephon when the panicle primordium was 5-15 mm long, and for practical purposes this was translated into a timing determined by the number of days elapsed since seeding (DAS).

In the following two years (1984/85) mostly demonstration work was conducted in Ecuador and over a great number of trials an average of 20 % (range 12-39 %) increase was obtained, corresponding to 875 kg rice grain/ha in the local conditions.

In the trials where ethephon was applied in transplanted rice, the yield increase was on average 24 % (range 11-39 %) whilst in the direct seeded crop only 15 % (range 12-17 %). Subsequently it has been confirmed that transplanted rice responded better to the ethephon treatment.

In the concurrently conducted small plot trials some of the yield components were also recorded, and the following data obtained with the var. INIAP 415 is a typical example :

TABLE 3

Yield Components	Treatments	
	Ethephon 0.24kg ai/ha	Untreated control
Panicles/m2	311.00	262.00
Grains/panicle	105.72	102.88
Panicle length cm	23.10	21.78
TGW	30.10	27.70

The yield increase obtained in this trial was 12 % due to the significantly increased number of panicles and perhaps also the slight increase in grain numbers and TGW.

A registration approval was issued for this use of ethephon in Ecuador in 1985.

Similar trials were conducted in the Philippines with three rates of ethephon (0.24, 0.48 , 0.72 kg ai/ha) applied 53-60 days after seeding. This timing was aimed to coincide with panicle initiation. The average yield results obtained at four trial sites are presented in the following table.

TABLE 4

Ethephon rates g a.i./ha	Yields obtained at the 4 trial sites with the indicated varieties T/ha			
	IR 58	IR 36	IR58	SELECT 183
Untreated control	5.50b	4.59b	4.75b	4.87b
240	6.12ab	5.15ab	6.00a	6.52a
480	6.87a	5.46a	6.12a	5.75a
720	6.37ab	5.22a	5.75a	6.25a

(* figures followed by different letters denote
highly significant statistical differences)

In all four trials significant yield increases of 11-34 % were obtained corresponding to 560-1650 kg/ha of rice.

The effect of various yield components was also investigated and the results obtained with the var.IR 64 are as follows :

TABLE 5

Ethephon G ai/ha	Height (cm)	N* Productive tillers	Panicle Length (CM)	N* Filled grains	N* unfilled grains	Weight 1000 seeds (G)	Grain yield Tons/ha
240	95.5 a	25.9 b	23.9 b	68.9 a	8.7 b	29.1	4.47 b
360	98.4 a	27.5 a	24.9 a	70.7 a	5.9 c	29.2	4.92 a
0	86.2 b	21.7 c	22.4 c	56.6 b	25.3 a	29.1	3.78 c

Applied once at 45 days after sowing (maximum tillering).
Solution used was 200 litres/ha.

(* figures followed by different letters denote highly significant statistical differences)

It is interesting to note that plant height was increased through ethephon treatment. The TGW remained unchanged, therefore the 18-30 % yield increase was due to the increase in productive tillers and the size of panicles (and perhaps weight).

Similar results were obtained in Indonesia with the varieties IR 36, IR 48, IR 64 and Cisadane.

On the South-American continent, the influence of ethephon upon rice yield and its components was thoroughly investigated with the variety Araure-1. Having a growth cycle of 120-140 days, its panicle primordia initiation occurred at 65 days after sowing and this was selected as the preferred timing of application for it gave better results than the 55 DAS and 75 DAS treatments. The rates tested were 0.24, 0.36, 0.48, kg ai/ha, alongside untreated controls.

The treatments had no significant effect upon plant height, flag leaf length, panicle length or TGW.

The average tiller number (TN) panicle number (PN) number of grains per panicle (GPP) and the yield were all significantly increased by the ethephon treatment to various degrees depending on timings and rates. The results obtained at the best timing of 65 DAS were as follows:

TABLE 6

Ethephon rates	PN	TN	GPP	YIELD Kg/ha
0.00	359	389	141	6337
0.24	430	458	158	7100 (12%)
0.36	470	484	165	7600 (20%)
0.48	540	564	180	8538 (35%)

Under local conditions, the highest rate of 0.48 kg ai/ha appeared to be the most effective : based on these results the farmers net income in Venezuela could be increased by 18,31 and 57 % respectively with increasing rates of ethephon.

In the United States the trial work has been aimed at selecting the best timing of application with a rate of 425 gr ai/ha. The earlier timings of 35-40 days after emergence proved to be more effective as shown by the following results obtained in Louisiana.

TABLE 7

Application stage	Spikelets per 10 panicles	Panicles per SQ.FT	Yields kg/ha	% of untreated control
Untreated control	1175	74	6312.5	100.0
35 DAE	1321	83	7434.8	117.8
42 DAE	1439	85	7335.5	116.2
49 DAE	1230	84	7119.8	112.8

The increases in the number of spikelets in the panicles and number of panicles were not statistically significant, though these factors in combination probably accounted for the significant yield increases. Indeed the other yield components were not altered by the ethephon treatment.

DISCUSSION

It appears from the trial results (of which the above examples are typical) that in most cases TGW is not, or only slightly, increased by the ethephon treatment and is not a major contributor to the yield increases. Neither is the number of tillers which is only occasionally increased.

What seems to be responsible for the yield benefits is a higher number of panicles (fertile tillers) and the number of well filled grains they bear. Occasionally the size and weight of panicles is also increased and at the same time the number of sterile spikelets (unfilled grains) is decreased.

Various theories could be put forward to explain the influence of ethephon upon rice yield and its components but the most likely hypothesis is that of Aufhammer (2). According to this, a relatively strong apical dominance (i.e. that of the main stalk) exists within a cereal plant at certain stages of its development, meaning that only a few tillers will be fertile and even these will be retarded compared with the main shoot. A similar type of dominance would also exist within the central region of each ear or panicle resulting in slower grain development at the top and the bottom of these organs.

Although it depends on the hormone balance (homeostasis) in the plant, apical dominance is primarily an auxin (IAA) mediated process. Therefore, all factors influencing auxin activity will eventually have an effect upon apical dominance. Ethylene is known to antagonize both auxin synthesis and transport and possibly stimulates some of the enzymes catalysing its deactivation. Therefore by applying ethephon (ethylene) at the time of panicle primordium initiation in the main stalk which corresponds with increased auxin synthesis in its apical meristem, the dominance over the tillers will be restricted to some extent enabling some of them to develop a panicle when normally they would not be able to do so : the results obtained are very much in agreement with this somewhat simplified physiological scheme.

The remaining, but all important question is, how to apply ethephon in order that the desired effect is consistently obtained.

Whilst the appropriate rate is reasonably well defined, the correct timing of application probably remains the critical point. Indeed the beginning of panicle primordium differentiation is not a visible occurrence. For this reason it has been attempted to correlate it to well identifiable phenological stages such as tillering which yielded acceptable results with a great number of, but not all, varieties.

A further problem is that panicle initiation occurs first in the main culms, the dominance of which over the tillers is supposed to be reduced by a timely treatment. This is however not easy in a field of rice which is composed of plants at different development stages. This situation is alleviated though, since ethephon remains in and releases ethylene into the rice plant for 2-3 weeks. This might also explain why relatively early applications made probably before panicle primordium differentiation were effective whilst treatments made after panicle initiation were not. It seems therefore that once the dominance of main culms (the correlative inhibition stimulus) has taken place it is not possible to reverse it, at least with the relatively low ethephon rates used to date.

The work conducted so far has concentrated on the determination of panicle primordium differentiation time for the most important rice varieties grown in the major areas and has attempted to connect this to easily identifiable parameters such as time interval since sowing or emergence or transplanting, heat units (DD 50), or morphological changes.

REFERENCES

1) RAO K.P. - PROCEEDINGS 1987 CALIFORNIA WEED CONTROL CONFERENCES

2) W.AUFHAMMER (1981) : ROLE OF PLANT GROWTH REGULATORS IN WHEAT YIELD, BPGRG Monograph n° 6-1981 131-140

ACKNOWLEDGMENTS

The results presented in this report have been obtained in trials conducted by RHONE-POULENC's Technical Services or through contract work by independent cooperators. The author wishes to express his thanks to all the colleagues and friends who contributed to the development of this data.

PACLOBUTRAZOL: CONTROL OF LODGING IN JAPANESE PADDY RICE

PAUL FRENCH
ICI Agrochemicals, Jealott's Hill Research Station
Bracknell, Berkshire RG12 6EY

HITOSHI MATSUYUKI AND HIROSHI UENO
ICI Japan Agrochemicals Division, P O Box 411, Tokyo 100, Japan

ABSTRACT

Paclobutrazol [(2RS, 3RS)-1-(4-chlorophenyl)-4,4-dimethyl-2-(1H-1,2,4-triazol-1-yl)pentan-3-ol], previously coded PP333, is a plant growth regulator developed by ICI Agrochemicals. This paper describes official and in-house trials conducted on rice throughout Japan. Results demonstrate reliable stem shortening and a reduction in lodging. This generally increased yield and improved work efficiency at harvest. Factors that may influence activity and the effect of variable farmer use patterns are also discussed eg. temperature, soil type and water management practices. The 0.6% granule formulation now on sale in Japan is recommended at 20-30 kg product/ha (120-180 g ai/ha), applied 15-10 days before heading. The product has low toxicity and crop residues so does not present a hazard to the farmer, or consumer.

INTRODUCTION

In 1986 Takano [1] clearly described the diversity of Japanese agricultural and the regional variations. Since that publication government policies have brought about a further reduction in the area planted to rice [2]. In turn this has prompted an increase in the area planted to the "quality" cultivars Koshihikari and Sasanishiki. Both of these cultivars, introduced over 20 years ago, have grown in popularity because of their outstanding flavour, quality and high yield potential; even though they are prone to lodge and have low resistance to disease and insect pests. Rice farmers in Japan can rely on agrochemicals to overcome these latter problems but have previously had no effective chemical to control lodging; a regular phenomena because of the high incidence of wind, rain and even typhoons, just before harvest.

Ohta [3] clearly demonstrated the cost and quality benefits that can be achieved when rice lodging is prevented. This paper describes how the chemical dwarfing agent paclobutrazol can be used to reduce the risk of lodging. Lever et al [4] first described the chemical, physical and growth regulatory properties of this triazole. Later Dalziel and Lawrence [5] detailed the mode of action while a previous paper by the present authors [6] described early field trial results on rice.

Since the spring of 1989, paclobutrazol has been commercially available in Japan for use in rice paddies as a 0.6% ai granular formulation called "Smarect G".*

Application should be made 15-10 days before heading at 20-30 kg product per hectare (ie. 120-180 g ai/ha). The higher rate is preferred in most situations.

MATERIALS AND METHODS
Chemical name (IUPAC): (2RS, 3RS)-1-(4-chlorophenyl)-4,
4-dimethyl-2-(1H-1,2,4-triazol-1-yl)
pentan-3-ol.

Common name (BSI approved): paclobutrazol
Code number: PP333
Structural formula:

* is a trademark of Imperial Chemical Industries PLC.

Formulation: 0.6% ai granule

Application rate: 20-30 kgs product/ha
                      (120-180 g ai/ha)
                      The preferred rate for most conditions is
                      180 g ai/ha.

Application timing: 15-10 days before heading.

Crop: Japonica rice.

Experimental design: Small plot randomised block, replicated 2 or 3
                      times.

Assessments: Stem and/or internode length, lodging and yield.

RESULTS

1. The effect of rate and timing on stem length, lodging and yield.

Early trials concentrated on establishing the optimum rate and timing
of application, [2],[6]. The dose response was found to be relatively
flat, with optimal stem shortening and subsequent lodging reduction
following paclobutrazol application at rates between 120 and 240 ai/ha
(Figure 1).

Figure 1. Paclobutrazol dose response on rice stem length and subsequent
          lodging control; var Koshihikari.

Concurrent studies showed that greater treatment benefit resulted from applications made 15-10 days before heading (DBH), which gave maximum reduction of the third internode, (Figure 2).

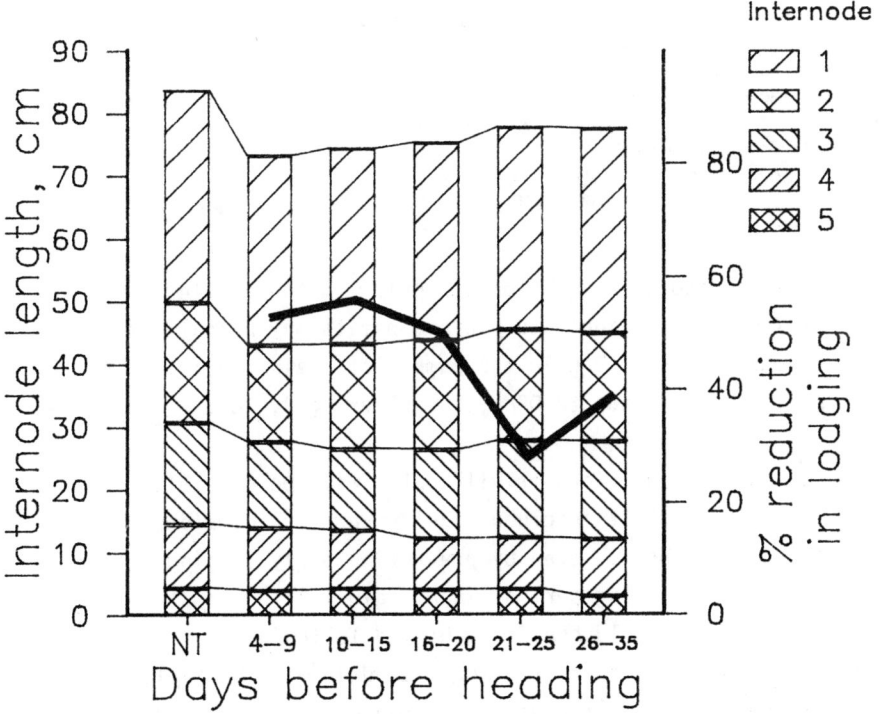

Figure 2. Effect of application timing on internode length and lodging; var Koshihikari, application rate 180 g ai/ha, mean of trials 1983-88.

By this growth stage rice has already completed tillering and the embryonic panicle is formed. So paclobutrazol application at this timing cannot interfere with these components of yield. In fact paclobutrazol treatment delays the on-set of lodging, hence increases the percentage of well filled grains and yield. (Table 1 and Figure 3).

TABLE 1

The typical effect of paclobutrazol on time of lodging

Location and treatment	Final plant height (%)	% lodging at various days after heading				Yield as percentage of untreated (%)
		20	30	40	Harvest	
Toyama 1988 (cv. Koshihikari)						
'Smarect' 18 DBH 180 g ai/ha	88	0	0	40	70	113
Untreated	100	50	60	80	100	100
Kagoshima 1988 (cv. Koshijiwase)						
'Smarect' 15 DBH 180 g ai/ha	94	0	0	10	20	110
Untreated	100	10	36	60	70	100

If paclobutrazol is used to reduce lodging prior to the earlier recommended timing (15 DBH) adequate yield benefits may not always result. For example a reduction in the number of grains/panicle can occur when application is made 20 DBH, or earlier, [2], [6].

In very severe weather conditions it will not always be possible to totally eliminate lodging with paclobutrazol. However, considerable yield benefit can still be achieved by preventing early lodging, through the early post heading period, when grain filling is still in progress (Table 1). The yield benefit achieved following paclobutrazol treatment was further demonstrated in a large field trial programme (60 trials) carried out throughout Japan between 1984 and 1988, (Figure 3).

In 46 out of 60 trials a yield increase was recorded. By examining the detailed yield component results it is possible to attribute the higher yield to an increase in the percentage of ripened grains. However increases in grain weight are not the only harvest benefits. In the normal, part-time farming situation that predominates in Japan, efficient use of time is essential. Table 2 clearly illustrates the overall benefit of paclobutrazol treatment : a large reduction in lodging, reduced harvesting time as well as weight increases and improved grain quality. The latter is particularly important and can result in the farmer being paid a premium of over 10% for his harvest.

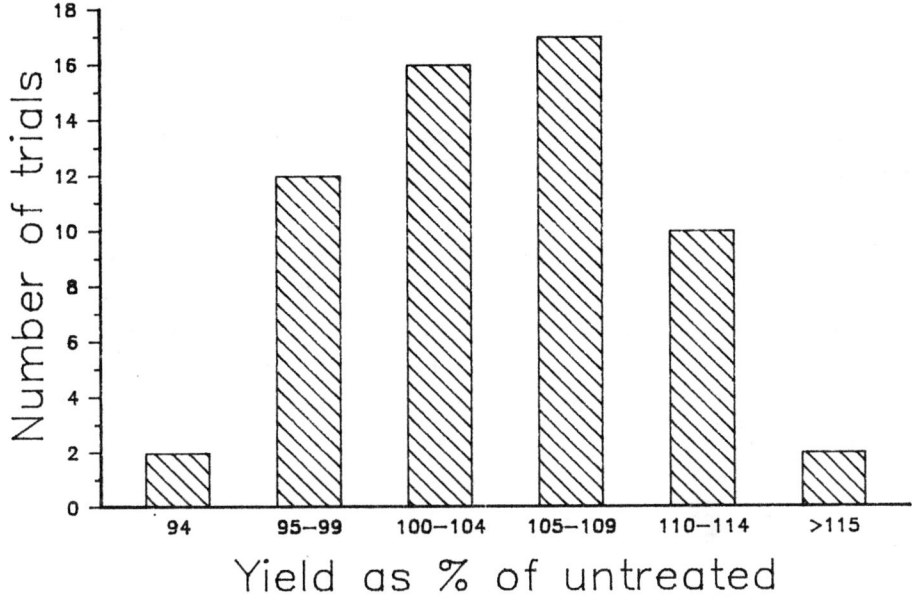

Figure 3. The influence of paclobutrazol on yield, applied at 180g ai/ha, 18-8 DBH. var Koshihikari, mean of trials 1984-88.

TABLE 2
Harvest benefits resulting from paclobutrazol treatment

Treatment	Applic time	lodging index*	time to harvest 0.1 ha**	yield from 0.1 ha	quality grade
Untreated		42.1	191 mins	572 kg	B
180g ai paclobutrazol	17 DBH	4.5	112 mins	651 kg	A

   * Coefficient of lodging degree x area lodged - max coefficient
 ** Using a 2-row combine harvester (std. commercial machine).

2. Factors influencing activity.

Paclobutrazol granules applied to paddy water are rapidly dispersed. Dissolution of the active ingredient into the water phase is then followed by adsorption onto paddy soil over the next few days. This facilitates uptake into the plants via the roots, although small amounts may also be taken in through the stem [2], [4], [5] and [6]. Hence uptake can be affected by factors such as; temperature, soil-type, crop variety and water management practices.

Growth chamber experiments conducted at the Hokuriku Agricultural
Experimental Station (MAFF), Japan, showed some changes in activity
with water temperature. At a mean of 16 or 20°C, paclobutrazol
performed as expected, giving a mean 25% reduction in the length of
the third internode. But at a mean temperature of 24°C activity
increased, reducing this internode length by 45%.

In trials throughout Japan, paclobutrazol showed a high degree of
consistency on the important paddy soils ie. sandy loams, loam and
clay loams (Table 3). On two less important paddy soils, some change
in response was detected. On clay soils activity increased, such that
an application of 120 g ai/ha produced a stem length reduction equal
to that achieved with 180 g ai/ha on loam soils. Conversely, on the
black volcanic soils 180 g ai/ha was needed to match the 120 g ai/ha
effect achieved on loam soils.

TABLE 3
The effect of paclobutrazol dose rate and soil type on stem length

Soil type	No trials	dose rate g ai/ha	mean stem length (% of untreated)
sandy loam	6	120	90
	24	180	88
loam	4	120	92
	15	180	89
clay loam	15	120	92
	44	180	89
clay	2	120	87
	4	180	84
black volcanic	12	180	92

paclobutrazol applied 8-18 DBH, mean of trials 1981-88 on cv. Koshihikani
(black volcanic soil data includes 2 trials with var. Tsugaru-otome).

It is difficult to accurately compare the response of major rice
varieties to paclobutrazol within a single trial because the maturity
times of each variety will differ. However a trial carried out at ICI
Japan Agricultural Research Station (JARS) in 1988, attempted to
compare eleven major varieties, with a range of genetic parentage,
(Figure 4). Overall no major differences in stem shortening were
detected following an application of 240 g ai/ha.

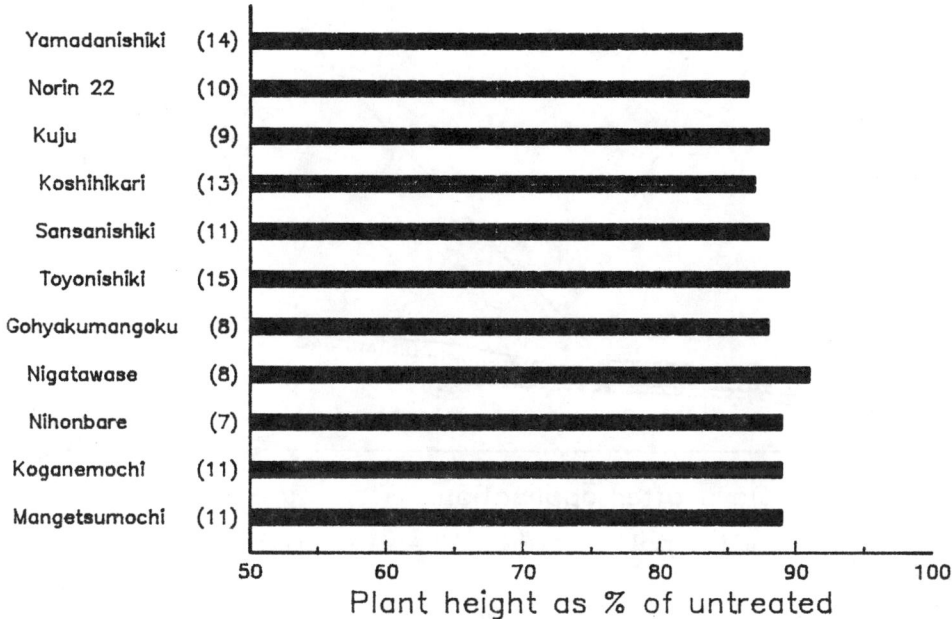

Figure 4. The effect of paclobutrazol on different rice varieties.
*(14) indicates DBH at application.

Water management practices vary greatly with locality, soil type, cultivar and even between neighbouring farmers. So a trial was set up at JARS in 1987 to compare the influence of different water management regimes on paclobutrazol activity.

Variations from almost zero water to 3 or 8 cms depth of water showed no appreciable difference in activity, when losses (percolation and overflow) were controlled. But when the percolation rate was increased to 3 cm/day, activity increased. Conversely, with no perculation and water losses only occurring from overflowing, activity slightly decreased. (Figure 5)

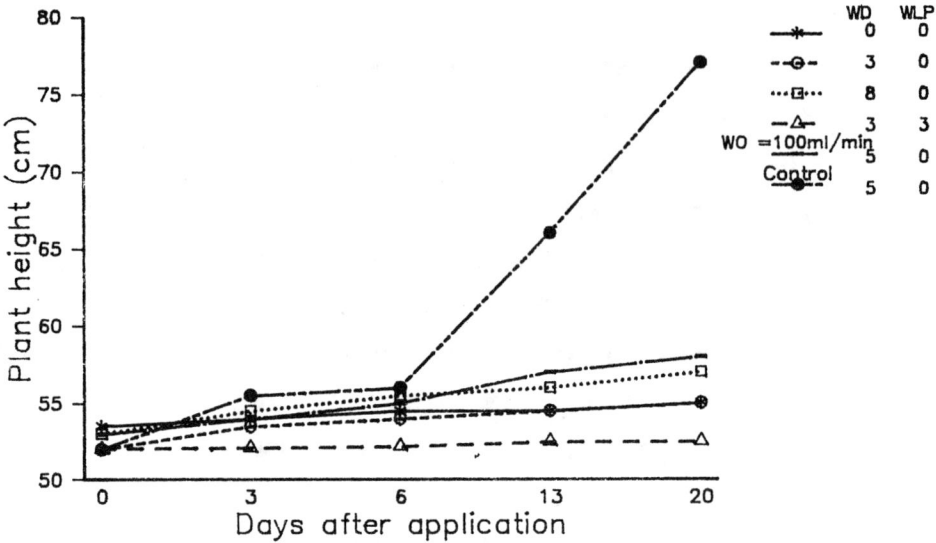

Paclobutrazol applied at 180 g ai/ha : WD = water depth, WLP = water loss by percolation, WO = water overflow

Figure 5. Effect of different water depths at treatment, and water loss following treatment, on paclobutrazol activity.

3. The effect of previous paclobutrazol applications on subsequent rice crops.

Unlike most other major rice producing countries, climate limits Japanese rice growers to a single crop each year, largely grown without crop rotation.

Studies were set up in several localities throughout Japan to establish what effect, if any, continuous or variable paclobutrazol use patterns would have on rice growth. Data from all the locations gave similar results; those obtained from Wakayama are displayed below. (Table 4)

These results show that annual applications over 5 years did not cause any reduction in stem length, over and above the reduction produced by a single application, while rice left untreated in the next cropping year grew to a normal height.

TABLE 4

Paclobutrazol responses over 5 years on var Koshihikari following
application at 180 g ai/ha (Wakayama).

| 'Smarect' application | Final plant height (% of untreated) | | | | |
	1984	1985	1986	1987	1988
180g ai/ha applied each year 1984 to 1988	88	91	88	82	88
180g ai/ha 1988 only	-	-	-	-	89
180g ai/ha 1984 only - no successive retreatment	88	100	-	-	-

DISCUSSION AND CONCLUSIONS

One to three weeks prior to heading, Japanese rice farmers can usually
predict fairly accurately if their crop is likely to lodge. This
overlaps with the optimum, recommended application time for paclobutrazol
granules, 15-10 days before heading. When applied at 120-180 g ai/ha in
granules, to flooded paddy fields, paclobutrazol is readily taken up by
rice plants, mainly through the roots. This effectively suppresses the
growth of the upper three internodes, subsequently reducing the bending
moment (height x panicle weight). Paclobutrazol treatment also induces a
more erect plant form and hence increase light interception [2]. A
combination of these effects results in yield and quality improvements,
largely as a result of an increase in the percent ripened grains.
Harvesting work rate is also improved as a direct consequence of reduced
lodging.

Several factors have been identified that will cause minor changes in
crop response eg. temperature, soil type and water percolation rate.
However, these have been taken into account in the product use
recommendations, which range from 120 to 180 g ai/ha. For example, the
lower recommended rate will be sufficient in paddies situated in warmer
southern regions with a high percolation rate. Although paclobutrazol
has been shown to be equally effective on all major Japanese rice
varieties, the lodging risk will vary with locality and variety, so
farmers should adjust application rates based on label and local research
station advice.

Paclobutrazol is broken down in soil by microbial activity [2], [4], its rate of decomposition being dependent upon soil and climatic conditions. Under anaerobic field conditions, paclobutrazol and its major decomposition product, the ketone, has been shown to break down in 1-10 months [2], [4], [5]. In field trials thoughout Japan, it has been demonstrated that paclobutrazol can be applied annually to rice, with no deleterious effects. Further studies, replanting a large range of other crops into treated fields are now in progress. So far, no adverse effects have been detected.

In conclusion paclobutrazol offers Japanese rice farmers an effective and reliable product to reduce lodging, aid harvesting efficiency and increase grain recovery and quality.

ACKNOWLEDGEMENTS

The authors wish to thank the National and Prefectural Research Stations of Japan, and JAPR (Japan Association for the Advancement of Phyto-Regulators) for their help and co-operation. We also wish to thank numerous colleagues in ICI Agrochemicals and our paclobutrazol distributors in Japan: Nihon Nohyaku Co. Ltd., Takeda Chemical Industries Ltd., and Ishihara Sangyo Kaisha Ltd., for their assistance and co-operation.

REFERENCES

1. Takano, M., History, geography and economics of Japanese Agriculture. Chemistry and Industry, 1986, 2, pp. 46-51.

2. Ueno, H., and Imanishi, K., Smarect : Development and Characteristic, 1989, Nohyaku, 36, pp46-53 (in Japanese).

3. Ohta, Y., Development and availability of plant growth regulators in cost-down techniques of rice Weeds and Weeds Control. Kanzatsu., 1984, 21, pp19-22 (in Japanese).

4. Lever, B.G., Shearing, S.J. and Batch, J.J., PP333 a new broad spectrum growth retardant. Proceedings 1982 British Crop Protection Conference - Weeds 3-10.

5. Dalziel, J., and Lawrence, D.K., Biochemical and biological effects of Kaurene oxidase inhibitors, such as paclobutrazol. British Plant Growth Regulator group, Monograph 11, 1984, pp. 43-53.

6. Ueno, H., French, P.N., Kohli, A., and Matsuyuki, H., Paclobutrazol :
Control of Rice lodging in Japan, Proceeding 11th International
Congress of Plant Protection. Manila. 1987.

Other Literature referred to :-

JAPR Annual Report on Plant Growth Regulators for Summer crops., 1983,
551-596., 1984, 467-506., 1985, 428-467., 1986, 449-520., 1987, 510-
551., 1988, 609-701 (in Japanese).

JAPR Annual Report of Demonstration Trials on Herbicides and Plant
Growth Regulators for Rice. 1989.

Macmillan, J., Goldsmith, I.R., and Hood, K.A., Inhibition of
Gibberellin Biosynthesis in Gibberella fujikuroi by PP333.

Hane, M., Tamori, T., and Yamamoto, Y., Effects of anti-lodging agents
on organo and yield components of rice plants. Proceedings Hokuri
Crop Science Soc., 1983, 19, pp21-22.

Nishiyama, I., Chemical control of lodging in direct sown paddy rice.,
179th Nation Conference of Crop Science Society of Japan, Tokyo, 1985,
pp190-191 (in Japanese).

THE IRAC RICE-WORKING-GROUP -
EFFORTS TOWARDS PEST MANAGEMENT IN RICE

W. Knauf
Biologische Forschung, Hoechst AG; Frankfurt; BRD

In recent years, industry has stepped up its efforts to tackle the challenge of possible resistance against insecticides in agriculture [1]. Pesticide producers used to react individually with different or even contradictionary strategies as a result. The industry is now endeavouring to achieve general agreement between the different companies and to harmonize its efforts in view of the following facts:

- growing numbers of efficient products in a single chemical group (e.g. pyrethroids or phosphate esters) and therefore an increasing risk of cross resistance.

- signs of developing resistance to older products which have been established on the market for a long time and should be kept there because of their safe use and their well-known properties.

- illegal use or abuse of unregistered products in at least some countries, with no application schemes supervised by the authorities and the producers.

- lowered chances of finding new insecticides as substitutes for older products because of more stringent registration requirements.

- concern about the build-up of resistance to very new products before a sufficient return on development investments can be achieved.

- growing number of data produced to verify resistance but generated by different methods.

- better knowledge about the mode of action of the different insecticides used and a better understanding of insect-specific detoxification and resistance mechanisms.

It was, therefore, felt that industry should take the initiative in coordinating ideas and efforts and to make its combined experience available for the avoidance of possible resistance problems [2].

In 1984, GIFAP set up the IRAC (Insecticide Resistance Action Committee) to be composed of experts from the producing industry [3].

This turned out to be very effective.
After an initial period, during which all participants were brought to the same level of knowledge about the problem, a "common list of terms" was established and various subgroups were formed.

Each subgroup of IRAC was made responsible for a specific crop or market segment (with one exception e.g. the IGREG = Insecticide Growth Regulator Efficacy Group). This was done also for rice.

Rice, which is the basic diet of millions of people around the world, is liable to infestation by a number of pests which have to be controlled during cultivation.

The following important pests are involved:

- Homoptera (e.g. hoppers)
- Lepidoptera (e.g. stemborers)
- Coleoptera (e.g. rice water weevil)
- Heteroptera (minor importance)

A survey performed by the rice group in 1986 identified a growing resistance potential in hoppers and to some extent in stemborers.

At the same time a number of new products with new chemical structures entered the market (e.g. Trebon, Applaud) and industry was keen to monitor the sensitivity status of pests (hoppers) from the very beginning of market penetration.

In agreement with the Central IRAC Panel it was decided to create standardized methods for measuring resistance, which were to be more reliable and practical than existing methods.

This was done first with a hopper-testing method (IRAC method no. 5), which was carefully established by intensive discussion and trials between and within the participating companies. The contribution of the East Asians (e.g. Japan) was especially helpful and resulted in a very practical approach.

The reasons not to use the existing method were:

- the existing method used topical application which was only acceptable for contact insecticides.

- there was no practicable modification to enable work to be carried out with treated plants, although this is the way the hopper normally comes into contact with the product.

- the method used technical equipment (e.g. microapplicator) which was unsuitable for field use and too expensive.

The new method was checked in a ring test by several laboratories and appears to fullfil the following criteria:

- it is easy to handle in the field and in the laboratory.

- it uses living rice plants as the natural environment for hopper populations.

- it combines tests either with contact or stomach poisons and considers intoxication via the gaseous phase to some extent.

- it needs no special technical equipment.

- it is flexible in using material from available sources.

- it can also be used in longterm tests, e.g. for insect growth regulators and all other known types of insecticides.

The method is described in detail and will be published elsewhere by GIFAP. Detailed information can be obtained from the GIFAP Headquarters in Brussels. All methods elaborated by the different working groups will be published and recommended for in tracing insect sensitivity to insecticides (see attachment, IRAC Method No. 5). If this is accepted by the researchers, the authorities or other industrial laboratories, then we will have taken a real step towards to cooperation and "speaking the same language". Those taking part in the discussion considered it very important to avoid a scattered list of results produced by different methods and thus not comparable with one another.

We are persuaded that this method is an economic, scientific and practical way to bring together some baseline data from all "hot spots" in the rice-growing areas. This would enable either the growers, the authorities, producers or scientists to compare data during the course of time in the same area and the data from different areas as well. It would also give early warning of developing resistance and enable appropriate controlling schemes to be employed.

This can only be successful if industry agrees to undertake longterm efforts. This also involves using company personnel to obtain data of this kind, and at least the baseline data should be published. Results from research institutes can then be used for direct comparison, if published likewise.

This also has the following advantages:

- an opportunity for producers to locate different sensitivity levels for their own products in different areas to optimize product positioning.

- comparable data base to counter rumours on resistance which might have an adverse influence on official recommendations.

- possibility of monitoring sensitivity drifts over several-year periods for own products.

- avoidance of misunderstandings in discussions dealing with "resistance".

- clear definitions of resistance levels.

- chance for small laboratories to work with an accepted method.

At the moment the IRAC - Rice working group is also working on a new method for borers which will be published once it has been verified in a ring test.

Above all, the members of the groups are aware that all efforts up to now are only the first step towards the formation of strategies with the aim of

- combating existing resistance
- avoiding resistance for already marketed and future products

Which general strategy (e.g. tank mix, subsequent applications within one season with products having different modes of action, subsequent spraying schemes from one season to the other) will turn out to be the best is not yet clear. It will be necessary to cooperate on trials with different spraying regimes and to monitor the sensitivity status regularly. This will require time and patience.

But we are convinced that decisions can be reached more easily if they are based on data from monitoring programmes of this kind.

As an example: It would help the users (and the producers) to crosscheck lists with data on cross-resistance for field populations to screen out suitable candidates as possible combination partners.

To sum up the evolution of the IRAC working groups up to the present, we feel that cooperation will be reached and will ultimately enable us to lengthen the lifespan of older but useful products without creating resistance, or will permit us to introduce new molecules and to achieve real pest management in rice.

REFERENCES

1. C.N.E. Ruscoe:
   The Attitude and Role of the Agrochemical Industry towards Pesticide Resistance
   in: Combating Resistance to Xenoliotics - Biological and Chemical Approaches; M.G. Ford, D.W. Holloman, B.P.S. Khambay and M.R. Sawicki Editors, pp 26-36 (1987)

2. Davies, R.A.H.:
    Insecticide Resistance; an Industry Viewpoint.
    Proceedings of the British Crop Protection Conference,
    2, 593-600 (1984)

3. Voss, G.:
    The Insecticide Resistance Action Committee of GIFAP:
    Objectives and Operation.
    Proceedings 17th International Congress on Entomology, Hamburg
    (1984), abstract volume

SOMACLONAL VARIATION FOR DISEASE RESISTANCE IN RICE (ORYZA SATIVA L.)

Q. J. Xie, M. C. Rush, and J. Cao
Department of Plant Pathology and Crop Physiology,
Agricultural Experiment Station,
Louisiana State University Agriculture Center,
Baton Rouge, Louisiana 70803, USA

ABSTRACT

Variation in resistance to rice blast and sheath blight was observed in somaclones regenerated from U. S. long-grain cultivars. Among 2,100 R2 somaclonal lines inoculated and screened for sheath blight resistance, three lines regenerated from the susceptible cultivar Labelle showed a high level of resistance. The sheath blight resistant somaclone SC 86-20001 has good agronomic characteristics when compared to the previously available resistant cultivars. The resistance of the three somaclonal lines to sheath blight was stable after 4 years of testing in the greenhouse and in the field. The inheritance of sheath blight resistance in SC 86-20001-5 was controlled by a single recessive gene. Sheath blight resistance of 86-20001-33 was controlled by two independently inherited recessive genes. Selection among F3 lines from crosses of the above lines with cultivar Lemont gave lines with resistance, good plant type, and high yield potential.

INTRODUCTION

Disease resistance provides one of the most economical and historically useful approaches to suppressing plant disease. Unfortunately, two major problems have been encountered in breeding disease resistant cultivars. First is the difficulty in finding suitable sources of resistance. An example would be the difficulty in finding sources of resistance to rice sheath blight. The second problem is the loss of resistance, as with rice blast, caused by the development of new pathogenic races of the pathogen. However, recent progress in the development of tissue culture techniques has provided the plant pathologist and breeder a novel approach to solve these problems.

Many research examples have demonstrated that genetic variation in disease resistance can be generated by subjecting cultivars or breeding lines to somatic cell culture. These variants can be used as new germplasm for disease resistance or, in rare cases, be released as new cultivars (1, 2, 3, 4, 5, 6. 7, 8, 9, 10, 11, 12, 13, 14, 15, 16). Genetic variation produced in tissue culture has a large advantage over variation from other sources; i.e., it is generated in vitro. Even in a single petri dish or flask, the millions of continuously dividing cells provide a population for selection much larger than the plant populations normally produced in conventional breeding programs. However, in vitro screening is applicable only if a screening procedure is available for a specific pathogen. Carlson (4) first demonstrated that disease resistance could be generated by tissue culture and selected by screening in vitro. Plants resistant to certain pathogens have been regenerated and selected among a few plant species by using his techniques. Examples are maize plants with resistance to T-toxin (1), sugarcane plants with resistance to eyespot caused by Helminthosporium sacchari (9), and rice plants with resistance to brown spots caused by Helminthosporium oryzae (10). A review focused on the selection for disease resistance in tissue culture has recently been published by Daub (17). However, the application of in vitro screening techniques on a large scale has been limited to pathogens that produce host-specific toxins. This has been considered a prerequisite for in vitro screening. Nevertheless, it has been demonstrated that unselected variation in disease resistance generated through somaculture is an excellent source of novel disease resistance, especially where sources of resistance are not readily available, even when the resistance has to be selected by conventional screening procedures.

Recently, rice somaculture systems have been successfully developed (2, 3, 18, 19, 20, 21, 22, 23, 24, 25, 26, 27, 28, 29, 30). Many genetic variations have been expressed in plants regenerated from somaculture (2, 3, 9, 14, 28, 31, 32, 33). In rice these include differences in morphological characteristics; such as color of apiculus, hull, leaf or sheath, awn production, and changes in pubescence of leaves. Variation has also been observed in physiological characteristics including tolerance to salt and to aluminum toxicity, as well as in important agronomic characteristics such as plant height, plant type, tillering ability, panicle shape, grain type and quality, grain weight, kernel numbers, sterility, days to heading, and yielding ability. Somaclonal variation for resistance to major rice diseases, such as rice blast, sheath blight, and brown spot disease have been observed and reported (3, 52).

Rice blast, caused by Pyricularia oryzae Cav., rice bacterial leaf blight, caused by Xanthomonas campestris pv. oryzae (Ishiyama 1922) Dowson; and rice sheath blight, caused by Rhizoctonia solani Kuhn, are the three most important nonviral rice diseases in the world. In the United States, especially in Louisiana, the major diseases on rice are sheath blight and blast. Recently, bacterial leaf blight was also observed in Texas and Louisiana. Although most U.S. cultivars have resistance genes to the major races of P. oryzae, the potential of the pathogen to produce new virulent races, or of a previously minor virulent race becoming prevalent, has been a continuing threat to rice production. Hence, a continous supply of new genes for resistance to rice blast has to be made available to the rice breeder. Recent research conducted in our laboratory with rice somaculture has shown that variation in resistance to rice blast can be generated through somaclonal variation (15). In 1986, among 2,000 somaclonal lines sceened in the field, 93 lines showed variation for blast resistance. Some of the lines were more susceptible than the parent cultivars and some lines were significantly more resistant. Fifty lines were evaluated by M. A. Marchetti in a blast nursery grown under controlled conditions. Four somaclonal lines showed much higher blast resistance than their parent cultivars (3, 14, 15). The inheritance of this resistance has yet to be investigated.

Rice sheath blight has been epidemic for many years in the southern rice area of the U. S., and recently this disease has become a serious problem in rice cultivation in Louisiana, Arkansas, and Texas, especially on the U.S. long-grain cultivars (35). No commercial cultivar in the world is highly resistant to this disease. In fact, all of the U.S. long grain cultivars, including Labelle and Lamont, are highly susceptible to sheath blight. The problem in breeding rice cultivars resistant to sheath blight is the difficulty in finding suitable sources of resistance. A few noncommercial cultivars are highly resistant, but the inheritance of this resistance is complicated (36). The agronomic characteristics of these cultivars, for example Tetep and Taducan, are extremely poor. It is difficult to use these sources of resistance in the breeding program and rapidly develop resistant cultivars with good agronomic characteristics. However, there is an urgent need for new long-grain rice cultivars with resistance to sheath blight. A new source of resistance with acceptable agronomic characteristics would be very valuable in breeding for sheath blight resistance. A major objective of our research in rice somaculture is to generate and identify somaclones which are highly resistant to sheath blight and have the characteristics of a typical U.S. long-grain cultivar. This resistance would be most useful if it is controlled by one or a few major genes. Locating such a source of resistance would certainly be a major contribution to the rice breeding program in the United States. The following discussion will focus on our attempts to find acceptable somaclonal resistance to sheath blight.

PLANT REGENERATION AND IDENTIFICATION OF RESISTANT SOMACLONES

The U.S. long-grain cultivars Labelle and Lemont were used in this experiment. The method of regenerating rice somaclones was developed by Cao (2) and Cao et al (3). The explant source for calli were pieces of immature panicle cut from 1 - 5 cm panicles removed from the boot using sterile technique and plated on MS (Murashige-Skoog) medium with 4 ppm 2,4-D. Explants were cultured in the dark at 28 C. Plants were regenerated from calli on MS medium without 2,4-D and with 0.5 ppm IAA (indole-3-acetic acid) and 0.8 ppm BA (6-benzylaminopurine). Calli were cultured on regeneration medium with 16 h light /8 h dark at 28 C. Plants forming on calli were subcultured every month until the seedlings were 5 to 7.5 cm long with a root system well enough developed to be transplanted into the greenhouse. More than 3,000 R1 green plants were regenerated from the two cultivars and transplanted to the greenhouse. The R2 generation plants produced from seed from the R1 plants were grown in the field to evaluate for variation.

The screening of the somaclones for sheath blight resistance started with R2 generation lines. In 1985, 700 lines were inoculated with R. solani and screened in the field for sheath blight resistance at the Louisiana State University Rice Research Station at Crowley, LA. Two somaclones from Labelle showed a high level of resistance to sheath blight. The agronomic characteristics of these lines were poor with the plants having characteristics described for polyploid or aneuploid plants (2). In 1986, 1,400 lines were screened in the field and one somaclone from Labelle was selected with sheath blight resistance and acceptable agronomic characteristics, such as resistance to lodging, glabrous leaf, typical long-grains, and good plant type. The three somaclones along with the parent cultivar Labelle and the resistant cultivars Tetep and Taducan were inoculated with R. solani in the greenhouse under controlled conditions. All three somaclonal lines showed resistance comparable to Tetep and Taducan (3, 14, 15). In 1987, forty panicle lines from SC 86-20001, the resistant somaclone with the best agronomic characteristics, were inoculated with R. solani and screened for sheath blight resistance in the field, along with the cultivar Labelle and six panicle rows of susceptible somaclone SC 86-20002. All the rows were rated at maturity for sheath blight using a 0 to 9 scale with 0 equal to no disease and 9 equal to all plants dead at maturity (37). All of the panicle lines of 86-20001 were significantly more resistant to sheath blight than the parent cultivar Labelle and the panicle lines from the susceptible somaclone SC 86-20002 (Table 1). However, there were differences in sheath blight resistance among the SC 86-20001 lines.

Table 1

Distribution of R3 somaclone panicle rows for sheath blight resistance in a 1987 field test

Cultivar or somaclone	Number of panicle rows with each disease rating based on 0-9 scale									
	0	1	2	3	4	5	6	7	8	9
Labelle							1	1	2	
SC 86-20001	9	14	7	8	1					
SC 86-20002										6

In 1988, the three somaclonal lines SC 86-20001-5, SC 86-20001-8, SC 86-20001-33, and SC 86-20001-41, selected from the forty SC 86-20001 panicle rows; the commercial cultivars Labelle and Lemont, (the most widely grown cultivar in Louisiana), and the highly resistant cultivar Taducan were tested again in the field for sheath blight resistance. The cultivars and lines were planted in plots with six rows in each plot, a row spacing of 25 cm and a row length of 1.5 m. A randomized complete block design with three replications was adopted. The plots were each inoculated uniformly at the booting stage with 250 ml of inoculum (R. solani isolate no. LR 172) grown on a sterilized mixture of rice grains and rice hulls (2:1), and disease ratings were made when the rice was mature. The same 0-9 disease rating system was used. The results clearly indicated that the resistance of the somaclonal lines was significantly higher than that of Labelle and Lemont cultivars, and the resistance levels of the two panicle lines SC 86-20001-33 and SC 86-20001-8 were as high as the resistant cultivar Taducan (Table 2). The agronomic characteristics of these two lines were much more acceptable than those of Taducan. However, significant differences in sheath blight resistance did exist among the selected panicle lines from SC 86-20001. The selected panicle lines from SC 86-20001 and the other two somaclones from Labelle were tested for 3 consecutive years in the field and in one greenhouse test for sheath blight resistance. All of the

Table 2

Comparison of sheath blight resistance among four selections from the sheath blight resistant somaclone SC 86-20001, the resistant cultivar Taducan, and two susceptible cultivars

Cultivar or line	R3 row rating (1987)	R4 plot rating (1988)
SC 86-20001-33	0.0	0.8
Taducan	-	1.0
SC 86-20001-8	0.0	1.3
SC 86-20001-41	4.0	4.3
SC 86-20001-5	3.0	4.8
Lemont	-	8.7
Labelle	7.3	9.0
$LSD_{0.05}$		0.2

somaclones showed stable resistance.

INHERITANCE OF SHEATH BLIGHT RESISTANCE FROM THE SELECTED SOMACLONES

Preliminary research has been conducted on the mode of inheritance of some rice somaclonal variants (2, 3, 32, 38, 39, 40 41), but no studies have addressed the inheritance of disease resistance. Smith and Murakishi (33) reported on the inheritance of somaclonal variation for resistance in tomato to tomato mosaic virus, and it appeared that the mode of resistance was complicated. Although the inheritance of most somaclonal variants has been relatively simple and often controlled by one or a few major genes, the inheritance of some variants is very complicated. One typical example would be that the variations were highly heritable from generation to generation, but they disappeared after the somaclones were crossed back to the parent cultivars. There were no F2 plants with the characteristics of the variants (39, 40). There was no satisfactory explanation for this phenomenon, and this kind of variation has little or no value in a breeding program if the gene(s) in control of the variant characteristic cannot be expressed when crossed with other cultivars. Information concerning the inheritance of rice somaclonal variation, especially variation in disease resistance, is still limited. However, a clear understanding of the inheritance of somaclonal variation is certainly of importance in both theoretical and applied areas of somaculture and plant breeding.

Two sheath blight resistant lines were selected for research on the inheritance of somaclonal resistance to sheath blight. Both were single panicle selections from the somaclone SC 86-20001. Line SC 86-20001-33 had the highest resistance level. The second line SC 86-20001-5, had a lower resistance level but better plant type when compared to SC 86-20001-33. Two cultivars were used to make crosses with the somaclonal lines. One was the parent cultivar Labelle, which is very susceptible to sheath blight, the other was Lemont, which is the most widely grown cultivar in the United States. Lemont has excellent yielding ability but is very susceptible to sheath blight. In 1987, reciprocal crosses were made between SC 86-20001-5 and Lemont and between SC 86-20001-33 and Labelle. The cross between 86-20001-33 and Lemont was made in 1988 in the greenhouse. The F1 plants from these crosses, the parent cultivars, and the parent somaclonal lines were inoculated with R. solani isolate LR-172 in a greenhouse test. The disease resistance ratings of the F1 plants were intermediate between the resistant somaclonal lines and the susceptible cultivars according to this test.

In 1988, the F2 populations from reciprocal crosses between SC 86-20001-5 and Lemont and the parent lines and the Lemont cultivar were space planted in 1.5 m rows with a row spacing of 25 cm in the field at the Rice Research Station at Crowley, LA. Plants were inoculated twice, 45 days after seeding and at the boot stage, with R. solani at 50 ml inoculum for each 1.5 m row. The disease rating was made at maturity and was based on the 0 - 5 scale of Guo and Cheng (42), with 0 equal to no disease, 1 equal to lesions below the fourth leaf, 2 equal to lesions below the third leaf, 3 equal to lesions up to third leaf, 4 equal to lesions up to the penultimate leaf, and 5 equal to lesions up to flag leaf. This rating system was easy to use in the field, but it was not as satisfactory a system as the 0 to 9 scale we adopted later. The same rating applied to two different plants did not distinguish the significant differences that were present in percent tissue diseased, the type of lesion, and lesion size and color. Based on this rating system, the distributions of resistance of SC 86-20001-5, Lemont and their F2 progenies to sheath blight were determined (Figures 1-1 and 1-2). The distribution of ratings of SC 86-20001-5 plants was from 0 to 3, with an average rating of 1.7. All Lemont plants were rated at 5. The distribution of the cross with SC 86-20001-5 as female parent showed a bi-modal distribution (Figure 1-1), which indicated major gene action, but its reciprocal cross showed a mono-modal distribution (Figure 1-2). Using the 0 - 5 rating system, there were significant differences in the amount of tissue diseased in the same rating when the rating was above 2. For example, plants of SC 86-20001-5 rated 3 had much less diseased tissue compared to most plants of the F2's rated 3. Also the lesion size and color showed the same tendency. Whether the plants rated 3 belonged to the resistant or susceptible class was decided by the disease resistance of their offspring, F3 panicle rows rated by using the 0 to 9 scale. After this adjustment, both reciprocal crosses showed a bi-modal distribution. The Chi-square estimates of goodness of fit of the F2 progenies of SC 86-20001-5 and Lemont to proposed genetic ratios are listed in Table 3. Based on the bi-modal distributions and the results of Chi-square analysis, the inheritance of sheath blight resistance in somaclone SC 86-20001-5 was controlled by a single recessive gene with incomplete dominance of susceptibility.

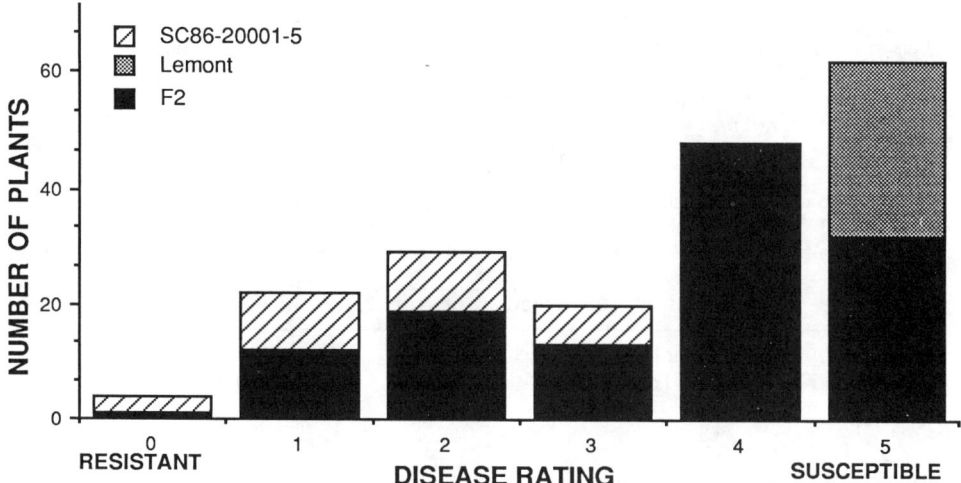

Figure 1-1. Distribution of F2 plants from the cross SC86-20001-5 X Lemont
for resistance to sheath blight.

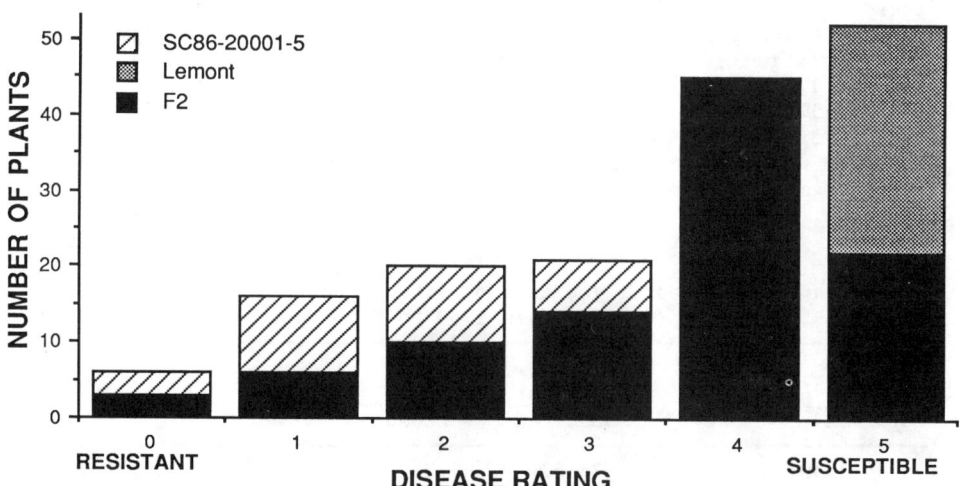

Figure 1-2. Distribution of F2 plants from the cross Lemont X SC86-20001-5
for resistance to sheath blight.

In 1989, F3 lines from the reciprocal crosses of SC 86-20001-5 and Lemont were tested for sheath blight resistance in the field again. Seeds from single plants harvested from both F2 populations in the previous year were planted in two sets. In both sets, 1.5 m rows with a 25 cm row spacing were used. In the first set, only F3 lines, SC

Table 3

Chi-square estimates of goodness of fit of the F2 progenies of SC 86-20001-5 by Lemont evaluated in 1988 to the expected genetic ratios

Crosses	Ratio (S:R)[*]			x^2	P
	Observed	Expected			
SC86-20001-5 X Lemont	85:40	94:31	(3:1)	3.488	0.10-0.05
Lemont X SC86-20001-5	74:26	75:25	(3:1)	0.054	0.99-0.97

[*]S:R = susceptible to resistant.

86-20001-5 and Lemont were planted. In the second set, each F3 row was separated by the susceptible cultivar Labelle to ensure uniform disease development across the F3 lines. The same inoculation technique was used. All ratings made in 1989 were based on the 0 to 9 scale. The rating of F3 rows was based on the reaction of apparently susceptible plants where there was segregation in disease resistance within the row.

The distributions of ratings received by F3 rows and their parents are shown in Figures 2-1 and 2-2. The ratings of SC 86-20001-5 rows ranged from 0 to 4. The Lemont rows received ratings of 8 or 9. Both reciprocal crosses showed a bimodal distribution, with most lines rated in susceptible class. Based on the distribution of parent ratings, F3 rows rated 0 to 5 were considered resistant, and rows rated 6 to 9 were considered susceptible. The chi-square test of goodness of fit to the proposed 3:1 ratio (Table 4) confirmed the F2 results from 1988, which was based on the 0 to 5 rating scale. Both year's evaluations indicated that the resistance of 86-20001-5 was controlled by a single recessive

Table 4

Chi-square estimates of goodness of fit of the F3 progenees of SC 86-20001-5 by Lemont evaluated in 1989 to the expected genetic ratios

Crosses	Ratio (S:R)[*] —			x^2	P
	Observed	Expected			
SC86-20001-5 X Lemont	85:37	92:31	(3:1)	1.847	0.25-0.10
Lemont X SC86-20001-5	72:28	75:25	(3:1)	0.480	0.50-0.25

[*]S:R = susceptible to resistant.

gene. Additional F2 plants from the cross of SC 86-20001-5 and Lemont were tested for resistance to sheath blight in 1989, and rated on the 0 to 9 scale. The bimodal distribution of the F2 plants (Figure 4-2) and chi-square test of goodness of fit (Table 5) confirmed that a single

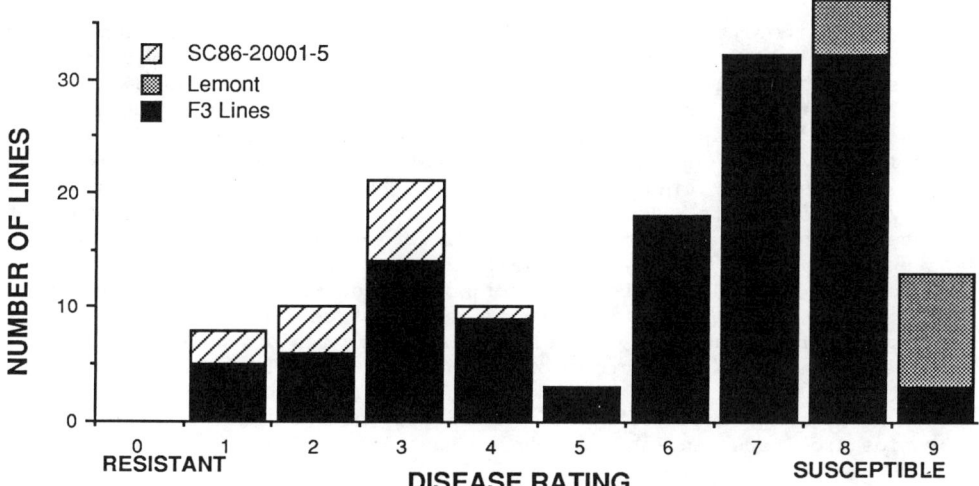

Figure 2-1. Distribution of F3 lines from the cross SC86-20001-5 X Lemont for resistance to sheath blight.

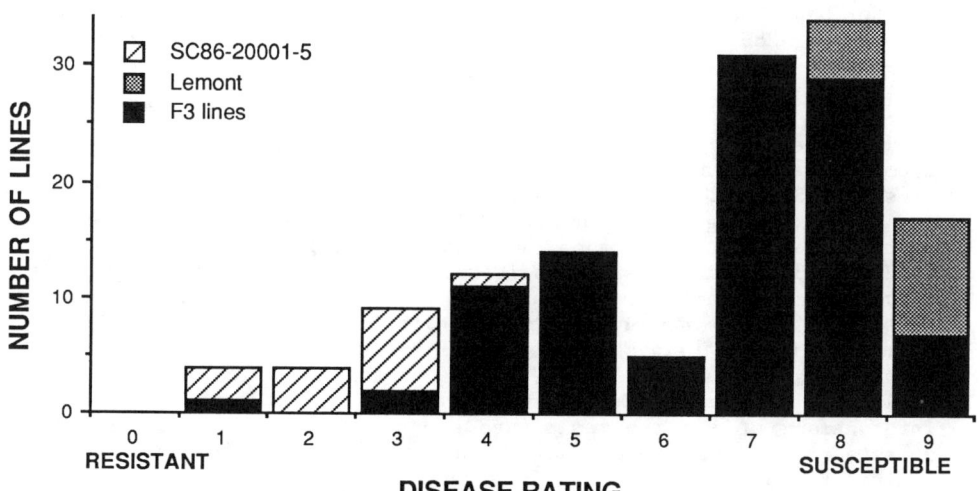

Figure 2-2. Distribution of F3 lines from the cross Lemont X SC86-20001-5 for resistance to sheath blight.

recessive gene played the major role in the sheath blight resistance of SC 86-20001-5.

The inheritance resistance in the most resistant somaclonal line, SC 86-20001-33, was investigated in 1989. F2 populations from reciprocal crosses between SC 86-20001-33 and Labelle, a cross of SC 86-20001-33 and Lemont, and the parent lines were planted and inoculated as described previously. Ratings were based on the 0 to 9 scale. The rating distributions of the parent line, Labelle and the F2 progenies are are shown in Figures 3-1 and 3-2. The rating distribution of SC 86-20001-33 plants was from 0 to 5, and the rating distribution of Labelle plants was from 6 to 9. The cross of 86-20001-33 X Labelle showed a bimodal distribution with the peaks at 3 and 6. The cross of Labelle X 86-20001-33 showed a tri-modal distribution with peaks at 1, 3, and 6. The poly-modal segregation suggested that major genes were involved in the control of the inheritance of resistance of 86-20001-33 to sheath blight. The cross of SC 86-20001-33 X Lemont also showed a bimodal distribution with peaks at 3 and 6 (Figure 4-1). Chi-square estimates of goodness of fit of the F2 progenies to the proposed genetic ratios are listed in Table 5, where plants rated 0 - 5 were considered

Table 5

Chi-square estimates of goodness of fit of the sheath blight reactions of F2 progenies evaluated in 1989 to the expected genetic ratios

| Crosses | Ratio (S:R)[*] | | | x^2 | P |
	Observed	Expected			
SC86-20001-33 X Labelle	57:56	64:49	(9:7)	1.547	0.25-0.10
Labelle X SC86-20001-33	160:143	170:133	(9:7)	1.450	0.25-0.10
SC86-20001-33 X Lemont	77:68	82:63	(9:7)	0.583	0.50-0.25
Lemont X SC86-20001-5	157:64	166:55	(3:1)	1.847	0.25-0.10

resistant and 6 - 9 susceptible based on the parent reactions. Results from the reciprocal crosses of SC 86-20001-33 and Labelle and from the cross of SC 86-20001-33 X Lemont (Figure 4-1) appeared to fit a 9:7 ratio, which suggested that two pairs of independent recessive genes controlled the inheritance of sheath blight resistance of SC 86-20001-33. If this assumption is true, the plants would be susceptible only if both dominant susceptible genes were present. In the F2 population, the proportion of plants with both dominant genes should be 9 out of 16. The plants will be resistant when they have either one or both recessive resistant genes, the proportion of which should be 7 out of 16 in the F2 population. There were insufficient data to explore the interactions of the two resistance genes because of the environmental effects on disease development and subsequent ratings and because of the limited F2 population studied. However, it appeared that there might be some additive gene effects between the two resistant genes based on the data available. The direct evidence was the third modal peak appearing at rating 1 in the F2 rating distribution of the Labelle X SC 86-20001-33 cross where the F2 population size was relatively large (Figure 3-2). Apparently the plants with both resistant genes reacted

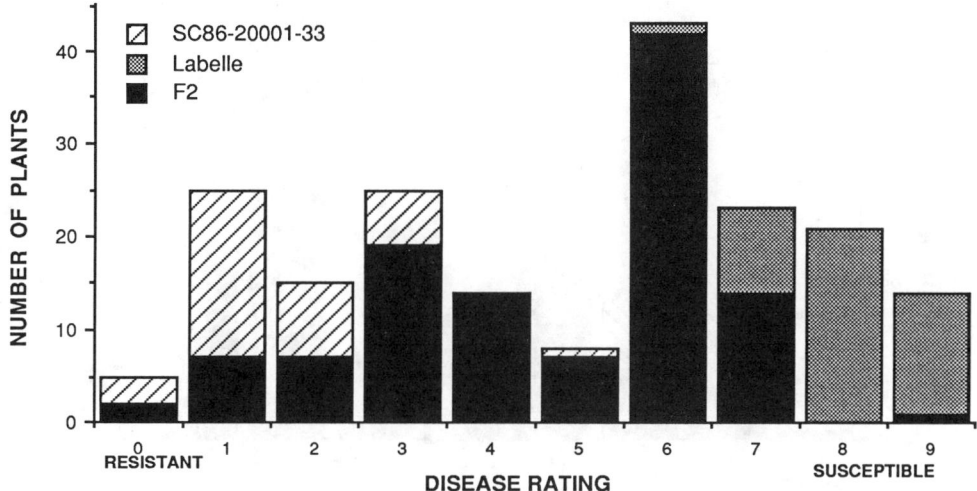

Figure 3-1. Distribution of F2 plants from the cross SC86-20001-33 X Labelle
for sheath blight resistance.

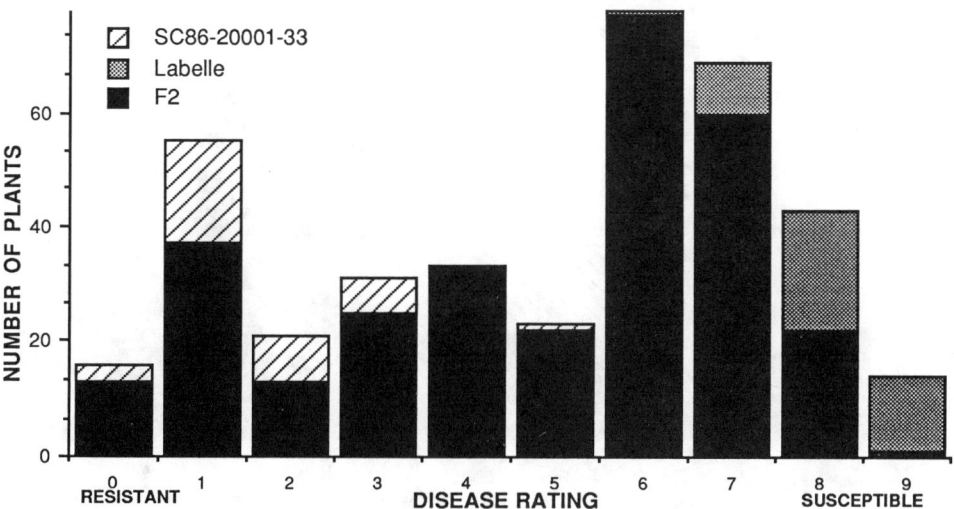

Figure 3-2. Distribution of F2 plants from the cross Labelle X SC86-20001-33
for sheath blight resistance.

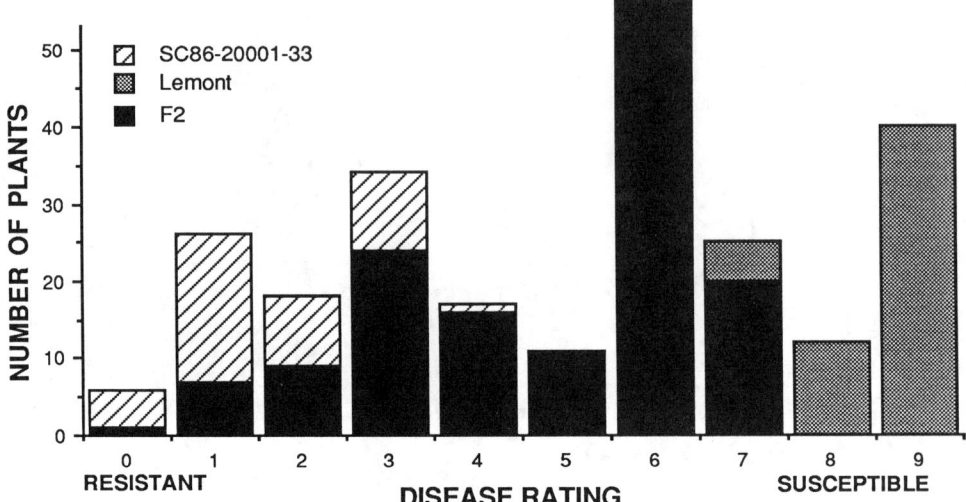

Figure 4-1. Distribution of F2 plants from the cross SC86-20001-33 X Lemont for sheath blight resistance.

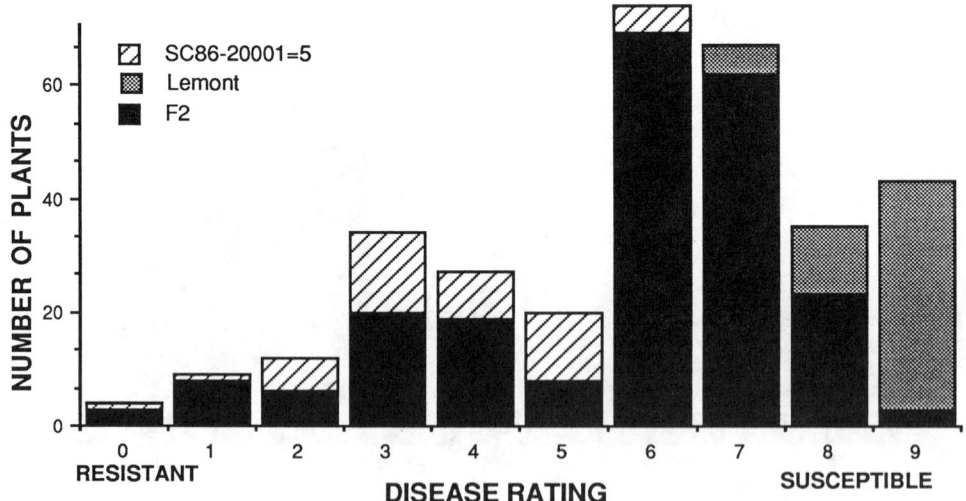

Figure 4-2. Distribution of F2 plants from the cross SC86-20001-5 X Lemont for sheath blight resistance.

with disease ratings of about 1. Plants with one gene were probably distributed around rating 4. However, the big environmental effect and the modifying genes involved would make it extremely difficult to distinguish the different classes among the plants rated within 0 to 5. The indirect evidence was that somaclone SC 86-20001-33, that had two resistance genes, was more resistant to sheath blight than somaclone SC 86-20001-5, that only had one resistance gene. Based on the available data, the conclusion could be made that two pairs of independently inherited recessive genes were involved in the control of sheath blight resistance of SC 86-20001-33. The rating distribution of the F2 population and the chi-square test of goodness of fit for the cross of SC86-20001-33 X Lemont fit the proposed 9:7 ratio. This further confirmed the assumption that two independently inherited recessive genes were controlling sheath blight resistance in SC 86-20001-33.

ROLE OF RICE SOMACULTURE IN DISEASE RESISTANCE

The development of rice somaculture has offered the rice breeder and pathologist a novel approach to reduce the losses caused by disease in rice. Current research shows that new disease resistant germplasm can be developed through somaculture. This is an especially valuable tool when sources of resistance to a disease are not available. The two lines SC 86-20001-5 and SC 86-20001-33, generated from the sheath blight susceptible cultivar Labelle, not only have good resistance to sheath blight and acceptable agronomic characteristics, the resistance is also highly heritable when crossed with susceptible cultivars. The preliminary results from selection among F3 lines were very promising as far as obtaining lines with resistance, good plant type, and high yield potential. However, a major difficulty with somaclonal variation in rice is that not all cultivars show somaclonal variation at acceptable frequencies. The cultivar Labelle is highly variable for many characteristics (2, 3, 14). However, some cultivars are tissue culture stable and seldom produce variants in somaculture. The cultivars Tetep and Taipei 309 have been demonstrated to have this characteristic in our laboratory. If the aim of somaculture is to generate variation for use in a breeding program, higher variation frequencies may need to be induced before the procedures can be adopted into a breeding program on a large scale. Two approaches might be useful in solving this problem. First, incorporate mutagens into tissue culture procedures. There have been significant increases in somaclonal variation frequency in tissue culture stable varieties in our laboratory by incorporation of EMS (ethyl methanesulfonate) into somaculture media (unpublished data). A second method is to use tissue from hybrids as the explant material for somaculture. Recently conducted research has determined how variation was generated in somaculture by culturing F1s from lines with specific genetic markers. This research suggested that the somaclonal variation from F1s from crosses made with these lines was much higher than the variation rate in somaclones from the parent lines (unpublished data).

Theoretically, the efficiency of obtaining somaclonal variants for disease resistance would be greatly increased by in vitro selection for resistance to a pathogen in calli, cell suspensions, or protoplasts. With increased variation frequency and stringent in vitro selection procedures, somaculture would be a major element in breeding for disease resistance in the future. In vitro selection would not only help plant pathologists and rice breeders to eliminate most of the tedious work of screening progeny in the greenhouse and field, but would also greatly increase the chances of obtaining variation for resistance, as the population of cells used for selection is many times larger than the population of plants that can be screened in the greenhouse and field. However, there has been only one successful report of in vitro selection for disease resistance in rice, and the success was based on the availability of a specific toxin for that disease (10). For the three major rice diseases, blast, sheath blight, and bacterial leaf blight, some "toxins" are available (43, 44, 45, 46, 47, 48, 49, 50, 51, 52). Whether these "toxins" can be effectively incorporated into an in vitro selection scheme and give reliable results has yet to be investigated. Bouharmont (53) reported that one toxic compound 3-methyl-thiopropionic acid, isolated from X. campestris pv. oryzae was not active at the cell level. At present, it appears that enough toxins are produced by rice pathogens that a successful in vitro selection technique can be established for several diseases. Besides the use of toxin, a technique for co-culture of the pathogen and host is being developed in our laboratory, which may have potential for use in an in vitro selection system. This would be useful where toxin is not available or is too expensive. These techniques will not detect whole plant resistance.

The success of rice protoplast culture (54, 55, 56, 57, 58, 59, 60) has opened the door for transformation of rice with foreign DNA. This would certainly increase the opportunities for enriching rice germplasm for disease resistance. Recent success in regeneration of green plants from protoplasts of the U.S. long-grain cultivar Labelle in our laboratory provides a method for transformation of U.S. cultivars with foreign DNA that can be used directly in U.S rice breeding programs.

Finally, it should be remembered that tissue culture is only one of the many tools that may be used in combating the disease problems in crop production. Even though tissue culture may offer a novel source of variation and an easier method for selection of resistance, its application in breeding programs is still in its infant stage. M. E. Daub (17) pointed out in her excellent review that "tissue culture is neither a fast nor an easy way of developing disease-resistant plants", and "using tissue culture to generate resistance to disease for which good resistant varieties already exist is not a profitable use of time". The successful application of tissue culture for breeding disease resistance relies not only on the development of tissue culture techniques, but also on the development of knowledge about the specific disease in question, the pathogen involved, and the interaction between the pathogen and the host at both the cellular and whole plant level.

REFERENCES

1. Brettell, R.I.S. and Thomas, E., Reversion of Texas male-sterile cytoplasm maize in culture to give fertile, T-toxin resistance plants. Theor. Appl. Genet., 1980, 58, 55-58.

2. Cao, J. Improvement of rice through somaculture. Ph. D. Dissertion. Louisiana State University, Baton Rouge. 184p. 1986.

3. Cao, J. Rush, M.C., Nabors, M.W., Xie, Q.J., Croughan, T.P. and Nowick, E., Development and inheritance of somaclonal variation in rice. In Biological Nitrogen Fixation Associated with Rice Production, Oxford and PIBH Publishing, New Delhi, India, 1990. (in press).

4. Carlson, P.S. Methionine sulfoximine-resistant mutants of tobacco. Science, 1973, 180, 1366-68.

5. Evans, D.A., Sharp, W.R. and Medinafilho, H. P., Somaclonal and gametoclonal variation. Am. J. Bot., 1984, 71, 759-74.

6. Hartman, C.L., McCoy, T.J.,and Knous, T.R., Selection of alfalfa (Medicago sativa) cell lines and regeneration of plants resistant to the toxin(s) produced by Fusarium oxysporum f. sp. medicaginis. Plant Sci. Lett., 1984, 34, 183-94.

7. Heinz, D.J., Krishnamurthi, M., Nickell, L.G. and Maretzki, A., Cell, tissue, and organ culture in sugarcane improvement. In Applied and Fundamental Aspect of Plant Cell, Tissue and Organ Culture, ed. J. Reinert, Y. P. S. Bajaj, Berlin:spring-verlag. 1977, pp. 3-27.

8. Larkin, P.J. and Scowcroft, W.R., Somaclonal variation-a novel source of variability from cell cultures for plant improvement. Theor. Appl. Genet., 1981, 60, 197-214.

9. Larkin, P.J. and Scowcroft, W.R., Somaclonal variation and eyespot toxin tolerance in sugarcane. Plant Cell Tissue Organ Cult., 1981, 2, 111-22.

10. Ling, D.H., Vidhyaseharan, P., Borromeo, E.S., Zapata, F.J., and Mew, T.W., In vitro screening of rice germplasm for resistance to brown spot disease using phytotoxin. Theor. Appl. Genet., 1985, 71, 133-35.

11. Liu, M.C. and Chen, W.H., Tissue and cell culture as aids to sugarcane breeding. I. Creation of genetic variation through callus culture. Euphytica, 1976, 25, 393-403.

12. Liu, M.C. 1987. Sugarcane improvement through somaclonal variation In Abstracts International Congress of Plant Tissue Culture Tropical Species, ed. A.A. Zerda, Bogota, Colombia, 1987, pp. 3-4.

13. Miller, S.A., Williams, G.R., Medina Filho, H. and Evans, D.A., A somaclonal variant of tomato resistant to race 2 of Fusarium oxyspoum f. sp. lycopersici. Phytopathol., 1985, 22, 215-45.

14. Xie, Q.J., Rush, M.C. and Cao, J. Somaclonal variation for rice improvement. Ann. Prog. Rpt., Rice Sta., LA Agr. Expt. Sta., LSU Agr. Center, 1987, 79, 237-41.

15. Xie, Q.J., Rush, M.C., Massaquoi, R. and Cao, J., Rice tissue culture: Potential for rice improvement through modified disease resistance. Phytopathol., 1987, 77, 1723.

16. Thanutong, P., Furusawa, I. and Yamamoto, M., Resistant tobacco plants from protoplast-derived callus selected for their resistance to Pseudomonas and Alternaria toxins. Theor. Appl. Genet., 1983, 66, 209-15.

17. Daub, M.E., Tissue culture and the selection of resistance to pathogens. Ann. Rev. Phytopathol., 1986, 24, 159-86.

18. Abe, T. and Futsuhara, Y., Efficient plant regeneration by somatic embryogenesis from root callus tissues of rice (Oryza sativa L.). J. Plant Physiol., 1985, 121, 111-18.

19. Abe, T. and Futsuhara, Y., Varietal difference of plant regeneration from root callus tissue in rice. Japan. J. Breed., 1984, 34, 147-55.

20. Abe, T. and Futsuhara, Y., Plant regeneration from suspension culture of rice (Oryza sativa L.). Japan. J. Breed. 1982, 36(1), 1-6.

21. Abrigo, E.M., Novero, A.U., Coronel, V.P., Cabuslay, G.S., Blanco, L.C., Parao, F.T. and S. Yoshida. Somatic cell culture at IRRI. Inter-Center Seminar on IARCs and Biotechnology. International Rice Research Institute, Los Banos, Philippines. 23-27 April, 1984.

22. Heyser, J.W., Dykes, T.A., Demott, K.J. and Nabors, M.W., High frequency, long-term regeneration of rice from callus culture. Plant Sci. Let., 1983, 29, 175-82.

23. Kawata, S.I. and Ishihara, A., The regeneration of rice plant, Oryza sativa L., in the callus derived from the seminal root. Proc. Japan Acad., 1968, 44(6), 549-53.

24. Kavikishor, P.B. Energy and osmotic requirement for high frequency regeneration of rice plants from long-term cultures. Plant Science, 1987, 48, 189-94.

25. Ling, D.H., Chen, W.Y. and Ma, Z.R., Somatic embryogenesis and plant regeneration in interspecific hybrid of Oryza. Plant Cell Reports, 1983, 2, 169-71.

26. Ling, D.H., Chen, W. Y. and Ma, Z. R., Direct development of plantlets from immature panicles of rice in vitro. Plant Cell Reports, 1983, 2, 172-74.

27. Marassi, M.A. and Rapela, M.A., Tissue culture in Argentine rice (O. sativa L.) improvement. In Abstracts International Congress of Plant Tissue Culture Tropical Species, ed. A.A.Zerda, Bogota, Colombia, 1987, pp. 67.

28. Nabors, M.W., Heyser, J.W., Dykes, T.A. and Demott, K.J., Long-duration, high-frequency plant regeneration from cereal tissue cultures. <u>Planta,</u> 1983, 157, 358-91.

29. Ye, H.C. Studies on cell suspension culture and plant regeneration in rice. <u>Acta Botanica Sinica,</u> 1984, 26(1), 52-59.

30. Zimmy, J. and Lorz, H., Plant regeneration and initiation of cell suspension from root-tip derived callus of <u>Oryza sativa L.</u> (rice). <u>Plant Cell Reports,</u> 1986, 5, 89-92.

31. Nabors, M.W. and Dykes, T.A., Obtaining cereal cultivars with increased tolerance to salt, drought and acid stressed soils through tissue culture. Inter-Center Seminar on IARCs and Biotechnology. International Rice Research Institute, Los Banos, Philippines. 23-27 April, 1984.

32. Oono, K. 1983. Genetic variability in rice plants regenerated from cell cultures. In <u>Cell and Tissue Culture Techniques for Cereal Crop Improvement,</u> Science Press. Beijing, China, 1983, pp.95-104.

33. Smith, S.S. and Murakishi, H.H., Genetic characteristics of somaclonal resistance to tomato mosaic virus (TOMV). <u>Phytopathol.,</u> 1987, 77, 1705.

34. Marchetti, M.A., Potential impact of sheath blight on yield and milling quality of short-statured rice lines in the Southern United States. <u>Plant Disease,</u> 1983, 67(2), 162-65.

35. Lee, F.N. and Rush, M.C., Rice sheath blight: a major rice disease. <u>Plant Disease,</u> 1983, 67, 829-32.

36. Sha, X.Y. Genetic studies on resistance to sheath blight (<u>Thanatephorus cucumeris</u> (Frank) <u>Donk</u>) in eight Indica rice varieties. M. S. Thesis. Nanjing Agricultural University, Nanjing, P. R. China, 1987, 49p.

37. Hoff, B.J., Rush, M.C., McIlrath, M.O. and Morgan, A., Disease nurseries. <u>Ann. Prog. Rpt.,</u> Rice Sta., LA Agr. Expt. Sta., LSU, Agr. Center, 1976, 68, 142-49.

38. Fukui, K. Sequential occurence of mutations in a growing rice callus. <u>Theor. Appl. Genet.,</u> 1983, 65, 225-30.

39. Nowick, E.M., Cao, J. and Rush, M.C., Genetic basis of selected somaclonal variants in the rice cultivar `Lemont'. <u>SABRAO,</u> 1988, 20, 1-10.

40. Oono, K., Putative homozygous mutations in regenerated plants of rice. <u>Mol. Gen. Genet.,</u> 1985, 198, 377-84.

41. Sun, Z.X., Zhao, C.H., Zheng, K.L., Qi, X. F. and Fu, Y.P., Somaclonal genetics of rice, <u>Oryza sativa L.</u> <u>Theor. Appl. Genet.,</u> 1983, 67, 67-73.

42. Guo, C.J. and Chen, Z.Y., Study of the pathogenesis differentiation of pathogen of rice sheath blight and the technique of evaluation of resistance. Scientia Agricultura Sinica, 1985, (5), 50-57.

43. Angadi, C.V., In vitro production of slime toxin in bacterial blight infected rice plants. Phytopath. Z., 1978, 92, 193-201.

44. Angadi, C.V., Extra-cellular slime of Xanthomonas oryzae in bacterial leaf blight of rice. Phytopath. Z., 1978, 93, 170-80.

45. Canonica, L., Fieecchi, A., Kienle, M.G. and Scala, A., Isolation and constitution of Cochliobolin B. Tetrahedron Lett. 1966, 13, 1329-33.

46. Canonica, L., Fiecchi, A., Kienle, M.G., Ranzi, B.M. and Scala, A., The biosynthesis of Cochliobolins A and B. Tetrahedron Lett., 1966, 26, 3035-39.

47. Choi, J.E., Matsuyama, N. and Wakimoyo, S., Biological and chemical properties of slime polysaccharide of Xanthomonas campestris pv oryzae. Ann. Phytopath. Soc. Japan, 1982, 58, 1-8.

48. Egawa, H., Yoshii, K. and Ueyama, A., Phenylacetic acid, a metabolite of Xanthomans oryzae (Uyeda et Ishiyama) Dowson in culture affecting upon the depressive growth of young roots of rice seedlings. Ann. Phytopath. Soc. Japan, 1986, 34, 46-50.

49. Iwaski, S., Nozoe, S., Okuda, S., Sato, Z. and Kozaka, T., Isolation and structural elucidation of a phytotoxic substance produced by Pyricularia oryzae Cavara. Tetrahedron Lett., 1969, 45, 3977-80.

50. Misaki, A., Kirkwood, S., Scaletti, J.V. and Smith, F., Structure of the extracellular polysaccharide produced by Xanthomonas oryzae. Canad. J. Chem., 1962, 40, 2204-13.

51. Purushothaman, D. and Prasad, N.N., Isolation of toxin from Xanthomnas oryzae. Phytopath. Z., 1972, 75, 178-80.

52. Vidhyasekaran, P., Borromeo, E.S. and Mew, T.W., Host-specific toxin produced by Heminthosporium oryzae. Phytopathol., 1986, 76, 261-66.

53. Bouharmont, J. and Dekeyser, A., In vitro culture for mutation selection in rice (Oryza sativa L.) (abstract). In Abstracts International Congress of Plant Tissue Culture Tropical Species, ed. A.A. Zerda, Bogota, Colombia, 1987, pp 32.

54. Abdullah, R., Cocking, E.C. and Thompson, J.A., Efficient plant regeneration from rice protoplasts through somatic embryogenesis. Biotechnology, 1986, 4, 1987-90.

55. Fujimura, T., Sakurai, M., Akagi, H., Negishi, T. and Hirose, A., Regeneration of rice plants from protoplasts. Plant Tissue Culture Letters, 1985, 2, 74-75.

56. Kyozuka, J., Hayashi, Y. and Shimamoto, K., High frequency plant regeneration from rice protoplasts by novel nurse culture methods. Mol. Gen. Genet., 1987, 206, 408-13.

57. Thompson, J.A., Abdullah, R. and Cocking, E.C., Protoplast culture of rice (Oryza sativa L.) using media solidified with agarose. Plant Science, 1986, 47(2), 123-33.

58. Thompson, J.A., Abdulian, R., Chen, W. H. and Cartland, K.M.A., Enhanced protoplast division in rice (Oryza sativa L.) following heat shock treatment. J. Plant Physiol., 1987, 127, 367-70.

59. Toriyama, K. and Hinata, K., Cell suspension and protoplast culture in rice. Plant Science, 1985, 41, 179-83.

60. Yamada, Y., Yang, Z.Q. and Tang, D.T., Plant regeneration from protoplast-derived callus of rice (Oryza sativa L.). Plant Cell Reports, 1988, 5, 85-88.

CAN INSECT PEST PROBLEMS IN RICE BE APPROACHED BY USING BACILLUS THURINGIENSIS CRYSTAL PROTEIN GENES ?

E.GÖBEL, M.PEFEROEN and A.REYNAERTS
Plant Genetic Systems N.V.,
J. Plateaustraat 22, 9000 Gent, Belgium

SUMMARY

During recent years several procedures have been developed to introduce foreign genes into plant cells and to subsequently regenerate fertile plants. In combination with an increasing understanding of plant gene regulation and the processing and targeting of proteins this has provided the opportunity to obtain genetically modified plants. Traits of agronomic importance can be introduced into crop plants, one example being the engineering of plants resistant to insect attack through the expression of insecticidal proteins. At present the preferential source of such insecticidal proteins is the bacterium Bacillus thuringiensis (B.t.).

Bacillus thuringiensis is a gram positive bacterium and produces crystalline inclusions upon sporulation. These inclusions contain insecticidal crystal proteins which are toxic to the larvae of specific groups of insects. Crystals, when ingested by the insect larvae, dissolve in the alkaline environment of the midgut. Most insecticidal proteins are produced by the bacteria as larger protoxins. They are proteolytically activated by midgut proteases to smaller active proteins which bind to specific receptors in the midgut epithelium and disturb the integrity of the brush border membrane. The larvae stop feeding and die.

There exists a wide variety of B.t. insecticidal crystal proteins , each of which are very specific for certain classes of insects. Four major pathotypes have been described: strains toxic to Lepidoptera, Lepidoptera/ Diptera, Diptera and Coleoptera. Even within the group of Lepidoptera-specific toxins different types can be identified. A classification of the different types of insecticidal proteins is proposed by Höfte and Whiteley, 1989 (1). The specificity of the crystal proteins is determined by the presence of binding molecules (receptors) in the brush border membrane of the midgut (2).

In recent years several research institutes have introduced genes encoding insecticidal proteins into various plant species. At PGS the first experiments towards engineering insect resistance in plants by transforming these plants with a B.t. gene were performed with tobacco(3). A B.t. insecticidal protein, CryIA(b), was selected which is active against

Lepidoptera such as <u>Manduca</u> <u>sexta</u> (tobacco hornworm) and <u>Heliothis</u> <u>virescens</u> (tobacco budworm).
A gene encoding the protoxin CryIA(b) was cloned and the region coding for the toxic fragment was identified. T-DNA mediated gene transfer was used to produce transgenic tobacco plants. The presence of CryIA(b) protein in leaves of transgenic tobacco plants was shown by ELISA assays. In most transgenic tobacco plants CryIA(b) protein represented less than 0.01 % of total protein. However, CryIA(b) levels above 0.004 % of total protein were sufficient to cause 100 % mortality in <u>Manduca</u> <u>sexta</u> first instar larvae in <u>in vitro</u> assays using detached tobacco leaves. Plants with the highest expression level (0.01 % - 0.02 %) were fully protected from insect feeding in greenhouse assays. Several years of field testing in North Carolina, US, in collaboration with Rohm and Haas have confirmed protection of transgenic tobacco plants against feeding damage by both <u>Manduca</u> <u>sexta</u> and <u>Heliothis</u> <u>virescens</u> under field conditions.

Transformation of tomato and potato with chimaeric <u>cryIA(b)</u> genes resulted in resistant plants as well; greenhouse assays using <u>Manduca</u> <u>sexta</u> as an indicator insect showed protection of transgenic plants against insect feeding. Furthermore, leaves and tubers of transgenic potato plants proved to be protected against <u>Phthorimaea</u> <u>operculella</u>, potato tuber moth, an important potato pest causing damage by both leaf and tuber mining.

In order to further explore the diversity of <u>Bacillus</u> <u>thuringiensis</u> strains, we have established an intensive isolation and screening program. Using selective media , <u>B.t.</u> strains are isolated from different sources and screened for the presence of crystalline inclusions. Strains containing inclusions are characterized further by SDS - PAGE and by ELISA assays. A representative number of newly isolated strains is selected for evaluation in insect bioassays. A collection of more than 5000 strains has been established and we currently isolate approx. 200 <u>B.t.</u> strains each month. New strains with improved activity against Lepidoptera in general and against <u>Heliothis</u> and <u>Spodoptera</u> species in particular have been isolated. In addition, several strains with new crystal proteins active against Coleoptera have been identified.

The results obtained with tobacco, tomato and potato indicate that insect resistance can ˙be engineered into plants by using <u>cry</u> genes of <u>Baccilus</u> <u>thuringiensis</u>. Can this technology also be applied to an important food crop such as rice ? Some of the requirements to be met are as follows:
1) Availability of an efficient and reproducible technique for the transfer of foreign genes into rice and production of fertile transgenic plants,
2) Identification of important pest insects,
3) Selection and characterization of <u>B.t.</u> insecticidal proteins with the appropriate toxicity against these insects,
4) Understanding of gene expression in transgenic rice.

In general cereal crops are more recalcitrant with respect to manipulation in cell and tissue culture than many dicotyledonous plant species.
A basic prerequisite for the genetic transformation of rice is the regeneration of fertile plants from protoplasts, as alternative methods for reproducible and efficient transformation of tissues or multicellular structures and subsequent plant regeneration have not yet been developed. Within the last few years plant regeneration from protoplasts has been obtained in japonica-type and indica-type rice in several laboratories (4), (5), (6), and the production of transgenic rice plants after direct

gene transfer into protoplasts has also recently been achieved (7), (8), thus opening up the possibility to transfer potentially useful genes into the rice genome.

In view of this a collaborative research project has been started between PGS and IRRI (project leader at IRRI: Dr.Litzinger), funded by the Rockefeller Foundation, on the control of rice insect pests using B.t. insecticidal protein genes. The objectives of this project are: 1) to isolate new B.t. strains with high activity against major rice insect pests, and at a later stage 2) to clone and characterize B.t. genes encoding the highly active insecticidal proteins, and 3) to engineer these genes, either singly or in combination, into rice in order to obtain insect resistant rice plants.

At this stage of the project we are focussing on the isolation of new B.t. strains toxic to some of the major lepidopteran pest insects in rice:
> Chilo suppressalis (Striped stemborer)
> Scirpophaga incertulas (Yellow stemborer)
> Scirpophaga innotato (White stemborer)
> Sesamia inferens (Pink stemborer)
> Cnaphalocrocis sp. (Leaffolder)
> Narasmia sp. (Leaffolder)

From different locations in the Philippines more than 250 samples of grain dust, soil and dead insects have been collected and processed for selective isolation of B.t. strains. From these samples a total of 2,038 B.t. strains was isolated and crystal proteins were characterized. The strains produced either single crystal proteins or mixtures of several different crystal proteins, and many isolates produced combinations of insecticidal proteins not previously observed. Representative strains are now being analyzed for insecticidal activity in insect bioassays on different target insects at IRRI and at PGS. Preliminary results from these assays indicate that B.t. crystal proteins do affect several rice insect pests: some of the strains tested show high activity against Cnaphalocrocis sp. and Narasmia sp. (leaffolders), others are highly toxic to the stemborer Chilo suppressalis. This screening program will continue throughout the project.

Further evaluation of the specific activity of crystal proteins active against different insects will allow the identification of insecticidal proteins most suitable to confer resistance to insects in rice.

The first steps towards an alternative system in rice insect pest management are being made. However, even in transgenic dicotyledonous plant species such as tobacco, tomato and potato expressing insecticidal proteins, a number of improvements remain to be made before such systems can be used commercially, concerning e.g. transformation efficiency, gene expression level and regulation of gene expression. In particular, as progress in transformation of cereals is much slower compared to many dicotyledonous plant species, little is known about the integration of introduced genes and their expression. It is not clear e.g. if the promoters successfully used in transformation of many dicotyledonous plant species will function in rice or other cereals. Thus, a good deal of fundamental research remains to be done. The progress currently being made in cell and tissue culture and transformation of rice suggests that rice will develop into a model system for cereals and that progress will be rapid.

SELECTED REFERENCES

1. Höfte, H. and Whiteley, H., Insecticidal crystal proteins of <u>Bacillus</u> <u>thuringiensis</u>. <u>Microbiological</u> <u>Reviews</u>, 1989, 53, 242-255.

2. Van Rie, J., McGaughey, W.H., Johnson, D.E., Barnett, B.D. and Van Mellaert, H., Mechanism of insect resistance to the microbial insecticide <u>Bacillus</u> <u>thuringiensis</u>. <u>Science</u>, 1990, 247, 72-74.

3. Vaeck, M., Reynaerts, A., Höfte, H., Jansens, S., De Beuckeleer, M., Dean, C., Zabeau, M., Van Montagu, M. and Leemans, J., Transgenic plants protected from insect attack. <u>Nature</u>, 1987, 327, 33-37.

4. Fujimura., T., Sakurai, M. ,Akagi, H., Negishi, T. and Hirose, A., Regeneration of rice plants from protoplasts. <u>Plant</u> <u>Tissue</u> <u>Culture</u> <u>Lett</u>., 1985, 2, 74-75.

5. Kyozuka, J., Hayashi, Y. and Shimamoto, K., High frequency plant regeneration from rice protoplasts by novel nurse culture methods. <u>Mol</u>. <u>Gen</u>. <u>Genet</u>., 1987, 206, 408-413.

6. Kyozuka, J., Otoo, E. and Shimamoto, K., Plant regeneration from protoplasts of indica rice: genotypic difference in culture response. <u>Theor</u>. <u>Appl</u>. <u>Genet</u>., 1988, 76, 887-890.

7. Toriyama, K., Arimoto, Y., Uchimiya, H. and Hinata, K., Transgenic rice plants after direct gene transfer into protoplasts. <u>Bio/Technology</u>, 1988, 6, 1072-1074.

8. Zhang, W. and Wu, R., Efficient regeneration of transgenic plants from rice protoplasts and correctly regulated expression of the foreign gene in the plants. <u>Theor</u>. <u>Appl</u>. <u>Genet</u>., 1988, 76, 835-840.

RICE BIOTECHNOLOGY: PROGRESS AND PROSPECTS

GARY H. TOENNIESSEN
Rockefeller Foundation, New York

ABSTRACT

Considerable progress has been made in the development of cellular biology and molecular genetic techniques that can be applied to the genetic improvement of rice. Tissue culture techniques such as anther culture, embryo rescue, and use of somaclonal variants have contributed to the release of new rice varieties. These technologies are of proven benefit to rice breeding and through research are becoming applicable to a broader range of rice cultivars and breeding objectives. Most molecular genetic techniques are still at an early stage of development but progress has been more rapid with rice than any other cereal. Species specific probes and genetic maps of rice have been produced and increasingly are being applied in breeding. Further development of these tools into a map-based system for cloning rice genes is underway. Regeneration of fertile plants from protoplasts has been achieved for both japonica and indica rice. This is the basis of rice genetic transformation systems that now exist in several laboratories. Transgenic rice plants containing alien marker genes are scheduled to be field tested this year. Several experiments are underway to introduce potentially useful cloned genes into rice. Numerous other research projects are aimed at identifying, constructing and cloning a wide variety of genes from plants, microbes and animals which might instill useful traits if introduced into the rice genome. This paper reviews the progress to date and describes several promising research projects.

INTRODUCTION

Rice is the most important food crop of the developing world and rice genetic improvement through breeding has proven to be an effective mechanism for delivering the benefits of science and technology to hundreds of millions of poor people. A system with regional, national, and international components is now in place for producing improved rice varieties and delivering them to the farmers that need them. Biotechnology can significantly strengthen rice breeding programs and help produce new varieties to stabilize higher yields, improve the efficiency of production, expand the area and population base receiving benefits, and further increase maximum yield potentials, thereby preventing future food shortages from occurring as demand for rice increases because of population growth and economic development. Biotechnology will enable breeders to do their work more quickly and efficiently and will help them to attain breeding goals not feasible using conventional techniques. Biotechnology will complement not replace breeding. In fact, a strong breeding program is a prerequisite to the application of these new technologies for rice genetic improvement.

ANTHER CULTURE

Anther culture of rice has been feasible for over twenty years. While first developed in Japan [1] it is used most extensively in China where over 100 varieties have been developed via anther culture. The major advantages of anther culture are reduction in the time required to obtain fixed lines and increased selection efficiency [2]. Instant homozygous lines result when the single set of chromosomes double in plants derived from anther culture. The gamete genotype including recessive genes is expressed. Early generation selection becomes feasible due to the additive effect of the doubling and the elimination of dominance.

Anther culture is most advantageous in situations where only one generation/year is feasible with conventional breeding and where recombination is not important. Examples are long duration or photosensitive cultivars and traits that are highly seasonal such as certain stress and disease tolerances.

Outside China anther culture has not been widely used in part due to low regeneration frequencies of indica rice. Efficiencies of 5% or greater are normally needed for anther culture to be used as a routine procedure in breeding; that is 5 green plants per 100 anthers plated [3]. Even in China where 10% efficiencies are common with japonica rice, efficiencies of only 3% are considered high for indica rice [4]. The occurrence of albino plants is a serious problem but little work has been done on its cause or solution. Various factors in addition to genotype, including physiological status of donor plant, pretreatment of anthers, developmental stage of anthers, callusing and differentiation media, and temperature are known to influence regeneration frequency [5].

Anther culture has contributed to improved pest management. At the Shanghai Academy of Agricultural Sciences anther culture was used to transfer genes for resistance to brown planthopper biotype I from indica rice to japonica rice [4]. Two lines have been released having strong BPH I resistance plus all the original favorable characteristics of the japonica line. Farmers no longer need to use insecticide and can now raise edible fungi in the rice fields simultaneously with rice cultivation. At the International Center for Tropical Agriculture (CIAT) in Colombia anther culture is being used to transfer genes for resistance to blast and other diseases from upland germplasm, which responds very well to anther culture, to irrigated breeding material [6]. Anther culture provides the opportunity to fix recessive genes and to compare the blast behavior of these lines with lines obtained through standard procedures. The goal is to identify and accumulate minor genes for blast resistance in the hope that their cumulative effect will provide a more durable resistance than that provided by major genes.

Utilization of anther culture in more breeding programs and for a broader range of breeding objectives should be feasible and beneficial. CIAT is currently conducting a cost/benefit analysis of anther culture to help national rice breeding programs decide under what circumstances it has most promise. Further theoretical and empirical research aimed at improving the anther culture response of indica rice is needed and if successful would help advance application of this potentially valuable breeding tool.

WIDE HYBRIDIZATION

The primary goal of wide hybridization is the introgression into breeding lines of a small chromosome segment from a wild species which contains one or more genes instilling a useful new trait. There are 22 <u>Oryza</u> species, only two of which are cultivated. The genus contains six known genomes, A-F, combined into tetraploid species as well as diploids. As indicated in Table 1 the wild <u>Oryza</u> species have many genetic traits that would be of value in cultivated rice including resistance to diseases and pests and to abiotic stresses [7]. Occasionally the technique results in unanticipated but useful new variability of unknown origin. Wide hybridization can significantly extend the genetic variability available to rice breeders.

Despite its potential value wide hybridization has not been extensively used for rice genetic improvement. By definition sexual hybridization between <u>Oryza</u> <u>sativa</u> and other <u>Oryza</u> species does not normally occur. It is a painstaking and time consuming process to overcome the inherent barriers. A variety of techniques are needed to stimulate pollination, to nurture the abortion-prone embryo through cell divisions to form a mature F1 hybrid plant, to overcome sterility problems, to stimulate and detect introgression of alien chromosome segments, and eventually to produce a fertile plant having only the desired introgression(s). Both tissue culture and molecular genetics can facilitate this process.

At the International Rice Research Institute (IRRI) studies of pollen tube growth in selected interspecific crosses have shown that the incompatibility mechanisms operating in the stigma/style/ovary vary depending on the cross combination. The type of incompatibility operating in a given cross needs to be established to decide the most appropriate approach to overcome it [8].

TABLE 1

Agronomically Important Characteristics Identified Among
the Wild Relatives of Rice (Oryza sativa)

SPECIES	2n	GENOME	CHARACTERISTICS
O. nivara	24	AA	Grassy stunt virus resistance
O. rufipogon	24	AA	Source of cytoplasmic male sterility, Tolerance to stagnant flooding
O. glaberrima	24	AA	GLH resistance, early vegetative vigour
O. barthii	24	AA	Bacterial blight resistance
O. longistaminata	24	AA	Floral characteristics for out-crossing
O. punctata	24,48	BB,BBCC	BPH,WBPH,GLH resistance
O. officinalis	24	CC	BPH,WBPH,GLH resistance
O. eichingeri	24	CC	BPH,WBPH,GLH resistance
O. minuta	48	BBCC	BPH,WBPH,GLH, blast and bacterial blight resistance
O. australiensis	24	EE	BPH resistance, drought tolerance
O. brachyantha	24	FF	Rice whorl maggot and stem borer resistance
O. ridleyi	48	----	Rice whorl maggot resistance

Once fertilization is achieved the immature hybrid embryo may need to be rescued through excision and germination under artificial conditions. Embryo rescue has been used to raise interspecific F1 hybrids between <u>Oryza</u> <u>sativa</u> and at least ten wild <u>Oryza</u> species having useful traits. Repeated embryo rescue may be required with the F2 plants and even with more advanced generations.

Once mature, the hybrid plants are often sterile or of low fertility due to irregularities that occur during the formation of gametes. Doubling the chromosomes with colchicine may improve fertility. Anther culture of some F1 hybrids may help by providing homozygous diploid substitution and addition lines [9]. Back crossing to the crop plant can enhance fertility and dilute out the undesirable attributes derived from the wild species, hopefully while retaining the desirable ones.

Finally, if there is little or no meiotic recombination, exchanges between the chromosomes of the cultivated rice and those of the wild species need to be stimulated. Passing the hybrid through a tissue culture cycle may help since this often causes chromosome breakage and rearrangements. Molecular probes made from repetitive DNA sequences can be used to test for the presence of alien chromosomes or chromosome segments in the hybrid plants. The ideal probe is one whose DNA sequence is highly repeated and dispersed throughout the genome of the wild rice species but absent from that of cultivated rice. For example, thousands of copies of the ribosomal genes of rice are dispersed throughout the genome and different <u>Oryza</u> species have different size spacer sequences in the tandem repeats of these genes [10]. Thus spacer sequences can be used as species specific probes and for studies of evolution. Zhao et al [11] have identified repetitive DNA sequences specific to the AA, CC, EE, and FF genomes which can be used similarly. Ongoing research should soon provide DNA probes which can be used to follow the transfer of alien chromatin from any wild species used in a wide cross with cultivated rice. Using <u>in</u> <u>situ</u> hybridization of these probes and scanning electron microscopy it is also possible to visualize the alien chromatin on the rice chromosomes [12].

One of the more important accomplishments of wide hybridization to date is the transfer of a gene for resistance to grassy stunt virus from Oryza nivara, a wild rice from India, to IRRI breeding lines. It is providing good resistance in numerous lines used throughout Asia and has saved farmers hundreds of millions of dollars in crop loss and pesticide costs. Advanced generation progenies with genes for resistance to brown planthopper and white-backed planthopper from Oryza officinalis have been sent by IRRI to national programs for field testing[8]. Hybrids have been made at IRRI with Oryza minuta in hope of transferring genes for resistance to blast and bacterial blight. Similar hybrids have been produced and are being tested in Korea and China. The male sterility gene used in China for hybrid rice seed production came from a wild relative. Hybrid seeds are now planted on over 10 million hectares and give an average of 15% yield increase.

RFLP MAPS AND MARKERS

Classical genetic maps have long been used in rice breeding as have cytological markers and biochemical markers such as isozymes. All have shortcomings when compared to RFLP maps and markers that are currently being developed for rice and other crops.

RFLP (Restriction Fragment Length Polymorphism) analysis will be most useful to breeders in following the inheritance of genes that are difficult, expensive, or time consuming to score and in identifying the major genetic components of quantitative traits. The major advantages of RFLP analysis in comparison to morphological markers are:

1.) RFLP markers are inherited in mendelian fashion and behave in a codominant manner, allowing the genotype of any locus to be determined in any breeding scheme. Inheritance of recessive genes tagged with RFLP markers can be followed as easily as dominant genes.

2.) Most RFLP markers are phenotype-neutral since most polymorphisms occur in non-transcribed regions of DNA. Using markers it should be possible to pyramid resistance genes, thereby producing varieties with more durable resistance.

3.) The level of allelic variation is high. For any gene it should be possible to identify one or more tightly linked markers.

4.) RFLP markers are free of epistatic effects so any number of markers can be monitored in a single population.

5.) RFLP markers are detectable in all plant tissues at all stages of plant development. Analysis can be conducted on a small piece of leaf material at the seedling stage.

6.) Nuclear, mitochondrial and chloroplast genomes can be analyzed.

7.) Polygenic traits can be disasgregated into monogenic components.

Rice is particularly well suited for RFLP mapping techniques. It is a true diploid with twelve chromosomes (2N = 24) containing only 5.8×10^5 kilobase pairs per haploid genome (maize has 7.2×10^6 kb). There is ample polymorphism in rice DNA and it is highly recombinogenic compared to other plants. One centimorgan in rice equals approximated 250 kb compared to more than 500 kb in tomato and 750 kb in potato [13]. The DNA content per map unit in rice, the most important food crop in the world, is only 2-3 times greater than

<u>Arabidopsis thaliana</u>, the weed some propose as the ideal plant for molecular genetics.

Restriction fragment length polymorphism is a complex term for a common phenomenon. At the DNA level a polymorphism is simply a difference between two plants in terms of their DNA sequence at a particular locus. Many such polymorphisms have evolved in all higher eucaryotic organisms including rice. Polymorphisms may be detected by cutting genomic DNA with a restriction enzyme, separating the resulting DNA fragments by length on a electrophoretic gel, probing the gel with a labelled fragment of DNA complementary in sequence to the DNA at or adjacent to the polymorphic locus and noting where on the gel the probe hybridizes to its complement. A combination of the right probe and the right restriction enzyme gives a reproducible detectable difference between the two plants in the position of labelled probe on the gel. This difference can be used as an indicator of the parental origin of DNA at that locus in the progeny of a cross between these plants. In practice selecting and matching the right probe and the right enzyme is an empirical process. There are procedures for increasing the probability of getting the right ones but it is still a matter of trial and error. If a combination works you keep it, if it doesn't work discard it.

For rice the first and most advanced RFLP map resulted from collaboration between scientists at Cornell University and IRRI [14]. IRRI had available cytogenetic stocks, segregating populations and isolines that greatly facilitated the molecular biology conducted at Cornell. Plant materials were often lyophilized at IRRI and sent to Cornell for DNA analysis.

In making the map at Cornell, DNA fragments from cultivar IR36 were selected as useful probes by trial and error. IRRI had available a set of primary trisomics in IR36 background which allowed the probes to be assigned to their respective chromosomes due to the dosage effect. Linkage analysis between probes was conducted on 53 plants grown at Cornell from parental and F2 seed of a cross previously made at IRRI. The result is the saturated genetic map of rice which now has over 300 markers covering all twelve chromosomes.

The Cornell map and markers have been widely distributed and efforts are now underway throughout the world to link the markers to genes for important qualitative and quantitative traits. The availability at IRRI and elsewhere of isolines for particular genes greatly facilitates linkage study. In theory isolines differ from each other due to the presence or absence of only the target gene and a small piece of flanking DNA. Segregating populations of near isogenic lines can be scored at IRRI and other breeding institutions and leaf material sent to Cornell or other RFLP labs for linkage analysis. Using this procedure a number of genes for resistance to white backed planthopper, green leafhopper, brown planthopper, bacterial leaf blight and blast are currently being linked to markers [15].

An attempt is also being made to use RFLP markers to map the genes responsible for complex quantitative traits in rice such as drought tolerance and salt tolerance. The RFLP map allows the quantitative trait loci (QTLs) of tolerant lines to be located on the rice chromosomes and tightly linked to markers. The QTLs can then be pyramided into modern varieties via RFLP selection. It may be possible to use different sources of tolerance and to incorporate complementary QTLs into a single variety providing higher levels of stress tolerance than exist in the original lines. Moreover, by disaggregating such complex traits into components, RFLP mapping enables scientists to determine which physiological and biochemical factors are major contributors to the trait. This in turn will help the molecular geneticists to know what genes they will need to isolate, clone, and transfer to help instill the desired trait via genetic engineering.

The RFLP map can also help the molecular geneticists to do the actual isolation and cloning of genes for which a phenotype is known but gene product and/or molecular function is not. To accomplish this scientists are first developing a physical map of the rice genome based on the RFLP map [16], and a complete library of large rice DNA fragments contained in yeast artificial chromosomes (YACs) [17]. The physical map is developed using pulsed-field gel electrophoresis (PFGE) to separate high molecular weight DNA. Since the DNA content per map unit in rice is relatively small, once 1000 markers are distributed on the map it should be possible to use PFGE to separate DNA fragments containing two markers. Any gene shown to be bracketed between the markers in linkage analysis can then be isolated by chromosome walking using the YAC clones.

The RFLP map of rice will be of significant value to rice breeders in the near term and to genetic engineers over the long term.

PROTOPLAST REGENERATION AND GENETIC TRANSFORMATION

Over twenty years ago rice was the first cereal for which regeneration of whole plants from callus tissue culture was reported [18]. Approximately five years ago, after many years of effort and some suggestions that cereal protoplasts were not totipotent, several laboratories reported successful regeneration of fertile rice plants from protoplasts [19,20,21,22]. This again was a first for the major cereals. All the initial reports were with japonica lines but within a few years regeneration of fertile plants from indica rice was also achieved [23,24]. These breakthroughs resulted more from persistent effort and hard work than from any key discovery or single solution to the regeneration problem. In fact, the protocols used are quite variable with regard to genotype, source of explant, callus media, regeneration media, and the need for nurse cultures or undefined additives. About all that can be generalized is that the efficient regeneration of rice requires the right genotype and a proper balance between the nutrient medium, the 2,4D concentration, and the age and development stage of the explant material [25]. Hodges' group has listed the following as critical factors in their successful protoplast regeneration of indica varieties IR54 and IR52 [26].

1.) Use cell suspension cultures that are highly embryogenic and regenerable as a source of protoplasts.

2.) Optimize conditions to obtain high yields of viable protoplasts (in excess of 10^6 protoplasts per gFW cells).

3.) Use nurse cells.

4.) Select small embryogenic calli at an early stage of protoplast callus formation and transfer to regeneration medium.

5.) Use cytokinin in the regeneration medium.

Once protoplast regeneration was achieved for rice it did not take long before the same laboratories and others reported the production of transgenic rice plants via protoplast uptake of DNA followed by regeneration [27, 28, 29, 30]. The DNA uptake was usually mediated by treatments that increase the permeability of the cell membrane such as electroporation and polyethylene glycol. Over a dozen laboratories in the United States, Japan, China, and Europe have now established the capability to genetically engineer rice plants via these techniques and experiments are underway to introduce potentially useful genes. It should be realized, however, that these protocols remain somewhat unreliable (sometimes they work, sometimes they do not), the efficiencies are low and the copy number and location of the alien gene is unpredictable. Much further research is needed to fine tune these protocols into more reliable and more efficient tools for rice genetic engineering.

Research is also continuing on a variety of other methods for genetic transformation of rice. Progress has been reported using <u>Agrobacterium</u> vectors [31, 32], particle bombardment [33] and transfer of DNA down pollen tubes [34] but at this time protoplast transformation and regeneration is the only reproducible technique for engineering fertile transgenic rice plants.

USEFUL GENES FOR RICE IMPROVEMENT

Now that the genetic engineering of rice plants is technically feasible attention is shifting to identifying, constructing and cloning genes that can instill new traits of agronomic, nutritional or commercial value when engineered into rice. The coding sequence for these genes can come from any source; rice, wildrice, other plants, microbes, animals, or chemical synthesis. The regulatory sequences however will need to function in rice. In fact many of the more sophisticated and powerful uses of genetic engineering will involve genes that are turned on or off in particular cells, tissues and/or organs at particular stages of development, and/or in responses to particular environmental stimuli such as insect feeding.

<u>Resistance Genes</u>. Plants have evolved a wide array of physiological and biochemical mechanisms for protecting themselves from microbial infection and insect attack, or for tolerating such infection. The host-pest relationship is genetically complex involving many genes in both organisms that have co-evolved over time. In most cases, classical genes for resistance will be regulatory genes controlling the expression of groups of other genes responsible for the physiological and biochemical functions which constitute the resistance trait, such as the hypersensitive response. Research is underway to isolate these classical genes for resistance from rice so that the structure and function of gene and gene product can be studied at the molecular level [35, 36]. The goal is to understand gene regulation as much as it is to enhance resistance. No such gene has yet been

isolated and it is likely to be some time before modifications in such genes can be engineered.

In the meantime efforts to enhance pest resistance in rice will utilize RFLP markers to pyramid genes for resistance, wide hybridization to incorporate resistance genes from wild relatives, and the engineering into rice of single genes from both alien sources and rice which code for proteins that directly inhibit microbial infection and insect attack.

Insect Resistance. Over 100 species of insects attack rice with 40 regarded as major pests. Losses of 10-30% of total yield from insect damage are not uncommon. Due to high costs, insecticides often are not a control strategy available to poor farmers in developing countries, and when they are health and environmental risks are inherent. Breeding for resistance and use of biocontrol are attractive alternatives to insecticides and both can be enhanced by genetic engineering. Most modern varieties have already incorporated the genes for insect resistance readily available via classical breeding but there are many insects against which effective resistance genes are not available and the emergence of new biotypes can lead to breakdown of resistance. The identification and judicious exploitation of exotic genes that can provide insect resistance will be one of the more significant near-term contributions of genetic engineering to rice production and utilization.

Initially most exotic insect resistance genes are likely to be insecticidal proteins where the gene product itself provides the toxicity and biosynthetic pathways are not involved. Scientists have only just begun to tap this pool of potential genes for resistance to insect pests. Endotoxins from Bacillus thuringiensis and protein inhibitors of insect digestive enzymes are the prime targets of current research. At IRRI and in Thailand, China, the Philippines, India and no doubt elsewhere researchers are identifying B.t. strains that are particularly effective against insect pests of rice. Once an effective strain is found the endotoxin gene can be isolated and incorporated into rice. The Chinese have already reported the production of transgenic rice plants containing a B.t. toxin gene [37]. Genes for inhibitors of insect digestive enzymes are likely to originate from wild rices and other crops. Cowpea trypsin inhibitor, potato proteinase inhibitors I and II, soybean trypsin inhibitor, arrowhead root trypsin inhibitor, wheat amylase inhibitors, papain inhibitors and corn trypsin inhibitor are being tested against digestive enzymes of specific rice pests [38, 39].

In utilizing insecticidal gene products that are produced endogenously in plants, considerable care and further research will be needed to control expression of the gene and to have the gene product delivered to the correct site. If the target insect attacks leafs and/or stems then the insecticide should be produced in the appropriate organ and not others. If the target insect is a phloem feeder then a leader sequence is likely to be needed to deliver the toxin to the phloem. If the target insect attacks the grain the gene product needs to be expressed in the endosperm and consideration will need to be given to human toxicity, heat liability, and other factors influencing suitability for human consumption. Promoters which are induced in response to insect feeding could play an important role.

Control of gene expression to prolong the usefulness of the gene is another important factor. Engineering constitutive expression of B.t. toxin genes at high levels throughout the life of all plants of a crop would be foolish. If the toxin were effective tremendous selection pressure would be placed on the insect population to evolve a new biotype resistant to the toxin. Concepts of evolutionary biology and ecology need to be brought to the genetic engineering process [40]. For example, if the plants or plant parts producing the toxin constitute only a portion of the pest's diet the selection process will be minimal. Combining genes coding for a toxin and genes coding for a repellant will offer longer lasting resistance than either approach alone [41]. Genetic engineering will bring new tools to pest control and if used wisely will enable other tools such as integrated pest management to be used more effectively.

Disease Resistance. In the developing world the principal strategy for controlling rice diseases is breeding for resistance. Research on the molecular genetics of the pathogens is leading to new tools which can strengthen breeding for disease resistance and allow existing sources of resistance to be used more effectively.

For the blast fungus, Magnoporthe grisea, and the bacterial blight pathogen, Xanthomonas campestris pv oryzae, scientists are developing race specific probes which enable more precise monitoring of race changes in pathogens and greater economy and precision in the deployment of resistant cultivars. For the blast fungus, a probe based on a family of repeated DNA sequences can distinguish various races and indicate location of origin [42]. Additional probes based on an RFLP map of Magnoporthe grisea are under development. [43].

Avirulence genes from Xanthomonas campestris pv oryzae are being cloned [44] and should provide DNA and/or serological probes which are race specific. A better understanding of avirulence gene function will also result and may allow rice to be engineered for enhanced elicitation of the resistance response.

Disease resistance genes that may be introduced or modified in rice via genetic engineering are being cloned from both rice and alien sources. Genes for chitinases and glucanases, which are major components of the lytic defense response, and genes for key enzymes in the production of phytoalexins have been cloned from rice [45]. Strategies to enhance resistance by modifying expression of such genes are under study. Many organisms produce anti-microbial products and exotic genes for disease resistance may come from some of these sources. In one screening procedure used in China, extracts from candidate plants and microbes are tested against rice pathogens. If there is an inhibitory effect, the extract is treated with proteases to determine if proteins are responsible for the inhibition. If they are, the proteins are purified and used to isolate the corresponding genes. Proteins inhibitory to the blast and bacterial blight pathogens have been identified using this method [46].

One of the most promising developments in plant biotechnology involves the use of exotic genes for introducing new sources of resistance to infection by plant pathogenic viruses. A highly successful strategy is termed coat protein-mediated protection. In several cases it has been demonstrated that transgenic plants expressing the coat protein gene of a pathogenic virus are resistant to infection by that virus, and that the resistance holds up under field conditions [47]. Moreover, the resistance extends to related viruses and works simultaneously with coat protein genes from more than one viral pathogen [48, 49]. Additional promising genes for resistance to plant viruses include: antisense genes, genes for ribozymes (catalytic RNA that cleaves viral RNA), genes for satellite RNA, genes for antibodies, and genes which inhibit vector transmission or cell to cell transmission of the virus.

The genomes of several important viral pathogens of rice are currently being characterized at the molecular level. This will allow cloning of candidate exotic genes for resistance utilizing one or more of the strategies noted above. For example, the coat protein gene from one of the viruses that causes tungro has been cloned and experiments aimed at transferring it into rice are underway. Transgenic rice plants containing this gene may well be available for field testing within the next few years.

Stress Tolerance The ability of rice plants to tolerate a stress such as drought results from the cumulative effect of component physiological and biochemical functions. Manipulation of such a trait via genetic engineering will require an understanding of and ability to manipulate the genetic determinants of the components. RFLP mapping can help in identifying the important genetic components and in following their inheritance in a breeding program. Genes induced in rice by stress can be cloned [50, 51, 52] and study of their regulation and function should provide additional insights.

Having the ability to genetically transform rice makes it possible to test the effect of candidate exotic genes for stress tolerance. If root structure and function are important determinants of drought tolerance, then addition of Ri plasmid genes which alter root phenotype might be beneficial. If osmotic adjustment plays a significant role in salinity tolerance then adding biosynthetic pathways for osmo-protective compounds like betaine might be useful [53]. At a minimum such research will help in determining the factors that do play an important role.

While genetic engineering of complex traits like drought tolerance is not yet feasible, scientists are generating the knowledge base which eventually should make this possible.

CONCLUSIONS

Over the past several years substantial progress has been made in rice biotechnology. New varieties with genetic improvements derived through anther culture and wide hybridization have reached the field and are being welcomed by farmers. Rice genetic maps and markers based on DNA polymorphisms have reached the point of application and are being welcomed by breeders. Genetic engineering of rice has been proven feasible and some scientists are now choosing rice as a model system for research in plant molecular biology.

The prospects for the future of this technology are bright. Rice plants containing useful foreign genes are likely to be field tested within two years. A variety of candidate genes, in addition to those for pest resistance and stress tolerance noted above, is being investigated. The objectives include improving the nutritional value and digestibility of rice storage proteins, biosynthesis of vitamin A precursors in rice grain, increasing yield through heterosis provided by apomictic seed production and/or new sources of male sterility, increasing starch biosynthesis and transport to the endosperm, improving seedling vigor, generating greater lodging resistance, and enhancing nitrogen uptake and assimilation.

The resulting transgenic rice plants will need to be tested in the field and evaluated by breeders. Surprises, both positive and negative, will surely occur. Over the coming decade the range of useful genetic variability available for rice improvement will be substantially increased and the precision and efficiency of rice breeding will be improved.

REFERENCES

1. Niizeki, H. and Oono, K., Induction of haploid rice plants from anther culture. Proc. Jpn. Acad., 1968, 44, 554-557.

2. Chen Y., Anther and pollen culture of rice. In Haploids of Higher Plants in Vitro, ed. H. Hu and H. Yang, Springer-Verlag, New York, 1986, pp 3-25.

3 Pulver, E.L. and Jennings, P.R., Application of anther culture to high volume rice breeding. In Rice Genetics: Proceedings of the International Rice Genetics Symposium, International Rice Research Institute, Manila, 1986, pp. 811-820.

4. Zhang, Z. Report of the Progress on Rice Biotechnology at SAAS in 1989. Shanghai Academy of Agricultural Sciences.

5. Raina, S.K., Tissue culture in rice improvement: status and potential. Advances in Agronomy, 1989, 42, 339-398.

6. Use of Anther Culture in Rice Breeding at CIAT: Proposal and Progress Report for 1989. Centro Internacional de Agricultura Tropical, Cali.

7. Progress Report on the Wide Hybridization Program, International Rice Research Institute, March, 1988.

8. Sitch, L.A., Jena, K.K., Dalmacio, R.D., Elloran, R., Romero, G.O., Amonte, A.D., Leung, H. and Khush, G.S., Wide hybridization for rice improvement. Presented at the Third Annual Meeting of the Rockefeller Foundation's International Program on Rice Biotechnology, March 1989.

9. Zhou, Q. and Hu, H., Transferring alien genes and creating new types of rice by anther culture. Presented at the Annual Meeting of the Cooperative Program on Biotechnology for Rice Improvement in China. November, 1989.

10. Cordesse, F., Second, G. and Delseny, M., Ribosomal gene spacer length variability in cultivated and wild rice species. Theor. Appl. Genet., 1990, 79, 81-88.

11. Zhao, X., Wu, T., Xie, Y. and Wu, R., Genome-specific repetitive sequences in the genus Oryza. Theor. Appl. Gent., 1989, 78, 201-209.

12. Aswidinnour, H., Dallas, J.F., Dillé, J.E., Gustafson, J.P., McIntyre, C.L. and Sears E.R., Molecular determination of species-specific repetitive DNA sequences in rice. Presented at the Third Annual Meeting of the Rockefeller Foundation's International Program on Rice Biotechnology, March, 1989.

13. Tanksley, S.D. and Coffman, W.R., Genetic and Physical Mapping of the Rice Genome: Proposal Submitted to the Rockefeller Foundation. September, 1989.

14. McCouch, S.R., Kochert, G., Yu, Z.H., Wang, Z.Y., Khush, G.S., Coffman, W.R. and Tanksley, S.D., Molecular mapping of rice chromosomes. Theor. Appl. Genet., 1988, 78, 815-829.

15. Op. cit. 13

16. Op. cit. 13

17. Ecker, J.R., Large DNA Cloning Methods for Rice: Proposal to the Rockefeller Foundation. September, 1989.

18. Nishi, T., Yamada, Y. and Takahashi, E., Organ redifferentiation and plant regeneration in rice callus. Nature, 1968, 219, 508-509.

19. Abdullah, R., Cocking, E.C. and Thompson, J.A. Efficient plant regeneration from rice protoplasts through somatic embryogenesis. Bio/technology, 1986, 4, 1087-1090.

20. Toriyama, K., Hinata, K. and Sasaki, T. Haploid and diploid plant regeneration from protoplasts of anther callus in rice. Theor. Appl. Genet., 1986, 73, 16-19.

21. Coulibaly, M.Y. and Demarly, Y., Regeneration of plantlets from protoplasts of rice, Oryza sativa L. Z. Planzenzüchtg, 1986, 96, 79-81.

22. Yamada, Y., Yang, Z.Q. and Tang, D.T., Plant regeneration from protoplast derived callus of rice (Oryza sativa). Plant Cell Rept., 1986, 5, 85-88.

23. Lee, L., Schroll, R.E., Grimes, H.D. and Hodges, T.K. Plant regeneration from indica rice (Oryza sativa L.) protoplasts. Planta, 1989, 178, 325-333.

24. Kyozuka, J., Otoo, E. and Shimamoto, K. Plant regeneration from protoplasts of indica rice: genotype differences in culture response. Theor. Appl. Genet. 1988, 76, 887-890.

25. Hodges, T.K., Peng, J.Y., Lee, L. and Koetje, D.S., In vitro culture of rice: transformation and regeneration of protoplasts. Proceedings of the Stadler Symposium, 1990 (In Press).

26. Op cit. 25

27. Yang, H., Zhang, H.M., Davey, M.R., Mulligan, B.J. and Cocking, E.C., Production of Kanamycin resistant rice tissues following DNA uptake into protoplasts. Plant Cell Rept., 1988, 7, 421-425.

28. Toriyama, K., Arimoto, Y., Uchimiya, H. and Hinata, K., Transgenic rice plants after direct gene transfer into protoplasts. Bio/technology, 1988, 6, 1072-1074.

29. Zhang, W. and Wu, R., Efficient regeneration of transgenic plants from rice protoplasts and correctly regulated expression of the foreign gene in the plants. Theor. Appl. Genet., 1988, 76, 835-840.

30. Shimamoto, K., Terado, R., Izawa, T. and Fugimoto, H., Fertile transgenic rice plants regenerated from transformed protoplasts. Nature, 1989, 338, 274-276.

31. Raineri, D.M., Bottino, P., Gordon, M.P. and Nester, E.W., Agrobacterium-mediated transformation of rice (Oryza sativa L.). Bio/Technology, 1990, 8, 33-38.

32. Li, B.J., Quyang, X.Z. and Zu, Y., Studies on the introduction of foreign genes into rice culture cells of Oryza sativa indica using Agrobacterium Ti plasmid system. Presented at the Annual Meeting of the Cooperative Program on Biotechnology for Rice Improvement in China. November, 1989.

33. Wang, Y.C., Klein, T.M., Fromm, M., Cao, J., Sanford, J.C. and Wu, R., Transient expression of foreign genes in rice, wheat and soybean cells following particle bombardment. Plant Molecular Biology, 1988, 11, 433-439.

34. Luo, Z.X. and Wu, R., A simple method for the transformation of rice via the pollen-tube pathway. Plant Molecular Biology Reporter, 1988, 6, 165-174.

35. Lamb, C.J., Molecular mechanisms for induction of rice defence responses to microbial diseases. Annual Report to the Rockefeller Foundation, 1989.

36. Op. cit. 13

37. Yang, H., Guo, S.D., Li, J.X., Chen, X.J. and Fan, Y.L., Transgenic rice plants produced from protoplasts following direct uptake of Bacillus thuringiensis endotoxin protein gene. Presented at the Annual Meeting of the Cooperative Program on Biotechnology for Rice Improvement in China. November, 1989.

38. Bennett, J., Enhancement of Insect Resistance in Rice with Special Reference to Gall Midge: Proposal to the Rockefeller Foundation. January, 1990.

39. Reeck, G., Inhibitors of Digestive Enzymes of Insect Pests of Rice: Proposal to the Rockefeller Foundation. September, 1989.

40. Gould, F., Evolutionary biology and genetically engineered crops. BioScience, 1988, 38, 26-33.

41. Op cit. 40.

42. Hamer, J.E., Farrall, L., Orbach, M.J., Valent, B. and Chumley, F.G., Host species-specific conservation of a family of repeated DNA sequences in the genome of a fungal plant pathogen. Proc. Natl. Acad. Sci. 1989, 86, 9981-9985.

43. Leong, S.A. and Holden, D.W., Molecular genetic approaches to the study of fungal pathogenesis. Annual Rev. Phytopathol., 1989, 27, 463-481.

44. Kelemu, S. and Leach, J.E., Cloning and characterization of an avirulence gene from Xanthomonas campestris pv. oryzae. Mol. Plant-Microbe Interaction, 1990 (In Press).

45. Op cit. 35

46. Chen, Z.L., Annual Progress Report to the Rockefeller Foundation from the National Laboratory for Plant Genetic Engineering, Beijing University. January, 1990.

47. Beachy, R.N., Nelson, R.S., Regester III, J., Fraley, R.T. and Tumer, N., Transformation to produce virus-resistant plants. In Genetic Improvements of Agriculturally Important Crops, ed. R.T. Fraley, N.M. Frey, and J. Shell, Cold Spring Harbor Laboratory Current Communication, 1988, pp. 47-53.

48. Beachy, R.N. and Stark, D.M., Protection against potyvirus infection in transgenic plants: evidence for broad spectrum resistance. Bio/Technology, 1989, 7, 1257-1262.

49. Lawson, C., Kaniewski, W., Haley, L., Rozmann, R., Newell, C., Sanders, P. and Tumer, N.E., Engineering resistance to mixed virus infection in a commercial potato cultivar: resistance to potato virus X and potato virus Y in transgenic Russet Burbank. Bio/Technology, 1990, 8, 127-134.

50. Mundy, J. and Chua, N.H., Abscisic acid and water-stress induce the expression of novel rice gene. EMBO, 1988, 7, 2279-2286.

51. Hahn, M. and Walbot, V., Effects of cold-treatment on protein synthesis and mRNA levels in rice leaves. Plant Physiol., 1989, 91, 930-938.

52. Claes, B., Dekeyser, R., Villarroel, R., Van Den Bulcke, M., Bauw, G., Van Montagu, M. and Caplan, A., Characterization of a rice gene showing organ-specific expression in response to salt stress and drought. The Plant Cell, 1990, 2, 19-27.

53. The Potentials of Biotechnology for Improving Grain Yield of Rice under Water Limited Conditions. Conference Report of the Rockefeller Foundation. 1989.